Lecture Notes in Physics

261

Conformal Groups and Related Symmetries Physical Results and Mathematical Background

Proceedings of a Symposium Held at the
Arnold Sommerfeld Institute for Mathematical Physics (ASI)
Technical University of Clausthal, Germany
August 12–14, 1985

Edited by A. O. Barut and H.-D. Doebner

Springer-Verlag Berlin Heidelberg GmbH

Editors

A. O. Barut
Department of Physics, University of Colorado at Boulder
Boulder, Colorado 80309, USA

H.-D. Doebner
Arnold Sommerfeld Institute for Mathematical Physics
Technical University Clausthal
D-3392 Clausthal-Zellerfeld, FRG

ISBN 978-3-662-14482-4 ISBN 978-3-540-47219-3 (eBook)
DOI 10.1007/978-3-540-47219-3

Originally published by Springer-Verlag Berlin Heidelberg New York in 1986
Softcover reprint of the hardcover 1st edition 1986

2153/3140-543210

PREFACE

This volume contains contributions presented at an International Symposium on Conformal Groups and Conformal Structures held in August 1985 at the Arnold Sommerfeld Institute for Mathematical Physics in Clausthal. We hope that the wide range of subjects treated here will give a picture of the present status of the importance of the conformal groups, other related groups and associated mathematical structures (such as superconformal algebra, Kac-Moody algebras), and spin structures. Symmetry, with group theory and algebras as its mathematical model, has always played a crucial and significant role in the development of physical theories. One of the prime reasons for the interest in the conformal group is that it is perhaps the most important of the larger groups containing the Poincaré group. It opens the door to applications far beyond the standard kinematical framework provided by the local symmetries of flat space-time.

It is stimulating to recognise the progress which has occurred in the last 15 years by comparing these proceedings with those of a similar conference held in 1970 (A.O. Barut, W.E. Brittin: De Sitter and Conformal Groups and Their Applications, Colorado University Press 1971). The emphasis has changed and numerous new fields have appeared which are mathematically and physically associated with the conformal group. The great interest shown in this conference and the material presented in this volume indicates that the field centred around conformal symmetry is very much alive and active.

The material is organised into six chapters:

I. Symmetries and Dynamics
II. Classical and Quantum Field Theory
III. Conformal Structures
IV. Conformal Spinors
V. Lie Groups, -Algebras and Superalgebras
VI. Infinite-Dimensional Lie Algebras

The papers range from direct physical applications (e.g. P. Magnollay and Dj. Šijački) to the presentation of mathematical methods and results (e.g. V.G. Kac) with likely future influence on particle phyiscs. We have also included articles with a bias towards fundamental questions using symmetry to reinforce parts of the foundations of physics and of space-time structure (e.g. C.F. v. Weizsäcker and also P. Budinich).

Some of the developments during recent years, and hence some of the contributions, have utilized conformal symmetry in combination with e.g. differential geometric and algebraic structures, as in string theory (e.g. Y. Ne'eman). There are also research reports based on applications of groups related to the conformal group (e.g. $\overline{SL}(4,R)$). The extended lectures by I.T. Todorov on "Infinite Dimensional Lie Algebras in Conformal QFT Models" aims to give new results combined with a review as an introduction to an important and fast-growing subject. Furthermore, some articles present reviews in a new and updated context. We have also included the material of some of the invited speakers who did not have the opportunity to present it at the conference.

To give this volume special value to postgraduate students and to physicists and mathematicians who want to enter the field, we asked for contributions which contain some introductory and review sections.

ACKNOWLEDGEMENTS

We wish to express our gratitude to the following organisations and persons for generous financial support and other assistance making the conference symposium and this volume possible:

- Der Niedersächsische Minister für Wissenschaft und Kunst, Hannover
- Alexander von Humboldt-Stiftung, Bonn
- Technische Universität Clausthal, specifically Rector Prof. Dr. K. Leschonski

Dr. V.K. Dobrev (Sofia/Clausthal) and Dr. W. Heidenreich (Clausthal) were scientific secretaries of the symposium. Dr. W. Heidenreich assisted furthermore in the preparation of the proceedings.
We want to thank Springer-Verlag, Heidelberg, for their kind assistance in matters of publication.

Last but not least we want to thank the members, co-workers and especially the students of the Arnold Sommerfeld Institute and the Institute of Theoretical Physics A who made the symposium run so smoothly and efficiently.

H.-D. Doebner
A.O. Barut

TABLE OF CONTENTS

IV. CONFORMAL SPINORS

V. LIE GROUPS, -ALGEBRAS AND SUPERALGEBRAS

VI. INFINITE-DIMENSIONAL LIE ALGEBRAS

I. SYMMETRIES AND DYNAMICS

FROM HEISENBERG ALGEBRA TO CONFORMAL DYNAMICAL GROUP

A. O. Barut

Department of Physics

Campus Box 390

University of Colorado

Boulder, CO. 80309-0390

ABSTRACT

The basic algebraic structures in the quantum theory of the electron, from Heisenberg algebra, kinematic algebra, Galilean, and Poincaré groups, to the internal and external conformal algebras are outlined. The universal role of the conformal dynamical group from electron, H-atom, hadrons, to periodic table is discussed.

I. Introduction

The postulates of quantum theory can be expressed most concisely as the representation theory of the symmetry groups and dynamical groups of physical systems. And the analytical methods and specific calculations in quantum theory are performed most economically in terms of the representations of the Lie and envelopping algebras and their matrix elements. In these notes I give an outline of the developments of the group theoretical ideas and methods mainly for the electron, but of course also applicable for other quantum systems. With an audience of both mathematicians and physicists in mind, I hope this presentation will be elementary and self-consistent, although some may find the text to be a bit too mathematical, others to concise in physics.

II. The Heisenberg Algebra h_n and Kinematical Algebra k_n

The algebraic quantum theory goes back to the initial work of Heisenberg, and the Born-Jordan-Heisenberg formulation of quantum mechanics.

For a mechanical Hamiltonian system of n-degrees of freedom with n generalized coordinates q_i, and n conjugate momenta p_i, $i = 1, 2, ... n$, we have the Heisenberg algebra h_n defined by the commutation relations:

$$[q_i, q_j] = 0 \quad , [p_i, p_j] = 0 \; ; \quad i, j = 1,2, \dots n$$

$$h_n: \quad [q_i, p_j] = i\hbar \, \delta_{ij} \, J \quad , \quad [J_i \, q_i] = 0, \; [J, p_j] = 0 \tag{1}$$

Here we have introduced, for purpose of later generalization, an operator J which in h_n has been choosen to be the identity operator. This can be done as long as, as is well known, p's and q's are not finite-dimensional matrices.

Originally Heisenberg introduced p_i, q_i as matrices in the energy basis of the quantum system. With the advent of transformation theory and Hilbert space formulation, eqs. (1) are general operator relations independent of basis.

The Heisenberg algebra h_n can be extended to a kinematical algebra k_n with the inclusion of SO(n)-rotation elements $\ell_{ij} = - \ell_{ji}$. The additional commutation relations to eqs. (1) are

$$[q_i, \ell_{jk}] = i\hbar \, (\delta_{ik} \, q_j - \delta_{ij} \, q_k)$$

$$[p_i, \ell_{jk}] = i\hbar \, (\delta_{ik} \, p_j - \delta_{ij} \, p_k)$$

$$k_n': \quad [\ell_{ij}, \ell_{k\ell}] = i\hbar \, (\varepsilon_{ik} \, \ell_{j\ell} + \delta_{j\ell} \, \ell_{ik} - \delta_{jk} \, \ell_{i\ell} - \delta_{i\ell} \, \ell_{jk}) \tag{2}$$

$$[J, \ell_{ij}] = 0$$

The dimension of $k_n = h_n + k_n'$ is $\frac{1}{2}(n+1)(n+2)$, the same as that of the Lie algebra of SO(n+2) or SO(n,2).

Any representation of h_n can be extended to a representation of k_n by the following realization of ℓ_{ij}:

$$\ell_{ij} = q_i \, p_k - q_k \, p_i \tag{3}$$

derived from the physical meaning of ℓ_{ij} as the components of orbital angular momentum. In this case k_n is just a derived algebra from h_n, a Lie algebra in the enveloping algebra of h_n. For this type of representations of k_n, the representations of k_n remain irreducible for the subalgebra h_n; conversely representations of h_n are automatically extended to the representations of k_n.

But there are other representations of k_n. For example we can set

$$\ell_{ij} = q_j \, p_i - q_i \, p_j + S_{ij} \tag{4}$$

where S_{ij} are the spin operators. We can then enlarge our dynamical system by the inclusion of the commutation relation of S_{ij}, $[S_{ij}, S_{k\ell}]$, or just keep the algebra k_n, independent of the realizations (3) or (4), and consider all its representations.

Sofar the kinematic algebra k_n describes the quantum system at a fixed time t. They can be realized also as differential operators acting on a time-dependent wave function $\psi(q,t)$ (Schrödinger representation), or they can be given a time-dependence $q = q(t)$, $p = p(t)$, acting or a time-independent Hilbert space (Heisenberg representation). Since the Hamiltonian system is characterized by a Hamiltonian H and the time evolution of the system by a unitary operator $U(t-t_0) = e^{-i\hbar H(t-t_0)}$, we have a quantum dynamical system of 2n-dimensions:

$$\dot{q}_j = \frac{i}{\hbar}[H, q_j]$$

$$\dot{p}_j = \frac{i}{\hbar}[H, p_j] \quad , \quad j = 1, 2, \ldots n \quad . \tag{5}$$

Because we are interested in the generalization of the operator J in eq. (1), it is important to note that if one postulates quantum mechanics first by eqs. (5), instead of eqs. (1), the most general Heisenberg commutation relations compatible with (5) are of the form[1]

$$[q_i, p_j] = i\hbar \delta_{ij} F \tag{6}$$

where F can be a function of the Hamiltonian.

A nonrelativistic quantum system must also show the symmetry under Galilean transformations of space and time if it is a system existing in space-time. For this purpose we introduce the total momentum of the system $\vec{P}$. [If q_i are the cartesian coordinates, then $\vec{P} = \vec{p}_1 + \ldots + \vec{p}_n$, otherwise $\vec{P}$ is related to p_i, q_i in a more complicated way]. Similarly, the system will have a total angular momentum $\vec{J}$, also a function of p's and q's. The introdution of the generators $\vec{M}$ of velocity (or boost) transformations is more subtle. They have explicit time-dependence in addition to their time evolution

$$\vec{M} = \sum_j \vec{M}_j = \sum_j (tp_i - m_j q_j) \quad , \tag{7}$$

for Cartesian coordinates q_j. The ten operators $P_0 = H$, $\vec{P}$, $\vec{J}$ and $\vec{M}$ are the generators of the Galilean group G . The representations of the Galilean group G cannot completely characterize our dynamical system of 2n degrees of freedom; the system is composite, it has a lot of internal degrees of freedom; the representation of G will be highly reducible. Irreducible representations of symmetry group apply to elementary systems.[2] In the purely geometric definition of the Galilean algebra we have

$$[\vec{M}, P_i] = 0 \tag{8}$$

But in the quantum mechanical realization (7) we have[3]

$$[M_i^{(j)}, P_k] = ih\, m^{(j)} \delta_{ik} \tag{9}$$

or, more generally,

$$[M_i, P_k] = ih \mathcal{M} \delta_{ik} \tag{9'}$$

where $\mathcal{M}$ is a mass operator. This is another instance, like eqs. (1) and (6), where we obtain new operators J, F, $\mathcal{M}$ in generalizing the simple commutation relations. The mathematical interpretation of (9) instead of (8) is that quantum theory uses actually <u>projective representations</u> (or ray representations) of symmetry groups, because an overall phase of the wave function is not observable; a state is characterized only by ray in Hilbert space. Equivalently, quantum mechanial representations are <u>extensions</u> of the geometrical representations of symmetry groups and algebras.

III. SO(n+2) and Compact Quantum Systems

Let us now see the position of the algebras h_n and k_n within the Lie algebra of SO(n+2) or SO(n,2). We denote the generators of SO(n+2) by $J_{AB} = -J_{BA}$; A, B = 1, ... n+2. They satisfy

$$[J_{AB}, J_{CD}] = i(g_{AC} J_{BD} + g_{BD} J_{AC} - g_{BC} J_{AD} - g_{AD} J_{BC}) \tag{10}$$

Let

$$J_{ij} = \frac{1}{\hbar} S_{ij} \;,\quad J_{i,n+1} = \frac{1}{\lambda} Q_i \;,\quad J_{i,n+2} = \frac{\lambda}{2\hbar} P_i \;,\quad J_{n+2,n+2} = \frac{1}{2} J \tag{11}$$

where for dimensional reasons we have introduced an "elementary length" λ, and in view of the following applications, new coordinates, and momenta Q_i, P_i.

Explicitly the antisymmetric set of generators are

$$\begin{pmatrix} 0 & S_{12} & S_{13} & \cdots & S_{1n} & Q_1 & P_1 \\ & 0 & S_{23} & \cdots & S_{2n} & Q_2 & P_2 \\ & & & \cdots & \cdots & \cdots & \cdots \\ & & & 0 & S_{n-1,n} & Q_{n-1} & P_{n-1} \\ & & & & 0 & Q_n & P_n \\ & & & & & 0 & J \\ & & & & & & 0 \end{pmatrix} \tag{11'}$$

To the Heisenberg algebra h_n corresponds now the algebra[4]

$$[Q_i, Q_j] = i\frac{\lambda^2}{\hbar}S_{ij} \quad ; \quad [P_i, P_j] = 4i\frac{\hbar}{\lambda^2}S_{ij}$$

$$H_n: \quad [Q_i, P_j] = i\hbar\,\delta_{ij}\,J \quad ; \quad [Q_i, J] = i\frac{\lambda^2}{\hbar}P_i \tag{12}$$

$$[P_i\ , J] = 4i\frac{\hbar}{\lambda^2}Q_i \quad ; \quad i,j = 1 \ldots\ldots n$$

The differences between h_n and H_n are that now the coordinates and momenta among themselves do not commute, and J also does not commute with Q_i and P_i. However, the extended kinematical algebra k_n' of eq. (2) remains the same:

$$[Q_i, S_{jk}] = i\hbar\,(\delta_{ik}\,Q_j - \delta_{ij}\,Q_k)$$

$$[P_i, S_{jk}] = i\hbar\,(\delta_{ik}\,P_j - \delta_{ij}\,P_k)$$

$$[S_{ij}, S_{kl}] = i\hbar(\delta_{ik}\,S_{jl} + \delta_{jl}\,S_{ik} - \delta_{jk}\,S_{il} - \delta_{il}\,S_{jk})$$

$$[J, S_{ij}] = 0 \quad . \tag{13}$$

In contrast to the Heisenberg algebra (1) - (2), the new algebra (12) - (13) now admits finite-dimensinal representations for Q_i, P_j, and S_{ij}. We shall see in fact that such systems actually occur in nature, namely as the internal structure of the electron and other relativistic spinning particles. In particular, the fundamental spinor representations of SO(n+2) comes as close as possible to the Heisenberg commutation relations in that J is traceless, has unique square and eigenvalues ± 1. The dimension of this representation is 2^p, where $p = 1/2(n+1)$ for n odd and $p = 1/2$, for n even, in which case there are two inequivalent representations. These representations coincides with the representations of Clifford algebras and are related with some realizations of superalgebras. The passage from SO(n+2) to k_n is via the <u>contraction</u> of the Lie algebra.[4] We define, starting from SO(n+2),

$$\tilde{q}_i \equiv \varepsilon_1\,Q_i\ ,\ \tilde{p}_i \equiv \varepsilon_2\,P_i\ ,\ \tilde{J} \equiv \varepsilon_1\varepsilon_2 J\ ,\ \tilde{\ell}_{ij} \equiv S_{ij} \tag{14}$$

and then obtain

$$[\tilde{q}_i, \tilde{q}_j] = i \frac{\lambda^2}{\hbar} \varepsilon_1^2 \tilde{\ell}_{ij} \quad , \quad [\tilde{p}_i, \tilde{p}_j] = 4i \frac{\hbar^2}{\lambda^2} \varepsilon_2^2 \tilde{\ell}_{ij}$$

$$[\tilde{q}_i, \tilde{p}_j] = i\hbar \, \delta_{ij} \, J \quad , \quad [\tilde{q}_i, \tilde{J}] = - i \frac{\lambda^2}{\hbar^2} \varepsilon_1^2 \tilde{p}_i$$

$$[\tilde{p}_i , \tilde{J}] = 4i \frac{\hbar}{\lambda^2} \varepsilon_2^2 \tilde{q}_i \tag{15}$$

There are two routes now. Either we let first $\varepsilon_1 \to 0$ and then ε_2, or vice versa. The intermediate algebra when one ε is set equal to zero and not the other, is interestingly, the euclidian algebra e(n+1) in (n+1) - dimensions.

All these relations show that the dynamical systems corresponding to (12), (13) are natural counterparts of the usual Heisenberg systems and should be also important. We recall here that <u>finite quantum systems</u> were first introduced by Weyl.[5] Weyl also treated the passage from Heisenberg algebra to the Heisenberg group, i.e. group whose infinitesimal generators are p_i and q_i, and recognized that the unitary representations of the Heisenberg group can be considered as ray representations of infinite abelian groups. Similarly the fundamental spinor representations of SO(n+2) can be considered as ray representations of finite abelian groups:[6] n commuting parity like operators Γ_i with

$$\Gamma_i^2 = 1 \quad , \quad \Gamma_i\Gamma_j = \Gamma_j\Gamma_i \quad ; \quad i,j = 1, 2, \ldots. n \tag{16}$$

have a projective representations of dimension $2^{n/2}$ or $2^{(n-1)/2}$ which is a Clifford algebra or the fundamental representation of SO(n+2). It is an open problem, as far as I know, to have a general theory of the relation between the projective rpresentations of finite groups and the corresponding Lie algebra representations.

The Heisenberg algebra can be transformed, as is well-known, into the boson algebra. In our case the new boson algebra maybe defined by[4]

$$A_i = \frac{1}{\lambda} Q_i + i \frac{\lambda}{2\hbar} P_i \quad , \quad A_i^+ = \frac{1}{\lambda} Q_i - i \frac{\lambda}{2\hbar} P_i \tag{17}$$

then we find the following commutation relations

$$[A_i, A_j] = 0 \quad , \quad [A_i^+, A_j^+] = 0 \quad , \quad [A_i^+ , J] = 2A_i^+ \quad ,$$

$$[A_i , J] = -2A_i \quad , \quad [A_i, A_j^+] = \delta_{ij} J + 2 \frac{i}{\hbar} S_{ij} \tag{18}$$

This system is naturally associated with a dynamical system

$$H = \frac{\hbar\omega}{2n} [A_i^+ A_j - A_i A_j^+] = \frac{h\omega}{2} J \tag{19}$$

with oscillator equations

$$\dot{A}_i = - i\omega A_i \quad , \quad \dot{A}_i^\dagger = i\omega A_i^\dagger \tag{20}$$

The double commutators are

$$[[A_i , A_j^+], A_k] = 2(\delta_{ij} A_k + \delta_{jk} A_i - \delta_{ik} A_j)$$

$$[[A_i , A_j^+], A_k^+] = 2(-\delta_{ij} A_k^+ + \delta_{jk} A_i^+ - \delta_{ik} A_j^+)$$

$$[[A_i , A_j^+] , [A_k , A_\ell^+]] = 2(\delta_{ik} [A_\ell , A_j^+] - \delta_{\ell j} [A_i , A_k^+]$$

$$+ \delta_{jk} [A_i , A_\ell^+] - \delta_{i\ell} [A_k , A_j^+]) \tag{21}$$

It is interesting to compare the system (21) with another finite system associated with the Hamiltonian

$$H = \frac{\hbar\omega}{n-1} (A_i^+ A_i + A_i A_i^+) \tag{22}$$

and satisfying the relations of the Lie superalgebra $s\ell(\ell,n)$

$$[\{A_i^+ , A_j\} , A_k] = - \delta_{ik} A_j + \delta_{ij} A_k$$

$$[\{A_i^+ , A_j\} , A_k^+] = \delta_{jk} A_i^+ - \delta_{ij} A_k^+$$

$$[\{A_i^+ , A_j\} , \{A_k^+ , A_\ell\}] = \delta_{jk} \{A_i^+ , A_\ell \} - \delta_{i\ell}\{A_k^+ , A_j\}$$

$$\{A_i , A_j \} = 0 \quad , \{A_i^+ , A_j^+\} = 0 \tag{23}$$

Only integer spin representations of SO(n) – subalgebra of $s\ell(\ell,n)$ occur here, whereas the system (21) also allows half-integer spins.

IV. Dynamics in the New Coordinates Q_i, P_i

We can now formulate dynamical problems with our new canonical coordinates Q_i, P_i satsifying the commutation relations (12) and (13) assuming a Hamiltonian. They provide novel type of finite (and infinite) quantum dynamical systems, and, by going over to the corresponding Poisson brackets, classical dynamical systems as well. Some of the problems of quantum dynamical systems, such as quantum chaos, maybe studied on such simple finite systems with their unusual phase space. Even a one-dimensional system of a free particle is a nontrivial interesting dynamical system:[8] We have in this case the commutation relations

$$[Q, P] = ih\, J, \quad [Q,J] = - i \frac{\lambda^2}{\hbar} P \,, \quad [P,J] = i \frac{\gamma^2\hbar}{\lambda^2} Q \tag{24}$$

and as the Hamiltonian of "free particle" we may choose

$$H = \frac{1}{2m} P^2 \quad . \tag{25}$$

The algebra (24) is isomorphic to so(3). If we diagonalize P in an irreducible $(2j + 1)$-dimensional representation of SO(3) with spectrum $\{-j, \ldots, j\}$, then the spectrum of energy is given by $E = aj^2$, $a(j-1)^2$, ..., 0 (j integer). The spectrum of an "oscillator" with $H = \alpha P^2 + \beta Q^2$ is a difficult problem of α and β are arbitrary.

The Heisenberg equations for $H = \alpha P^2 + \beta Q^2$ are highly nonlinear

$$\dot{P} = - \beta(QJ + JQ) \,, \; \dot{Q} = \alpha(PJ + JP) \,, \; \dot{J} = (\beta \frac{\lambda^2}{\hbar^2} - \alpha \frac{\gamma^2}{\lambda^2})(PQ + QP) \tag{26}$$

compared to the ordinary oscillator $\dot{p} = aq$, $\dot{q} = bp$.

Actually such a dynamical system occur in nature, namely in the internal motion of the relativistic Dirac electon, a dynamics called the Zitterbewegung.[9] It is possible to identify in the rest frame of the electron $(p = 0)$, operators Q_i and P_i as well as S_{ij} and J, i = 1, 2, 3, which precisely satisfy the commutation relations (12) and 13). In this case they have been extracted from the Dirac matrices, hence they are 4×4-matrices. The "Hamiltonian" representing the internal energy is in this case just J so that Heisenberg equations are linear oscilator equations

$$\dot{Q}_j = \frac{1}{m} P_j \quad , \quad \dot{P}_j = -\frac{4\hbar}{\lambda^3} Q_j \quad , \quad \dot{J} = 0 \quad . \tag{27}$$

The Zitterbewegung is just this oscillation of the charge of the electron around its center of mass.

For the massless neutrino we obtain an internal dynamics again with the same algebra (12) and (13) but everywhere δ_{ij} replaced by $\tilde{\delta}_{ij}$ and S_{ij} replaced by $\tilde{S}_{ij}$ where

$$\tilde{\delta}_{ij} = \delta_{ij} - \frac{p_i p_j}{\vec{p}^2} \quad , \quad \tilde{S}_{ij} = S_{ij} - \frac{p_i p_k}{p^2} S_{kj} - \frac{p_k p_j}{p^2} S_{ik} \tag{28}$$

which means that the internal motion takes place on an hypersurface perpendicular to $\vec{p}$, and that it has effectively two degrees of freedom.[10]

V. Relativistic Systems

There are different approaches to the dynamics of a single relativistic particle which are all at the end equivalent. But the relativistic dynamics of two or more interacting particles is more subtle.

Continuing the line of our developments in the previous Sections, we can still start from the Heisenberg algebra (1), the angular momentum algebra (2) and the realization of angular momentum given by (4) including spin. Instead of the nonrelativistic Galilean algebra we must now realize the Poincaré algebra with the generators $P_0 = H$, $\vec{\pi}$, $\vec{J}$ (angular momentum), and again the boost operators $\vec{M}$ satisfying the commutation relations of the Poincaré Lie algebra:

$$[\pi_i, \pi_j] = 0 \quad , \quad [\pi_i, H] = 0 \quad , \quad [J_i, H] = 0$$

$$[J_i, J_j] = \varepsilon_{ijk} J_k \quad , \quad [J_i, \pi_j] = \varepsilon_{ijk} \pi_k$$

$$[J_i, M_j] = \varepsilon_{ijk} M_k \quad , \quad [M_j, H] = \pi_j$$

$$[M_i, M_j] = -\varepsilon_{ijk} J_k \quad , \quad [M_i, \pi_j] = \delta_{ij} H \tag{29}$$

Conversely if one starts from an irreducible representation of the Poincaré group with generators $J_{\mu\nu}$ and P_μ there are no position operators q_μ. How do we introduce them? Under certain additional criteria and using imprimitivity theorems one can introduce position operators.[11] For example, for a spinless particle, they can be defined as differential operators on the carrier space of an irreducible representation

$$(q_k \psi)(p) = i \left(\frac{\partial}{\partial p_k} - \frac{p_k}{2p_o^2}\right) \psi(p) \tag{30}$$

or, for a spinning particle, by

$$(q_k \psi)(p) = \left\{i \left(\frac{\partial}{\partial p_k} + \frac{\gamma_k}{2p_o}\right) - i \frac{(\gamma p))_k + (\Sigma x p)_k p_o}{2p_o^2 (p_o + m)} - i \frac{p_k}{p_o^2}\right\} \psi(p) \tag{30'}$$

However, for a system like the Dirac electron, we have a reducible representation of the Poincaré group and the above position operator does not really apply. For a single spin 1/2-irreducible representation of mass m given by

$$U^{m,\, 1/2}_{(a,\Lambda)} \psi(p) = e^{ipa} D^{(1/2,0)} (\Lambda) \psi(L_\Lambda^{-1} p) \tag{31}$$

and acting on functions $\psi(p)$ over the mass hyperboloid ($p^2 = m^2$, $p^0 > 0$), parity operator is not defined and there is no four-vector current operator. We double the space by

$$U^{m,\, 1/2}_{(a,\Lambda)} \psi(p) = e^{ipa} [D^{(1/2,0)} \otimes D^{(0,1/2)}] \psi(L_\Lambda^{-1} p) \tag{31'}$$

We can work in this doubled space but at the end we have to reproject on two physical components by the projection operator $\pi = \begin{pmatrix} 1 & \\ & 0 \end{pmatrix} = 1/2(\gamma_0 + I)$. This projection operator in an arbitrary frame is the Dirac equation[11]

$$(\gamma^\mu p_\mu - m) \psi (p) = 0 \quad , \quad p^0 > 0 \tag{32}$$

The other half-space describes the antiparticle

$$(\gamma^\mu p_\mu + m) \psi(p) = 0 \quad , \quad p^0 > 0 \; . \tag{33}$$

(The solutions of (32) for $p_0 < 0$ coincide with those of (33) for $p^0 > 0$).

Now for the Dirac electron-positron complex it is more convenient to introduce two position operators and not one. One is a center of mass coordinate, the other a relative cordinate, and their sum is the coordinate x that appears in the Dirac equation, and x is the position of the charge, because the electromagnetic field couples locally to x. The fact that center of mass position and the charge position do not coincides indicate an internal structure which shows itself in the spin degrees of freedom. In contrast to the representation (4) or (29), spin is a dynamical variable, hence any spinning system must have a larger set of basic dynamical variables than the Heisenberg algebra of p's and q's. Alternatively we can speak of an external Heisenberg algebra and an additional internal Heisenberg algebra. And it turns out that the former satisfy eqs. (1), but the latter the new Heisenberg algebra (12), as we have already mentioned.

In this Section we shall give the covariant version of the new internal Heisenberg algebra (12).

It turns out that both quantum Dirac theory of the electron and a recently proposed classical relativistic model for the spinning electron lead exactly to the same internal algebra, the latter in terms of the Poisson brackets, the former, of course, in terms of commutators. The classical theory is based on the Lagrangian

$$L = -\frac{\lambda}{2i}(\dot{\bar{z}}z - \bar{z}\dot{z}) + p_\mu(\dot{x}^\mu - \bar{z}\gamma^\mu z) + e A_\mu \bar{z}\gamma^\mu z \tag{34}$$

Here $z(\tau)$ is a complex c-number spinor, $z(\tau) \epsilon C^4$, representing the internal spin degrees of freedom, τ = an invariant parameter. The dynamical system (34) is a Hamiltonian system with a covariant "Hamiltonian" (relative to τ)

$\mathcal{H} = \gamma^\mu \bar{z}\gamma_\mu z \equiv \gamma^\mu v_\mu$; and $(x_\mu, \pi_\mu = p_\mu - e A_\mu)$ and $(z, i\bar{z})$ are conjugate pairs.

One can elimnate z, $\bar{z}$ in favor of the spin variables $S_{\mu\nu}$ and obtain the dynamical system

$$\dot{x}_\mu = v_\mu , \quad v_\mu = 4S_{\mu\nu}\pi^\nu$$

$$\dot{\pi}_\mu = eF_{\mu\nu}v^\nu , \quad \dot{S}_{\mu\nu} = \pi_\mu v_\nu - \pi_\nu v_\mu \tag{35}$$

with the Poisson algebra

$$\{x_\mu, \pi_\nu\} = g_{\mu\nu} \quad , \quad \{\pi_\mu, \pi_\nu\} = e\, F_{\mu\nu}$$

$$\{v_\mu, v_\nu\} = 4\, S_{\mu\nu} \quad , \quad \{S_{\alpha\beta}, v_\nu\} = g_{\alpha\nu}\, v_\beta - g_{\beta\nu}\, v_\alpha$$

$$\{S_{\alpha\beta}, S_{\gamma\delta}\} = g_{\alpha\gamma}\, S_{\beta\delta} - g_{\beta\gamma}\, S_{\alpha\beta} - g_{\alpha\delta} S_{\beta\gamma} + g_{\beta\delta} S_{\alpha\gamma} \tag{36}$$

Note that momentum and velocity, π_μ and v_μ, are independent dynamical variables even for a free particle ($A_\mu = 0$). [A similar situation occurs if the radiation reaction force of the classical electron is taken into account].[13] For a free particle we now separate internal and external coordinates as follows. Let $x_\mu = X_\mu + Q_\mu$, hence $v_\mu = \dot{X}_\mu + \dot{Q}_\mu$. Then we set $\dot{X}_\mu = p_\mu/m$ which is the velocity a particle of momentum p_μ and mass m. Then we can interprete Q_μ as the relative coordinate and $P_\mu = m\dot{Q}_\mu$ as the relative or internal velocity and x_μ as the position of the charge. Similarly, the total angular momentum $J_{\mu\nu}$ can be decomposed either as $J_{\mu\nu} = L_{\mu\nu} + S_{\mu\nu}$ (orbital and spin angular momentum of the charge), or as $J_{\mu\nu} = L_{\mu\nu} + \Sigma_{\mu\nu}$ (orbital angular momentum of the center of mass and that of internal motion). Then the internal algebra generated by Q_μ, L_μ, $\Sigma_{\mu\nu}$ and $\mathcal{H}$ is closed and is the covariant form of the algebra (12) - (13) (or (28)):[12]

$$\{Q_\mu, Q_\nu\} = m^{-2}\, \Sigma_{\mu\nu}$$

$$\{P_\mu, P_\nu\} = 4m^2\, \Sigma_{\mu\nu} \, , \quad \{P_\mu, \mathcal{H}\} = -4m^2 Q_\mu$$

$$\{Q_\mu, P_\nu\} = -\, g_{\mu\nu} m^{-1} \, , \quad \{Q_\mu, \mathcal{H}\} = m^{-1} P_\mu$$

$$\{Q_\mu, \Sigma_{\alpha\beta}\} = (\tilde{g}_{\mu\alpha}\, Q_\beta - \tilde{g}_{\mu\beta} Q_\alpha)$$

$$\{P_\mu, \Sigma_{\alpha\beta}\} = (\tilde{g}_{\mu\alpha}\, P_\beta - \tilde{g}_{\mu\beta} P_\alpha)$$

$$\{\Sigma_{\alpha\beta}, \Sigma_{\gamma\delta}\} = \tilde{g}_{\alpha\gamma}\, \Sigma_{\beta\delta} + \tilde{g}_{\beta\delta}\, \Sigma_{\alpha\gamma}\, \tilde{g}_{\alpha\delta}\, \Sigma_{\beta\gamma} - \tilde{g}_{\beta\gamma} \Sigma_{\alpha\delta}$$

$$\{\Sigma_{\mu\nu}, \mathcal{H}\} = 0 \tag{37}$$

where

$$\tilde{g}_{\mu\nu} = g_{\mu\nu} - \frac{p_\mu p_\nu}{m^2}$$

$$\Sigma_{\mu\nu} = S_{\mu\nu} - \frac{p_\mu p^\alpha}{m^2} S_{\alpha\nu} - \frac{p_\nu p^\alpha}{m^2} S_{\mu\alpha} \tag{38}$$

Equations (38) show that the internal motion, in spite of the covariant 4-dimensional form, is actually three-dimensional and takes place on a 3-dimensional hyper-space in Minkowski space perpendicular to p_μ.

In the quantum case, also we can derive the equations of internal motion inside the electron in a covariant form in the proper-time formalism, generalizing the eqs. (12), (13), and (27). In order to do this we write the Dirac equation in a five-dimensional form $\psi(x_\mu,\tau)$, where τ is an invariant parameter -conjugate to mass m. The "Hamiltonian" with respet to τ is

$$\mathcal{H} = \gamma^\mu p_\mu \tag{39}$$

It is then possible to solve the quantum Heisenberg equations in covariant form. Again setting the charge coordinate X_μ equal to

$$x_\mu = X_\mu + Q_\mu \tag{40}$$

where X_μ is the center of mass coordinate and Q_μ the internal coordinate, and setting $P_\mu = m\dot{Q}_\mu$, $\dot{X}_\mu = \mathcal{H}^{-1}p_\mu$, we not only find the explicit time-dependences $Q_\mu(\tau)$, $P_\mu(\tau)$, but also the internal algebra generated by Q_μ, P_μ, $S_{\mu\nu}$ and . The result is exactly the equations (37) and (38) with the only difference that the Poisson bracket $\{\ \}$ is replaced by the commutator [ ] and a factor i appears everywhere on the right hand side of eqs. (37).[14] This correspondance constitute the canonical quantizaion of the classical electron theory to the Dirac electron. I believe this solves one of the outstanding problems of relativistic quantum theory, namely the precise classical counterpart of the Dirac electron and the nature of the phase space of the quantum spin. We may recall that Dirac discovered his equation, "by chance", as he put it,[15] and not by quantization of an existing classical model. Ever since, the physical meaning of the Dirac matrices has been rather mysterious. We can now directly relate them to the internal oscillatory degrees of freedom z and $\bar{z}$. In fact, the real and imaginary parts of z and $\bar{z}$ describe real oscillations of the charge around the center of mass and spin corresponds to the orbital angular momentum of these internal oscillations. One of the dynamical equations (35):

$$\dot{x}_\mu = v_\mu = \bar{z}\gamma_\mu z \tag{41}$$

relates the velocity of the charge to an internal velocity $\bar{z}\gamma_\mu z$ anologous to a rolling condition of a ball on an inclined plane.

Another noteworthy feature of the classical model is that mass m does not enter into the basic Lagrangian (34) as a fundamental parameter. It appears rather later as the value of the constant of motion = $\bar{z}\gamma_\mu z P^\mu$. Hence it can be modified by external interactions or by self-interaction. This is also true in the covariant formulation of the Dirac electron: mass is the eigenvalue of the constant of the motion $\mathcal{H} = \gamma^\mu \pi_\mu$. The Lagrangian (34) has however besides charge e, a fundamental constant λ of dimension of action which in quantized form becomes the Planck's constant h.

A second independent form of the quantization of the classical model of the electron is via the path integral formalism. It was also an outstanding unsolved problem how to obtain the quantum theory of discrete spin of the electron by a path integration based on a continuous classical action."[16] Since we have now an action (34) which is in one-to-one correspondance with the Dirac electron in canonical formalism, we can evaluate the path integrals not only in the (x_μ, p_μ) space, but also in the ($z,\bar{z}$)-space. Indeed, the quantum propagator can be obtained in a rather straightforward way not only for a free electron,[17] but also for an electron in an external field and for several interacting particles.[18] We have now a direct passage from classical particle trajectories to Feynman diagrams of perturbative quantumelectrodynamics.

VI. Further Generalizations of the Universal Role of the Conformal Dynamical Group

Having obtained the classical or the quantum algebra algebra (37) from the theories of the electron, we can now consider other representations or theories of the electron, we can now consider other representations or realizations of this algebra, than just the four-dimensional realization for the electron. We then obtain a family of compact quantum systems representing relativistic systems in their center of mass frame. The corresponding relativistic wave equations can be obtained by boosting these systems

$$v_\mu = p_\mu/m + \frac{1}{m} P_\mu \quad .$$

In the limit we get the infinite-dimensional representations of the algebra (37) and we shall now show that these describe composite relativistic objects, like H-atom or hadrons or nuclei.

If we disregard for a moment the restrictions (38), the algebra (37) can be made to be isomorphic to the Lie algebra of dimension 15 of the conformal group. [Here P_μ, Q_μ are combinations of the standard generators "P_μ" and "K_μ" of the conformal group]. However, because of the restrictions (38) not all of the 15 generators are independent and we have effectively the Lie algebra of SO(3,2).[14] The electron theory has in addition the observables $\bar{z}\gamma_5 z$ and $\bar{z}\gamma_\mu\gamma_5 z$ which are decoupled, but should be included in a full theory. These observables restore again the dynamical group SO(4,2). The electron theory (34) in fact is more concisely formulated in 5 (or 6)-dimensions because of the existence of 5 anticommuting γ-matrices. The 5-velocity $v_a = (\bar{z}\gamma_\mu z,\ i\bar{z}\gamma_5 z)$ satisfies $v_a v^a = 1$. And with $S_{\mu 5} = -\frac{1}{2}\bar{z}\gamma_\mu\gamma_5 z$, $S_{55} = 0$, we can write the electron equation in the form

$$m\ddot{x}_a = e F_{ab}\dot{x}^b + S_{ab} x^b \tag{42}$$

Quantummechanically also, the proper-time electron equation is more concisely written in the 5-dimensional form.

The physical interpretation of the conformal algebra (37) in the case of infinite-dimensional unitary representations is well-known. In this case P_μ, Q_μ, $\tilde{S}_{\mu\nu}$ are bona-fide relative coordinates of the constituents of a composite system in the center of mass frame.[19] For example, in H-atom, they are realized by the relative coordinates $\vec{r}$, $\vec{p}$ of the electron-proton system. Again a covariant wave equation for the moving atom may be obtained by boosting the system. The full algebraic framework of a moving relativistic system consists of the internal algebra plus the external Poincaré algebra which itself maybe generalized to a conformal algebra of space-time.[20] We should emphasize the physical difference between the two realizations of the same conformal algebra, one as the usual space-time interpretation, the other entirely different internal dynamical interpretation.

The appearance of the conformal dynamical group SO(4,2) in the dynamics of the 2-body problem maybe traced to electromagnetic interactions and to the zero mass of the exchanged photons. It is due this fact that the relative four vector coordinate $r_\mu = x_{1\mu} - x_{2\mu}$ satisfies $r_\mu r^\mu = 0$, and this condition then determines the realization of the conformal group in momentum space used in the relativistic Coulomb problem.[19,20] This is completely dual to the conformal group in coordinate space when the masslessness condition $P_\mu P^\mu = 0$ is satisfied.

Finally, I may add the remarkable role, which is surely not accidental, of the conformal dynamical group in the symmetry of the Periodic Table of elements which enhances its universality.[21]

REFERENCES

1. E. P. Wigner, Phys. Rev. 77, 711 (1950).
2. A. O. Barut, in Lectures in Theoretical Physics, Vol. IXB, (Gordon & Breach, 1967), p. 273.
3. V. Bargmann, Ann. of Math. 59, 1 (1954).
4. A. O. Barut and A. J. Bracken, J. Math. Phys. 26, 2515 (1985).
5. H. Weyl, The Theory of Gorups and Quantum Mechanics (Dover, New York, 1950), p. 272-280.
6. A. O. Barut and S. Komy, J. Math. Phys. 7, 1903 (1966); A. O. Barut, J. Math. Phys. 7, 1908 (1966).
7. T. D. Palev, J. Math. Phys. 23, 1778 (1982).
8. A. J. Bracken, (to be published).
9. A. O. Barut and A. J. Bracken, Phys. Rev. D23, 2454 (1981); D24, 3333 (1981).
10. A. O. Barut, A. J. Bracken, and W. D. Thacker, Lett. Math. Phys. 8, 472 (1984).
11. A. O. Barut and R. Raczka, Theory of Group Representations and Applications, Second Edition, 1980, (PWN-Warsaw).
12. A. O. Barut and N. Zhangi, Phys. Rev. Lett. 52, 2009 (1984).
13. A. O. Barut, in Differential Geometric Methods in Physics, Lecture Notes in Math. Vol. 905 edit. H. Doebner (Springer, 1982), p. 90; and in Quantum Optics, Relativity and Theory of Measurement, edit. P. Meystre (Plenum, 1983); p. 155.
14. A. O. Barut and W. D. Thacker, Phys. Rev. D31, 1386; 2076 (1985).
15. P. A. M. Dirac, The Relativistic Electron Wave Equation, Proc. European Conference on Particle Physics, Budapest 1977, p. 17.
16. R. P. Feynman and A. R. Hibbs, Quantum Mechanis and Path Integrals (McGraw Hill, N.Y., 1965), p. 34-36; L. S. Schulman, Techniques and Applications of Path Integration (Wiley, N.Y., 1981).
17. A. O. Barut and I. H.Duru, Phys. Rev. Lett. 53, 2355 (1984).
18. A. O. Barut and I. H. Duru, J. Math. Phys.

19. A. O. Barut, in Groups, Systems, and Many-Body Physics (Vieweg Verlag, 1980), edit. P. Kramer *et al*, p. Ch. VI.

20. A. O. Barut and G. Bornzin, J. Math. Phys. **15**, 1000 (1974).

21. A. O. Barut, in Prof. Rutherford Centennary Symposium, edit. B. Wybourne, (Univ. of Canterbury Press, 1972); p. 126.

$\overline{SL}(4,R)$ DYNAMICAL SYMMETRY FOR HADRONS

Dj. Šijački

Institute of Physics
P.O. Box 57, Belgrade, Yugoslavia

The double covering group $\overline{SL}(4,R)$ of the SL(4,R) group is proposed as a dynamical symmetry for hadron resonances. It is suggested that the spectrum of baryon and meson resonances, for each flavour, corresponds to a set of infinite-component field equation projected states of the spinor and tensor unirreps of $\overline{SL}(4,R)$ respectively. $\overline{SL}(4,R)$ is a geometrical space-time originated symmetry, presumably resulting from QCD, with possible connection to the affine gauge gravity and/or extended object picture of hadrons. The comparison with experiment seems very good.

Introduction

We have proposed[1)] recently that the complete spectrum of resonances for each baryon and meson flavour can be determined by infinite-component fields[2,3)] corresponding respectively to spinor and tensor infinite-dimensional unitary irreducible representations[4)] (unirreps) of the $\overline{SL}(4,R)$ group, i.e. the double covering of the SL(4,R) group. The suggested model makes use of the recent results about the $\overline{SL}(4,R)$ multiplicity-free unirreps, with the field theory serving as a guiding principle in actual assignement of hadronic states and making a contact with observations.

According to QCD, the observed spectrum of hadrons represents the set of stable and metastable solutions of the Euler-Largrange equations for a second-quantized action, constructed from quark and gluon fields. The parallels are with Chemistry, where the elements and compounds, with their excited states,are known to represent the solutions of Schrödinger's equation, with nuclei, photons and electrons as constituents. In each of these cases however, it has not been possible to use the fundamental dynamical model for actual calculation beyond the relevant "hydrogen atom" level. In hadron physics, the experimental exploration of the hadron spectrum goes on even though theory has moved away to the constituent level, except for the "bag model" approximate calculations. Our model may arise as a geometrical-symmetry of the QCD equations, in the same sense that the Nuclear Shell Model is belived to be generated by meson exchanges between nucleons. Alternatively, it is also possible that the success of the $\overline{SL}(4,R)$ scheme be due to an additional interaction component which is generally not included in the $SU(3)^{color}$ setting. Such a component might involve extensions of gravity such as might arise from an GA(4,R) gauge,[5-8] or from a string-like generalized treatment incorporating the bag model. An evolving confined lump (the bag) would indeed be represented by an $\overline{SL}(4,R)$ 4-measure,[9-11] just as the evolving string is given by that of $\overline{SL}(2,R)\simeq SU(1,1)$, the 2-measure spanned by the spinning string.

In contradistinction to leptons, which appear as point-like objects and whose space-time structure is completly determined by the Poincaré group, the strongly interacting particles, the hadrons, show additional structure. Hadrons of a given flavour (the same internal quantum numbers) lie on practically linear trajectories in the Chew-Frautschi plot (J vs. m^2). Furthermore, particles belonging to the same trajectory satisfy the $\Delta J=2$ rule. The seemingly infinite number of equally spaced hadron states of Regge trajctory were interpreted as excitations of a single physical object and classified by means of the unitary irreducible representations (unirreps) of the noncompact SL(3,R) group.[12] A minimal fully relativistic extension of the SL(3,R) model is given by the SL(4,R) spectrum generating symmetry,[13] with the six Lorentz J and nine shear T generators. By adding the dilation invariance, another important feature in hadronic interactions, one arrives at the general linear group GL(4,R). Finally, together with the translations one obtains the general affine group GA(4,R).

Several, manifestly relativistic, extended object models have been proposed either to explain quark confinement or with a built in confinement of them. According to the bag model, a strongly interacting particle is a finite region of space-time to which the fields are confined in a Lorentz invariant was by endowing the finite region with a constant energy per unit volume B. Strong interactions are descibed by the following action integral

$$A = \int dt \int d^3x \left[\mathcal{L}_{QCD}(\text{quarks, gluons}) - B\right].$$

The second term is invariant for fixed time with respect to the SL(3,R) transformations. In general, the second part of the bag action is invariant under the SL(4,R) group, which contains as subgroups the Lorentz group and SL(3,R). The dynamics of a hadron described by say a spheroidal bag are rotationally invariant giving rise to the conserved bag internal orbital angular momentum L, and to a good quantum number K which is due to the rotational invariance about the bag symmetry axis $R(\pi)$. The wave function of such a bag is of the form

$$\chi_K(g)D^L_{KM}(\alpha,\beta,\gamma) + (-)^{L+K}\chi_{-K}(g)D^L_{-KM}(\alpha,\beta,\gamma),$$

where α, β, γ are Euler angles and q are the remaining coordinates. The states with K=O can be labeled by the eigenvalue r of R, where $r=(-)^L$ and therefore the allowed values of L are L=0,2,4,... for K=0, r=1 and L=1,3,5,... for K=0, r=-1. When $K\neq 0$ there is only a constraint $L \geq K$, i.e. L=K,K+1,K+2,... These values of L are exactly those of the SL(3,R) unirreps. For K=0, group theoreticaly one has the Ladder unirreps, while phenomenologically one has the states belonging to the same Regge trajectory. It turns out that the SL(4,R) unirreps desribe both the orbital (SL(3,R) unirrep) and the radial excitations of a hadronic bag.

The success of the dual string models indicates strongly the importance of considering hadrons as extended objects. These models are based on the $\overline{SL}(2,R)$ group. Dual amplitudes, as well as the Virasoro and the Neveu-Schwarz-Ramond gauge algebras can be directly constructed by making use of the infinite dimensional $\overline{SL}(2,R)$ representations. The string model can be generalized to the 3-dimensional model of a lump, i.e. to a region of 3-space embedded in space-time. It is parametrized by 4 internal coordinates y^μ $\mu=0,1,2,3$. The first one y^o plays the role of the proper time, while the remaining three y^i can be thought of as labeling the points belonging to the lump. The coordinates $x^a(y^\mu)$ locate the lump in the embedding space-time as internal coordinates. In analogy with the relativistic action of a point particle or of a free string we take the relativistic action for a free lump to be proportional to the volume of space-time generated by the evolution of the lump, i.e.

$$A = -\alpha^{-2}\int_{y_1^o}^{y_2^o} dy^o \int_V d^3y\left[-\det(g_{\mu\nu})\right]^{1/2},$$

where $g_{\mu\nu}=\eta_{ab}(\partial x^a/\partial y^\mu)(\partial x^b/\partial y^\nu)$ is the metric induced on the submanifold of space-time generated by the lump from the embedding flat space-time. V is the volume of the lump and α has the dimension M^{-2}. If we perform variations for which initial and final positions of the lump are not kept fixed, but only actual motions of the lump are allowed, we can compute the momentum P_a, the angular momentum M_{ab} and the shear T_{ab} currents of the lump. The $\mu=o$ integrated components of these operators over the lump volume V generate the $SA(4,R)=T_4 \circledS SL(4,R)$ group.

$\overline{GA}(4,R)$ and $\overline{SL}(4,R)$ unirreps

From the Particle Physics point of view, one is interested in a unified description of both bosons and fermions. This would require the existence of respectively tensorial and (double valued) spinorial representations of the GA(4,R) group. Mathematically speaking, one is interested in the corresponding single valued representations of the double covering $\overline{GA}(4,R)$ of the GA(4,R) group, since its topology, is given by the topology of its (double connected) linear compact subgroup SO(4).

The $\overline{GA}(4,R)$ group is a semidirect product of the group of translations in four dimensions (Minkowski space-time), and of the double covering $\overline{GL}(4,R)$ of the general linear GL(4,R) group, i.e.

$$\overline{GA}(4,R) = T_4 \circledS \overline{GL}(4,R).$$

The GL(4,R) group can be split into the one-parameter group of dilations, and the SL(4,R) group. The latter is a group of volume preserving transformations in the Minkowski space-time. The maximal compact subgroup of SL(4,R) is SO(4). The universal covering, i.e. the double covering, group we denote by $\overline{SL}(4,R)$ and its maximal compact subgroup is $\overline{SO}(4)$ which is isomorphic to $SU(2) \otimes SU(2)$. The $\overline{SL}(4,R)$ group is physically relevent since it has (infinite-dimensional) spinorial unitary irreducible representations (unirreps) which are double-valued unirreps of SL(4,R). The SL(4,R) group has as a subgroup the Lorentz group SO(3,1), and correspondingly $\overline{SL}(4,R)$ has as a subgroup $\overline{SO}(3,1) \simeq SL(2,C)$.

The Lorentz group is generated by the angular momentum and the boost operators J_i and K_i, $i = 1,2,3$ respectively. We write them as J_{ab}, $a,b = 0,1,2,3$, where $J_{ab} = -J_{ba}$. The remaining nine generators form a symmetric second rank shear operator T_{ab}, $a,b = 0,1,2,3$, i.e. $T_{ab} = T_{ba}$ and $\mathrm{tr}T_{ab} = 0$. The commutation relations of the SL(4,R) algebra are given by the following relations

$$[J_{ab}, J_{cd}] = -i(\eta_{ac}J_{bd} - \eta_{ad}J_{bc} - \eta_{bc}J_{ad} + \eta_{bd}J_{ac}$$

$$[J_{ab}, T_{cd}] = -i(\eta_{ac}T_{bd} + \eta_{ad}T_{bc} - \eta_{bc}T_{ad} - \eta_{bd}T_{ac}$$

$$[T_{ab}, T_{cd}] = i(\eta_{ac}J_{bd} + \eta_{ad}J_{bc} + \eta_{bc}J_{ad} + \eta_{bd}J_{ac}),$$

where n_{ab} is the Minkowski metric $\eta_{ab} = \mathrm{diag}(+1,-1,-1,-1)$.

$$J_{ab} = \frac{1}{2}(Q_{ab} - Q_{ba}), \quad T_{ab} = Q_{(ab)} = \frac{1}{2}(Q_{ab} + Q_{ba}) - \frac{1}{4} n_{ab} Q^{c}{}_{c},$$

and the dilation generator is $D = \frac{1}{4} n_{ab} Q^{c}{}_{c}$. The T_{ab} and D operators form together a 10-component symmetric (not traceless) tensot

$$Q_{\{ab\}} = \frac{1}{2}(Q_{ab} + Q_{ba}).$$

The translation generators P_a together with the $\overline{GL}(4,R)$, generators Q_{ab} fulfil the $\overline{GA}(4,R)$ commutation relations

$$[Q_{ab}, Q_{cd}] = i\eta_{bc}Q_{ad} - i\eta_{ad}Q_{cb},$$

$$[Q_{ab}, P_c] = -i\eta_{ac}P_b,$$

$$[P_a, P_b] = 0.$$

Owing to the $\overline{GA}(4,R)$ semidirect product structure it is rather straightforward to write down its (unitary irreducible) representations. The general recipe for constructing a semidirect product group representations is well known (Wigner, Mackey,...). There are two important ingredients to be determined: i) The orbits of the translations and

ii) The corresponding little groups (which are the subgroups of $\overline{GL}(4,R)$).

When the orbit is $\hat{T}_4-\{0\}$, $\hat{T}_4$ being the character group of the translation subgroup, the corresponding little group is $\tilde{T}_3 \circledS \overline{SL}(3,R)$. The $\tilde{T}_3$ subgroup is generated by $Q_{oi} = 1/2(J_{oi}+T_{oi})$, $i = 1,2,3$ which commute mutualy. Now, the first possibility is to represent the whole little group trivially, and we obtain the scalar state $\phi(p)$. The second possibility is to represent $\tilde{T}_3$ trivially, and the remaining $\overline{SL}(3,R)$ subgroup linearly. The $\overline{SL}(3,R)$ unitary irreducible representations are infinite dimensional and can be both spinorial and tensorial.[14] These unirreps determine the Regge trajectory spin content.

Let us consider now the $\overline{GA}(4,R)$ representations on fields. The $\overline{GL}(4,R)$ generators $Q^a{}_b$ can be split into the orbital and intrinsic parts

$$Q^a{}_b = \hat{Q}^a{}_b + O^a{}_b,$$

where the orbital part is of the form $O^a{}_b = x^a p_b$, $a,b = 0,1,2,3$. The $\overline{GA}(4,R)$ commutation relations listed above are now supplemented by the following relations

$$[\hat{Q}_{ab}, O_{cd}] = 0,$$

$$[\hat{Q}_{ab}, P_c] = 0.$$

The linear $\overline{GA}(4,R)$ representations on fields are of the form

$$(a,\bar{A}) \; : \; \Psi(x) \to D(\bar{A})\Psi(A^{-1}(x-a)),$$

$$D(\bar{A}) = \exp(-i\xi^a{}_b Q^b{}_a),$$

and $D(\bar{A})$ is a representation of the intrincis $\overline{GL}(4,R)$ component. It is obvious from the above expression for the $\overline{GA}(4,R)$ representations on fields, that the essential part is given by the $\overline{GL}(4,R)$, i.e.$\overline{SL}(4,R)$ unirreps.

In the physical applications we will only make use of the multiplicity free $\overline{SL}(4,R)$ unirreps. These unirreps contain each representation (j_1,j_2) of its $\overline{SO}(4) \simeq SU(2) \otimes SU(2)$ maximal compact subgroup at most once. The complete set of these representations is given as follows.[4]

<u>Principal series</u>: $D^{pr}(0,0;e_2)$, and $D^{pr}(1,0;e_2)$, $e_2 \in R$, with the $\{(j_1 j_2)\}$ content given by $j_1 + j_2 \equiv 0 \pmod 2$, and $j_1 + j_2 \equiv 1 \pmod 2$ respectively.

<u>Supplementary series</u>: $D^{supp}(1,0;e_1), 0<|e_1|<1$, with the $\{(j_1,j_2)\}$ content given by $j_1 + j_2 \equiv 1 (\text{mod } 2)$.

<u>Discrete Series</u>: $D^{disc}(j_0,0)$ and $D^{disc}(0,j_o), j_o = \frac{1}{2},1,\frac{3}{2},2,\ldots$, with the $\{(j_1,j_2)\}$ content given by $j_1 - j_2 \geq j_o$, $j_1 + j_2 \equiv j_o (\text{mod } 2)$, and $j_2 - j_1 \geq j_o$, $j_1 + j_2 \equiv j_0 (\text{mod } 2)$ respectively.

<u>Ledder series</u>: $D^{ladd}(0;e_2)$, and $D^{ladd}(\frac{1}{2};e_2), e_2 \varepsilon R$, with the $\{(j_1,j_2)\}$ content fiven by $\{(j_1,j_2)\} = \{(j,j)|j = 0,1,2,\ldots\}$, and $\{(j_1,j_2)\} = \{(j,j)|j = \frac{1}{2},\frac{3}{2},\frac{5}{2},\ldots\}$ respectively.

The $D^{ladd}(\frac{1}{2})$, $D^{disc}(\frac{1}{2},0)$ and $D^{disc}(0,\frac{1}{2})$ unirrep $\{(j_1,j_2)\}$ -content is illustrated in Fig.1.

<u>Space-time description of leptons and hadrons</u>

The $\overline{GA}(4,R)$ symetry approach to Particle Physics, allows one to have rather different space-time field description of leptons and hadrons. For leptons we make use of the $\overline{GA}(4,R)$ representations, which are linear when restricted to the Poincaré subgroup.[7)] In this way leptons are essentialy given by the Dirac field, with the nonlinear symmetry realizer given by the metric field.

In the case of hadrons, the non-trivial part of the field structure is, as shown, determined by the $\overline{SL}(4,R)$ unirreps. Furthermore, there are additional constraints coming from the appropriate field equations. For the tensorial (meson) representations, the simplest choices are either $D^{ladd}(0;e_2)$ or $D^{ladd}(\frac{1}{2};e_2)$ with a Klein-Gordon-like (infinite-component) equation[1,3)] for the corresponding manifield $\phi(x)$

$$(\partial_\mu \partial^\mu + m^2)\phi(x) = 0.$$

Spinor (baryon) manifields obey a first-order equation[2)] with infinite X_μ matrices generalizing Dirac's, except for the requirement of anti-commutation. The X_μ behave as a Lorentz 4-vector $(\frac{1}{2},\frac{1}{2})$ and we are forced to use the reducible pair of $\overline{SL}(4,R)$ unirreps $D^{disc}(\frac{1}{2},0)\otimes D^{disc}(0,\frac{1}{2})$ for the menifield $\Psi(x)$,

$$(iX_\mu \partial^\mu - \kappa)\Psi(x) = 0.$$

The X_μ operators only connect the $|j_1-j_2|= \frac{1}{2}$ states across the two unirreps (see Fig.1). These are thus the only physical (propagating) states in $\Psi(x)$, all others decouple. For the $\Delta(1232)$ system we use the same pair of intrinsic, unirreps now adjoining a n explicit 4-vector index as in the Rarita-Schwinger field,

$$(iX_\mu \partial^\mu - \kappa)\Psi_\rho(x) = 0.$$

The physical $\overline{SO}(4)$ multiplets projected out by Lorentz-invariance in the equations are thus

$D^{ladd}(o,e_2), \phi_{glueballs}$ if any: $\{j_1,j_2\} = \{(0,0)(1,1)(2,2)\dots\}$,

$D^{ladd}(\frac{1}{2},e_2), \Phi:\{(j_1,j_2)\} = \{(\frac{1}{2},\frac{1}{2})(\frac{3}{2},\frac{3}{2})(\frac{5}{2},\frac{5}{2}),\dots\}$,

$$D^{disc}(\tfrac{1}{2},0)\oplus D^{disc}(0,\tfrac{1}{2}), \Psi:\{(j_1,j_2)\} = \{(\tfrac{1}{2},0),(\tfrac{3}{2},1),(\tfrac{5}{2},2),\dots\}\oplus \oplus\{(0,\tfrac{1}{2}),(1,\tfrac{3}{2}),(2,\tfrac{5}{2}),\dots\},$$

$$\left[D^{disc}(\tfrac{1}{2},0)\oplus D^{disc}(0,\tfrac{1}{2})\right]\otimes(\tfrac{1}{2},\tfrac{1}{2}), \Psi_\rho:\{(j_1,j_2)\} = \{(1,\tfrac{1}{2}),(2,\tfrac{3}{2}),(3,\tfrac{5}{2}),\dots\}\oplus \oplus\{(\tfrac{1}{2},1),(\tfrac{3}{2},2),(\tfrac{5}{2},3),\dots\}.$$

For a given $\overline{SO}(4)$ representation (j_1,j_2), the total (spin) angular momentum is $\vec{j} = \vec{j}^{(1)} + \vec{j}^{(2)}$, and $\vec{N} = \vec{j}^{(1)} - \vec{j}^{(2)}$. The vector operator $\vec{N}$ connects different spin values, and is an odd operator under parity. Thus the J^P content of a (j_1,j_2) $\overline{SO}(4)$ representation is

$$J^P = (j_1 + j_2)^P,\ (j_1 + j_2-1)^{-P}, (j_1 + j_2-2)^P,\dots,(|j_1 - j_2|).$$

The $\overline{SL}(3,R)$ subgroup[9] unirreps determine the Regge trajcetory states of a given $\overline{SL}(4,R)$ unirrep. In decomposing an $\overline{SL}(4,R)$ unirrep v.r.t. the $\overline{SL}(3,R)$ ones,[14] it is convenient to introduce a quantum number n defined by[13]

$$j_1 + j_2 = J + n.$$

Since the different spin values of an $\overline{SL}(3,R)$ unirrep are connected by an even parity quadrupole operator T_{ij}, all unirrep states have the same parity. The $\overline{SL}(4,R)$ ladder unirreps contain an infinite sum of $\overline{SL}(3,R)$ ladder unirreps,[13] cf. Fig.2.

The $\overline{SL}(4,R)$ unirrep wave equation projected states are in an excellent agreement with the observed hadronic states.[1] We find a striking match between the (J^P,mass) values and the wave equation-projected $\overline{SL}(4,R)$ unirrep states. Moreover, a remarkably simple mass formula fits these infinite systems of hadronic states. For N and Λ (and the higher spin Δ) resonances we write

$$m^2 = m_o^2 + \frac{1}{\alpha_f'}(j_1 + j_2 - \tfrac{1}{2} - \tfrac{1}{2}n)$$

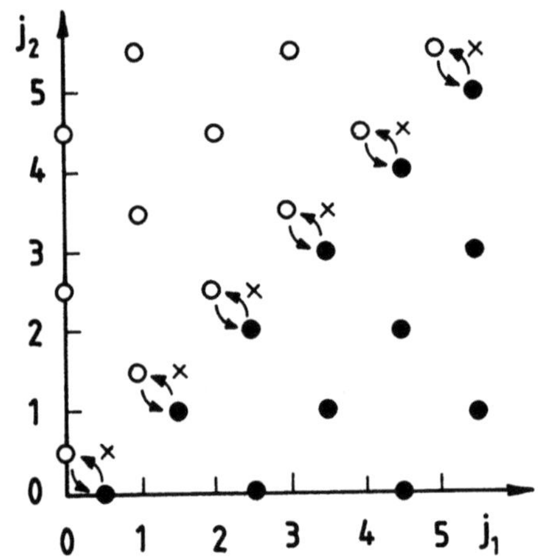

Figure 1. $\{(j_1,j_2)\}$content of representations $D^{ladd}(1/2)$: croses, $D^{disc}(1/2,0)$: solid dots, $D^{disc}(0,1/2)$: open dots; arrows connect the wave equation projected states.

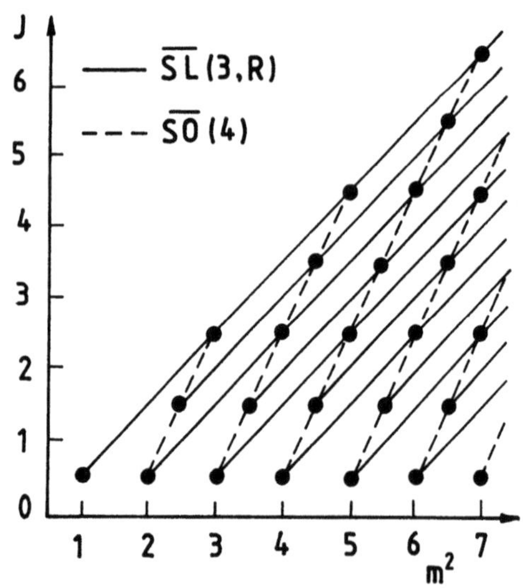

Figure 2. Physical $D^{disc}(1/2,0)$ or $D^{disc}(0,1/2)$ states accordind to the mass formula with $m_o=(\alpha')^{-1}$ = 1 GeV.

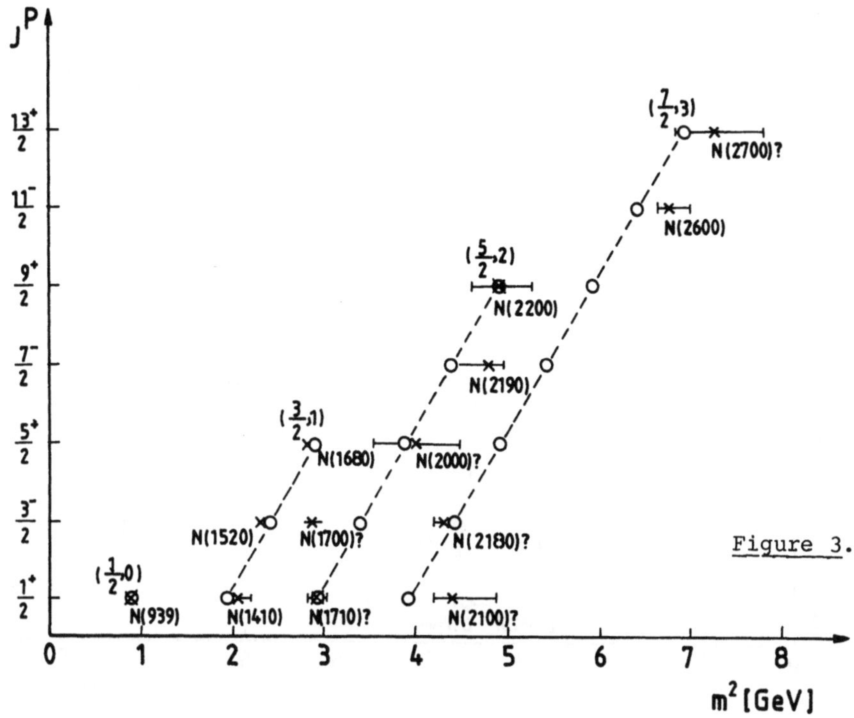

Figure 3.

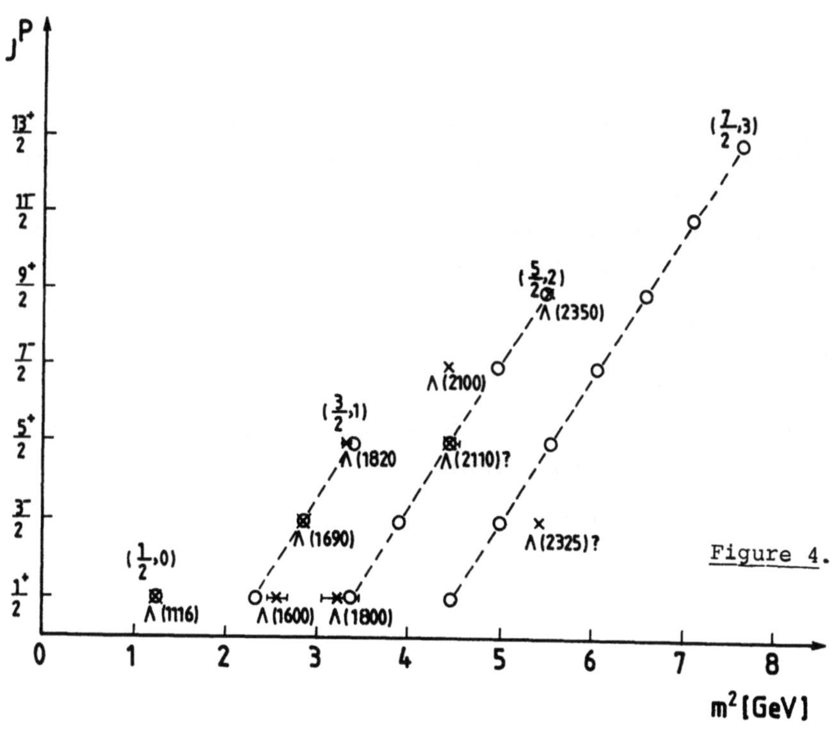

Figure 4.

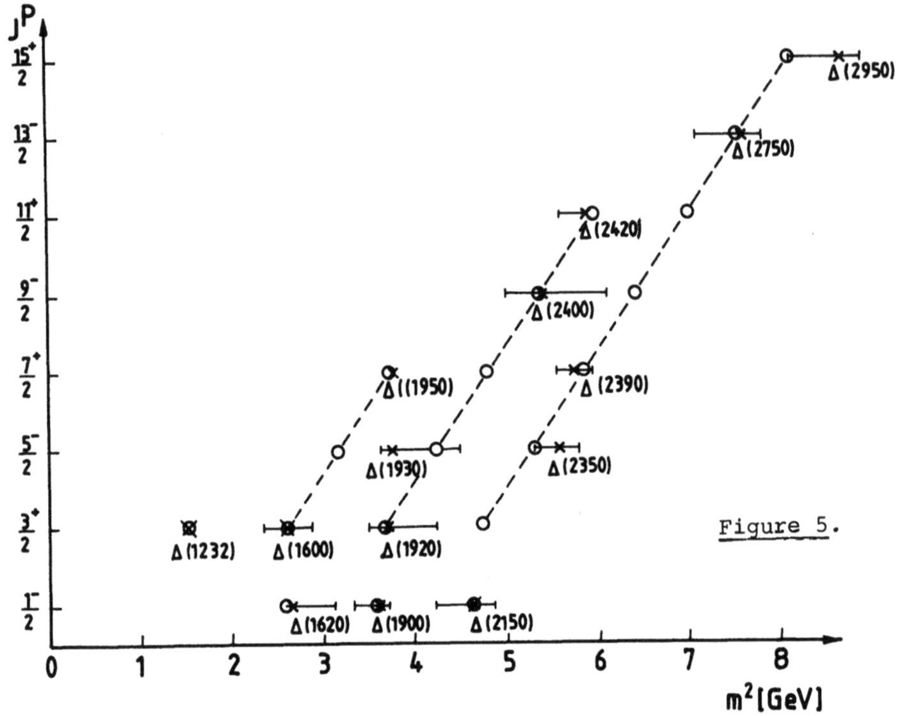

Figure 5.

where m_o is the mass of the lowest lying state, and α'_f is the slope of the Regge trajectory for that flavour (see Fig.2). We illustrate this mass formula for the best known system of N, Λ and Δ resonances (Fig.3, Fig.4 and Fig.5 respectively). For the average masses of the (j,j) unirreps of mesons belonging

to $D^{ladd}(\frac{1}{2},e_2)$,

$$m^2 = m_o^{\ 2} + \frac{1}{\alpha'_f}(j - \frac{1}{2})$$

while (at least) for the lowest $\overline{SO}(4)$ states $(\frac{1}{2},\frac{1}{2})$ of opposite parity we find the following mass formula

$$m^2(0^-) + m^2(1^+) = m^2(0^+) + m^2(1^-).$$

References

1. Y. Ne'eman and Dj. Šijački, Phys. Lett. 157B(1985)267.
2. A. Cant and Y. Ne'eman, J. Math. Phys. (in press).
3. Y. Ne'eman and Dj. Šijački, Phys. Lett. 157B(1985)275.
4. Dj. Šijački and Y. Ne'eman, J. Math. Phys. 26(1985)2457.
5. F. W. Hehl, G. D. Kerlick and P. von der Heyde, Phys. Lett. 63B(1976)446.
6. F. W. Hehl, E. A. Lord and Y. Ne'eman, Phys. Lett. 71B(1977)432; Phys. Rev. D 17(1978)428.
7. Y. Ne'eman and Dj. Šijački, Proc. Natl. Acad. Sci. USA 76(1979)561; Ann. Phys. (N.Y.) 120(1979)292.
8. Dj. Šijački, Phys. Lett. 109B(1982)435.
9. Dj. Šijački, Lecture Notes in Phys. 79(1978)394.
10. Dj. Šijački, Ann. Isr. Phys. Soc. 3(1980)35.
11. Y. Ne'eman and Dj. Šijački, Tel-Aviv Univ. preprint TAUP 699-78.
12. Y. Dothan, M. Gell-Mann and Y. Ne'eman, Phys. Lett. 17(1965)148.
13. Dj. Šijački, Ph. D. Thesis, Duke Univ. (1974).
14. Dj. Šijački, J. Math. Phys. 16(1975)289.

A NEW QUANTUM RELATIVISTIC OSCILLATOR AND THE HADRON MASS SPECTRUM

P. Magnollay
Center for Particle Theory
The University of Texas at Austin
Austin, Texas 78712

Abstract

We construct a new quantum relativistic oscillator (QRO) based on the spectrum generating group SO(3,2). We show that all the features of the three-dimensional harmonic oscillator are recovered in the non-relativistic limit. This QRO gives a classification of hadrons that leads to the linear Regge trajectories; moreover, the Regge slopes are SU(3) invariant.

Introduction

A QRO must satisfy essentially three criteria:

1) The hamiltonian must be a Lorentz scalar.
2) The usual three-dimensional harmonic oscillator should be recovered in the limit $C \to \infty$.
3) The mass levels of the QRO should correspond to elementary particles (irreps of the Poincaré group) in order to apply it to particle physics.

The simplest way to build a model satisfying these criteria is to use the formalism of constrained hamiltonian mechanics.[1] The covariant hamiltonian is defined from a constraint on the mass operator and, following Dirac, the constraint is set equal to zero only after the equations of motion have been calculated. From the above point 3), it is clear that the symmetry of the QRO must be larger than the Poincaré symmetry and we therefore extend the Poincaré symmetry into a relativistic symmetry.[2] We choose the following relativistic symmetry:

$$P_{J_{\mu\nu},\hat{P}_\mu} \circledS SO(3,2)_{S_{\mu\nu},\Gamma_\mu} \qquad (1)$$

where $\hat{P}_\mu = P_\mu/M$ and $M^2 = P_\mu P^\mu$.

The semi-direct product indicates that the intrinsic spin $S_{\mu\nu}$ (which is not the physical spin) and the operator Γ_μ transform respectively as a second rank tensor and as a vector operator under the action of the total angular momentum $J_{\mu\nu}$. Once the constraint on the mass operator is imposed, P_μ no longer commutes with SO(3,2). After the constraint is imposed we assume that $\hat{P}_\mu$ still commutes with SO(3,2) and therefore the symmetry (1) remains. Notice that the hamiltonian is invariant only under the Poincaré group and that the Lie algebra of (1) is exactly the algebra defining spin-½ particles in the Dirac theory. In the Dirac theory, there is no problem with the constraint (Dirac equation) because the four-dimensional representation of SO(3,2) is a degenerate representation (only one mass and one spin). The dynamics of our relativistic oscillator is very reminiscent of the "zitterbewegung" of the electron.[3)]

Quantum Relativistic Oscillator

Our QRO has two positions[4)]: Y_μ, the center of mass position operator and Q_μ, the particle position operator (canonical conjugate to P_μ). The difference $d_\mu = Y_\mu - Q_\mu$ is what we call the internal relativistic coordinate and is defined by[5)]

$$d_\mu = S_{\mu\nu} \frac{P^\nu}{M^2} \tag{2}$$

The total angular momentum $J_{\mu\nu}$ can be decomposed in terms of the intrinsic spin $S_{\mu\nu}$ or the physical spin $\Sigma_{\mu\nu}$ in the following ways:

$$J_{\mu\nu} = Q_\mu P_\nu - Q_\nu P_\mu + S_{\mu\nu} = Y_\mu P_\nu - Y_\nu P_\mu + \Sigma_{\mu\nu} \tag{3}$$

The two spin tensors are therefore related by the equation

$$\Sigma_{\mu\nu} = S_{\mu\nu} - d_\mu P_\nu + d_\nu P_\mu \tag{4}$$

The spin can also be described by the Pauli-Lubanski vector $\hat{W}_\mu = \frac{1}{2} \varepsilon_{\mu\nu\rho\sigma} \hat{P}^\nu \Sigma^{\rho\sigma}$ and the definition (2) of d_μ is the only one consistent in order to have:

$$-\hat{W}_\mu \hat{W}^\mu = \tfrac{1}{2} \Sigma_{\mu\nu} \Sigma^{\mu\nu} \tag{5}$$

Our construction of d_μ only allows spacelike oscillations and the physical spin tensor $\Sigma_{\mu\nu}$ has only three independent components. Finally the hamiltonian of an QRO is taken to be:

$$\mathcal{H} = \phi[P_\mu P^\mu - \frac{1}{\alpha'} \hat{P}_\mu \Gamma^\mu] \tag{6}$$

The coefficient ϕ is a Lagrange multiplier and is found to be $-\frac{1}{2M}$ when the evolution parameter τ associated to $\mathcal{H}$ is the proper time of the center of mass. The parameter $\frac{1}{\alpha'}$ has units of $(\mathrm{GeV})^2$. The constraint on the mass operator is imposed when $\mathcal{H} = 0$ on the physical states. Using the canonical commutation relation $[Q_\mu, P_\nu] = -ig_{\mu\nu}$ and the Heisenberg equations of motion to get the proper time derivatives of the observables, we obtain:

$$\dot{J}_{\mu\nu} = 0 \qquad \dot{\Sigma}_{\mu\nu} = 0 \qquad \dot{Y}_{\mu} = \hat{P}_{\mu}$$
$$\dot{\Gamma}_{\mu} = -\frac{1}{2\alpha'} d_{\mu} \qquad \dot{d}_{\mu} = \frac{1}{2\alpha' M^2} (\Gamma_{\mu} - \hat{P}_{\mu}(\hat{P}\cdot\Gamma)) \tag{7}$$

Taking the second derivative of d_{μ}, one finds the standard oscillator equation:

$$\ddot{d}_{\mu} + \frac{1}{4\alpha'^2 M^2} d_{\mu} = 0 \tag{8}$$

We then define the canonical conjugate momentum $\pi_{\mu} = 2M\dot{d}_{\mu}$ of d_{μ} and obtain the following relativistic Heisenberg commutation relations[5]:

$$[d_{\mu},d_{\nu}] = -\frac{i}{M^2}\Sigma_{\mu\nu} \qquad [\pi_{\mu},\pi_{\nu}] = -\frac{i}{\alpha'^2 M^2}\Sigma_{\mu\nu} \qquad [d_{\mu},\pi_{\nu}] \overset{\mathcal{C}}{=} -i\check{g}_{\mu\nu} \tag{9}$$

The constraint $\mathcal{H} = 0$ has been used in order to get the last equality and $\check{g}_{\mu\nu} = g_{\mu\nu} - \hat{P}_{\mu}\hat{P}_{\nu}$.

Representation Space of the Relativistic Symmetry

In order to choose a basis of the representation space of the relativistic symmetry (1), we choose the following complete set of commuting observables: $\hat{P}_{\mu}$, M^2, $\hat{P}_{\mu}\Gamma^{\mu}$, $\frac{1}{2}\Sigma_{\mu\nu}\Sigma^{\mu\nu}$ and $\Sigma_{12}^{(R)}$ (taken in the rest frame) with the eigenvalues $\hat{p}_{\mu}$, m^2, μ, $j(j+1)$ and j_3 respectively. The basis vectors are therefore labeled by $|\hat{p}_{\mu},m^2,\mu,j,j_3\rangle$. We use irreducible, unitary, multiplicity-free representations of SO(3,2) that are characterized by the minimum value μ_{min} of Γ_o (or $\hat{P}_{\mu}\Gamma^{\mu}$) and by the minimum value S of the spin appearing in the representation. The quadratic and quartic Casimir operators are then functions of μ_{min} and S and are given by[7]:

$$\text{eigenvalue of } C_{(2)} = -R = (\mu_{min} - \tfrac{3}{2})^2 + S(S+1) - \tfrac{9}{4}$$
$$\begin{cases} \mu_{min} \geq S + \frac{1}{2} & \text{for } S = 0,\frac{1}{2} \\ \mu_{min} = S + 1 & \text{for } S \geq 1 \end{cases}$$
$$\text{eigenvalue of } C_{(4)} = S(S+1)[-R - (S-1)(S+2)] \tag{10}$$

The spectrum of Γ_o is discrete and μ takes the values $\mu_{min} + n$ with $n = 0,1,2,\cdots$. By applying a Lorentz boost on the states at rest, we induce a representation of the full relativistic symmetry (1). Every subspace corresponding to an irreducible representation of SO(3) × SO(2) (maximal compact subgroup) determined by a particular set (n,j) now becomes an irreducible representation D(m(n),j) of the physical Poincaré group[8] (with $m^2(n) = \frac{1}{\alpha'} n$). The complete representation space of our QRO is therefore a discrete direct sum of irreducible representations of the Poincaré group so that every mass level of the oscillator corresponds to an elementary particle. These spaces are written as:

$$\sum_{\substack{n=0,1,2,\cdots \\ j=0,2,4,\cdots n \ (n \text{ even}) \\ j=1,3,5,\cdots n \ (n \text{ odd})}} \oplus \ D(m(n),j) \qquad \text{when} \quad S = 0 \tag{11}$$

and

$$\sum_{\substack{n=0,1,2,3\cdots \\ j=S,S+1,\cdots n+S}} \oplus \ D(m(n),j) \qquad \text{when} \quad S = \tfrac{1}{2},1,\tfrac{3}{2},2,\cdots \tag{12}$$

For $\mu_{min} = S + 1$ and $S = 0,\frac{1}{2},1,\frac{3}{2},2\cdots$ the space (12) is also an irreducible representation of a relativistic symmetry based on SO(4,2)[9] (instead of SO(3,2)). Even though electromagnetic decays seem to indicate that a larger symmetry such as SO(4,2) is needed, we will restrict ourselves to SO(3,2) since our QRO is not yet coupled to interactions.

Non Relativistic Limit

The non-relativistic limit is taken by contracting the Poincaré group and SO(3,2).[10] The contraction parameter $\frac{1}{c}$ goes to zero and we follow a sequence of irreducible representations of the symmetry (1) along which the operators S_{io}, Γ_i, Γ_o and P_o go to infinity, but such that the following limits are finite:

$$\frac{S_{io}}{mc} \to \overset{\circ}{\xi}_i \,, \quad \frac{\Gamma_i}{\alpha' mc} \to \overset{\circ}{\pi}_i \,, \quad \frac{\Gamma_o}{\alpha'(mc)^2} \to \mathbf{1} \,, \quad \frac{P_o}{mc} \to \mathbf{1} \tag{13}$$

The Galilean mass m and the non-relativistic hamiltonian H are defined by the expansion of P_o:

$$P_o = mc + \frac{H}{c} + O(\frac{1}{c^3}) \tag{14}$$

In order to have a well defined representation space at any step of the contraction, we establish a one to one correspondence between the value of c and the value of the quadratic Casimir oeprator by imposing the condition:

$$-R = (mc)^4 \alpha'^2 \tag{15}$$

This condition is of course consistent with the contraction (13). The results of the contraction are presented in the table (I). The Poincaré group contracts to the Galilei group, SO(3,2) contracts to the group of the three-dimensional harmonic oscillator (it contains the Heisenberg group and SO(3) as subgroups). The relativistic Heisenberg commutation relations contract into the usual Heisenberg commutation relations. The constraint $\mathcal{H} = 0$ gives the hamiltonian of the three-dimensional harmonic oscillator.

TABLE I. Contraction of the QRO.

$$
\begin{aligned}
P_{J_{\mu\nu}}, P_\mu &\rightarrow G_{J_i}, G_i, M, H, P_i \\
SO(3,2)_{S_{\mu\nu}, \Gamma_\mu} &\rightarrow SO(3)_{S_{ij}} \otimes H(3)_{\xi_i, \pi_j, \mathbf{1}} \\
d_o &\rightarrow 0 \\
\pi_o &\rightarrow 0 \\
d_i &\rightarrow \overset{\circ}{\xi}_i = \xi_i^{(\infty)} \\
\pi_i &\rightarrow \overset{\circ}{\pi}_i + P_i = \pi_i^{(\infty)} \\
\Sigma_{io} &\rightarrow 0 \\
\Sigma_{ij} &\rightarrow S_{ij} + \xi_i^{(\infty)} P_j - \xi_j^{(\infty)} P_i = \Sigma_{ij}^{(\infty)} \\
[d_\mu, d_\nu] = -\frac{i}{M^2}\Sigma_{\mu\nu} &\rightarrow [\xi_i^{(\infty)}, \xi_j^{(\infty)}] = 0 \\
[\pi_\mu, \pi_\nu] = -\frac{i}{\alpha'^2 M^2}\Sigma_{\mu\nu} &\rightarrow [\pi_i^{(\infty)}, \pi_j^{(\infty)}] = 0 \\
[d_\mu, \pi_\nu] \cong -i\check{g}_{\mu\nu} &\rightarrow [\xi_i^{(\infty)}, \pi_j^{(\infty)}] = i\delta_{ij} \\
\mathcal{H} = 0 &\rightarrow H = \frac{\vec{P}^2}{2m} + \frac{\vec{\pi}^{2(\infty)}}{4m} + \frac{1}{4m\alpha'^2}\vec{\xi}^{2(\infty)}
\end{aligned}
$$

and eigenvalue of $C_{(4)} = S(S+1)[-R-(S-1)(S+2)]$

$\rightarrow$ eigenvalue of $\vec{\tilde{S}}^2 = S(S+1)$

where $\tilde{S}_i = S_i - \varepsilon_{ijk}\overset{\circ}{\xi}_j\overset{\circ}{\pi}_k$ and $S_i = \frac{1}{2}\varepsilon_{ijk}S_{jk}$.

The number of constituents of our QRO is not determined (it could be a three-dimensional volume oscillating harmonically) but if we want to get two constituents in the non-relativistic limit, we are free to rescale the operators d_μ and π_μ in such a way that the right reduced mass appears in the expression of H. The contraction of the quartic Casimir operator leads to the definition of a new spin operator $\tilde{S}_i$. (This operator can also be written as $\tilde{S}_i = \Sigma_i^{(\infty)} - \varepsilon_{ijk}\xi_j^{(\infty)}\pi_k^{(\infty)}$) which is the difference between the physical spin and the internal orbital angular momentum of of the particle. The parameter S characterizing the SO(3,2) representations can therefore be interpreted as being the total intrinsic spin of the constituents of the oscillator. For $S = 0$, we get the usual three-dimensional harmonic oscillator with no intrinsic spin and the total spin is purely orbital; there is also a one to one correspondence between the states of the representation space (11) and the states of a non-relativistic oscillator.

Application to the Hadron Mass Spectrum (Regge trajectories)

Intuitively, an oscillator is not enough to describe a hadron because it could also perform rigid rotations; we therefore complete our QRO by adding the term $\frac{\lambda^2}{2} \Sigma_{\mu\nu}\Sigma^{\mu\nu}$ corresponding to a quantum relativistic rotator.[11)] The general mass formula then becomes:

$$m^2 = \frac{1}{\alpha'} n + \lambda^2 j(j+1) + m_o^2 \tag{16}$$

We now have to assign states of the different representations of SO(3,2) to the different hadrons. Because of the analogy between the minimum spin S and the total intrinsic spin of the quarks, the mesons with $CP = +1$ enter the representation with $S = 1$, the mesons with $CP = -1$ enter the representation with $S = 0$ and the nucleons, for instance, enter the representation with $S = \frac{1}{2}$. The strange mesons which are not eigenstates of CP enter the same representation as their SU(3) counterparts; the K^* mesons and their excitations are in the same representation ($S = 1$) as the ρ and ω mesons and the K mesons and their excitations are in the same representation ($S = 0$) as the π and η mesons. These assignments are of course not unique and there is a part of subjectivity in our classification. We have performed a χ^2 computer fit to determine the slopes $\frac{1}{\alpha'}$ and λ^2; the results are presented in table II.

TABLE II. Fits of Hadrons.

family	π ,	K ,	ρ-ω ,	K^* ,	ϕ ,	N ,
number of particles	6	7	19	6	6	12
value of S	0	0	1	1	1	$\frac{1}{2}$
$\frac{1}{\alpha'}$ in $(GeV)^2$	0.82	0.84	1.03	1.11	1.00	1.05
λ^2 in $(GeV)^2$	0.24	0.22	0.02	0.02	0.05	0.01
χ^2	2.0	1.6	2.0	0.8	2.0	7.4

In figures (1) and (2), we give the representation spaces and the different particle assignments for the π and ρ-ω families respectively.

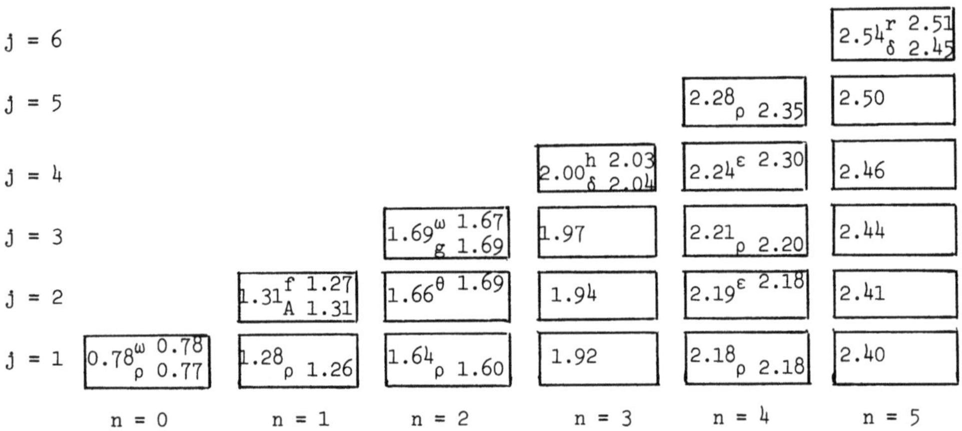

Fig. 1 ρ-ω family. The predicted masses in GeV appear on the left side of each box. The spin-7 meson M(2.75) is not displayed here. The particles with I = 1 are listed below the particles with I = 0 ; they are all included in the fit.

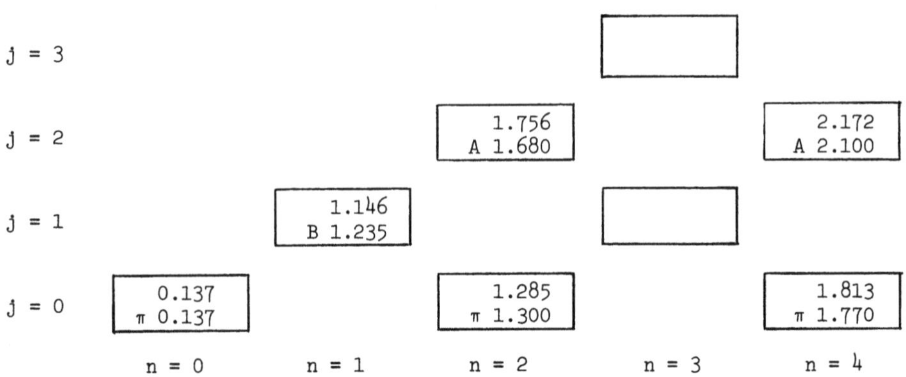

Fig. 2 π family. The predicted masses in GeV appear at the top of each box. The states of this representation of SO(3,2) are identical to the states of the three-dimensional harmonic oscillator.

The above fits have three important consequences. First, the rotational excitation bands of the mass spectrum are much smaller than the radial excitation bands since λ^2 is in general much smaller than $\frac{1}{\alpha'}$. This result is analogous to the situation in molecular and nuclear physics. Second, the mass formula (16) leads to linear Regge trajectories for the subspace n = j - s (subspace corresponding to the Regge trajectory) and an approximate mass formula can be written for this subspace as:

$$m^2 \cong \frac{1}{\alpha'} j + \tilde{m}^2 \qquad j = s, s+1, \cdots \tag{17}$$

Third, the slopes $\frac{1}{\alpha'}$ and λ^2 seem to be SU(3) invariant. The monet $\rho-\omega$, K^* and ϕ has the same mass squared splittings in the radial and rotational bands. If we interpret the ι (1440) and η (1.275) as being the first radial excitations of the η' and η mesons respectively, the monet π, η, K and η' then also has the same slopes $\frac{1}{\alpha'}$ and λ^2 within a small uncertainty.

The situation is not as clear for the baryons because the Ξ family is very incomplete. Moreover the Σ family splits into two subfamilies; half the Σ particles enter the octet and the other half enter the decuplet. There is, anyway, some good hope to find the right particle assignment to verify the SU(3) invariance of the slopes.

References

1) P. A. M. Dirac, Proc. R. Soc. London A328, 1 (1972).
2) P. Budini and C. Fronsdal, Phys. Rev. Letters 14, 968 (1965).
3) A. O. Barut and W. Thacker, Phys. Rev. D31, 2076 (1985).
4) A. Bohm et al., Phys. Rev. D31, 2304 (1985).
5) M. M. L. Pryce, Proc. R. Soc. London A195, 62 (1948).
6) A. Bohm et al., Phys. Rev. Letters 53, 2292 (1984).
7) J. B. Ehrman, Ph.D. thesis, Princeton University (1954).
8) A. Bohm et al., Phys. Rev. D28, 3032 (1983).
9) A. O. Barut and A. Bohm, J. Math. Phys. 11, 2938 (1970).
10) A. Bohm et al., Phys. Rev. D32, 2828 (1985).
11) R. R. Aldinger et al., Phys. Rev. D29, 2828 (1984).
12) R. R. Aldinger et al., Phys. Rev. D28, 3020 (1983).

PATH INTEGRAL REALIZATION OF A DYNAMICAL GROUP

A. Inomata
Department of Physics, State University of New York at Albany
Albany, New York 12222, USA

and

R. Wilson
Department of Mathematics, University of Texas
San Antonio, Texas 78285, USA

Abstract

A way to realize a dynamical group in terms of a path integral is illustrated by using the Poschl-Teller oscillator.

I. Introduction

"No definite connection is known at the present time between the use of De Sitter and conformal groups as dynamical groups or spectrum generating groups and their use as 'space-time-scale' groups," said Barut[1] at the Boulder Conference on De Sitter and Conformal Groups in 1970, and added, "The conformal group in the SO(4,2)-interpretation has been found to describe the Dirac particle, the H atom and a model for proton....." The SO(4,2) symmetry underlying electrodynamics and other massless field theories is a geometrical symmetry which is directly related to space-time transformations. The same SO(4,2) group structure appears to be significant at the dynamical level of composite systems as well. Fifteen years after the Boulder Conference, we still have no clear understanding of their connection. What we are certain of is that the SO(4,2) dynamical scheme works rather well in quantizing composite systems.[2] Recently, it has also been found that the dynamical group idea is effective in reducing a nontrivial path integral into a soluble path integral.[3-5]

The dynamical group SO(4,2) contains SO(3)×SO(2,1) or SO(3)×SO(3) as a subgroup. The SO(3) commonly involved in the above two choices is a geometrical group representing the spherical symmetry in space, whereas the remaining subgroup SO(2,1) or SO(3) is nongeometrical and to generate the energy spectrum of a composite system. A path integral for a system of SO(2,1) has been considered,[3] which is basically equivalent to a Schrödinger equation reducible to a confluent hypergeometric equation. In this paper, we wish to propose a way to realize the dynamical group SO(3) in terms of a path integral. For illustration, we choose the Pöschl-Teller oscillator[6] in which interest has recently been revived.[7] First, we demonstrate the SO(3) structure of the oscillator.[7,8] Then, we convert Feynman's path integral for the one-dimensional system into a path integral on S^3. This path integral gives us not only the correct energy spectrum but also the properly normalized wave functions.

II. Dynamical Group of the Pöschl-Teller Oscillator

The Hamiltonian for the Pöschl-Teller oscillator is [6]

$$H = \frac{1}{2M} p^2 + \tfrac{1}{2} V_0[\kappa(\kappa-1)\csc^2 ax + \lambda(\lambda-1)\sec^2 ax] \tag{2.1}$$

where $V_0 = a^2\hbar^2/M$, $\kappa > 1$, $\lambda > 1$, $x \in [0, \pi/2a]$, and a is a constant. Here, following ref. 8, we demonstrate that this one-dimensional oscillator has the SO(3) dynamical symmetry.

The Schrödinger equation for (2.1) can be written as [8]

$$\{\frac{d^2}{d\theta^2} - \frac{1}{4}[(m+g+\tfrac{1}{2})(m+g-\tfrac{1}{2})\csc^2\tfrac{1}{2}\theta + (m-g+\tfrac{1}{2})(m-g-\tfrac{1}{2})\sec^2\tfrac{1}{2}\theta] + \Lambda\}\,\Psi_m = 0 \tag{2.2}$$

where we have set $\kappa = m + g + \frac{1}{2}$, $\lambda = m - g + \frac{1}{2}$, $\Lambda = ME/2a^2\hbar^2$ and $\theta = 2ax \in [0,\pi]$. Now we introduce three operators,

$$L_1 = -i\cos\phi\,\cot\theta\frac{\partial}{\partial\phi} - i\sin\phi\frac{\partial}{\partial\theta} - i\frac{\cos\phi}{\sin\theta}\frac{\partial}{\partial\psi} + \tfrac{1}{2}i\sin\phi\cot\theta \tag{2.3}$$

$$L_2 = -i\sin\phi\,\cot\theta\frac{\partial}{\partial\phi} + i\cos\phi\frac{\partial}{\partial\theta} - i\frac{\sin\phi}{\sin\theta}\frac{\partial}{\partial\psi} - \tfrac{1}{2}i\cos\phi\cot\theta \tag{2.4}$$

$$L_3 = -i\frac{\partial}{\partial\phi} \tag{2.5}$$

which form an SO(3) algebra,

$$[L_i, L_j] = i L_k \qquad \text{cyclic in } i, j, k. \tag{2.6}$$

Let $\Phi_{\ell m}$ be a simultaneous eigenstate of the Casimir operator $C_m = \vec{L}^2$ and L_3, so that

$$C_m\,\Phi_{\ell m} = \ell(\ell+1)\,\Phi_{\ell m} \tag{2.7}$$

$$L_3 \, \Phi_{\ell m} = m \, \Phi_{\ell m} \tag{2.8}$$

where $\ell = 0, 1, 2, \ldots; \; m = \pm 1, \pm 2, \ldots \pm \ell$. Using (2.3)-(2.8), we can express the wave equation (2.2) as

$$[C_m - \Lambda + \tfrac{1}{4}] \, \Psi_m = 0. \tag{2.9}$$

If we identify Ψ_m with $\Phi_{\ell m}$, we immediately get

$$\Lambda - \tfrac{1}{4} = \ell(\ell + 1), \quad \text{or} \quad E = (2a^2h^2/M)(\ell + \tfrac{1}{2})^2.$$

Since $\ell = |m| + n$ $(n = 0, 1, 2, \ldots)$, we obtain the well-known energy spectrum[6,9]

$$E_n = (a^2\hbar^2/2M)(2n + \kappa + \lambda)^2. \tag{2.10}$$

Thus, we see that although the system is one-dimensional there is an underlying SO(3) symmetry in its dynamics.

III. Path Integral Realization of the Pöschl-Teller Oscillator

As we have seen above, the spectrum generating group of the Pöschl-Teller oscillator is SO(3), which is independent of the apparent space symmetry of the Hamiltonian (2.1). Since SO(3) is locally isomorphic to SU(2) and the group manifold of SU(2) is homeomorphic to S^3, we attempt to realize the oscillator in a path integral on S^3.

The Lagrangian for this system is given by

$$L = \tfrac{1}{2} M \dot{x}^2 - \tfrac{1}{2} V_0 [\kappa(\kappa - 1) \csc^2 ax + \lambda(\lambda - 1) \sec^2 ax], \tag{3.1}$$

for which Feynman's path integral may be expressed as

$$K(x'', x'; \tau) = \lim_{N\to\infty} \int \prod_{j=1}^{N} \exp[\tfrac{i}{\hbar} S_j] \prod_{j=1}^{N} [\tfrac{M}{2\pi i\hbar\varepsilon}]^{\frac{1}{2}} \prod_{j=1}^{N-1} dx_j \tag{3.2}$$

where

$$S_j = \frac{M}{2\varepsilon}(\Delta x_j)^2 - \tfrac{1}{2}\varepsilon V_0 [\kappa(\kappa - 1) \widetilde{\csc}^2 ax_j + \lambda(\lambda - 1) \widetilde{\sec}^2 ax_j]. \tag{3.3}$$

In the above, we have used the notations: $x_j = x(t_j)$, $t' = t_0$, $t'' = t_N$, $t_j - t_{j-1} = \varepsilon$, $\tau = t'' - t'$, $\widetilde{\csc}^2 \theta_j = \csc\theta_j \csc\theta_{j-1}$, etc. As before, we set $\theta_j = 2ax_j$ and rewrite (3.2) as

$$K(x'',x';\tau) = 2a\lim_{N\to\infty} \int \prod_{j=1}^{N} \exp[\tfrac{i}{\hbar}S_j] \prod_{j=1}^{N} [\frac{M}{8a^2\pi i\hbar\varepsilon}]^{\frac{1}{2}} \prod_{j=1}^{N-1} d\theta_j \tag{3.4}$$

The short time action (3.3) has also to be expressed in terms of θ-variables. As is well-known, terms of $O(\varepsilon^{1+\delta})$ with $\delta > 0$ are unimportant in a short time action, and $(\Delta\theta_j)^2 \sim \varepsilon$, so that $[-\alpha(\Delta\theta)^2 + \beta(\Delta\theta)^4]$ is equivalent to $[-\alpha(\Delta\theta)^2 + (3\beta/4)\alpha^{-2}]$ for large α. Hence,

$$S_j = \frac{M}{a^2\varepsilon}(1 - \cos\tfrac{1}{2}\Delta\theta_j) - \frac{a^2\hbar^2}{8M}\varepsilon - \tfrac{1}{2}\varepsilon V_0[\kappa(\kappa-1)\widetilde{\csc}^2\tfrac{1}{2}\theta_j + \lambda(\lambda-1)\widetilde{\sec}^2\tfrac{1}{2}\theta_j] \tag{3.5}$$

Next, using the asymptotic formula valid for z large and p an integer,[4]

$$\int_0^{2\pi} \exp[ip\alpha - z(1-\cos\alpha)]\, d\alpha \simeq (2\pi/z)^{\frac{1}{2}}\exp[-(p^2 - \tfrac{1}{4})/2z], \tag{3.6}$$

we generate two angular variables α and β from the last two terms of (3.5) as

$$\exp[\frac{iV_0\kappa(\kappa-1)\varepsilon}{\hbar\widetilde{\cos}^2\frac{1}{2}\theta_j}] = [\frac{M\widetilde{\cos}^2\frac{1}{2}\theta_j}{2\pi i a^2\hbar\varepsilon}]^{\frac{1}{2}} \int_0^{2\pi} \exp[ip\alpha_j + \frac{iM}{a^2\hbar\varepsilon}\widetilde{\cos}^2\tfrac{1}{2}\theta_j(1-\cos\alpha_j)]\, d\alpha_j ,$$

$$\exp[\frac{iV_0\lambda(\lambda-1)\varepsilon}{\hbar\widetilde{\sin}^2\frac{1}{2}\theta_j}] = [\frac{M\widetilde{\sin}^2\frac{1}{2}\theta_j}{2\pi i a^2\hbar\varepsilon}]^{\frac{1}{2}} \int_0^{2\pi} \exp[iq\beta_j + \frac{iM}{a^2\hbar\varepsilon}\widetilde{\sin}^2\tfrac{1}{2}\theta_j(1-\cos\beta_j)]\, d\beta_j ,$$

where $p = \kappa - \frac{1}{2} = m + g$ and $q = \lambda - \frac{1}{2} = m - g$ are assumed to be positive integers. Substituting of these results into (3.4) yields

$$K(x'',x';\tau) = a(\sin\theta'\sin\theta'')^{\frac{1}{2}} \lim_{N\to\infty} \int \prod_{j=1}^{N} \exp[\tfrac{i}{\hbar}\tilde{S}_j] \prod_{j=1}^{N} [\frac{M}{8\pi i\hbar a^2\varepsilon}]^{3/2} \prod_{j=1}^{N-1} (2\sin\theta_j d\theta_j d\alpha_j d\beta_j)(4\,d\alpha'' d\beta'') \tag{3.7}$$

where

$$\tilde{S}_j = \frac{M}{a^2\varepsilon}(1 - \cos\tfrac{1}{2}\Omega_j) - \frac{\hbar^2 a^2\varepsilon}{8M} + \hbar(\kappa - \tfrac{1}{2})\alpha_j + \hbar(\lambda - \tfrac{1}{2})\beta_j \tag{3.8}$$

with $\cos\frac{1}{2}\Omega_j = \widetilde{\cos}^2\frac{1}{2}\theta_j\cos\alpha_j + \widetilde{\sin}^2\frac{1}{2}\theta_j\cos\beta_j$. The newly introduced variables α_j and β_j may be converted into Euler angles ϕ_j and ψ_j by

$$\alpha_j = \tfrac{1}{2}(\Delta\psi_j + \Delta\phi_j), \qquad \beta_j = \tfrac{1}{2}(\Delta\psi_j - \Delta\phi_j) \tag{3.9}$$

and

$$\int_0^{2\pi} d\alpha_j \int_0^{2\pi} d\beta_j = \tfrac{1}{2}\int_0^{2\pi} d\phi_j \int_{-2\pi}^{2\pi} d\psi_j . \tag{3.10}$$

As a result, we obtain

$$K(x'',x';\tau) = 2a(\sin\theta'\sin\theta'')^{\frac{1}{2}} \exp[-ia^2\hbar\tau/8M]$$

$$\times \int_0^{2\pi} d\phi'' \int_{-2\pi}^{2\pi} d\psi'' \exp[\tfrac{1}{2}i(\kappa-\lambda)\phi'' + \tfrac{1}{2}i(\kappa+\lambda-1)\psi'']\tilde{K}(\theta'',\phi'',\psi'';\theta',0,0;\tau) \tag{3.11}$$

where

$$\tilde{K}(\theta'',\phi'',\psi'';\theta',0,0;\tau) = \lim_{N\to\infty} \int \prod_{j=1}^{N} \exp[\tfrac{i}{\hbar}\hat{S}_j] \prod_{j=1}^{N} [M/8\pi i\hbar a^2\varepsilon]^{3/2} \prod_{j=1}^{N-1} (\sin\theta_j d\theta_j d\phi_j d\psi_j) \tag{3.12}$$

with

$$\hat{S}_j = (M/a^2\varepsilon)(1 - \cos\tfrac{1}{2}\Omega_j) \tag{3.13}$$

and $\cos\frac{1}{2}\Omega_j = \tilde{\cos}^2\frac{1}{2}\theta_j \cos\frac{1}{2}(\Delta\phi_j + \Delta\psi_j) + \tilde{\sin}^2\frac{1}{2}\theta_j \sin\frac{1}{2}(\Delta\phi_j - \Delta\psi_j)$. The path integral (3.12) is nothing but an SU(2) path integral, which can be evaluated and has been given by[5]

$$\tilde{K}(\Omega;\tau) = (16\pi^2)^{-1} \sum_{J=J_0}^{\infty} (2J+1)\, C^1_{2J}(\cos\tfrac{1}{2}\Omega) \exp[-\frac{2i\hbar a^2\tau}{M} J(J+1) - \frac{3i\hbar a^2\tau}{8M}] \tag{3.14}$$

where $J_0 = \max\{\frac{1}{2}|\kappa-\lambda|, \frac{1}{2}|\kappa+\lambda-1|\}$, and

$$C^1_{2J}(\cos\tfrac{1}{2}\Omega) = \sum_{\mu,\nu=-J}^{J} e^{-i\mu\phi''} e^{-i\nu\psi''} d^{J*}_{\mu,\nu}(\cos\theta')\, d^{J}_{\mu,\nu}(\cos\theta''). \tag{3.15}$$

Finally, substituting (3.14) into (3.11) and completing the integration, we arrive at

$$K(x'',x';\tau) = 2a(\sin 2ax' \sin 2ax'')^{\frac{1}{2}} \sum_{n=0}^{\infty} (2n+\kappa+\lambda)$$

$$\times \exp[-(i\hbar a^2/2M)(2n+\kappa+\lambda)^2\tau]\, d^{J*}_{\mu,\nu}(2ax')\, d^{J}_{\mu,\nu}(2ax'') \tag{3.16}$$

where $J = n+\frac{1}{2}(\kappa+\lambda-1)$, $\mu = \frac{1}{2}(\kappa-\lambda)$, $\nu = \frac{1}{2}(\kappa+\lambda-1)$ and the θ-variable has been transformed back into the x-variable. From (3.16), we can immediately read off the energy spectrum and the normalized energy eigenfunctions,

$$E_n = (a^2\hbar^2/2M)(2n+\kappa+\lambda)^2 \tag{3.17}$$

$$\Psi_n = [a(2n+\kappa+\lambda)\sin 2ax]^{\frac{1}{2}}\, d^{n+\frac{1}{2}(\kappa+\lambda-1)}_{\frac{1}{2}(\kappa-\lambda),\, \frac{1}{2}(\kappa+\lambda-1)}(2ax). \tag{3.18}$$

The wave functions (3.18) are consistent with, but differs by a phase factor from the result obtained by Nieto.[9]

IV. Conclusion

By taking the Pöschl-Teller oscillator as an example, we have explicitly constructed a path integral which accommodates a dynamical group. Apparently, the one-dimensional Hamiltonian (2.1) is not scale-invariant. However, the spectrum generating group SO(3) SO(4,2) can be realized as a path integral on S^3. Thus, we conclude that a certain dynamical group can be realized in terms of a path integral and that the dynamical group idea is helpful in solving Feynman's path integral for a non-trivial system such as the Pöschl-Teller oscillator.

References

1. A.O. Barut, in De Sitter and Conformal Groups and Their Applications, Lectures in Theoretical Physics Vol. XIII, eds. A.O. Barut and W.E. Brittin,(Colorado University Press, Boulder, 1971) p. 20.

2. See, e.g., A.O. Barut, Dynamical Groups and Generalized Symmetries in Quantum Theory (University of Canterbury, Christchurch, 1972).

3. R. Ho and A. Inomata, Phys. Rev. Lett. 48, 231 (1982);
A. Inomata, Phys. Lett. 101A, 253 (1984);
P.Y. Cai, A. Inomata and R. Wilson, Phys. Lett. 96A, 117 (1983).

4. A. Inomata and M.A. Kayed, Phys. Lett. 108A, 9 (1985);
A. Inomata and M.A. Kayed, J. Phys. A18, 235 (1985).

5. G. Junker and A. Inomata, in Path Integrals from meV to MeV, eds. M.C. Gutzwiller, A. Inomata, J.R. Klauder and L. Streit, (World Scientific Publishing Co. Singapore, 1986).

6. G. Pöschl and E. Teller, Z. Phys. 83, 143 (1933); S. Flügge, Practical Quantum Mechanics (Springer-Verlag, New York, 1974) p. 89.

7. Y. Alhassid, F. Gürsey and F. Iachello, Phys. Rev. Lett. 50, 873 (1983);
A. Frank and K.B. Wolf, Phys. Rev. Lett. 52, 1737 (1984).

8. A.O. Barut, A. Inomata and R. Wilson, "A New Method of Algebraization by Factorization," to be published.

9. M.M. Nieto, Phys. Rev. A17, 1273 (1978).

POLYNOMIAL IDENTITIES ASSOCIATED WITH DYNAMICAL SYMMETRIES

Mircea Iosifescu and Horia Scutaru

Department of Theoretical Physics, Central Institute of Physics

Bucharest, P.O.Box MG-6, ROMANIA

The existence of relations between the generators of a linear representation of a Lie algebra is a well-known fact and has been pointed out in connection with various physical problems (e.g. /1/-/7/).

During the last fifteen years, several methods for the construction of such identities have been deduced /8/-/14/.These methods have in common the fact that they are aimed at determining the polynomial identities satisfied by a given linear representation. The idea which underlies several of them (e.g. /10/, /12/) is the following :

Given two representations ρ_Λ and ρ_Ω of an n-dimensional reductive Lie algebra L acting in the linear spaces V_Λ and V_Ω, respectively, the operator

$$A = \sum_{i=1}^{n} \rho_\Lambda(e_i) \otimes \rho_\Omega(e^i) \qquad (1)$$

can be defined ; $\{e_i\}_{i=1}^n$, $\{e^i\}_{i=1}^n$ are bases in L dual with respect to a nondegenerate bilinear form on L. The operator A commutes with the direct product of the two representations ρ_Λ and ρ_Ω of L. Its eigenvalues label thus the components into which the direct product $\rho_\Lambda \otimes \rho_\Omega$ decomposes. The number of nonequivalent irreducible components of $\rho_\Lambda \otimes \rho_\Omega$ is equal to the degree of the minimal polynomial $p(A)$ satisfied by A. The polynomial relations satisfied by one of the representations, ρ_Λ say, are obtained by taking the matrix elements of the relation $p(A) = 0$ with respect to the basis vectors in V_Ω.

A different approach of the problem has been derived in /15/, /16/. In /15/ a method allowing the determination of all the identities satisfied by the generators of a Poisson bracket realization (PBR) of a Lie algebra has been pointed out ; these polynomial identities are obtained by equating to zero the vec - tors belonging to subrepresentations of the symmetric part of di-rect powers of the adjoint representation $(\mathrm{ad}^{\otimes k})_s$. The isomor-phism existing between the extensions of the adjoint action to the symmetric algebra S(L) and to the algebra of polynomials de-fined on L^* allows one to conclude that the equations provided by the polynomial identities (written in L^*) are invariant under the co-adjoint action and describe thus an intrinsic pro-perty of the Lie algebra L. Thus for a sufficiently large set of powers {k} these equations will provide all the equations which can be satisfied by the generators of a PBR. As the symme-trizer intertwins the extensions of the adjoint action to S(L) and to the universal enveloping algebra U(L), the symmetrization of the identities obtained for the Poisson bracket realization of a Lie algebra L leads to identities for the linear represen - tations of L.

Let us observe that the identities satisfied by the gene-rators of a realization of a dynamical (or of an invariance) Lie algebra of a Hamiltonian system are constraints imposed on the Hamiltonian system by its symmetry.

The procedure described above has been applied in /15/ to the explicit determination of the second-degree polynomial identities which characterize the PBR's of the $B_2 \sim C_2$ and $D_3 \sim A_3$ Lie algebras.

In a subsequent work /16/ (cf. also /17/) the second-degree irreducible tensors $t_{L,\sigma} \in S(L)$ which transform under subrepresentations of $(ad \otimes ad)_s$ of the Lie algebra L have been determined for the following semisimple Lie algebras: $A_n (n \geq 3)$, $B_n (n \geq 2)$, $C_n (n \geq 2)$, $D_n (n \geq 5)$. The corresponding tensor operators, $T_{L,\sigma}$ are obtained from $t_{L,\sigma}$ by symmetrization. The equations $T_{L,\sigma}(\rho) = T_{L,\sigma}(x_1, x_2, \ldots, x_n) = 0$ (x_i = generators of the representation ρ) obtained in this way provide the second-degree polynomial identities satisfied by the representations ρ of Ω.

These polynomial identities contain an amount of information concerning the linear representations ρ (the generators of) which satisfy them. This has been proved, for instance, in papers /18/, /19/ in which for a class of linear representations of the Lie algebras so (2n,R) and sp(2n,R) which appear in Gross-Neveu models /20/ and in the study of collective motion in nuclei /21/, respectively , boson realizations of the Holstein-Primakoff-type have been obtained from the corresponding polynomial identities.

One of the purposes of the present note is to obtain specific information concerning the representations ρ (of the Lie algebra L) which satisfy the polynomial identities $T_{L,\sigma}(\rho) = 0$. This is illustrated by the subsequent theorem in which the representations ρ which satisfy the second-degree polynomial identities $T_{L,\sigma}(\rho) = 0$ deduced in /16/ are determined for the Lie algebras sp(2n,C) and so(2n,C).

Before stating the theorem, let us remind that, for the semisimple Lie algebras of types $C_n (n \geq 2)$ and $D_n (n \geq 5)$, the Clebsch-Gordan series of $(ad \otimes ad)_s$ are /22/: For algebras of type $C_n (n \geq 2)$ (cf.explanation to Table 1):

$$(ad \otimes ad)_s = (0) \oplus (\Lambda_2) \oplus (4\Lambda_1) \oplus (2\Lambda_2) \qquad (2)$$

For algebras of type D_n $(n \geq 5)$:

$$(ad \otimes ad)_s = (0) \oplus (2\Lambda_1) \oplus (\Lambda_4) \oplus (2\Lambda_2) \qquad (3)$$

The expressions of the tensor operators $T_{L,\sigma}$, transforming under the subrepresentations σ of $(ad \otimes ad)_s$ (for $L = C_n, D_n$) pointed out above, are given in the Appendix.

Theorem. Let $T_{L,\sigma} \in U(L)$ be the second-degree tensor operator which transforms under the subrepresentation σ of $(ad \otimes ad)_s$ of the Lie algebra L. Let L be one of the Lie algebras sp(2n,C) and so(2n,C). The finite-dimensional representations ρ of L on (the states of) which $T_{L,\sigma}$ vanishes, i.e. for which $T_{L,\sigma}(\rho) = 0$, are those contained in Table 1:

Table 1

Lie algebra L	*Representation σ under which $T_{L,\sigma}$ transforms*	*Representation ρ for which $T_{L,\sigma}(\rho) = 0$*
sp(2n,C)	(Λ_2)	$(k\Lambda_n)$
	$(4\Lambda_1)$	(Λ_1)
	$(2\Lambda_2)$	-
so(2n,C)	$(2\Lambda_1)$	$(k\Lambda_{n-1})$
		$(k\Lambda_n)$
	(Λ_4)	$(k\Lambda_1)$
	$(2\Lambda_2)$	$(\Lambda_{n-1})(\Lambda_n)$

In Table 1 Λ_i denotes the highest weight of the fundamental representation (Λ_i) $(i = 1,2,\ldots,n)$ of the Lie algebra L under consideration.

The proof is based on the observation that, for a finite-dimensional linear space V_ρ , in order to have

$$T_{(\Lambda)}(x_1, x_2,\ldots,x_n)\, v = 0 \tag{4}$$

for any component of $T_{(\Lambda)}$ and any $v \in V_\rho$, it is sufficient to prove that

$$T_{(\Lambda)}(x_1, x_2,\ldots,x_n)v_\rho = 0 \tag{5}$$

for any component of $T_{(\Lambda)}$ where v_ρ is the vector corresponding to the highest weight of the representation ρ . To do that,the polynomial identities $T_{L,\sigma} = 0$ have been expressed in a Cartan-Weyl basis and the action of the generators of the Cartan sub - algebra and of the raising operators upon v_ρ have been taken into account.

Besides the results summarized in Table 1, Eqs. (5) provide also information concerning the highest-weight vector v_ρ. Detailed proofs have been given in /23/, /24/.

It has been proved /25/ that for the semisimple Lie algebras C_n and D_n the following direct products decompose into a direct sum of two nonequivalent irreducible representations :

For Lie algebras of type C_n:

$$(\Lambda_1) \otimes (m\Lambda_n) = (\Lambda_1 + m\Lambda_n) \oplus (\Lambda_{n-1} + (m-1)\Lambda_n) \tag{6}$$

For Lie algebras of type D_n:

$$\begin{aligned}
(m\Lambda_1) \otimes (\Lambda_{n-1}) &= (m\Lambda_1 + \Lambda_{n-1}) \oplus ((m-1)\Lambda_1 + \Lambda_n) \\
(m\Lambda_1) \otimes (\Lambda_n) &= (m\Lambda_1 + \Lambda_n) \oplus ((m-1)\Lambda_1 + \Lambda_{n-1}) \\
(\Lambda_1) \otimes (m\Lambda_{n-1}) &= (\Lambda_1 + m\Lambda_{n-1}) \oplus ((m-1)\Lambda_{n-1} + \Lambda_n) \\
(\Lambda_1) \otimes (m\Lambda_n) &= (\Lambda_1 + m\Lambda_n) \oplus (\Lambda_{n-1} + (m-1)\Lambda_n)
\end{aligned} \tag{7}$$

where m are arbitrary positive integers. Okubo's method /12/, applied to the direct products (6) and (7), proves the consistency of the statements of the theorem with the sets of equations (6) and (7).

A second purpose of this note is to draw attention upon a mapping /26/ which presents similarities with the moment map /27/, /28/ and the properties of which make it useful in the study of G-manifolds and, in particular, for the determination of polynomial identities which characterize a Poisson Bracket Realization. Applications of this mapping to the conformal algebra will be pointed out.

Let L be a Lie algebra of a Lie group G, $\{x_i, i=1,\dots,n\}$ be its generators and $[x_i, x_j] = \sum_{k=1}^{n} c_{ij}^k x_k$ be its structure relations.

Let M be a manifold endowed with a Poisson algebra A and with a G-action on A /24/. The Poisson product in A is denoted by $\{\ ,\ \}$. Let $R : L \to A$ be a PBR of L

$$R : x_i \to f_{x_i} \in A, \qquad R : [x_i, x_j] \to \{f_{x_i}, f_{x_j}\} \tag{8}$$

which is supposed to have the equivariance property

$$f_{Ad(g)x_i}(m) = f_{x_i}(g^{-1}m) \tag{9}$$

Let V be a G-module on which G acts by the linear representation U and let π be the corresponding representation of L on V : $\pi(x_i) = \frac{d}{dt} U(\exp t\, x_i)$

Definition : The mapping $K : m \in M \to K(m) \in \operatorname{End} V$ is defined by

$$K(m) = \sum_{i=1}^{n} f_{x_i}(m)\, \pi(x^i) \quad (m \in M) \tag{10}$$

where x^i is the element dual to x_i with respect to a nondegenerate bilinear Ad-invariant form (,) on L.

The mapping K has the following equivariance property /26/ :

Theorem. For any $g \in G$ and any $m \in M$

$$K(g.m) = U(g)\, K(m)\, U(g^{-1}) \tag{11}$$

where g.m denotes the action of $g \in G$ upon $m \in M$.

The equivariance property (11) of the mapping K allows the determination of the algebraic identities satisfied by the generators of the PBR (8) which have been used in the definition of K(m). Indeed, relation (11) keeps valid if K is replaced by a polynomial in the indeterminate K. Thus if a polynomial relation $P(K(m)) = 0$ can be proved to be valid for a given point $m \in M$, it will be valid for all the points which are on the G-orbit through m. Hence the algebraic relations which result by taking the matrix elements of the relation $P(K(m)) = 0$, characterize the restriction of the PBR to the G-orbit through m.

Let us observe that K(m) always satisfies a polynomial relation $P(K(m))=0$ (the Cayley-Hamilton theorem); this leads to a proof for the existence of polynomial relations between the generators of a PBR.

It is essential to remark that the equivariant mapping $K : M \to \mathrm{End}V$ reduces the nonlinear problem of the classification of G-orbits on M to the linear problem of the classification of G-orbits in EndV, which can be solved more easily and for which some general solutions exist in the literature. This will become evident from the examples new to be mentioned.

We shall give two examples of mappings K which allow one to exploit the existing classifications of G-orbits on EndV in order to obtain polynomial identities for the PBR defined on $\mathcal{M}$. In these examples (in which G will be the conformal group) the G-modules are fundamental modules of G and the polynomial identities satisfied by the operators from the image of K are of second degree; the manifold $\mathcal{M}$ is the dual space L^* of L endowed with the Kirillov-Kostant-Souriau bracket and with the co-adjoint action of G on L^*.

Let $L = so(4,2) \sim su(2,2)$. There exists a classification of the orbits of SU(2,2) in $\mathrm{End}V_{\Lambda_1}$ /30/ (V_{Λ_1} is called the twistor space) and a classification of the orbits of SO(4,2) in $\mathrm{End}V_{\Lambda_2}$ /31/, /32/. Hence, for the Lie algebra so(4,2) $\sim$ su(2,2), we can give two examples of K-mappings , K_i (i=1,2) which differ by the image space $\mathrm{End}V_{\Lambda_i}$ (i = 1,2).

In both cases we shall use for the so(4,2) generators the following vectorial notations :

$L_i = M_{jk}$(i,j,k = cyclic permutations of 1,2,3), $A_i = M_{i4}$
$B_i = M_{i5}$, $C_i = M_{i6}$ (i = 1,2,3), $B_4 = M_{45}$, $C_4 = M_{46}$, $M = -M_{56}$, the generators $M_{ij} = -M_{ji}$ (i,j=1,...,6) being supposed to satisfy the Lie relations

$$[M_{ij},M_{kl}] = -g_{ik}M_{jl} - g_{jl}M_{ik} + g_{il}M_{jk} + g_{jk}M_{il} \qquad (12)$$

with $g_{11}=g_{22}=g_{33}=g_{44}=-g_{55}=-g_{66}=1$ and $g_{ij}=0$ for $i \neq j$.
The corresponding dual elements of these generators, with respect to the Cartan-Killing bilinear form, are $L^i = L_i$, $A^i = A_i$ (i = 1,2,3), M = M, $B^j = -B^j$, $C^j = -C_j$ (j = 1,2,3,4).

The matrix $K_1(m)$ ($m \in L^*$) is defined by relation (10) in which the representation π is generated by

$$\pi(L_K) = \frac{1}{2}\begin{pmatrix} i\sigma_k & 0 \\ 0 & i\sigma_k \end{pmatrix}, \qquad \pi(A_k) = \frac{1}{2}\begin{pmatrix} 0 & -\sigma_k \\ \sigma_k & 0 \end{pmatrix} \qquad (k = 1,2,3)$$

$$\pi(B_k) = \frac{1}{2}\begin{pmatrix} -\sigma_k & 0 \\ 0 & \sigma_k \end{pmatrix}, \qquad \pi(C_k) = \frac{1}{2}\begin{pmatrix} 0 & \sigma_k \\ \sigma_k & 0 \end{pmatrix} \qquad (k = 1,2,3)$$

$$\pi(B_4) = \frac{1}{2}\begin{pmatrix} 0 & 1 \\ 1 & 0 \end{pmatrix}, \qquad \pi(C_4) = \frac{1}{2}\begin{pmatrix} 1 & 0 \\ 0 & -1 \end{pmatrix}, \quad \pi(M) = \frac{1}{2}\begin{pmatrix} 0 & -1 \\ 1 & 0 \end{pmatrix}$$

Denoting the coordinate functions of a point $m \in so(4,2)^*$ by the same letters as the corresponding generators of so(4,2) (cf./15/) and reminding that the coordinate functions generate a PBR with respect to the Kirillov-Konstant-Souriau bracket, we obtain for $K_1(m)$ the expression

$$K_1(m) = \frac{1}{2}\begin{pmatrix} -(\vec{B}+i\vec{L})\cdot\vec{\sigma} + C_4 I_2 & (\vec{A}+\vec{C})\cdot\vec{\sigma} + (B_4+M)I_2 \\ -(\vec{A}-\vec{C})\cdot\vec{\sigma} + (B_4-M)I_2 & (\vec{B}-i\vec{L})\cdot\vec{\sigma} - C_4 I_2 \end{pmatrix} \tag{13}$$

where by I_n we have denoted the n-dimensional unity matrix and by $\vec{X}\cdot\vec{Y}$ the scalar product between $\vec{X}$ and $\vec{Y}$.

Then, for the six-dimensional complex orbit nr. XXIV from /30/ we obtain that $K_1(m)$ satisfies the second-degree matrix equation

$$K_1(m)^2 + 2i\lambda K_1(m) + 3\lambda^2 I_4 = 0 \tag{14}$$

the matrix elements of which lead /26/ to polynomial identities obtained previously (/15/, relations (3.24-3.30)).

The second case, in the expression (10) of the matrix $K_2(m)$ the representation π will be the natural, six-dimensional representation of so(4,2): $\pi(M_{ij}) = M_{ij}$, where

$M_{ij} = e_{ij} - e_{ji} = -M_{ji} (1 \leq i \leq j \leq 4)$, $M_{56} = -e_{56} + e_{65} = -M_{65}$,

$M_{i\alpha} = e_{i\alpha} + e_{\alpha i} = M_{\alpha i}$ $(1 \leq i \leq 4;\ \alpha = 5,6)$ and $(e_{ij})_{kl} = \delta_{ik}\delta_{jl}$.

The matrix K_2 is

$$K_2(m) = \begin{vmatrix} 0 & -L_3 & L_2 & -A_1 & B_1 & C_1 \\ L_3 & 0 & -L_1 & -A_2 & B_2 & C_2 \\ -L_2 & L_1 & 0 & -A_3 & B_3 & C_3 \\ A_1 & A_2 & A_3 & 0 & B_4 & C_4 \\ B_1 & B_2 & B_3 & B_4 & 0 & -M \\ C_1 & C_2 & C_3 & C_4 & M & 0 \end{vmatrix} \tag{15}$$

and if we choose m such that $K_2(m)$ is of the form $\alpha M_{12} + \beta M_{34} + \gamma M_{56}$ $(\alpha = \beta = \gamma)$ in the notations of /32/ then

$$K_2(m)^2 = \alpha^2 I_6 \tag{16}$$

The matrix elements of this equation lead also to polynomial identities obtained in /15/.

The polynomial identities which result from equations (14) and (16) (and which are satisfied by the coordinate functions in $L^* = so(4,2)^*$) are equations of manifolds invariant under the co-adjoint action of G, to which correspond, through the mapping K, G-orbits in $\mathrm{End}V_{\Lambda_1}$ and $\mathrm{End}V_{\Lambda_2}$, respectively.

APPENDIX

A) *The second-degree tensors $t_{L,\sigma}$ for the Lie algebra sp(2n,C)*

Let $g_{ij}=\delta_{i,j+n}-\delta_{i+n,j}$ $(i,\ j=1,\dots,2n)$ and let

$$S_{ij}=\sum_{k=1}^{2n}(g_{ik}e_{kj}-g_{kj}e_{ki})\qquad(i,j=1,\dots,2n)$$

be the generators of the algebra sp(2n,C) $(S_{ij}=S_{ji})$ with the structure relations

$$[S_{ij},S_{kl}]=g_{kj}S_{il}-g_{il}S_{kj}-g_{ik}S_{jl}+g_{lj}S_{ki}$$

The expressions of the second-degree tensors t_σ, which transform under the representations σ of sp(2n,C) into which decomposes $(\mathrm{ad}\otimes\mathrm{ad})_s$ (formula (2)), are obtained by projection from the generic element $S_{pq}S_{rs}$. The corresponding tensor operators T_σ are obtained from t_σ by symmetrization i.e. by replacing products by anticommutators. We obtain

$$t_{(0)}=-\frac{1}{8n(2n+1)}(g_{ps}g_{qr}+g_{pr}g_{qs})\sum_{i,j,k,l}g_{ji}g_{lk}S_{li}S_{jk}$$

$$t_{(\Lambda_2)}=-\frac{1}{4(n+1)}\sum_{i,j}g_{ji}(g_{ps}S_{iq}S_{jr}+g_{pr}S_{iq}S_{js}+g_{qr}S_{ip}S_{js}+g_{qs}S_{ip}S_{jr}$$

$$-\frac{2n+1}{n+1}t_{(0)}$$

$$t_{(4\Lambda_1)}=\frac{1}{3}(S_{pq}S_{rs}+S_{ps}S_{rq}+S_{pr}S_{qs})$$

$$t_{(2\Lambda_2)}=\frac{1}{3}(2S_{pq}S_{rs}-S_{ps}S_{rq}-S_{pr}S_{qs})+\frac{n}{n+1}t_{(0)}$$

$$+\frac{1}{4(n+1)}\sum_{i,j}g_{ji}(g_{ps}S_{iq}S_{jr}+g_{pr}S_{iq}S_{js}+g_{qr}S_{ip}S_{js}+g_{qs}S_{ip}S_{jr}$$

B) *The second degree tensors $t_{L,\sigma}$ for the Lie algebra so(2n,C)*

The structure relations of so(2n,C) are relations (10) with $g_{ij} = \delta_{ij}$ and i, j=1,...,2n. The second-degree tensors t_σ, which transform under the representations σ of so(2n,C) into which decomposes $(\mathrm{ad} \otimes \mathrm{ad})_S$ (formula (3)), are obtained by projection from the generic element $M_{pq}M_{rs}$. The corresponding tensor operators T_σ are obtained from t_σ by symmetrization. Denoting

$$\mu_{ps} = \sum_{i=1}^{2n} M_{pi}M_{is}$$

we obtain

$$t_{(0)} = \frac{1}{2n(2n-1)} (\delta_{qr}\delta_{ps} - \delta_{pr}\delta_{qs}) \sum_{i,j=1}^{2n} M_{ij}M_{ji}$$

$$t_{(2\Lambda_1)} = \frac{1}{2(n-1)} (\delta_{qr}\mu_{ps} + \delta_{ps}\mu_{qr} - \delta_{pr}\mu_{qs} - \delta_{qs}\mu_{pr}) - \frac{2n-1}{n-1} t_{(0)}$$

$$t_{(\Lambda_4)} = \frac{1}{3}(M_{pq}M_{rs} + M_{ps}M_{qr} + M_{pr}M_{sq})$$

$$t_{(2\Lambda_2)} = \frac{1}{3}(2M_{pq}M_{rs} - M_{ps}M_{qr} - M_{pr}M_{sq})$$

$$- \frac{1}{2(n-1)}(\delta_{qr}\mu_{ps} + \delta_{ps}\mu_{qr} - \delta_{pr}\mu_{qs} - \delta_{qs}\mu_{pr}) + \frac{n}{n-1} t_{(0)}$$

ACKNOWLEDGEMENTS

Part of this work has been done while one of the authors (M.I.) enjoyed the hospitality of the Joint Institute of Nuclear Research, Dubna, U.S.S.R. It is his pleasure to express his gratitude to Professor V.G.SOLOVIEV and to Dr. I.N.MIHAILOV.

References

/1/ J.D.Louck, "Theory of Angular Momentum in N-Dimensional Space " Los Alamos Sci. Lab. Report La 2451 (1966).

/2/ H.Bacry, Nuovo Cim. 41A, 222 (1966).

/3/ C.E.Wulfman, Y.Takahata, J.Chem. Phys. 47, 488 (1967).

/4/ G.Györgyi, Nuovo Cim. 62A, 449 (1969).

/5/ A.O.Barut, A.Böhm, J.Math. Phys. 11, 2938 (1970).

/6/ L.P.Stanton, J. Math. Phys. 19, 1471 (1978); Phys. Rev. D8, 2446 (1973), D10, 1760 (1974).

/7/ I.V.Komarov, Teor. Mat. Fiz. 53, 296 (1982).

/8/ A.J.Bracken, H.S.Green, J. Math. Phys. 12, 2099 (1971).

/9/ H.S.Green, J. Math. Phys. 12, 2106 (1971).

/10/ K.C.Hannabuss "Characteristic Identities for Semisimple Lie Groups", Preprint, Mathematical Institute, Oxford, 1972.

/11/ S.Okubo, J.Math.Phys. 16, 528 (1975).

/12/ S.Okubo, J.Math.Phys. 18, 2382 (1977).

/13/ D.M.O'Brien, A. Cant, A.L.Carey, Ann.Inst.H.Poincaré, 26 405 (1977).

/14/ M.D.Gould "Characteristic Identities for Semisimple Lie Algebras ", Preprint Univ. of Adelaide.

/15/ M.Iosifescu, H.Scutaru, J. Math. Phys. 25, 2856 (1984).

/16/ M.Iosifescu, H.Scutaru, "Polynomial Constraints Associated to Dynamical Symmetries". Preprint, Bucharest FT-253 (1984).

/17/ M.Iosifescu, H.Scutaru, "Polynomial Identities Satisfied by Realization of Lie Algebras " in "Geometrodynamics Proceedings", A.Prâstaro, editor. World Scientific (1985) p. 173.

/18/ M.Iosifescu, H.Scutaru, "Boson Realizations from Quantum Constraints", to be published in J. Math. Phys. (1986) Preprint, Central Institute of Physics, Bucharest FT-255 (1984).

/19/ M.Iosifescu, H.Scutaru, Phys.Lett. 109A, 433 (1985).

/20/ N.Papanicolaou, Ann. Phys. (N.Y.) 136, 210 (1981).

/21/ J.Deenen, C.Quesne, J. Math. Phys. 23, 2004 (1982).

/22/ P.D.Jarvis, J. Phys. A, Math. Gen. 12, 1 (1979).

/23/ M.Iosifescu, H.Scutaru, "Representations which Satisfy Identities Associated with Tensor Operators. Second-Degree Tensor Operators which Transform under the Representations $(2\Lambda_1)$ of so(2n,C) and (Λ_2) of sp(2n,C)". Communication of the JINR, Dubna E4-85-525 (1985).

/24/ M.Iosifescu, H.Scutaru, "Representations which Satisfy Identities Associated with Tensor Operators. Second - degree Tensor Operators Transforming under the Representations (Λ_4), $(2\Lambda_2)$ of so(2n,C) and $(4\Lambda_1)$, $(2\Lambda_2)$ of sp(2n,C)" Communication of the JINR, Dubna E4-85-526 (1985).

/25/ M.Iosifescu, H.Scutaru, "Kronecker Products which Decompose into two Irreducible Representations", Communication of the JINR, Dubna E4-85-527 (1985).

/26/ M.Iosifescu, H.Scutaru, A Moment Map-like Mapping", Preprint , Bucharest, FT-254 (1984).

/27/ J.M.Souriau, "Structure des Systèmes Dynamiques", Dunod Paris (1970).

/28/ V.Guillemin, S.Sternberg, Ann. Phys. (N.Y.) 127, 220 (1980).

/29/ P.Kramer, Ann. Phys. (N.Y.) 141, 254 (1982) ; "Advances in the Theory of Collective Motion in Nuclei" in "Group Theoretical Methods in Physics", Lecture Notes in Physics, Nr. 201, Springer (1984).

/30/ A.L.Carey, K.C.Hannabuss, Rep. Math. Phys. 13, 199 (1978).

/31/ H.Zassenhaus, Can. Math. Bull. 1, 31, 101, 183 (1958).

/32/ E.G.Kalnins, W.Miller, Jr., J. Math. Phys. 18, 271 (1977).

DE - SITTER REPRESENTATIONS AND THE PARTICLE CONCEPT, STUDIED IN AN UR-THEORETICAL COSMOLOGICAL MODEL[x)]

Th. Görnitz and C. F. v. Weizsäcker
Arbeitsgruppe Afheld in der Max - Planck - Gesellschaft
Bahnhofplatz 4, D - 8130 Starnberg, Germany

Abstract: The theory of urs (basic two-valued observables) is used to describe particles in cosmic space-time. Cosmic position space is described as S^3, interpreted as a homogeneous space of SU(2). An expanding model of the universe is locally approximated by de Sitter spaces. Irreducible representations of the de Sitter group are explicitly constructed in ur theory. From these, Poincaré group representations in Minkowski space with well-defined rest mass are deduced by a special rule of contraction.

1. Ur - Theory and Cosmology

We use the terms a b s t r a c t q u a n t u m t h e o r y for the universal laws of quantum theory in Hilbert space, and c o n c r e t e q u a n t u m t h e o r y for the description of objects as they really exist in the world./1/ Abstract quantum theory includes the universal law of dynamics: the time dependence of states is described by a one-parameter unitary transformation group in the Hilbert space. Concrete quantum theory comprises the existence of particles in a 3,1-dimensional space-time with relativistically invariant interaction laws. We call u r h y p o t h e s i s the assumption that all state spaces occuring in concrete quantum theory can be resolved into tensor products of two-dimensional complex vector

x) Work supported in part by the Deutsche Forschungsgemeinschaft

spaces which are defined by one unique type of observable, called Ur - Alternative (e.g. original alternative) in German; the abstract object corresponding to this alternative is called the ur. We call u r t h e o r y the study of the premisses and consequences of the ur hypothesis.(/1/ ch. 9 and 10)

The ur theory will have to test two conjectures:

The s u f f i c i e n c y c o n j e c t u r e (SC): The ur hypothesis is sufficient for deducing the complete concrete quantum theory from abstract quantum theory.

The t r i v i a l i t y c o n j e c t u r e (TC): The ur hypothesis is trivial in the sense of being a necessary concequence of those postulates from which abstract quantum theory itself can be reconstructed.

SC, the sufficiency of the ur hypothesis, may seem to be a very daring assumption. In order to confirm it we would have to deduce from the quantum theory of an arbitrary number of urs

a) the existence of a 3,1-dimensional space-time

b) the existence and properties of all known particles and fields.

We suppose to have solved problem a) in /1/ chapter 9, as far as space can be described as flat or constantly curved. The basic idea is to interpret a symmetry group of the ur, SU(2), as defining a real 3-space in which it acts as an SO(3), and which is hypothetically treated in the ur theory as the position space of physics. The solution of problem b) cannot be easier than a unified theory of elementary particles (/1/, chapter 10). The present paper describes a special model, not contained in /1/, in which we can formulate an ur-theoretical approach towards problem b).

In any new fundamental theory there occurs a reversal of the historical order of some arguments. We use two well-known examples which will turn out to be relevant again for ur theory.

Astronomy, the most ancient exact natural science, was from antiquity down to Kepler a morphological theory of planetary orbits, be it around the earth or around the sun, completed by the morphological cosmology of a finite spherical universe. General laws prescribed the mathematical form of the orbits as built from circles or, finally, as ellipses. These shapes were different from those of terrestrical motions. Newton's mechanics was not a "great", i. e. additive, but a "radical", i. e. reductive unification. In astronomy it permitted for the first time to ask and answer the question why and to what approximation planets should have mathematically well-defined orbits at all. Historically the orbits were the way towards general mechanics; in the new theory mechanics was the reason why there had to be orbits.

Similarly, the quantum theory of the atom was a radical unification of mechanics and chemistry. Bohr's correspondence principle presupposed the good approximate macroscopic validity of classical physics and paved the way towards a consistent quantum theory. Quantum mechanics reversed the argument and explained classical mechanics as its limiting case.

Ur theory again pressuposes the good approximate validity of two earlier concepts: of the *visible universe* and of *particles*. The high degree of homogeneity and the systematic expansion of the visible universe has permitted to treat it approximately as one large physical object. Its history is described in the semi-empirical , semi-speculative science of cosmology. This science presupposes general relativity which was conceived as a theory of a locally defined field. On the other hand the concepts of mass-point particles and/or localisable fields were presupposed by most models of elementary particle theories; a string in a high-dimensional space is no more than a generalisation of the particle concept. An additive unification of the concepts of universe and particle has begun to exist in the description of the earlier phases of cosmic expansion. Both concepts presuppose the validity of the concept of space-time.

Since the ur theory claims to derive position space from the quantum state space of the binary alternative, it is essentially a radical unification of cosmology and particle theory. A single ur, containing no more than one bit of information, cannot possibly be localized in the universe. The simplest model of cosmic space in the ur theory is the largest homogeneous space of the group SU(2), that is the group itself, considered as a topological and metrical space. It is the S^3, the position part of an Einstein cosmos. In this space, one ur can be described as a spinor wave-function with a wavelength equal to the diameter of the universe (/1/, chapter 9, section 3b). If we assume this diameter to be 10^{27}cm, a particle can be localized down to the Compton wavelength of the electron by superposing 10^{37} ur wavefunctions. Thus the ur is essentially cosmic. The accuracy of the measurement of a small distance is limited by the available number of urs.

In this theory, space is not an objective ultimate entity like Newton's or Einstein's spaces. Its coordinatisation as S^3 is done by the group parameters of SU(2) which are no quantum observables. As far as its properties can be observed, it is rather a "surface" of the high-dimensional quantum state space of a large number of urs. Quantum theory is immensely richer in information than any classical theory in space-time.

On the other side, the concept of particle equally loses its apparent evidence. If a particle "consists" of 10^{37} urs, why should these stick together? This problem should not surprise us. The history of atomism teaches that nearly every particle which was considered elementary turned out to be composite sooner or later. Ur theory seems to be the most radical possible form of atomism; there is no smaller meaningful alternative than yes - no. Hence we may expect all objects to be divisible into urs in principible. This division is no longer spatial, but informationial. The question then is: what is the dynamics of the urs; how does it motivate them to keep together? This can be subdivided into two succesive questions:

1) How to describe the inertial motion of a free particle?
2) How to describe interaction?

The present paper is confined to question 1). The answer is given in principle by Wigner's definition: The state space of a free pointlike particle is the representation space of an irreducible representation of the Poincaré group. Thus, if we can construct such representations by urs, they will permit an interpretation as particles.

The problem is that ur theory does not yet fully determine the relevant relativistic group. In order to understand this problem we turn to TC, the conjecture that ur theory is trivial (/1/, chapter 9, section 2b). It is indeed trivial as far as we leave dynamics aside. It is logically trivial that any n-fold alternative can be represented within the Cartesian product of k binary alternatives with $2^k \geq n$. It is mathematically true that an n-dimensional vector space can be represented within the tensor product of k two-dimensional vector spaces. If n is countably infinite, so will be k . This decomposition is not unique; there are many different possibilities of defining the ur. The problem is wether the law of dynamics keeps all these differently defined urs or some of them invariant in time, such that they can be considered as physical objects (or, rather, "subobjects").

Certainly the ur hypothesis is not trivial in full abstract quantum theory which permits any self-adjoint operator as an Hamiltonian. In /1/, ch. 9, sec. 2b, we try to narrow down the basic postulates of quantum theory so as to make TC a necessary concequence. In the present paper we follow a different path, by way of a cosmological model.

We assume S^3, as defined by the symmetry group of the ur, to be a parametrisation of the cosmic position space. S^3 is compact. Hence we seem to have assumed a finite universe. As long as we may assume that the information content, hence the number of urs, in the universe is finite, a compact position space is indeed a natural description. (We should never forget that in ur theory space is not a basic entity, but, only a way of describing a quantum world, hence perhaps to some extent conventional.) If, however, we assume an infinite number of urs, we must represent the quantum state of the universe in an infinite-dimensional Hilbert space in which noncompact groups possess unitary representations. Then we can use unbounded world models.

Yet, in an infinite-dimensional Hilbert space we must distinguish between actual and virtual urs, i.e. between alternatives that can be decided, given a real situation, and alternatives that might only be decided by producing a different situation. A free particle in flat position space is an example. A discrete basis of its wave functions is given by all eigenfunctions of the total angular momentum, j, and of one of its components, m. The angular momentum is defined with respect to some position in space; let this be the observer's position. Then there will be an upper bound j_{max} for those wave functions which the observer will be able actually to observe; for $j > j_{max}$ the value of ψ in the volume accessible to the observer will, for him, be practically indistinguishable from zero. That means that this observer will only make use of a finite-dimensional part of the Hilbert space of the particle; a part which can be decomposed into the state spaces of a finite number of urs. If he wants to know more about the particle, he must move to another place, finding an additional finite number of decisions; and so on. The full representations of non-compact groups like spatial translations or Lorentz boosts is always done by virtual urs; we cannot actually walk indefinitely into space or accelerate a particle indefinitely.

For a cosmological working model/2/ we choose a cosmological (absolute) time coordinate t and a time-dependent total number $N_u(t)$ of urs in the universe. N_u is supposed to increase monotonously with t : the number of possible decisions in the world increases steadily with time. N_u will be a measure of the volume of the cosmic space, as measured by elementary particles. Thus our assumption describes the expansion of the universe. The semantic consistency of the model will only admit a test when we shall have understood how particles can be described in such a universe.

2. Particles

The concept of a pointlike particle is historicaly an abstraction from the concept of a body, or of its center of mass, neglecting the body's extension or inner dynamics. In the ur theory, this concept must be derived from more basic concepts. Wigner derived the free particle from the representation theory of the Poincaré group. The Wigner particles are characterised by two numbers: the spin s, and the mass m. In the ur theory, particles with arbitrary spin can be represented (/1/, ch. 9, sec. 3e). The determination of m remains an unresolved problem (/1/, ch. 10, sec. 6d).

In the ur-theoretical context it is plausible that the rest masses of real particles are cosmologically determined. A possible consideration might be the following: We measure cosmic dimensions by ponderable matter (and with the help of light). The mass of ponderable matter is mainly concentrated in nucleons. Let λ be the Compton wavelength of the nucleon, R_u the radius of the universe. Assume the number N_u of urs in the universe to be the volume of the universe, measured in nuclear volumes:

$$N_u \approx R_u^3 / \lambda^3 \tag{1}$$

In order to localize a nucleon in 3 dimensions we need

$$\gamma \approx 3\ R_u/\lambda \approx N^{1/3} \tag{2}$$

urs. If we assume this to be the number of urs "contained" in the nucleon, there would have to be $N^{2/3}$ nucleons in the world. With $N^{1/3} \approx 10^{40}$, this gives 10^{80} nucleons, not too far from the estimated empirical number. The real task of the theory would be to explain this "condensation" of urs by statistical considerations.

Our present aim is more limited. We search for a precise mathematical description of particles in our cosmological model. Wigner's construction presupposes a Minkowski space. It would seem natural at any space-time point to choose the locally tangential Minkowski space. This is what we will finally do. But we shall insert a locally approximating de Sitter space between the cosmological model and the Minkowski space. The reason for this intercalation is that a de Sitter space combines two properties, none of which ought to be lost. It contains an S^3 as its spatial part; by approximating the world model this can be identified with the S^3 prevailing in the model at the respective cosmological time t . We need this compact spatial volume in order to perform the program of determining the rest mass of the particle. And on the other hand it possesses a 10-parameter symmetry group which will permit us to define particles by the Wigner method which then can be translated into the usual Minkowski description by the procedure of contraction.

The free particles thus defined will then be the starting material for a theory of interaction (/1/, ch. 10) including the transition to the Riemannian space-time of general relativity (/1/, ch. 10, sec. 7; /2/).

3. How to build Particles out of Urs

The state space of an ur is C^2. The norm is conserved by $SU(2) \times U(1)$, and by complex conjugation. The latter can, according to Castell/3/, be linearly represented by introducing anti-urs, represented together with the urs in a common C^4. The state space for n urs is then the tensor product $C^{4^n} = \bigotimes_1^n C^4$. All states of any number of urs are then contained in the "tensor space" $T = \bigoplus_{n=0}^{\infty} C^{4^n}$. C^1 is the vacuum. Let r_i (i = 1...4) be a basis in C^4. Then the monomials

$r_{i_1} \ldots r_{i_n}$ form a basis in C^{4^n} with

$$\langle r_{i_1} \ldots r_{i_n} \mid r_{k_1} \ldots r_{k_m} \rangle = \delta_{nm} \cdot \delta_{i_1 k_1} \ldots \delta_{i_n k_n} \tag{3}$$

Now we define "pick-" and "stuff-" operators R_r and S_r resp. (see /1/, ch. 10, sec. 2b) by their action on the basis monomials:

$$S_r\, r_1 \ldots r_n = r r_1 \ldots r_n + r_1 r \ldots r_n + \ldots + r_1 \ldots r r_n + r_1 \ldots r_n r \tag{4}$$

$$R_r\, r_1 \ldots r_n = r_1 \ldots \not{r}_{i_1} \ldots r_n + r_1 \ldots \not{r}_{i_2} \ldots r_n + \ldots + r_1 \ldots \not{r}_{i_j} \ldots r_n \tag{5}$$

with $r_{i_k} = r$ and $r_k \neq r$ for all the other indices; $\not{r}_{i_j}$ means omit this vector from the monomial. With respect to the scalar product (3) R_r and S_r are adjoint operators: $R_r^+ = S_r$

Further on we define "trucking" - operators t_{rs} by

$$t_{rs}\, r_1 \ldots s_1 \ldots s_k \ldots r_n = r_1 \ldots r \ldots s_k \ldots r_n + \ldots + r_1 \ldots s_1 \ldots r \ldots r_n \tag{6}$$

for $s_i \equiv s$ and $r_i \neq s$

The operator t_{rr} only multiplies a monom with the number of vectors r contained in it.

The following commutation relations hold

$$\begin{aligned} &[S_i, S_j] = [R_i, R_j] = 0 \qquad && [R_r, t_{st}] = + R_t \delta_{rs} \\ &[R_r, S_s] = t_{sr} \quad \text{for } s \neq r \qquad && [S_r, t_{st}] = - S_s \delta_{rt} \\ &[R_r, S_r] = t_{rr} + \widehat{(n+1)} && \end{aligned} \tag{7'}$$

$\hat{n}$ denotes the operator multiplying a monom with the number of its factors.

Castell has shown how the conformal group SO(4,2) can be represented in the subspace $\bar{T}$ of T which consists of the symmetric tensors only. The operators acting in $\bar{T}$ which are used in these representations are

$$a_r^+ = S_r / \sqrt{n+1} \qquad ; \qquad a_r = R_r / \sqrt{n} \tag{8}$$

with the canonical commutation relations

$$[a_r, a_s^+] = \delta_{rs} \quad ; \quad [a_r, a_s] = [a_r^+, a_s^+] = 0 \tag{9}$$

They correspond to Bose statistics for the urs. In these representations, $s = (n_1 + n_2 - n_3 - n_4)/2$ with n_r designing the number of urs in state r, is the Casimir operator of SO(4,2) which describes the helicity of the respective particle; the whole space $\bar{T}$ contains just one representation for each value of s, describing a massless particle.

If we want to describe many-particle systems and particles with non-zero rest mass we must make use of nonsymmetric tensors. This raises the question wether urs are individually distinguishable. In principle one would prefer to assume them to be indistinguishable, since their distinction would be an additional alternative, not in accord with SC. If TC were correct, the version of the ur hypothesis turning out to be trivial would decide the question. We have so far considered two alternative answers.

The most general statistics for indistinguishable urs is parastatistics, especially para-Bose statistics (/1/, ch. 10, sec. 2d). It acts on a larger subspace $\tilde{T}$ of T which contains in every C^{4^n} one representative of every irreducible representation of the permutation group S_n, i.e. for every Young standard scheme. It permits representations of SO(4,2) and its subgroups with finite rest mass.

The present paper presents an other possibility. It works, in principle, in the full tensor space T. R_r and S_r are defined everywhere in T. If, following a proposal by Drühl, we define

$$R_r = - t_{or} \quad ; \quad S_r = t_{ro} \quad ; \quad \widehat{(n+1)} = - t_{oo} \quad , \tag{10}$$

then the relations (7') can be summarized into

$$\left[t_{rs}, t_{tu}\right] = t_{ru}\,\delta_{st} - t_{ts}\,\delta_{ru} \tag{7}$$

Commutation relations for arbitrary powers of these operators are given in the appendix.

As a "ground state" we define a normed vector Ω_N in T with

$$R_r\ \Omega_N = 0 \ ; \ t_{rs}\,\Omega_N = 0 \quad \text{for} \quad r \neq s \quad ; \ -t_{oo}\,\Omega_N = (N+1)\,\Omega_N \tag{11}$$

Ω_0 is the regular vacuum, a general ground state is given by

$$\Omega_{4n} = (4!)^{-n/2} \sum_{P_i} (-1)^{T_1} P_1(r_1 r_2 r_3 r_4) \ldots (-1)^{T_n} P_n(r_1 r_2 r_3 r_4) \tag{12}$$

P_k denotes a permutation of the quadruple of basis vectors and T_k counts the transpositions in it. The sum goes over all possible permutations in all the quadruples.

Over a ground state the stuff-operators generate a linear subspace. Its orthonormal basis vectors are given by

$$|l_1 l_2 l_3 l_4\rangle_N = \sqrt{\frac{(N+n)!}{(N+n+L)!\,l_1!\,l_2!\,l_3!\,l_4!}}\ S_1^{l_1}\, S_2^{l_2}\, S_3^{l_3}\, S_4^{l_4}\ \Omega_N \tag{13}$$

with $N = 4n$, $L = l_1+l_2+l_3+l_4$

On $|\,l_1 l_2 l_3 l_4\rangle_N \overset{!}{=} |\,L\rangle_N$ the pick- and stuff-operators act like

$$R_j\ |\,l_i, l_j, l_k, l_l\rangle_N = \sqrt{l_j}\ \sqrt{N+n+L}\ |l_i, l_j-1, l_k, l_l\rangle_N$$

$$S_j\ |\,l_i, l_j, l_k, l_l\rangle_N = \sqrt{l_j+1}\ \sqrt{N+n+L+1}\ |l_i, l_j+1, l_k, l_l\rangle_N \tag{14}$$

$$t_{ji}|l_i,l_j,l_k,l_l\rangle_N = \sqrt{l_j+1}\sqrt{l_i}\;|l_i-1,l_j+1,l_k,l_l\rangle_N$$

$$t_{jj}\;|\;L\rangle_N = (n+l_j)\;|\;L\rangle_N \qquad -t_{oo}\;|\;L\rangle_N = (N+L+1)\;|\;L\rangle_N$$

4. De Sitter representation for a given ground state

Given a ground state Ω_N, then, by modified pick- and stuff-operators, an irreducible unitary representation ν_{r,N^2} of the de Sitter-group /4/ can be constructed such that for different particles the ratio of their numbers of urs in the ground state corresponds to their mass ratio. If the spin r of the particle is not zero then we have (2r+1) vectors with minimal ur number which in this case is N + 2r.

Let the indicees a, b be equal to 1 or 2 and c, d to 3 or 4. Now we define the following operators

$$S_{ac}|L\rangle_N = S_{ca}|L\rangle_N = \sqrt{\frac{(L/2+1-r)(L/2+2+r)}{(l_1+l_2+1)(l_1+l_2+2)(l_3+l_4+1)(l_3+l_4+2)}}\cdot \sqrt{\frac{(L/2+1)(L/2+2)+N^2}{(N+n+L+1)(N+n+L+2)}}\cdot S_a\cdot S_c|L\rangle_N \tag{15}$$

$$R_{ac}|L\rangle_N = R_{ca}|L\rangle_N = \sqrt{\frac{(L/2-r)(L/2+1+r)}{(l_1+l_2)(l_1+l_2+1)(l_3+l_4)(l_3+l_4+1)}}\cdot \sqrt{\frac{L/2\,(L/2+1)+N^2}{(N+n+L)(N+n+L+1)}}\cdot R_a\cdot R_c|L\rangle_N \tag{16}$$

$$T_{ac}|L\rangle_N = \sqrt{\frac{((l_3+l_4-l_1-l_2)/2+r)((l_1+l_2-l_3-l_4)/2+1+r)}{(l_3+l_4)(l_3+l_4+1)(l_1+l_2+1)(l_1+l_2+2)}}\cdot 1/2\cdot\sqrt{(l_3+l_4-l_1-l_2)(2+l_3+l_4-l_1-l_2)+4N^2}\;t_{ac}|L\rangle_N \tag{17}$$

$$T_{ca}|L\rangle_N = \sqrt{\frac{((l_1+l_2-l_3-l_4)/2+r)((l_3+l_4-l_1-l_2)/2+1+r)}{(l_1+l_2)(l_1+l_2+1)(l_3+l_4+1)(l_3+l_4+2)}}\cdot 1/2\cdot\sqrt{(l_1+l_2-l_3-l_4)(2+l_1+l_2-l_3-l_4)+4N^2}\;t_{ca}|L\rangle_N \tag{18}$$

The generators of the wanted unitary irreducible representation of the SO(4,1) are (19)

$$M_1 = (t_{12}+t_{21}+t_{34}+t_{43})/2 \qquad P_1 = (t_{12}+t_{21}-t_{34}-t_{43})/2$$

$$M_2 = -i(t_{12}-t_{21}+t_{34}-t_{43})/2 \qquad P_2 = -i(t_{12}-t_{21}-t_{34}+t_{43})/2$$

$M_3 = (t_{11}- t_{22}+ t_{33}- t_{44})/2 \qquad P_3 = (t_{11}- t_{22}- t_{33}+ t_{44})/2$

These six operators form the SO(4) - subgroup and preserve the number of urs and of antiurs.

$$P_0 = (S_{14}- S_{32}+ R_{14}- R_{32}+ T_{31}+ T_{13}+ T_{42}+ T_{24})/2 \qquad (20)$$

$$N_1 = -i(S_{13}- S_{24}+ R_{24}- R_{13}+ T_{32}- T_{23}+ T_{41}- T_{14})/2$$

$$N_2 = -(S_{13}+ S_{24}+ R_{24}+ R_{13}+ T_{32}+ T_{23}- T_{41}- T_{14})/2$$

$$N_3 = i(S_{14}+ S_{23}- R_{14}- R_{23}+ T_{13}- T_{31}- T_{24}+ T_{42})/2$$

The Casimir-operators for this representation are

$$C_2 = P_0^2 - P_1^2 - P_2^2 - P_3^2 + N_1^2 + N_2^2 + N_3^2 - M_1^2 - M_2^2 - M_3^2 \qquad (21)$$

$$C_4 = (\vec{M}\cdot\vec{P})^2 - (P_0\,\vec{M} - \vec{P}\times\vec{N})^2 - (\vec{M}\cdot\vec{N})^2 \qquad (22)$$

with the eigenvalue equations

$$C_2 \mid l_1,l_2,l_3,l_4 \rangle_N = (N^2 - r\,(r+1) + 2) \mid l_1,l_2,l_3,l_4 \rangle_N \qquad (23)$$

$$C_4 \mid l_1,l_2,l_3,l_4 \rangle_N = - N^2\, r\,(r+1) \mid l_1,l_2,l_3,l_4 \rangle_N \qquad (24)$$

written in the pick- and stuff-operators C_2 has the form

$$\begin{aligned} 2\,C_2 = \; & S_{14}R_{14} + S_{13}R_{13} + S_{24}R_{24} + S_{23}R_{23} - (t_{11}- t_{22})^2 + \\ & + R_{14}S_{14} + R_{13}S_{13} + R_{24}S_{24} + R_{23}S_{23} - (t_{33}- t_{44})^2 + \\ & + T_{13}T_{31} + T_{14}T_{41} + T_{23}T_{32} + T_{24}T_{42} - 2(t_{12}t_{21}+ t_{21}t_{12}) \\ & + T_{31}T_{13} + T_{41}T_{14} + T_{32}T_{23} + T_{42}T_{24} - 2(t_{34}t_{43}+ t_{43}t_{34}) \end{aligned} \qquad (25)$$

We define (for i, k, l, m mutually different)

$$\begin{aligned} & (t_{ii} + t_{kk} - t_{ll} - t_{mm})T_{ik} + 2t_{il}T_{lk} + 2t_{mk}T_{im} \\ & \stackrel{!}{=} (\hat{l}_1 + \hat{l}_2 + \hat{l}_3 + \hat{l}_4 + 2)T_{ik} \end{aligned} \qquad (26)$$

Then C_4 can be written as

$$\begin{aligned} 8C_4 = & -(t_{11}+t_{22}+t_{33}+t_{44})^2\left(\{S_{13},R_{13}\}+\{S_{14},R_{14}\}+\{S_{23},R_{23}\}+\{S_{24},R_{24}\}\right) \\ & -(\hat{l}_1+\hat{l}_2+\hat{l}_3+\hat{l}_4+2)^2\left(\{T_{13},T_{31}\}+\{T_{14},T_{41}\}+\{T_{23},T_{32}\}+\{T_{24},T_{42}\}\right) \\ & + 4\{t_{12},t_{21}\} - 4\{t_{34},t_{43}\} + 2(t_{11}- t_{22})^2 - 2(t_{33}- t_{44})^2 \end{aligned} \qquad (27)$$

In the spin-zero-case there is $r = 0$ and $l_1+ l_2 = l_3+ l_4$, so all T_{ac} and T_{ca} vanish. For half-integer spin the representations are unitary only for $N^2 \geq 1/4$. Representations with integer spin are unitary also for $N^2 = 0$, but then they decompose into a direct sum of tree irreducible representations of the so-called discrete series $\pi^{\pm}_{r,q}$:

$$\lim_{N^2\to 0} \nu_{r,N^2} = \pi^{+}_{r,1} + \pi_{r,0} + \pi^{-}_{r,1} \qquad (28)$$

Castell's massless particles[3] all belong to the representations $\pi^{+}_{r,r}$ and $\pi^{-}_{r,r}$. In the limit (28) only the photon representations are of this type.

5. Transition to the Poincaré Group

We have introduced de Sitter space as an approximation to the cosmological model in order to interpret states from the tensor space of urs as states of a particle in de Sitter space. The particle was defined by its minimal ur number N. It turned out that our operators R and S defined irreducible representations of the de Sitter group characterized by the number N^2. These representations are localisable on the S^3 as considered as the position space in the de Sitter world[5]. To the degree to which we can neglegt the curvature of this S^3, hence approximate it by a flat space, or the de Sitter world by a Minkowski world, such a state can be considered as a localized state in a representation of the Poncaré group P . We shall consider the resulting Poincaré representation as the Wigner description of a free particle.

The transition from the de Sitter representation to the Poincaré representation is achived by a group contraction. It is well known that this contraction can be done in different ways, so as to give the Poincaré particle any value m of its rest mass. We consider N as the quantity in T which corresponds to the rest mass. Hence we shall carry out a contraction such that the ratio N'/N'' of two different particles is transformed into the ratio m'/m'' of their masses.

In the process of contraction a parameter λ which corresponds to the curvature scalar of the de Sitter space goes towards zero. Simultaneously the parameter N^2 which characterises the representation moves towards infinity. The rest mass m in the resulting Poincaré representation is given by

$$m^2 = \lim_{\lambda^2 \to 0\,;\, N^2 \to \infty} (\lambda^2 N^2) \tag{29}$$

We need a relation between λ and N in order to fix m. It is sufficient to postulate that this relation should be such that for two different particles the ratio N' / N'' is kept constant throughout the process of going to the limit; then we will achieve

$$m' / m'' = N' / N'' \tag{30}$$

We can e.g. abitrarily choose that the Planck-Wheeler mass should be $m_P = 1$ for all time. The number of urs in the Planck-Wheeler particle is $N_u^{1/2}$. If, as our cosmological model assumes, N_u depends on the cosmological time t, the number N of urs in the ground state of a particle whose mass is assumed to have at a given time a fixed value in units of the Planck mass will depend on cosmological time:

$$N / N_u^{1/2} = f(t) \tag{31}$$

It will, however, depend on the intended theory of rest masses in which way this condition will be specified.

Böhm and Moylan /6/ have shown that the representation space of an irreducible representation of SO(4,1) is the direct sum of the representation spaces of two irreducible representations of the Poincaré group, both with positive energy and equal mass m, but different by a charge-like quantum number. So, coming from the de Sitter group, the particle - antiparticle dualism is very natural. In their theory m is not fixed, this is done by our presciption.

We recapitulate our answer to the question 1., how to describe the inertial motion. Locally we have justified the Wigner description in Minkowski space; its empirical success justifies our calling the Minkowski coordinate x_0 the time. Through the local de Sitter space this identification leads back to the local time in the cosmological model. However, with increasing cosmological time t the local Minkowski space is replaced by another one; it is to be assumed that N and m will thus depend on t. This dependence will be determined by the assumed dependence of N_u on t in the cosmological model. Since the actual measurement of time will depend on the functions $N(t)$ and $m(t)$, t being the assumed cosmological time, it seems possible that the model contains no arbitrary function $N_u(t_m)$, if t_m means the time as locally measured. But this question is further to be studied.

A c k n o w l e d g e m e n t

One of us (Th.G.) thanks P. Moylan for interesting discussions and the Deutsche Forschungsgemeinschaft for financial support.

Literature

/1/ C. F. v. Weizsäcker, Aufbau der Physik, Hanser Verl., München, 1985
A series of three shorter english papers on the content of this book is under preparation
/2/ Th. Görnitz in L. Castell, C. F. v. Weizsäcker (Eds.) Quantum theory and the structures of time and space, Vol.6 Hanser Verl., München, 1985
/3/ L. Castell in L. Castell, M. Drieschner, C. F. v. Weizsäcker, Quantum theory and the structures of space and time, Vol.1 Hanser Verl., München, 1975
/4/ S. Ström, Arkiv för Fysik, 30, (1965), 455 - 472
/5/ K. C. Hannabuss, Proc. Camb. Phil. Soc. 70, (1971) 238-302
/6/ A. Böhm in Studies in Math. Phys., A. O. Barut (Ed.), Reidel, New York, 1973
P. Moylan, J. Math. Phys. 24, (1983) 2706 - 2721

A p p e n d i x

Some commutation relations for powers of pick- and stuff-operators

let be $r \neq s$

$$R_r^k\, t_{rs}^n = \sum_{j=o}^{\min(n,k)} \frac{n!\; k!}{(n-j)!\; j!\; (k-j)!}\, t_{rs}^{n-j}\, R_s^j\, R_r^{k-j} \qquad (A1)$$

$$t_{sr}^n\, S_r^k = \sum_{j=o}^{\min(n,k)} \frac{n!\; k!}{(n-j)!\; j!\; (k-j)!}\, S_r^{k-j}\, S_s^j\, t_{sr}^{n-j} \qquad (A2)$$

$$R_r^k\, S_s^n = \sum_{j=o}^{\min(n,k)} \frac{n!\; k!}{(n-j)!\; j!\; (k-j)!}\, S_s^{n-j}\, t_{sr}^j\, R_r^{k-j} \qquad (A3)$$

$$R_i^k\, S_i^n = \sum_{j=o}^{\min(n,k)} \frac{n!\; k!}{(n-j)!j!(k-j)!}\, S_i^{n-j} \left[\prod_{s=1}^{j} (t_{ii}-t_{oo}+n+k-j-s)\right] R_i^{k-j} \qquad (A4)$$

II. CLASSICAL AND QUANTUM FIELD THEORY

THE STRUCTURE OF LOCAL ALGEBRAS IN QUANTUM FIELD THEORY

(Talk presented at the International Symposium on Conformal Groups and Structures 1985 in Clausthal)

D. Buchholz
II. Institut für Theoretische Physik
Universität Hamburg, D-2000 Hamburg 50

It is the purpose of these notes to give an account of some recent work on the structure of local algebras appearing in the algebraic formulation of relativistic quantum physics. (For a review of this subject up to 1980 cf. [1].) There has been considerable progress on this problem in the last few years, both from the mathematical and the physical side, and one may say that we have reached now a satisfactory understanding of the properties of these algebras in generic cases.

Let me begin by recalling the general postulates of algebraic quantum field theory. The basic input in this setting is the assumption that one is given a mapping (a "net")

$$\mathcal{O} \longrightarrow \mathfrak{A}(\mathcal{O}), \tag{1}$$

assigning to each open, bounded region $\mathcal{O}$ of Minkowski space some von Neumann algebra [1] $\mathfrak{A}(\mathcal{O})$ on a separable Hilbert space $\mathcal{H}$. Each $\mathfrak{A}(\mathcal{O})$ is interpreted

1) A von Neumann algebra is a weakly closed *-algebra of bounded operators.

as the algebra generated by all observables which can be measured within $\mathcal{O}$, and $\mathcal{H}$ is the space of physical states. In view of this interpretation one is led to assume that

$$\mathfrak{A}(\mathcal{O}_1) \subset \mathfrak{A}(\mathcal{O}_2) \quad \text{if} \quad \mathcal{O}_1 \subset \mathcal{O}_2 , \tag{2}$$

and that locality holds, i.e.

$$\mathfrak{A}(\mathcal{O}_1) \subset \mathfrak{A}(\mathcal{O}_2)' \quad \text{if} \quad \mathcal{O}_1 \subset \mathcal{O}_2' . \tag{3}$$

Here $\mathcal{O}'$ denotes the spacelike complement of $\mathcal{O}$ and $\mathfrak{A}(\mathcal{O})'$ the algebra of all bounded operators commuting with the elements of $\mathfrak{A}(\mathcal{O})$. The space-time symmetry group $\mathcal{L}$ (i.e. the Poincaré group, possibly extended by conformal transformations) is assumed to act on $\mathcal{H}$ by a continuous, unitary representation $U(L), L \in \mathcal{L}$, and the unitaries $U(L)$ generate automorphisms inducing the symmetry transformations L on the local algebras,

$$U(L)\, \mathfrak{A}(\mathcal{O})\, U(L)^{-1} = \mathfrak{A}(L\mathcal{O}). \tag{4}$$

It is furthermore assumed that the generators of the space-time translations $U(x)$, $x \in \mathbb{R}^4$ satisfy the relativistic spectrum condition (positivity of energy), and that there is an (up to a phase unique) vector $\Omega \in \mathcal{H}$, representing the vacuum, for which

$$U(L)\, \Omega = \Omega , \quad L \in \mathcal{L} . \tag{5}$$

In the following we also assume that Ω is cyclic for the local algebras, i.e.

$$\overline{\bigcup_{\mathcal{O}} \mathfrak{A}(\mathcal{O})\, \Omega} = \mathcal{H}. \tag{6}$$

Since observables do not change the charge quantum numbers of a state this means that we restrict our attention to states lying in the same superselection sector as the vacuum. But this is no restriction of generality, since the structure of the local algebras $\mathfrak{A}(\mathcal{O})$ we are interested in, is the same in each superselection sector of the Hilbert space of all physical states.

In contrast to the more conventional formulations of the general postulates of quantum field theory, such as the Wightman axioms [2], one deals in the algebraic setting with algebras of bounded operators. This assumption is mathematically convenient, because there do not appear subtle "domain problems" in this setting. But it is also physically reasonable: since observables are to be represented by self-adjoint operators, one can proceed to the corresponding spectral resolutions, giving a family of orthogonal projections which contains the same information as the original operators. With reference to the Wightman framework one may thus think of $\mathfrak{A}(\mathcal{O})$ as the algebra generated by all bounded functions of some basic field(s) φ , ... smeared with real testfunctions f having support in $\mathcal{O}$ [2]. So from the algebraic point of view the fields are regarded as a collection of generators of the local algebras.

It has been emphasized by Haag [4], that for the physical interpretation of a model it is not necessary to know the physical meaning of each individual observable . All what is needed in order to determine e.g. the superselection structure, or the particle spectrum, or collision cross sections etc., is the correspondence (1) between space-time regions and local algebras.

In view of this fact it is natural to ask, which types of algebras $\mathfrak{A}(\mathcal{O})$ are suitable as carriers of this information. There exists an abundance of different (non-isomorphic) von Neumann algebras, and it is known that not all of them can appear as

---

2) For sufficient conditions on the unbounded field operators allowing a rigorous construction of the local algebras cf. for example [3].

elements of the net (1) due to the restrictions imposed by the general principles of relativistic quantum physics. What we want to outline here is the relatively new insight that for a large class of physically relevant models, which are distinguished by a "tame" high-energy behaviour, the structure of the local algebras is in fact unique (i.e. model-independent). Phrased differently: the internal structure of the local algebras is the same in interacting and free field theories, so these algebras are explicitly known. This result shows that the dynamics of a particular model enters only in the specific properties of the mapping $\mathcal{O} \longrightarrow \mathfrak{A}(\mathcal{O})$.

In order to substantiate this result we need various concepts from the theory of von Neumann algebras, which will be explained in the following.

1. Let $\mathcal{M} \subset \mathcal{B}(\mathcal{H})$ be any von Neumann algebra . The commutative algebra $\mathcal{M} \cap \mathcal{M}'$ is called the <u>center</u> of $\mathcal{M}$. A von Neumann algebra $\mathcal{M}$ with trivial center, i.e. $\mathcal{M} \cap \mathcal{M}' = \mathbb{C} \cdot 1$, is called a <u>factor</u>. So our first question is: do the local algebras $\mathfrak{A}(\mathcal{O})$ have a center?

A general answer to this question is not known. But it has been shown by explicit calculations that the local algebras are factors in many field theoretic models [5]. On the other hand there exist certain artificial models, where the local algebras associated with some given space-time region $\mathcal{O}$ do have a non-trivial center. It is noteworthy that these counterexamples violate the so-called time slice axiom, where one assumes that the inclusion (2) still holds if the region $\mathcal{O}_1$ is contained in the causal shadow $\mathcal{O}_2''$ of $\mathcal{O}_2$. This condition should be satisfied whenever there is a dynamical law with hyperbolic propagation in the model. So it seems that the local algebras are factors in these generic cases, and, to simplify the subsequent discussion, we will restrict our attention to such models.

2. According to Murray and von Neumann the factors $\mathcal{M}$ can be subdivided into various types by looking at the relative dimensions of the orthogonal projections

in $\mathcal{M}$ [6]. Based on the Tomita-Takesaki theory, a more refined classification of factors has been given by Connes [7]; in fact, this classification seems to be exhaustive as far as the von Neumann algebras appearing in physics are concerned. Following Connes one proceeds as follows: let $\mathcal{M} \subset \mathcal{B}(\mathcal{H})$ and let $\phi \in \mathcal{H}$ be any vector which is cyclic and separating for $\mathcal{M}$, i.e.

$$\overline{\mathcal{M}\phi} = \mathcal{H} \quad \text{and} \quad M\phi \neq 0 \ \text{for} \ M \in \mathcal{M}, M \neq 0 . \tag{7}$$

(We assume that such vectors exist. Note that the vacuum Ω is cyclic and separating for the local algebras, according to the Reeh-Schlieder theorem.) One then defines an anti-linear involution S_ϕ, setting

$$S_\phi \cdot M\phi = M^*\phi \ \text{for} \ M \in \mathcal{M}. \tag{8}$$

It is easy to see that S_ϕ is a closable operator, so the operator $S_\phi^* S_\phi = \Delta_\phi$ is a densely defined, positive invertible operator, called the <u>modular operator</u> associated with the pair $(\mathcal{M}, \phi)$. Amongst the remarkable properties of these modular operators following from the Tomita-Takesaki theory [8], we only mention that the unitaries Δ_ϕ^{it}, $t \in \mathbb{R}$ induce automorphisms of $\mathcal{M}$, i.e.

$$\Delta_\phi^{it} \mathcal{M} \Delta_\phi^{-it} = \mathcal{M} \ \text{for} \ t \in \mathbb{R} . \tag{9}$$

Looking at the spectrum $\mathrm{sp}\,\Delta_\phi$ of Δ_ϕ, Connes [7] invented an algebraic invariant of $\mathcal{M}$,

$$S(\mathcal{M}) = \bigcap_\phi \mathrm{sp}\,\Delta_\phi , \tag{10}$$

where the intersection is to be taken with respect to all states ϕ satisfying the condition (7). Connes was able to show that $S(\mathcal{M})$ has to be one of the following sets if $\mathcal{M}$ is a factor:

$$\{1\} \ , \ \{0,1\}$$

$$\{0\} \cup \{\lambda^n : n \in \mathbb{Z}\} \text{ for some } 0 < \lambda < 1 \tag{11}$$

$$\{\lambda \geq 0\}$$

The factors $\mathcal{M}$ for which $S(\mathcal{M})$ has the form given in the first line can still be subdivided into various types, but we do not need to discuss this here. In the latter two cases $\mathcal{M}$ is called a factor of type III_λ and III_1, respectively. Hence our second question: what is the type of the local algebras?

At first sight it might seem hopeless to answer this question since one must calculate the spectrum of an abundance of operators. But there is a useful result due to Connes [7], saying that if there is some $\phi \in \mathcal{H}$ such that the corresponding modular automorphisms (9) do not have any fixed point (apart from multiples of the identity), then $S(\mathcal{M}) = sp\,\Delta_\phi$. So in many cases it suffices to calculate the modular operator associated with a single vector.

Let us now turn back to quantum field theory. Assuming that the local algebras are generated by Wightman fields, Bisognano and Wichmann [9] have calculated the modular operators associated with $(\mathfrak{A}(W), \Omega)$, where W is a wedge-shaped region such as

$$W = \{ x \in \mathbb{R}^4 : \ x_1 > |x_0| \}. \tag{12}$$

They showed that the modular group Δ_W^{it} associated with $(\mathfrak{A}(W), \Omega)$ coincides with the unitary representation $U(\Lambda(t))$ of the Lorentz-transformations

$$\Lambda(t) = \begin{pmatrix} \mathrm{ch}\, 2\pi t & \mathrm{sh}\, 2\pi t & 0 & 0 \\ \mathrm{sh}\, 2\pi t & \mathrm{ch}\, 2\pi t & 0 & 0 \\ 0 & 0 & 1 & 0 \\ 0 & 0 & 0 & 1 \end{pmatrix}. \tag{13}$$

So, irrespective of the dynamics, one has

$$\Delta_W^{it} = U(\Lambda(t)) \quad , t \in \mathbb{R} , \tag{14}$$

which in view the completely different origin of these groups is a quite remarkable relation. From this relation it is now easy to deduce the type of the algebras $\mathfrak{A}(W)$: Since apart from multiples of the identity there does not exist any operator in $\mathfrak{A}(W)$ which is invariant under the Lorentz-transformations $U(\Lambda(t))$, it follows from (14) that $S(\mathfrak{A}(W)) = \mathrm{sp}\,\Delta_W = \mathbb{R}_+$, so $\mathfrak{A}(W)$ is a factor of type III_1 according to the classification of Connes. (That $\mathfrak{A}(W)$ is a factor follows also from general arguments [1].)

The modular operator corresponding to the algebras of other space-time regions could explicitly be calculated only in special models, however. For models of free, massless particles it was shown by Buchholz [10], that the modular group Δ_V^{it} associated with $(\mathfrak{A}(V), \Omega)$, where V is the light cone, is a representation of the dilations. Since there are no non-trivial fixed points in $\mathfrak{A}(V)$ under dilations it follows that $\mathfrak{A}(V)$ is also a factor of type III_1 . For the same restricted class of models Hislop and Longo [11] have been able to calculate the modular group $\Delta_{\mathcal{O}}^{it}$ associated with $(\mathfrak{A}(\mathcal{O}), \Omega)$, where $\mathcal{O}$ is a double cone. They could show that in this case the modular group is a representation of a 1-parameter subgroup $t \to K(t)$ of the conformal group which has a timelike generator and leaves $\mathcal{O}$ invariant. Again it follows that $\mathfrak{A}(\mathcal{O})$ is a factor of type III_1.

In all these cases the calculation of the modular group was possible because of its purely geometrical action on the local algebras. It was pointed out by Haag that under these circumstances the appearance of conformal transformations is no coincidence: since the causal structure of Minkowski space manifests itself in the spacelike commutation relations of the local algebras, any automorphism of these algebras having a purely geometrical meaning must respect this causal structure, and therefore correspond to a conformal transformation. This remark reveals the limitations of the above direct method for the calculation of the type of the local algebras: since in general only the Poincarè transformations are a symmetry of field-theoretic models, the modular automorphisms associated with bounded regions will in general not have a geometric interpretation, and it is therefore difficult to determine their action explicitly.

At this point the subject got stuck for some time. But it was recently realized by Fredenhagen [12] that one can determine the spectrum of the modular operators associated with bounded space-time regions by going to the scaling limit of the underlying model. Fredenhagen started from the simple geometric observation that if $\mathcal{O} \subset W$ (cf. equation (12)) is a double cone containing the origin in its closure, and if $\Lambda(t)$ are the Lorentz transformations introduced in relation (13), then one has for any λ with $0 < \lambda < 1$

$$\Lambda(t) \cdot \lambda\mathcal{O} \subset \mathcal{O} \quad \text{if} \quad 2\pi|t| \leq |\ln \lambda|, \tag{15a}$$

and consequently (cf. equation (2))

$$U(\Lambda(t))\, \mathfrak{A}(\lambda\mathcal{O})\, U(\Lambda(t))^{-1} \subset \mathfrak{A}(\mathcal{O}). \tag{15b}$$

Now from the work of Bisognano and Wichmann it is known that $U(\Lambda(t))$ coincides with the modular group Δ_W^{it} associated with $(\mathfrak{A}(W),\Omega)$. Moreover, since

$\mathfrak{A}(\mathcal{O}) \subset \mathfrak{A}(W)$ it follows from the very definitions of the modular operator $\Delta_{\mathcal{O}}$ associated with $(\mathfrak{A}(\mathcal{O}), \Omega)$ and Δ_W that for any $A, B \in \mathfrak{A}(\mathcal{O})$

$$(\Delta_W^{1/2} A\Omega, \Delta_W^{1/2} B\Omega) = (\Delta_{\mathcal{O}}^{1/2} A\Omega, \Delta_{\mathcal{O}}^{1/2} B\Omega), \tag{16}$$

so Δ_W is an extension (in the sense of bilinear forms) of $\Delta_{\mathcal{O}}$. Using this fact and equation (15) Fredenhagen could show that the unitaries $\Delta_{\mathcal{O}}^{it}$ and $U(\Lambda(t))$ act on the vectors $A\Omega$, $A \in \mathfrak{A}(\lambda\mathcal{O})$ in "almost the same manner", provided λ is sufficiently small. Namely, given any testfunction f and any $\varepsilon > 0$ there exists a λ, $0 < \lambda < 1$, such that for all $A \in \mathfrak{A}(\lambda\mathcal{O})$

$$\| \int dt\, f(t) \{ \Delta_{\mathcal{O}}^{it} - U(\Lambda(t)) \} A\Omega \|^2 \leq \varepsilon \{ \|A\Omega\|^2 + \|A^*\Omega\|^2 \}. \tag{17}$$

Hence if one chooses in this relation a function f_0 whose Fourier transform has support in the <u>complement</u> of the spectrum of the generator $-\ln \Delta_{\mathcal{O}}$ of $\Delta_{\mathcal{O}}^{it}$ one obtains

$$\| \int dt\, f_0(t)\, U(\Lambda(t)) A\Omega \|^2 \leq \varepsilon \{ \|A\Omega\|^2 + \|A^*\Omega\|^2 \}. \tag{18}$$

Note that this relation is a statement on the spectral properties of the Lorentz-transformations.

Now in theories where dilations are a symmetry it immediately follows from (18) that this relation does not only hold for $A \in \mathfrak{A}(\lambda\mathcal{O})$, but for all $A \in \mathfrak{A}(\mathcal{O})$, since the dilations commute with the Lorentz-transformations. So in this case one can put $\varepsilon = 0$ in (18). But since the spectrum of the generators of the Lorentz-transformations is $\mathbb{R}$, this implies that f_0 must be 0, which means that $\mathrm{sp}\, \Delta_{\mathcal{O}} = \mathbb{R}_+$.

The same conclusion can also be drawn if the underlying theory is not dilation invariant, but has some non-trivial scaling limit. The technical input needed is that there exists some Wightman field φ affiliated with the local algebras (cf.

footnote 2), for which the scaled Wightman distributions

$$N(\lambda)^n \cdot (\Omega, \varphi(\lambda x_1) \cdots \varphi(\lambda x_n) \Omega) \tag{19}$$

have a non-trivial limit as λ tends to 0, if the scaling factor $N(\lambda)$ is suitably chosen. It is expected that such fields exist in all renormalizable field theories having an ultraviolet fixed point. With this input Fredenhagen could calculate the Connes invariant of the local algebras $\mathfrak{A}(\mathcal{O})$ and show that they are factors of type III_1 (or, if the local algebras have a center, that only such factors appear in the central decomposition of these algebras). So we have learned from this argument that this specific structure of the local algebras is intimately connected with the conformal invariance of field theoretic models in the short distance limit.

3. The last concept which is needed for a complete characterization of the local algebras is the notion of hyperfiniteness. A von Neumann algebra $\mathcal{M}$ is said to be <u>hyperfinite</u> if there exists an increasing family of finite dimensional subalgebras $\mathcal{M}_1 \subset \mathcal{M}_2 \subset \cdots \subset \mathcal{M}_n \subset \cdots$ of $\mathcal{M}$ which generates $\mathcal{M}$. These hyperfinite von Neumann algebras are well studied.

For the problem at hand a recent result due to Haagerup [13] is of great interest: this result says that all hyperfinite factors of type III_1 are isomorphic. In view of this fact and the preceding results it is therefore natural to ask under which circumstances the local algebras in field-theoretic models are hyperfinite.

From the point of view of physics one would expect that such models should describe systems with a "reasonable" (i.e. not too large) number of degrees of freedom. But there is the problem of an appropriate characterization of this class of models. It has been proposed by Buchholz and Wichmann to distinguish these models by a nuclearity criterion [14]. According to this criterion the sets of vectors

$$e^{-\beta H}\ \mathcal{U}(\mathcal{O})\,\Omega\,, \quad \beta > 0\,, \tag{20}$$

where H is the Hamiltonian and $\mathcal{U}(\mathcal{O})$ the group of unitaries in $\mathfrak{A}(\mathcal{O})$, ought to be nuclear, i.e. any such set should be contained in the image of the unit ball in $\mathcal{H}$ under the action of some trace class operator. It was argued in [14] that this condition is satisfied whenever a model admits thermodynamical equilibrium states for all temperatures $\beta > 0$. This in turn is only possible if the particle spectrum of a model is such that the sum $\sum_i e^{-\beta m_i}$, where the m_i are the particle masses counted according to their multiplicity, is finite for any $\beta > 0$. So, roughly speaking, the nuclearity criterion characterizes models with a particle spectrum which does not grow too rapidly at high energies.

It has recently been shown by Buchholz, D'Antoni and Fredenhagen [15] that the local algebras are indeed hyperfinite in all models satisfying a (slightly strengthened) version of this nuclearity criterion. So summing up, we see that in all models exhibiting conformal invariance in the short distance limit and a reasonable particle spectrum at high energies, the local algebras $\mathfrak{A}(\mathcal{O})$ are hyperfinite factors of type III_1, (respectively direct integrals of such factors if the local algebras have a center). Disregarding the latter cases and making use of the result of Haagerup quoted before, this implies that for this physically relevant class of models the local algebras are all isomorphic, and thus model-independent. Hence, as far as the internal algebraic structure is concerned, one may think of any local algebra $\mathfrak{A}(\mathcal{O})$ corresponding to some double cone $\mathcal{O}$ as a fixed, concrete object: the Araki-Woods factor $\mathcal{R}_\infty$ [16].

One may expect that this very explicit information on the local algebras will be the key to further progress in the structural analysis of the local nets $\mathcal{O} \rightarrow \mathfrak{A}(\mathcal{O})$ appearing in quantum field theory.

References

1. R. Longo: Structure of von Neumann algebras of physics. In: Proceedings of Symposia in Pure Mathematics, Vol. 38/2, Providence: Am. Math. Socity 1982

2. R.F. Streater, A.S. Wightman: PCT, spin and statistics and all that, New York: Benjamin Inc. 1964

3. W. Driessler, S. Summers, E.H. Wichmann: On the connection between quantum fields and von Neumann algebras of local observables. Preprint (1985)

4. R. Haag: Lille Conference 1957. Les problèmes mathématiques de la théorie quantique des champs. CNRS, Paris (1959)

5. G.F. Dell'Antoni: Structure of the algebras of some free systems, Commun. Math. Phys. 9, 51 (1968)

6. S. Sakai: C*-algebras and W*-algebras, Berlin, Heidelberg, New-York: Springer 1971

7. A. Connes: Une classification des facteurs de type III. Ann. Sci. Ecole Normale Sup. 6, 133 (1973)

8. M. Takesaki: Tomita's theory of modular Hilbert algebras. In: Lecture Notes in Mathematics, Vol. 128, Berlin, Heidelberg, New-York: Springer 1970

9. J.J. Bisognano, E.H. Wichmann: On the duality condition for a hermitean scalar field, J. Math. Phys. 16, 985 (1975)

10. D. Buchholz: On the structure of local quantum fields with non-trivial interaction. In: Proc. of the Int. Conf. on Operator Algebras. Leipzig: Teubner Verlagsgesellschaft 1978

11. P.D. Hislop, R. Longo: Modular structure of the local algebras associated with the free massless scalar field theory, Commun. Math. Phys. 84, 71 (1982)

12. K. Fredenhagen: On the modular structure of local algebras of observables, Commun. Math. Phys. 97, 79 (1985)

13. U. Haagerup: Connes' bicentralizer problem and uniqueness of the injective factor of type III_1. Odense Universitet, preprint Nr. 10 (1984)

14. D. Buchholz, E.H. Wichmann: Causal independence and the energy level density of states in local quantum field theory. Universität Hamburg, preprint (1985)

15. D. Buchholz, C. D'Antoni, K. Fredenhagen: The universal structure of local algebras in quantum field theory, to appear

16. H. Araki, E.J. Woods: A classification of factors. Publ. Res. Inst. Math. Sci. A4, 51 (1968)

DOES SUPERGRAVITY ALLOW A POSITIVE COSMOLOGICAL CONSTANT?

M. F. Sohnius
The Blackett Laboratory
Imperial College of Science and Technology
London SW7 2BZ, England

In this talk, I shall report on work done in collaboration with P. van Nieuwenhuizen and K. Pilch [1] on the sign of the cosmological constant in supergravity models with unbroken supersymmetry. It has long been believed that supersymmetry rules out a positive cosmological constant, i.e. that a de Sitter background with $O(1,4)$ symmetry cannot sustain a supersymmetric theory. This is in stark contrast to the relative ease with which it is possible to incorporate a negative cosmological constant and its supersymmetric counterparts into supergravity [2], giving a theory with an anti-de Sitter background and $O(2,3)$ symmetry. Not too long ago, J. Lukierski and A. Nowicki [3] suggested that it should be possible to construct a field theory around the quaternionic superalgebra $UU_{\alpha}(1,1;1,\mathbb{H})$. The Lie subalgebra of this is locally isomorphic*) to $O(1,4)\times O(2)$ and we therefore have an obvious starting point for the construction of a supersymmetric theory with de Sitter background.

The non-supersymmetric (Einstein) theory of gravity allows the addition of a "cosmological term", $-\frac{1}{4}\lambda\sqrt{-g}$, to the action with either sign for λ, and there is no natural theoretical reason for λ to be positive, negative or indeed $\lambda=0$ as observation tells us it is to within very stringent limits (with $G=$ Newton's constant we have a bound for $G^2\lambda \lesssim 10^{-120}$). From the theorist's point of view, any restrictions are of great interest which are placed on λ in models with even the slightest chances of being related to the actual laws of nature, and the question must therefore be answered conclusively whether supergravity is as "one-sided" as believed in its preference of anti-de Sitter or Minkowski backgrounds over de Sitter ones. We think we provided this answer in our paper [1], and the answer is a cautious "yes, de Sitter supergravity does not work".

---

*) in this talk, I ignore distinctions between groups and their algebras and between different but locally isomorphic groups

We started with a classification of all superalgebras whose bosonic Lie subalgebras are of the form

$O(2,3) \times$ something (anti-de Sitter case)

$O(1,4) \times$ something (de Sitter case)

and found that only in the "normal case" of $O(2,3) \times$ compact group the algebraic structure is such that always $\{Q, Q^\dagger\} > 0$. These "good cases" are the well known (extended) supergravity theories with negative cosmological constant and a compact internal symmetry group. Since in a Hilbert space with positive definite metric an operator $\{Q, Q^\dagger\}$ is hermitian and positive definite, all other cases will necessarily lead to problems somewhere. We did not therefore find it surprising that in the simplest of the "bad cases" ($N=2$ de Sitter supergravity) the relative sign of the Einstein term and of the Maxwell term in the action

$$\mathcal{L}^{\text{de Sitter}} = \sqrt{-g}\,\left(\tfrac{1}{2} R - \tfrac{1}{4} \lambda + \ldots + \tfrac{1}{2} F^{\mu\nu} F_{\mu\nu} + \ldots \right)$$

came out wrong and predicts that either the graviton or the graviphoton must be a ghost (the unconventional normalization of the Maxwell field with a factor of $\frac{1}{2}$ instead of $\frac{1}{4}$ avoids an abundance of $\sqrt{2}$ later).

I begin the main body of the talk with short outlines of the general structure of Poincaré supergravity on the one hand and gravity with a cosmological constant on the other. I then proceed to describe the relatively easy way in which one can construct anti-de Sitter $N=1$ supergravity using tensor calculus techniques. After this, I present our result of the classification of de Sitter superalgebras, and finally I present the field theory constructed out of the simplest of those superalgebras which have an $O(1,4)$ space-time group. The result will be the lagrangian outlined above with its problem of giving rise to unitarity ghosts.

Poincaré supergravity is based on the algebra

$$\{ Q, \bar{Q}\} = 2\gamma^a P_a \quad ; \quad [Q, P_a] = [P_a, P_b] = 0 \quad , \tag{1}$$

which describes the relationship of supersymmetry (fermion-boson) transformations and translations of space and time. This algebra has the Lorentz group $O(1,3)$ as outer automorphism group, with

$$[Q, M_{ab}] = \tfrac{1}{2}\sigma_{ab} Q \quad ; \quad [P_a, M_{bc}] = i\eta_{ab} P_c - i\eta_{ac} P_b \quad . \tag{2}$$

In the case of supergravity, the parameters of all transformations are functions of space and time, according to the following table:

Generator:	Type of transformation:	Parameters:
P_a	general coordinate transformations	$\xi^a(x)$
Q_α	local supersymmetry transformations	$\bar{\zeta}^\alpha(x)$
M_{ab}	local Lorentz frame rotations	$\lambda^{ab}(x)$

As in all gauge theories, there is a gauge connection associated with each local transformation. The field quanta of these are the "particles" predicted by the model:

Generator:	Connection:	Particle:
P_a	vierbein $e_\mu^{\ a}$	graviton
Q_α	Rarita-Schwinger field $\bar{\psi}_\mu^{\ \alpha}$	gravitino
M_{ab}	Lorentz connection $\omega_\mu^{\ ab}$	—

No particle is associated with the Lorentz connection, since its own (algebraic) equations of motion express $\omega_\mu^{\ ab}$ in terms of $e_\mu^{\ a}$ and $\bar{\psi}_\mu^{\ \alpha}$. The relationship between metric and vierbein is the usual $g_{\mu\nu} = \eta_{ab}\, e_\mu^{\ a}\, e_\nu^{\ b}$ so that

$$\sqrt{-g} = e \equiv \det\left[e_\mu^{\ a}\right] \quad . \tag{3}$$

The lagrangian of $N=1$ supergravity,

$$\mathcal{L} = \tfrac{1}{2}\, e\, R + 2i\, \epsilon^{\mu\nu\rho\sigma}\, \bar{\psi}_\mu\, e_\nu^{\ a}\, \gamma_a \gamma_5\, \nabla_\rho \psi_\sigma \quad , \tag{4}$$

is a density under the supersymmetry transformations

$$\delta e_\mu^{\ a} = -2i\, \bar{\zeta}\, \gamma^a\, \psi_\mu \quad ; \quad \delta\psi_\mu = \nabla_\mu \zeta \tag{5}$$

where ∇_μ is a derivative which is covariantized with respect to local Lorentz transformations only (using $\omega_\mu^{\ ab}$). At the linearized level, the algebra of the transformations (5) is just (1), provided that the equations of motion are employed:

$$\begin{aligned} 0 &= R_\mu^{\ a} - \tfrac{1}{2} e_\mu^{\ a} R && \text{(Einstein equation)} \\ 0 &= \mathcal{R}^\mu \equiv \epsilon^{\mu\nu\rho\sigma}\, \gamma_5 \gamma_\nu\, \nabla_\rho \psi_\sigma && \text{(Rarita-Schwinger equation)} \; . \end{aligned} \tag{6}$$

Note that these are coupled equations since $\omega_\mu^{\ ab}$ and hence $R_{\mu\nu}^{\ \ ab}$ are functions of both $e_\mu^{\ a}$ and $\psi_{\mu\alpha}$.

In flat Minkowski space, two translations commute: $[\partial_\mu, \partial_\nu] = 0$. In a curved space-time situation, however, the commutator of two (covariant) translations is in general not zero but rather a combination of a further translation and a rotation. Somewhat loosely, I write this structure as

$$[\nabla_\mu , \nabla_\nu] = i\, T_{\mu\nu}{}^a P_a + \tfrac{i}{2} R_{\mu\nu}{}^{ab} M_{ab} \quad . \tag{7}$$

The equations of motion (6) allow the special solution $e_\mu{}^a = \delta_\mu{}^a$ and $\psi_\mu = 0$ in which case the right-hand side of (7) is zero and the commutator collapses to that of flat Minkowski space: a Minkowskian background is a solution of $N=1$ Poincaré supergravity. If, however, the equations of motion force $R \neq 0$, as e.g. in the presence of a cosmological term, then a flat metric is not a solution. Thus

$$\mathcal{L} = \sqrt{-g}\, (\tfrac{1}{2} R - \tfrac{1}{4} \lambda) \tag{8}$$

gives $R = \lambda$ because the Einstein equation is now

$$0 = R_\mu{}^a - \tfrac{1}{2} e_\mu{}^a R + \tfrac{1}{4} e_\mu{}^a \lambda \quad . \tag{9}$$

A solution is now given by the metric of a de Sitter (or anti-de Sitter) space with

$$R_{\mu\nu}{}^{ab} = \frac{1}{12} (\delta^a_\mu \delta^b_\nu - \delta^b_\mu \delta^a_\nu)\, R = \frac{1}{12} (\delta^a_\mu \delta^b_\nu - \delta^b_\mu \delta^a_\nu)\, \lambda \quad . \tag{10}$$

The algebraic relationship associated with eq.(7) is now

$$[P_a , P_b] = \frac{i\lambda}{12} M_{ab} \tag{11}$$

and the group generated by P_a and M_{ab} is

$O(1,4)$	(de Sitter group)	for	$\lambda > 0$
$O(2,3)$	(anti-de Sitter group)	for	$\lambda < 0$.

The question arises whether in the presence of supersymmetry the action can also be augmented by a further invariant to include a cosmological term:

$$\mathcal{L} = \mathcal{L}_{SG} - \tfrac{1}{4} e \lambda + \ldots . \tag{12}$$

The easiest way to answer this question is to extend both the lagrangian and the transformation laws to include auxiliary fields S, P and A_a:

$$\mathcal{L} = \ldots - \tfrac{4}{3} e\, (S^2 + P^2 + A_a A^a) \tag{13}$$

$$\begin{aligned} \delta\psi_\mu &= \ldots - \tfrac{i}{3} \gamma_\mu (S + \gamma_5 P - i \gamma^a \gamma_5 A_a) \zeta - \gamma_5 \zeta A_\mu \\ \delta(S + iP) &= \tfrac{i}{4} \bar{\zeta} (1 + i\gamma_5) \gamma^a \mathcal{R}_a \\ \delta A_a &= \tfrac{3}{4} \bar{\zeta} \gamma_5 (\mathcal{R}_a - \tfrac{1}{3} \gamma_a \gamma_b \mathcal{R}^b) \; . \end{aligned} \tag{14}$$

Now, in the presence of "off-shell supersymmetry", equations of motion come as whole supermultiplets. One of these is

$$\$ = \left[S, P; \tfrac{i}{4}\gamma^a \mathcal{R}_a ; -\tfrac{1}{4}R - \tfrac{4}{3}(S^2 + P^2 - \tfrac{1}{2}A_a A^a), \nabla_a A^a\right) , \tag{15}$$

a supermultiplet which vanishes on-shell: $\$ = 0$. This, together with other equations of motion which ensure that $A_a = 0$ on-shell, implies $R = 0$ and hence a Minkowskian background. It is now, however, relatively easy to add another invariant to the lagrangian (13), namely

$$\mathcal{L}_{cosm} = 2egS - ieg\bar{\psi}_\mu \sigma^{\mu\nu}\psi_\nu \tag{16}$$

(here g is an arbitrary real coupling constant and not $\det g_{\mu\nu}$). The relevant equation of motion is now not $\$ = 0$ but

$$\$ = (\tfrac{3}{4}g, 0; 0; 0, 0) \tag{17}$$

which implies $S = \frac{3}{4}g$ and

$$R = -\frac{16}{3}S^2 = -3g^2 \quad , \tag{18}$$

i.e., we have an anti-de Sitter background. Note how the sign of R comes out negative, irrespective of the sign of g. This is ultimately due to the sign of the S^2-term in the lagrangian which is fixed by the requirement of supersymmetry.

If we wish, we can eliminate the auxiliary fields, using their own equations of motion, and get

$$\mathcal{L} = \tfrac{1}{2}eR + 2i\,\epsilon^{\mu\nu\rho\sigma}\bar{\psi}_\mu \gamma_\nu \gamma_5 \nabla_\rho \psi_\sigma + \tfrac{3}{4}eg^2 - ieg\bar{\psi}_\mu \sigma^{\mu\nu}\psi_\nu \tag{19}$$

$$\begin{aligned} \delta e_\mu{}^a &= -2i\bar{\zeta}\gamma^a \psi_\mu \\ \delta\psi_\mu &= \nabla_\mu \zeta - \frac{ig}{4}\gamma_\mu \zeta \quad . \end{aligned} \tag{20}$$

To get the other sign for the cosmological constant, one would need terms of the type $\ldots + eX^2 + 2egX + \ldots$ in the lagrangian. Such terms are present in the coupling of matter to supergravity, but in those cases $\langle X \rangle \neq 0$ implies spontaneous breaking of supersymmetry. These difficulties have given rise to the suspicion that there is no de Sitter supergravity.

For a more systematic study of this problem, we backtrack and explore possible superalgebras which contain the group algebras of $O(1,4)$ or $O(2,3)$ as factors. We write these algebras in a 5-dimensional notation ($a,b = 0,1,2,3,5$):

$$[M_{ab}, M_{cd}] = i\eta_{bc}M_{ad} - i\eta_{ac}M_{bd} - i\eta_{bd}M_{ac} + i\eta_{ad}M_{bc} \quad . \tag{21}$$

Before enlarging this to a superalgebra, we must look at the properties of spinors in five dimensions:

$$[Q_\alpha, M_{ab}] = \tfrac{1}{2}(\sigma_{ab})_\alpha{}^\beta Q_\beta \qquad \text{with} \qquad \sigma_{ab} = \tfrac{i}{2}[\gamma_a, \gamma_b] \; . \tag{22}$$

The hermitian adjoint spinor $Q^\dagger$ will transform under the representation which is generated by $-\frac{1}{2}(\sigma_{ab})^*$. These representations are equivalent, i.e., a matrix D exists with

$$-(\sigma_{ab})^* = D^{-1}\sigma_{ab}D \quad . \tag{23}$$

This D has certain properties (for more details see the Appendix of ref.[4]); the most important here is

$$DD^* = \begin{cases} +1 & \text{for } O(2,3) \quad \text{(anti-de Sitter)} \\ -1 & \text{for } O(1,4) \quad \text{(de Sitter).} \end{cases} \tag{24}$$

This means that only in the anti-de Sitter case Majorana spinors with

$$Q_{\alpha i} = Q^c_{\alpha i} \equiv D_\alpha{}^\beta (Q_{\beta i})^\dagger \tag{25}$$

can be present: eq.(27) is only consistent for $DD^* = +1$. In the de Sitter case, the components of Q are linearly independent of those of $Q^\dagger$. This absence of Majorana spinors means that for each Q there is an independent Q^c or, in the usual way of counting, that N is even. Introducing a base $Q_{\alpha i}$ which contains all Q's and all Q^c's, we then necessarily have a condition

$$Q_{\alpha i} = E_i{}^j Q^c_{\alpha j} = E_i{}^j D_\alpha{}^\beta (Q_{\beta j})^\dagger \; . \tag{26}$$

This is consistent only if $EE^* DD^* = +1$, so that for the de Sitter case we get $EE^* = -1$. Indeed, we can always number the Q's in such a way that

$$E = \begin{bmatrix} 0 & 1 \\ -1 & 0 \end{bmatrix} \qquad \text{for } O(1,4) \quad . \tag{27}$$

The most general solution of the Jacobi identities for the extension of the algebra (21-22) into a full superalgebra is

$$\begin{aligned}
\{Q_{\alpha i}, Q_{\beta j}\} &= -\omega_{ij}(\sigma^{ab}C)_{\alpha\beta}M_{ab} + 2i\,C_{\alpha\beta}T_{ij} \\
[M_{ab}, T_{ij}] &= 0 \\
[Q_{\alpha i}, T_{jk}] &= i\omega_{ij}Q_{\alpha k} - i\omega_{ik}Q_{\alpha j} \\
[T_{ij}, T_{k\ell}] &= i\omega_{jk}T_{i\ell} - i\omega_{j\ell}T_{ik} - i\omega_{ik}T_{j\ell} + i\omega_{i\ell}T_{jk}
\end{aligned} \tag{28}$$

with C the (antisymmetric) charge conjugation matrix for five space-time dimensions. The symmetry properties of the structure constants ω_{ij} and of the internal symmetry generators T_{ij} are

$$\omega_{ij} = \omega_{ji} \quad ; \quad T_{ij} = -T_{ji} \quad . \tag{29}$$

The (symplectic) Majorana condition, eq.(26), implies reality conditions for the matrices ω and T,

$$\omega^{\dagger} = E\omega E^{-1} \quad \text{and} \quad T^{\dagger} = -ETE^{-1} \quad , \tag{30}$$

which are defined by

$$\begin{aligned} (\omega)_{ij} &= \omega_{ij} \quad ; \quad (\omega^{\dagger})_{ij} = (\omega_{ji})^{*} \\ (T)_{ij} &= T_{ij} \quad ; \quad (T^{\dagger})_{ij} = (T_{ji})^{\dagger} \end{aligned} \tag{31}$$

In the *anti-de Sitter case* where $E=1$, this implies that ω is hermitian and its symmetry means that it is real. If it is non-singular, it can be brought into the form

$$\omega = \text{diag}\,(1,\ldots,1,-1,\ldots,-1) \tag{32}$$

and the internal symmetry group, whose structure is the only open question at this point, is $O(p,q)$ with $p+q=N$. If ω has any zero eigenvalues, we have a group contraction of $O(p,q)$ and I must refer to the paper [1] for details.

In the *de Sitter case*, the T_{ij} will also generate some (complex) version of $O(N)$. Eq.(30) implies the following most general form for T:

$$T = \begin{bmatrix} a & ih \\ -ih^{\mathsf{T}} & a^{\dagger} \end{bmatrix} \qquad \text{where} \quad a = -a^{\mathsf{T}} \;;\; h = h^{\dagger} \tag{33}$$

and closer analysis will show that the $(\frac{N}{2})^2$ matrices

$$T_{\text{compact}} = \begin{bmatrix} 0 & ih \\ -ih^{\mathsf{T}} & 0 \end{bmatrix} \tag{34}$$

generate $U(\frac{N}{2})$, while the $\frac{1}{2}N(\frac{1}{2}N-1)$ matrices

$$T_{\text{non-compact}} = \begin{bmatrix} a & 0 \\ 0 & a^{\dagger} \end{bmatrix} \tag{35}$$

are all generators of non-compact transformations. That complex version of $O(N)$ which has $U(\frac{N}{2})$ as its maximal compact subgroup is called $O^{*}(N)$ and we conclude that the internal symmetry group of de Sitter superalgebras is $O^{*}(N)$ or some contraction thereof. The following table expresses the first few star-groups in terms of more familiar classical groups:

$$
\begin{aligned}
O^*(2) &= O(2)\\
O^*(4) &= SU(2)\times SU(1,1)\\
O^*(6) &= SU(3,1)\\
O^*(8) &= O(6,2)\,.
\end{aligned}
\tag{36}
$$

We find that for de Sitter superalgebras, the internal symmetry group is always non-compact. Such internal symmetries were ruled out by Coleman and Mandula [5] and hence in [6]. Tracking back, one finds that it was the requirement of a positive energy spectrum and the positivity of the metric in Hilbert space that lead to the conclusion that internal symmetry groups must be compact. Usually, these are fulfilled in supersymmetric theories where the relationship

$$\sum_{\alpha=1}^{4} \{Q_\alpha, (Q_\alpha)^\dagger\} \propto H \tag{37}$$

ensures positivity of the energy because the left-hand side is a positive definite operator if the Hilbert space is positive definite:

$$\langle x|\,\{Q, Q^\dagger\}\,|x\rangle = \big|Q^\dagger\,|x\rangle\big|^2 + \big|Q\,|x\rangle\big|^2 \geqslant 0 \tag{38}$$

($=0$ only if $Q=0$).

Let us therefore examine how eq.(37) comes out in our case. We choose a set of four γ-matrices which are purely imaginary (Majorana representation), and a fifth one

$$\gamma_5 = \begin{cases} \gamma_0\gamma_1\gamma_2\gamma_3 & \text{for } O(1,4)\\ i\,\gamma_0\gamma_1\gamma_2\gamma_3 & \text{for } O(2,3)\,. \end{cases}$$

which is real for $O(1,4)$ and imaginary for $O(2,3)$. We then have

$$C = \begin{cases} \gamma_0\gamma_5\\ -i\,\gamma_0\gamma_5 \end{cases} \quad\text{and}\quad D = \begin{cases} \gamma_5 & \text{for } O(1,4)\\ 1 & \text{for } O(2,3)\,. \end{cases}$$

This means that $\{Q,Q^\dagger\} = -\omega\,(\sigma^{ab}CD^{-1T})\,M_{ab} + 2i\,CD^{-1T}\,T$ and

$$\sum_{\alpha=1}^{4} \{Q_{\alpha i}, (Q_{\alpha j})^\dagger\} = \begin{cases} 0 & \text{for } O(1,4)\\ 8\,\omega_{ij}\,M_{05} & \text{for } O(2,3)\,. \end{cases} \tag{39}$$

We find what we suspected: the energy M_{05} is positive definite for the anti-de Sitter case and internal symmetry $O(N)$, indefinite for the anti-de Sitter case and $O(p,q)$, and something is wrong with the Hilbert space for the de Sitter case as well as for degenerate ω's in the anti-de Sitter case.

Let us find out what exactly goes wrong for $N=2$ de Sitter supergravity. There the algebra can always be brought into the following form (in four-dimensional notation, $a,b=0,1,2,3$):

$$[P_a, P_b] = i M_{ab} \quad ; \quad [Q_i, P_a] = \tfrac{i}{2}\gamma_a\gamma_5 Q_i \quad ; \quad [Q_i, T] = i\epsilon_{ij} Q_j$$
$$\{Q_i, Q_j\} = 2i\delta_{ij}\gamma^a C P_a + \delta_{ij}\sigma^{ab}\gamma_5 C M_{ab} - 2i\gamma_5 C \epsilon_{ij} T . \tag{40}$$

The Majorana condition is now

$$Q_{\alpha i} = i\epsilon_{ij}(\gamma_5 C)_{\alpha\beta}\bar{Q}_j^{\beta} . \tag{41}$$

The following set of transformation laws on vierbein, Maxwell field and Rarita-Schwinger field,

$$\delta e_\mu{}^a = 2i\bar{\zeta}_i\gamma^a\gamma_5\psi_{\mu j}\epsilon_{ij}$$
$$\delta A_\mu = -2i\bar{\zeta}_i\psi_{\mu i} \tag{42}$$
$$\delta\psi_{\mu i} = \nabla_\mu\zeta_i + \epsilon_{ij}\zeta_j A_\mu + \tfrac{1}{2}\gamma_\mu\gamma_5\zeta_i + \tfrac{i}{4}\epsilon_{ij}\sigma^{ab}\gamma_\mu\gamma_5\zeta_j F_{ab}$$

with

$$F_{ab} = e_a{}^\mu e_b{}^\nu(\partial_\mu A_\nu - \partial_\nu A_\mu + 2i\bar{\psi}_{\mu i}\psi_{\nu i})$$
$$\omega_\mu{}^{ab} = e_{\mu c}(K^{cab} + K^{bca} - K^{abc})$$
$$K_{ab}{}^c = e_a{}^\mu e_b{}^\nu(-\tfrac{1}{2}\partial_\mu e_\nu{}^c + \tfrac{1}{2}\partial_\nu e_\mu{}^c + i\bar{\psi}_{\mu i}\gamma^c\gamma_5\psi_{\nu j}\epsilon_{ij}) ,$$

close on "local translations", Maxwell transformations and Lorentz frame rotations:

$$[\delta^1, \delta^2] = \delta_P(\xi^a) + \delta_M(\alpha) + \delta_L(\lambda^{ab}) \tag{43}$$

with (partly field-dependent) parameters

$$\xi^a = -2i\bar{\zeta}_i^1\gamma^a\gamma_5\zeta_j^2\epsilon_{ij} \quad , \quad \alpha = 2i\bar{\zeta}_i^1\zeta_i^2$$
$$\lambda^{ab} = \bar{\zeta}_i^1\left[-2\epsilon_{ij}\sigma^{ab} - 2i\delta_{ij}(F^{ab} - i\gamma_5{}^*F^{ab})\right]\zeta_j^2 .$$

The P-transformation is itself a general coordinate transformation followed by field dependent gauge transformations. The Maxwell transformations are

$$\delta_M e_\mu{}^a = 0 \quad ; \quad \delta_M A_\mu = \partial_\mu\alpha \quad ; \quad \delta_M\psi_{\mu i} = -\alpha\epsilon_{ij}\psi_{\mu j} . \tag{44}$$

The algebra closes only modulo equations of motion. Working out all commutation relations, one finds that at the linearized level everything indeed reduces to the $UU_\alpha(1,1,;1,\mathbb{H})$ algebra (40).

We now construct a lagrangian which is a density under the transformations (42). The result is

$$\mathcal{L} = \tfrac{e}{2}R + 2i\epsilon^{\mu\nu\rho\sigma}\bar{\psi}_{\mu i}\gamma_\nu\epsilon_{ij}\nabla_\rho\psi_{\sigma j} + \tfrac{e}{2}F^{\mu\nu}F_{\mu\nu} + 2ie\bar{\psi}_{\mu i}(F^{\mu\nu} - i\gamma_5{}^*F^{\mu\nu})\psi_{\nu i}$$
$$- 3e - 2e\bar{\psi}_{\mu i}\sigma^{\mu\nu}\epsilon_{ij}\psi_{\nu j} + \psi^4\text{-term} \tag{45}$$

which gives rise to just the equations of motion necessary to close the algebra.

Two points must be noted about this lagrangian:

(1) there is a positive cosmological constant $\lambda = 12$, in agreement with eqs.(11) and (40),

(2) the Maxwell part of the action has the wrong sign.

Before we can say that $F^{\mu\nu}F_{\mu\nu}$ has the wrong sign, we must be sure that the lagrangian as a whole has the correct one. It is quite intricate to establish the correct sign for the Hilbert action since it defies straightforward canonical formalism. There is, however, a short-cut: it is solidly established that independent of any conventions, the Weyl field which appears in a conformal transformation $g_{\mu\nu} \to e^{2C} g_{\mu\nu}$ will always appear as a unitarity ghost in the action. In my conventions, the transformation law for the Hilbert action under conformal transformations is

$$\int d^4x \sqrt{-g}\, R \to \int d^4x \sqrt{-g}\, (R - 6\, \partial_\mu C\, \partial^\mu C - 6\, e_a{}^\nu\, \partial_\mu e_\nu{}^a\, \partial^\mu C)\, e^{2C} \tag{46}$$

so that a positive factor for $\sqrt{-g}\, R$ is the correct one, since then the "kinetic term for C" is that of a ghost. Hence we conclude that the sign for eR in (45) is correct, and that of $F^{\mu\nu}F_{\mu\nu}$ is wrong.

In conclusion, I may summarize our results as follows:

(1) de Sitter superalgebras exist. N is always even and the bosonic symmetry group is $O(1,4) \times O^*(N)$.

(2) Representations of these algebras on gauge fields have been constructed for $N = 2$ and probably exist for $N < 8$.

(3) An invariant density has been constructed for $N = 2$ and probably exists for $N < 8$.

(4) This density cannot be interpreted as a lagrangian, since it leads to unitarity ghosts (fields with the wrong sign for the kinetic part of the lagrangian). This is in agreeement with expectations raised by the structure of the algebra which cannot be realized on a positive definite Hilbert space.

<u>References</u>

[1] K. Pilch, P. van Nieuwenhuizen and M.F. Sohnius, Comm. Math. Phys. 98 (1985) 105
[2] P.K. Townsend, Phys. Rev. D15 (1977) 2802
P.K. Townsend and P. van Nieuwenhuizen, Phys. Lett. 67B (1977) 439
[3] J. Lukierski and A. Nowicki, unpublished, and private communication
[4] M.F. Sohnius, Phys. Rep. 128 (1985) 39
[5] S. Coleman and J. Mandula, Phys. Rev. 159 (1967) 1251
[6] R. Haag, J.T. Lopuszanski and M.F. Sohnius, Nucl. Phys. B88 (1975) 257

PHOTONS AND GRAVITONS IN CONFORMAL FIELD THEORY

W.F. HEIDENREICH
INSTITUT FÜR THEORETISCHE PHYSIK
TU CLAUSTHAL
W.GERMANY

1. INTRODUCTION

What is the spacetime symmetry group of the fundamental equations of physics? Several answers are possible: the group of motions of a spacetime, like the Poincaré- or the de Sitter groups; the conformal group of the spacetime; another type of symmetry group, like the diffeomorphisms in general relativity. For present day quantized theories the selection between these answers is easy: before symmetry breaking our theories contain only massless fields with conformally invariant interactions, for both, long range and short range forces. For long range forces we may expect that the observed massless particles appear in the fundamental equations as gauge fields without changes due to symmetry breaking or confinement. Therefore I will discuss here photons and gravitons in manifestly conformal field theory; photons as part of a reasonably well understood model, conformal gravitons as a possible step to quantum gravity.

2. PHOTONS IN CONFORMAL FIELD THEORY

According to Wigner photons are a direct sum of UIRs of the Poincaré-group with mass 0 and helicities $\lambda = +1$ and $\lambda = -1$. The massless helicity λ UIRs $D(\lambda)$ of the Poincaré-group can be extended uniquely to UIRs of the conformal group [1]. We label the irreducible lowest weight representations of SU(2.2) by the $U(1)xSU(2)xSU(2)$ quantum numbers (E_0, j_1, j_2) of the lowest weight. For helicity $\lambda \geq 0$ the extension is $D(1+\lambda, \lambda, 0)$, for $\lambda \leq 0$ it is $D(1+\lambda, 0, \lambda)$. So photons are the direct sum $D(2,1,0) \oplus D(2,0,1)$.

In field theory the positive energy solution space V of $\Box A_\mu = 0$ carries the tensor product of the four-dimensional and the massless scalar representation of the Poincaré-group. Its reduction

$$(1) \qquad D_4 \otimes D(0) = D(0) \rightarrow [D(+1) \oplus D(-1)] \rightarrow D(0)$$

is indecomposable [2]. The full tensor product space V has invariant subspaces $V \supset V_L \supset V_g$ which are not invariantly complemented. The Lorentz-condition $\partial^\mu A_\mu = 0$ projects on V_L. The subspace V_g of pure gauge fields $A_\mu = \partial_\mu \Lambda$ carries a $D(0)$, the photons $D(+1) \oplus D(-1)$ lie in the quotient space V_L/V_g. The quotient V/V_L carries a $D(0)$, the "scalar modes". The arrows in Eq. (1) denote "leaks" from scalar to physical to gauge states, that is if we act with group-elements on for example a physical state, we obtain in general a linear combination of physical and gauge states. Such an indecomposable resprentation (1) with invariant indefinite scalar product is called a Gupta-Bleuler triplet. It is the group theoretical structure connected with indefinite metric quantization [3].

The tensor product in Eq. (1) gives immediately the Pauli-Jordan commutation function as

$$[A_\mu^{(+)}(y), A_\nu^{(-)}(y')] = -i\eta_{\mu\nu} D^{(+)}(y-y')$$

and hence a Poincaré-invariant free field theory with field operator $A = A^{(+)} + A^{(-)}$.

Now we want to extend this free field theory to the conformal group. Simply extending the terms on the right hand side of Eq. (1) is impossible, as the UIRs $D(1,0,0)$ and $D(2,1,0)$ are not Weyl-equivalent, a necessary condition to appear in an indecomposable representation [4]. The next simple possibility is to extend the terms on the left side of Eq. (1). Replacing the vector D_4 by the conformal vector D_6 we get

$$(2) \qquad D_6 \otimes D(1,0,0) = D(1,1/2,1/2) \rightarrow [D(2,1,0) \oplus D(2,0,1) \oplus D(0,0,0)] \rightarrow D(1,1/2,1/2).$$

This tensor product can be calculated using minimal weight techniques, see appendix. $D(1,0,0)$ is the 1-dimensional trivial respresentation. A field theoretical realization of this tensor product is easily formulated in Dirac's conformal space [5], a compactification of Minkowski space. It is a "light cone" $x_a x^a = 0$, $a = 1, .., 6$ in R^6 with (4,2)-metric and points along rays identified, $x \mathrel{\hat{=}} \lambda x$, $\lambda \neq 0$. Its principal virtue is, that the conformal group $SO(4,2)/Z_2$ acts linearly on it. The positive energy solutions of

(3) $\partial_a \partial^a A_b = 0, \quad x_a \partial^a A_b = - A_b$

carry the full tensor product (2); transversality

(4) $x_a A^a = 0$

is the "Lorenz condition" which projects on physical and gauge fields; the pure gauge field $A_a = \partial_a \Lambda$ is a dipole ghost satisfying $\partial^2 \partial^2 \Lambda = 0$ [6]. The tensor product (2) immediately gives the conformal commutation functions as

(5) $[A_a^{(+)}(x), A_b^{(-)}(x')] = -i\eta_{ab} \, (x.x')_+^{-1},$

which clearly contains the physical (transverse) modes, according to the decomposition in Eq.(2). These field equations can straightforwardly be translated in usual flat coordinates yielding five potentials A_μ, A_+ [3], which satisfy

(6)
$$\Box A_\mu - \partial_\mu \, \partial.A + 1/2 \, \Box A_+ = 0,$$
$$\Box \partial.A = 0.$$

The auxiliary potential A_+ is necessary to accomodate the additional "scalar" modes. It too is a dipole ghost, $\Box\Box A_+ = 0$. The Lorentz condition (4) gives $A_+ = 0$. If it is imposed the Eqs.(6) become the well known Maxwell equations with conformal gauge fixing [7].

3. COUPLING TO A MASSLESS ELECTRON

If we start with classical electrodynamics in conformal space, coupled to an external current,

(7) $\partial^2 A_a = J_a, \quad x.A = 0, \quad x.J = 0,$

we get in flat space

(8)
$$\Box A_\mu - \partial_\mu \, \partial.A = J_\mu,$$
$$\Box \partial.A = -4J_B.$$

Putting $J_B = 0$ is not an invariant condition, unless $J_\mu = 0$ also. The gauge fixing $\partial.A = 0$ is not invariant by itself.

The necessary fifth component J_B of the current in conformal electrodynamics could be obtained by introducing an auxiliary dipole ghost [8]. Yet if working in conformal space there is a more natural possibility.

Consider a charged massless spinor with helicity +1/2. There are two different field descriptions for such an object [9]. One carries the "neutrino representation" D(3/2,1/2,0) irreducibly, the other one has gauge freedom [10]; it carries an indecomposable representation

$$(9) \qquad D(3/2,1/2,0) \rightarrow D(5/2,0,1/2).$$

If we form the usual conformal current [11] J_a and transform it to flat space coordinates we find that J_μ is the usual gauge invariant current of a charged massless helicity +1/2 field, while J_B is a "gauge current" of the form

$$J_B = (\text{pure gauge}) \cdot (\text{physical modes}).$$

The gauge freedom of massless electrons allows for a conformally invariant coupling.

For quantization of this electron field we need "scalar modes" belonging to a D(5/2,0,1/2) which leak into the representation (9). We have to relax the wave equations to enlarge the solution space. There are two possibilities: Either give up the homogeneity condition ($x\partial\Psi = -2\Psi$) in conformal space and keep the wave equation there [12] (but not in flat space!), or work with homogeneous functions, but give up the wave equation in conformal space (it becomes the "Lorentz-condition") [9]. The latter possibility amounts to using the tensor product of the 4-dimensional (semi-)spinor and an "anomalous" scalar representation,

$$(10) \qquad D(-1/2,1/2,0) \otimes D(2,0,0) = D(5/2,0,1/2) \rightarrow D(3/2,1/2,0) \rightarrow D(5/2,0,1/2),$$

which gives an extremely simple commutation relation

$$(11) \qquad [\Psi^{(+)}(x), \overline{\Psi}^{(-)}(x')] = (x\,x')_+^{-2}.$$

The extensions necessary to formulate conformal QED as compared to massless Poincaré QED concern the gauge sectors only. So the physical predictions should be the same. Yet there are some curious points: The Feynman propagators are not conformally invariant in Minkowski space itself, but only in a twofold (or higher) covering. There may appear cancellations in perterbation theory in such spaces [13].

And the role of the trivial representation in conformal QED (see Eq. (2)) and of the spinor gauge freedom is not fully understood.

Even so the power of the group theoretical techniques described bring a more than formal treatment of conformal spin 2 theories within reach.

4. GRAVITONS

The free positive energy solutions of linearized Einstein's field equations (with $h^{\mu}{}_{\mu} = 0$, $\partial^{\mu} h_{\mu\nu} = 0$)

$$\square h_{\mu\nu} = \kappa \, T_{\mu\nu} \tag{12}$$

carry the indecomposable Poincaré representation (see e.g. [15])

$$\begin{aligned} D(0) \to [D(+1) \oplus D(-1)] \to [D(+2) \oplus D(0) \oplus D(-2)] \\ \to [D(+1) \oplus D(-1)] \to [D(0)] \,. \end{aligned} \tag{13}$$

The helicity (± 2) states are conventionally interpreted as gravitons. The irreducible graviton representations can be uniquely extended to $D(3,2,0) \oplus D(3,0,2)$ of $SO(4.2)$. We want to extend the indecomposable structure (13), which is the one relevant for field theory. First we use the second order Casimir operator of $SO(4.2)$ to find the smallest tensors in conformal space which can describe conformal gravitons. For traceless fields with symmetry (m_1, m_2, m_3) and degree n, which satisfy the subsidiary conditions $x.T = 0$ and $\partial.T = 0$ it is

$$C_2 T = [n(n+4) - 2m + m_1 (m_1+4) + m_2 (m_2+2) + m_3^2] T.$$

Here m_i is the number of boxes in the ith row of a Young symmetrizer; $m = \Sigma m_i$. For massless particles with helicity λ we have

$$C_2 = 3(\lambda^2 - 1).$$

So for photons ($\lambda^2 = 1$) we would require $C_2 = 0$ and obtain the possibilities $\cdot$, $n = 0$ (scalar), $\square$, $n = -1$ (vector), and $\begin{smallmatrix}\square\\\square\end{smallmatrix}$, $n = -2$ (antisymmetric 2-tensor). For gravitons ($\lambda^2 = 4$, $C_2 = 9$) we get the possibilities

(14) $\begin{smallmatrix}\square\\\square\end{smallmatrix}$, $n = +1$, $\begin{smallmatrix}\square\square\\\square\end{smallmatrix}$, $n = 0$, and $\begin{smallmatrix}\square\square\\\square\square\end{smallmatrix}$, $n = -1$.

A symmetric 2-tensor cannot describe gravitons in conformal space! Looking closer shows that the antisymmetric 2-tensor describes pure gauge in a spin 2 conformal theory; a mixed 3-tensor ($\begin{smallmatrix}\square&\square\\ \square&\end{smallmatrix}$) with degree 0 is the smallest tensor to describe conformal gravitons. The corresponding field equations are

$$\text{(15)}\qquad \begin{aligned} &\Psi_{abc} = -\Psi_{bac},\ \Psi_{abc} + \Psi_{bca} + \Psi_{cab} = 0,\ \Psi_{ab}{}^{b} = 0,\\ &x\partial\Psi = 0, \quad \partial^2\partial^2\Psi = 0 \end{aligned}$$

Their positive energy solution space carries the tensor product of the dipole ghost $D(1,1/2,1/2)$ and the finite $D_{\begin{smallmatrix}\square&\square\\ \square&\end{smallmatrix}} = D(-2,1/2,1/2)$. It can (lengthily) be reduced to an indecomposable

$$\text{(16)}\qquad D(0,1/2,3/2) \rightarrow \left\{\begin{matrix} D(3,2,0)\oplus \\ D(1,0,2)\oplus \\ D(-1,0,1) \end{matrix}\right\} \rightarrow D(0,1/2,3/2) \oplus \begin{matrix}\text{helicity}\\ \text{conjugate}\end{matrix} \oplus \text{more.}$$

The subsidiary conditions $\partial\cdot\Psi = 0$ and $x^a\Psi_{a(bc)}$ project on the two Gupta-Bleuler triplets of conformal gravity, $x^a\Psi_{a[bc]} = 0$ is the "Lorentz-condition". $D(-1,0,1)$ is finite, the ghost $D(1,0,2)$ makes the theory nonunitary, unless it can be prevented from propagating. Conditions on the coupling to achieve this have been formulated [14].

Here we want to concentrate on the physical content of the theory. An equation which projects on the physical and gauge modes, $(D(3,2,0) \rightarrow D(0,1/2,3/2) \oplus (D(3,0,2) \rightarrow D(0,3/2,1/2))$ is in flat space coordinates

$$\text{(17)}\qquad \begin{aligned} &C_{\mu\nu\lambda\rho} \equiv \boxed{\begin{array}{c|c} \mu & \nu \\ \hline \lambda & \rho \end{array}}\ (\partial_\rho \Psi_{\mu\lambda\nu} - 3/2\ \Psi_{\mu\nu} H_{\lambda\rho}) = 0,\\ &H_{\mu\nu} \equiv -1/2\ \partial^\rho(\Psi_{\rho\mu\nu} + \Psi_{\rho\nu\mu}). \end{aligned}$$

From this follows (up to terms $\partial^\rho\Psi_{\rho[\mu\nu]}$ which can be eliminated by fixing the gauge)

$$\text{(18)}\qquad \Box\Psi_{\mu\nu\lambda} = 0,$$

and from that

$$\text{(19)}\qquad \Box H_{\mu\nu} = 0.$$

Due to the last equation, $H_{\mu\nu}$ was interpreted as the metric of Eq.(12) in ref.(14). Yet the physical meaning of the main field

remained obscure. We try an approach which is based on new results on the tensor product of finite and the massless scalar representation of the Poincaré group [15].
Consider the positive energy solution space of Eq.(18), which contains those of Eqs.(17,19). It carries an indecomposable representation

(20) $$\begin{matrix} & D(+2) & \to & D(+1) & \to & D(0) & \to & D(-1) \\ \nearrow & & \nearrow & & \nearrow & & \nearrow & \\ D(+1) & \to & D(0) & \to & D(-1) & \to & D(-2) & \end{matrix} \qquad \oplus \text{ helicity conjugate.}$$

There are two sets of helicity ($\pm$ 2) modes. Gauge fixing $\partial^{\mu}\Psi_{\mu[\nu\lambda]} = 0$ and the conformally invariant equation $C = 0$ project on

(21) $$\begin{matrix} D(+2) \to D(+1) & \searrow & & \nearrow & D(+1) \\ & & D(0) & & \\ D(-2) \to D(-1) & \nearrow & & \searrow & D(-1). \end{matrix}$$

So conformal gravitons are the "upper" spin 2 modes which are also carried by $H_{\mu\nu}$. The "lower" spin 2 modes are described by

$$\Psi_{\mu\nu\lambda} = \partial_{\mu}g_{\lambda\nu} - \partial_{\nu}g_{\lambda\mu}, \qquad \Box g = 0, \qquad g^{\nu}{}_{\nu} = 0.$$

This suggests two possible different interpretations of the fields in conformal gravity:

	Metric interpretation	Ricci interpretation
$g_{\mu\nu}$	?	metric
$_{\mu\nu\lambda}$	?	⊞-part of Christoffel
$H_{\mu\nu}$	metric	Ricci
$C_{\mu\nu\lambda\rho}$	?	Weyl

Although the metric interpretation gives a direct extension of linearized Einstein gravity, it remains disturbing that the other fields - specifically Ψ - do not play a physical role. In contrast for the Ricci interpretation all the fields can be identified with linearized geometrical objects. But it no longer is an extension of linearized Einstein's theory. Yet maybe this is what we should expect from a conformal theory. After all the coupled Einstein theory contains the dimensional coupling constant κ. If coupling to matter - even to massless particles only - we cannot expect to obtain a SO(4.2) invariant theory. The step from "massless" theory to phenomenological gravity requires symmetry breaking, as for short range forces.

But then, what is the non-linear theory whose linear approximation gives our linear conformal spin 2 theory? Weyl's theory $L = C^2$ is in linear approximation

$$\Box\Box h_{\mu\nu} = 0 \Rightarrow \Box R_{\mu\nu} = 0.$$

We may have got a unitarizable variant of linearized Weyl theory.

APPENDIX: Reduction of $D_6 \otimes D(1,0,0)$

The tensor product of finite and infinite-dimensional representations of the conformal group (or coverings) were one of the essential techniques used above. Here - as an example - Eq.(2) shall be derived. In conformal space $x^2 = \eta^{ab}x_a x_b = x_1^2 + x_2^2 + x_3^2 - x_4^2 + x_5^2 - x_6^2 = 0$, $x_a = \lambda x_a$, $\lambda \neq 0$, a basis of the Lie algebra so(4.2) acting on scalar fields is

(A1) $L_{ab} = -i(x_a\partial_b - x_b\partial_a)$.

It contains the Lie algebra of the maximal (essentially) compact subgroup $\tilde{U}(1) \times SU(2) \times SU(2)$ with basis L_{46}, L_{ij} $(i,j = 1,2,3,5)$, and step operators $L_i^{\pm} = L_{i6} \pm iL_{i4}$. They raise or lower the eigenvalues of energy L_{46} by one.
The state

(A2) $\varphi_o = x_+^{-1} \equiv (x_4 + ix_6)^{-1}$

satisfies $L_i^- \varphi_o = 0$, $L_{46}\varphi_o = \varphi_o$, $L_{ij}\varphi_o = 0$, and therefore carries a lowest $\tilde{U}(1) \times SU(2) \times SU(2)$-weight with quantum numbers (1,0,0). Acting with rasing operators L_i^+ on it gives a basis of the irreducible representation D(1,0,0). The first few states and their corresponding weights are

(A3) $x_i\, x_+^{-2} \in (2,1/2,1/2)$,
$(x_i\, x_j\text{-trace})x_+^{-3} \in (3,1,1)$, ...

All these states have degree of homogeneity $x\partial\varphi = -\varphi$ and satisfy the field equation $\partial^2\varphi = 0$.
The 6-dimensional representation is realized on z_a , with Lie algebra

(A4) $S_{ab} = -i(z_a \partial^z_b - z_b \partial^z_a)$.

Its lowest weight is $z_+ \equiv z_4 + iz_6 \in (-1,0,0)$, its highest weight is $z_- \equiv z_4 - iz_6 \in (1,0,0)$, and the other states z_i belong to $(0,1/2,1/2)$.

Now consider the following states in the tensor product $D_6 \otimes D(1,0,0)$ and their respective weights:

$$A_t = z_+x_+^{-1} \qquad \in (0,0,0)$$

$$A_s^j = z_jx_+^{-1} + z_+x_jx_+^{-2} \qquad \in (1,1/2,1/2)$$

$$A_g^j = z_jx_+^{-1} - z_+x_jx_+^{-2} \qquad \in (1,1/2,1/2)$$

$$A_p^{jk} = (z_jx_k - z_kx_j)x_+^{-2} \qquad \in (2,1,0) \oplus (2,0,1).$$

Acting with the step operators $M_i^{\pm} = L_i^{\pm} + S_i^{\pm}$ on them we get the following structure of invariant subspaces:

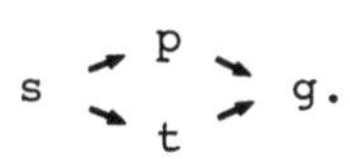

It means for example: lowering from A_s we get A_t, but not the reverse; raising from A_t we get A_p, but not the reverse, ... We can read off the indecomposable structure of Eq.(2). Comparing weight diagrams shows that all states of the tensor product appear in the decomposition.

REFERENCES

[1] E. Angelopoulos, M. Flato, Lett. Math. Phys., 2, 405 (1978).

[2] J. Bertrand, Nuovo Cimento A, 1, 1 (1971).

[3] B. Binegar, C. Fronsdal, W. Heidenreich, Journ. Math. Phys., 24, 2828 (1983).

[4] I.N. Bernshtein, I.M. Gel'fand, S.J. Gel'fand, Funct. Anal. Appl., 5, 1 (1971), G. Pinczon, J. Simon, Rep. Math. Phys., 16, 49 (1979).

[5] P.A.M. Dirac, Ann. Math., 37, 429 (1936).

[6] W. Heidenreich, Journ. Math. Phys., 25, 376 (1984).

[7] D.H. Mayer, Journ. Math. Phys., 16, 884 (1975), F. Bayen, M. Flato, Journ. Math. Phys., 17, 1112 (1976).

[8] G. Sotkov, D. Stoyanov, Journ. Phys., A16, 2817 (1983), V.B. Petkova, G.M. Sotkov, I.T. Todorov, Comm. Math. Phys., 97, 227 (1985).

[9] W. Heidenreich, Nuovo Cimento, A80, 220 (1984).

[10] L. Castell, Nuovo Cimento, A42, 160 (1977).

[11] H.A. Kastrup, Phys. Rev., 150, 1183 (1966).

[12] F. Bayen, M. Flato, C. Fronsdal, "Conformal invariance and gauge fixing in QED", UCLA 84/TEP/18.

[13] L. Castell, W. Heidenreich, Phys. Rev., D24, 98 (1981), B. Binegar, M. Flato, C. Fronsdal, S. Salamó, Czech. Journ. Phys., B32, 439 (1982).

[14] B. Binegar, C. Fronsdal, W. Heidenreich, Phys. Rev., D27, 2249 (1983).

[15] W. Heidenreich, "On solution spaces of massless field equations with arbitrary spin", to appear in Journ. Math. Phys.

ON CONFORMALLY COVARIANT ENERGY MOMENTUM TENSOR AND VACUUM SOLUTIONS

BO-WEI XU
DEPARTMENT OF PHYSICS
SHANGHAI JIAOTONG UNIVERSITY
SHANGHAI, CHINA

1. INTRODUCTION

The conformal group which includes Poincaré transformations is a generalized symmetry group of space-time. By using Noether's theorem to a conformally covariant theory, we can obtain canonical currents such as energy-momentum tensor $T_{\mu\nu}^{c}$, angular momentum tensor $J_{\mu\nu\lambda}^{c}$, dilatation tensor D_{μ}^{c} and special conformal tensor $K_{\mu\nu}^{c}$ [1]. The special conformal tensor, however, is not formally conserved. In order to get rid of this defect, the usual way is to add to $K_{\mu\nu}^{c}$ a term which does not modify the commutator of the conformal charges [2] [3] [4]. Although this technique for the canonical currents is acceptable, we feel this method is somewhat artificial. In these canonical currents, the energy-momentum tensor from which the other canonical currents are composed plays a fundamental role. But $T_{\mu\nu}^{c}$ is not conformally covariant. The usual symmetric energy-momentum tensor in general is not conformally covariant either. This situation is unnatural in conformal symmetric theories. We shall here propose a conformally covariant energy momentum tensor $\Theta_{\mu\nu}$, in terms of which we can redefine the angular momentum tensor, the dilatation tensor, and the special conformal tensor. The expressions of these tensors are simple, and all these currents are conserved formally [5]-[7]. The various energy momentum tensors mentioned above differ from each other by total divergence terms which do not contribute to the total energy momentum. In discussing the vacuum solutions of conformally symmetric theories, however, these energy momentum tensors are unequal, and the conformally covariant energy momentum tensor is useful in looking for the vacuum solutions of field equations [8]-[11]. In this paper we first review the general properties of conformal transformations, discuss the conformally symmetric theories, and give the expressions of $\Theta_{\mu\nu}$ for different fields. The vacuum solutions and vacuum state are discussed at the end.

2. CONFORMAL TRANSFORMATIONS

The conformal group includes the following transformations Poincaré transformations

(1) $$x'_\mu = \Lambda_\mu{}^\nu x_\nu + a_\mu ,$$

dilatation transformations

(2) $$x'_\mu = x_\mu ,$$

special conformal transformations

(3) $$x'_\mu = \frac{x_\mu + c_\mu x^2}{\Omega(x)} \qquad \Omega(x) = 1 + 2\,cx + c^2x^2 .$$

A set of fields $\phi_\alpha(x)$ belonging to a linear representation of Lorentz group behaves under the conformal transformations as [12]

(4) $$\phi'_\mu(x') = \left|\det\left(\frac{\partial x'}{\partial x}\right)\right|^{\frac{\ell}{4}} D_\mu{}^\alpha (\Lambda(x)\phi_\alpha(x)) ,$$

where ℓ is the conformal weight of the field. For special conformal transformations, we obtain from eqs. (3) and (4)

(5) $$\phi'_\mu(x') = \Omega^{-\ell} D_\mu{}^\alpha(\Lambda(x))\ \phi_\alpha(x) ,$$

where

(6) $$D_\mu{}^\alpha(\Lambda(x)) = g_\mu{}^\alpha + (c^\lambda x^\sigma - x^\lambda c^\sigma)\ I_{\lambda\sigma\mu}{}^\alpha ,$$

$I_{\lambda\sigma\mu}{}^\alpha$ denotes the spin matrices which satisfy the following commutation relations

(7) $$[I_{\mu\nu}, I_{\lambda\sigma}] = g_{\mu\sigma}I_{\nu\lambda} + g_{\nu\lambda}I_{\mu\sigma} - g_{\mu\lambda}I_{\nu\sigma} - g_{\nu\sigma}I_{\mu\lambda} .$$

3. CONFORMALLY COVARIANT THEORIES

Consider a Lagrangian which is Poincaré invariant as well as dilatation covariant, then under the special conformal transformations it transforms as [13]

$$(8) \qquad L'(\phi'_\mu, \partial'_\nu\phi'_\mu) = \Omega^4 L(\phi_\mu, \partial_\nu\phi_\mu) + 2\, C^\lambda R_\lambda ,$$

where

$$(9) \qquad R_\lambda = -\pi^{\sigma\alpha}\,(I_{\sigma\lambda\alpha}{}^{\beta}\,\phi_\beta + \ell\, g_{\lambda\sigma}\,\phi_\alpha)$$

$$(10) \qquad \pi^{\sigma\alpha} = \frac{\partial L}{\partial\, \partial_\sigma\phi_\alpha} .$$

We restrict ourselves to the cases where $R_\lambda = 0$ and $R_\lambda = \partial^\sigma R_{\sigma\lambda}$. Here $R_{\sigma\lambda}$ is some function of $\phi_\alpha(x)$, and has the conformal weight $\ell_R = -2$. The field equations will thus be conformally covariant.

Below we list the Lagrangian L for different fields. We confine our discussion to the kinematic terms of L only. The more general situation, which includes several fields and conformally covariant interaction terms, is a straight-forward extension of the same formalism.

Real scalar field

$$(11) \qquad L = -\frac{1}{2}\,\partial_\mu\phi\partial^\mu\phi \qquad\qquad R_{\mu\nu} = \frac{1}{2}\, g_{\mu\nu}\phi^2 ,$$

spinor field

$$(12) \qquad L = -\frac{1}{2}\,\bar\psi\gamma_\mu\overleftrightarrow{\partial}^\mu\psi \qquad\qquad R_{\mu\nu} = 0 ,$$

vector field

$$(13) \qquad L = -\frac{1}{4}\, F_{\mu\nu}F^{\mu\nu} \qquad\qquad R_{\mu\nu} = 0 ,$$

second rank symmetric tensor field [14] [15]

$$(14) \qquad L = -\frac{1}{2}(\partial_\sigma h_{\mu\nu})^2 + \frac{2}{3}\partial_\sigma h_{\mu\nu}\partial^\nu h^{\mu\sigma} + \frac{1}{6}(\partial_\sigma h)^2 - \frac{1}{3}\partial_\sigma h\partial_\mu h^{\mu\sigma}$$

$$R_{\mu\nu} = \frac{1}{3}\,[\frac{1}{2}\, g_{\mu\nu}(h_{\sigma\lambda})^2 + 2\, h_{\mu\lambda}h^\lambda{}_\nu - h\;\; h_{\mu\nu}]$$

second rank antisymmetric tensor field

$$(15)\qquad L = -\frac{1}{2}(\partial_\sigma A_{\mu\nu})^2 + \partial_\sigma A_{\mu\nu}\partial^\nu A^{\mu\sigma} + \partial_\sigma A_{\mu\nu}\partial^\mu A^{\sigma\nu}$$

$$R_{\mu\nu} = \frac{1}{2}[3\, g_{\mu\nu}(A_{\sigma\lambda})^2 - 4\, A_{\mu\lambda}\, A_\nu{}^\lambda].$$

The canonical currents using Noether's theorem are

$$(16)\qquad T^c_{\mu\nu} = g_{\mu\nu}\, L - \pi_\mu{}^\alpha \partial_\nu \phi_\alpha$$

$$(17)\qquad J_{\mu\nu\lambda}{}^c = x_\lambda\, T^c_{\mu\nu} - x_\nu T^c_{\mu\lambda} + \pi_\mu{}^\alpha I_{\nu\lambda\alpha}{}^\beta \phi_\beta$$

$$(18)\qquad D_\mu{}^c = x^\nu\, T^c_{\mu\nu} + \ell\, \pi_\mu{}^\alpha\, \phi_\alpha$$

$$(19)\qquad K^c_{\mu\nu} = x^2\, T^c_{\mu\nu} - 2\, x_\nu x^\lambda\, T^c_{\mu\lambda} - 2\pi_\mu{}^\alpha(\ell x_\nu\, \phi_\alpha - x^\lambda I_{\nu\lambda\alpha}{}^\beta\, \phi_\beta).$$

Inserting eqs. (11)-(15) into (16)-(19) respectively, we obtain explicit forms of canonical currents for different fields.

4. CONFORMALLY COVARIANT ENERGY MOMENTUM TENSOR

The conformally covariant energy momentum tensor has the form [5]-[7]

$$(20)\qquad \Theta_{\mu\nu} = T^c_{\mu\nu} - \frac{1}{2}\partial^\lambda\, [(\pi_\lambda{}^\alpha\, I_{\mu\nu\alpha}{}^\beta + \pi_\mu{}^\alpha I_{\nu\lambda\alpha}{}^\beta + \pi_\nu{}^\alpha\, I_{\mu\lambda\alpha}{}^\beta)\, \phi_\beta]$$
$$- \frac{1}{2}(\partial^2 R_{\mu\nu} + g_{\mu\nu}\, \partial^\lambda\partial^\rho\, R_{\lambda\rho}) + \frac{1}{2}(\partial_\mu\partial^\rho\, R_{\rho\nu} + \partial_\nu\, \partial^\rho\, R_{\rho\mu})$$
$$- \frac{1}{6}(\partial_\mu\, \partial_\nu - g_{\mu\nu}\, \partial^2)\, R_\lambda{}^\lambda$$

and

$$(21)\qquad \Theta'_{\mu\nu} = \Omega^4\, D_{\mu\nu}{}^{\lambda\sigma}\, \Theta_{\lambda\sigma}\,,$$

where

$$(22)\qquad D_{\mu\nu}{}^{\lambda\sigma} = g_\mu{}^\lambda\, g_\nu{}^\sigma + 2\,(c_\mu x^\lambda - x_\mu c^\lambda)\, g_\nu{}^\sigma + 2\, g_\mu{}^\lambda(c_\nu x^\sigma - x_\nu c^\sigma).$$

Furthermore, we can verify that $\Theta_{\mu\nu}$ has the properties

$$(23)\qquad \partial^\mu\Theta_{\mu\nu} = 0 \qquad \Theta_{\mu\nu} = \Theta_{\nu\mu} \qquad \Theta_\mu{}^\mu = 0\,.$$

Now one can redefine the other conformal currents in terms of $\Theta_{\mu\nu}$ as follows

$$(24) \qquad J_{\mu\nu\lambda} = x_{\lambda}\Theta_{\mu\nu} - x_{\nu}\Theta_{\mu\lambda}$$

$$(25) \qquad D_{\mu} = x^{\nu}\Theta_{\mu\nu}$$

$$(26) \qquad K_{\mu\nu} = x^2\Theta_{\mu\nu} - 2\, x_{\nu}\, x^{\lambda}\Theta_{\mu\lambda} \, .$$

It is easily seen that all these conformal currents of eqs. (24)-(26) are conserved due to eq. (23).

Substituting eq. (20) into eqs. (24)-(26), we have

$$\begin{aligned}(27) \qquad J_{\mu\nu\lambda} = {} & J_{\mu\nu\lambda}^{c} + \partial^{\rho}\{x_{\lambda}\, G_{[\rho,\mu]\nu} - x_{\nu}\, G_{[\rho,\mu]\lambda} \\ & + \tfrac{1}{2}\,(g_{\rho\lambda}\, R_{\mu\nu} - g_{\mu\lambda}R_{\rho\nu}) + \tfrac{1}{2}\,(g_{\mu\nu}R_{\rho\lambda} - g_{\rho\nu}R_{\mu\lambda}) \\ & + \tfrac{1}{6}\,(g_{\mu\lambda}\, g_{\nu\rho} - g_{\rho\lambda}g_{\nu\mu})R_{\sigma}{}^{\sigma}\}\end{aligned}$$

$$(28) \qquad D_{\mu} = D_{\mu}^{c} + \partial^{\rho}\{x^2\, G_{[\rho,\mu]\nu}\}$$

$$\begin{aligned}(29) \qquad K_{\mu\nu} = {} & K_{\mu\nu}^{c} + \partial^{\rho}\{x^2\, G_{[\rho,\mu]\nu} - 2\, x_{\nu}x^{\lambda}G_{[\rho,\mu]\lambda} \\ & + x_{\rho}R_{\mu\nu} - x_{\mu}R_{\rho\nu} + g_{\mu\nu}x^{\lambda}R_{\rho\nu} - g_{\;\nu}x^{\lambda}R_{\mu\lambda} \\ & + \tfrac{1}{3}\,(x_{\mu}g_{\rho\nu} - x_{\rho}g_{\mu\nu})R_{\sigma}{}^{\sigma}\} - 2\, R_{\mu\nu} \, ,\end{aligned}$$

where

$$(30) \qquad G_{[\rho,\mu]\nu} = -\tfrac{1}{2}\,(H_{\rho[\mu,\nu]} + H_{\mu[\nu,\rho]} + H_{\nu[\mu,\rho]})$$

$$\begin{aligned}(31) \qquad H_{\rho[\mu,\nu]} = {} & \pi^{\alpha}\, I_{\mu\nu\alpha}{}^{\beta}\, \phi_{\beta} + \tfrac{1}{2}\,(\partial_{\nu}R_{\rho\mu} - \partial_{\mu}R_{\rho\nu} + g_{\rho\mu}\,\partial^{\lambda}R_{\lambda\nu} \\ & - g_{\rho\nu}\partial^{\lambda}R_{\lambda\mu}) + \tfrac{1}{6}\,(g_{\rho\nu}\,\partial_{\mu} - g_{\rho\mu}\partial_{\nu})R_{\sigma}{}^{\sigma} \, .\end{aligned}$$

The conformal currents $\Theta_{\mu\nu}$, $J_{\mu\nu\lambda}$, and D_{μ} differ from the original canonical currents $T_{\mu\nu}^{c}$, $J_{\mu\nu\lambda}^{c}$, and D_{μ}^{c} only by the total divergence terms which do not contribute to total charges. As for $K_{\mu\nu}$ and $K_{\mu\nu}^{c}$, in addition to the total divergence term, there is another term $2\,R_{\mu\nu}$ which exactly cancels the nonconserved term in K_{μ}^{c} and keeps $K_{\mu\nu}$ conserved.

Substituting eqs. (11)-(15) into (20) respectively, we can write down the explicit expressions for different fields.
For scalar field

$$\Theta_{\mu\nu} = -\frac{1}{6} g_{\mu\nu}(\partial_\rho\phi)^2 + \frac{2}{3}\partial_\mu\phi\partial_\nu\phi - \frac{1}{3}\phi\,\partial_\mu\partial_\nu\phi \tag{32}$$

which is identical to the improved energy momentum tensor [16]
spinor field

$$\Theta_{\mu\nu} = \frac{1}{4}(\bar\psi\gamma_\mu\overleftrightarrow{\partial}_\nu\psi + \bar\psi\gamma_\nu\overleftrightarrow{\partial}_\mu\psi)\,, \tag{33}$$

vector field

$$\Theta_{\mu\nu} = -\frac{1}{4} g_{\mu\nu} F_{\rho\lambda}F^{\rho\lambda} - F_{\rho\mu}F_\nu{}^\rho, \tag{34}$$

second rank symmetric tensor field

$$\begin{aligned}\Theta_{\mu\nu} = {} & g_{\mu\nu}[-\tfrac{1}{18}(\partial_\lambda h_{\alpha\beta})^2 - \tfrac{1}{3}(\partial^\alpha h_{\alpha\beta})^2 + \tfrac{1}{3}\partial^\lambda h^{\alpha\beta}\partial_\alpha h_{\beta\lambda}\\
& + \tfrac{1}{18}(\partial_\lambda h)^2 + \tfrac{1}{3}\partial^\alpha h\partial^\beta h_{\beta\alpha}] + \tfrac{2}{3}\partial^\alpha h_\mu{}^\beta\partial_\alpha h_{\nu\beta}\\
& + \tfrac{4}{3}\partial^\alpha h_\mu{}^\beta\partial_\beta h_{\nu\alpha} - \tfrac{4}{3}\partial^\alpha h_{\mu\nu}\partial^\beta h_{\beta\alpha} + \tfrac{8}{9}\partial_\mu h^{\alpha\beta}\partial_\nu h_{\alpha\beta}\\
& - \tfrac{2}{9}\partial_\mu h\,\partial_\nu h - \tfrac{4}{3}\partial_\mu h^{\alpha\beta}\partial_\alpha h_{\beta\nu} - \tfrac{4}{3}\partial_\nu h^{\alpha\beta}\partial_\alpha h_{\beta\mu}\\
& + \tfrac{2}{3}\partial_\mu h_\nu{}^\alpha\partial^\beta h_{\beta\alpha} + \tfrac{2}{3}\partial_\nu h_\mu{}^\alpha\partial^\beta h_{\beta\alpha} + \tfrac{1}{6}\partial_\mu h\,\partial^\alpha h_{\alpha\nu}\\
& + \tfrac{1}{6}\partial_\nu h\partial^\alpha h_{\alpha\mu} - \tfrac{1}{6}\partial^\alpha h\,\partial_\mu h_{\nu\alpha} - \tfrac{1}{6}\partial^\alpha h\,\partial_\nu h_{\mu\alpha}\\
& + g_{\mu\nu}[h^{\alpha\beta}(\tfrac{1}{9}\partial^2 h_{\alpha\beta} - \tfrac{2}{3}\partial_\beta\partial^\lambda h_{\lambda\alpha} + \tfrac{1}{2}\partial_\alpha\partial_\beta h\\
& \qquad + \tfrac{1}{6} g_{\alpha\beta}\partial^\lambda\partial^\rho h_{\lambda\rho} - \tfrac{1}{9} g_{\alpha\beta}\partial^2 h)]\\
& + h_\mu{}^\alpha(\tfrac{1}{3}\partial^2 h_{\nu\alpha} - \tfrac{2}{3}\partial_\nu\partial^\beta h_{\beta\alpha} - \tfrac{1}{6}\partial_\nu\partial_\alpha h + \tfrac{2}{3}\partial_\alpha\partial^\beta h_{\beta\nu})\\
& + h_\nu{}^\alpha(\tfrac{1}{3}\partial^2 h_{\mu\alpha} - \tfrac{2}{3}\partial_\mu\partial^\beta h_{\beta\alpha} - \tfrac{1}{6}\partial_\mu\partial_\alpha h + \tfrac{2}{3}\partial_\alpha\partial^\beta h_{\beta\mu})\\
& + h^{\alpha\beta}(\tfrac{2}{3}\partial_\mu\partial_\alpha h_{\beta\nu} + \tfrac{2}{3}\partial_\nu\partial_\alpha h_{\beta\mu} - \tfrac{1}{9}\partial_\mu\partial_\nu h_{\alpha\beta} - \tfrac{4}{3}\partial_\alpha\partial_\beta h_{\mu\nu})\\
& + \tfrac{1}{6}h_{\mu\nu}\partial^2 h + \tfrac{1}{6}h\,\partial^2 h_{\mu\nu} - \tfrac{1}{6}h\,\partial_\mu\partial^\alpha h_{\alpha\nu} - \tfrac{1}{6}h\,\partial_\nu\partial^\alpha h_{\alpha\mu}\\
& + \tfrac{1}{9}h\,\partial_\mu\partial_\nu h\,,\end{aligned} \tag{35}$$

second rank antisymmetric tensor field

$$(36)\qquad \Theta_{\mu\nu} = g_{\mu\nu}\,[-\tfrac{13}{6}\,(\partial_\lambda A_{\alpha\beta})^2 + 3\,\partial^\alpha A^{\lambda\beta}\,\partial_\lambda\,A_{\alpha\beta} + (\partial^\alpha A_{\alpha\beta})^2$$
$$-\tfrac{5}{3}\,A^{\alpha\beta}\,\partial^2 A_{\alpha\beta} + 2\,A^{\lambda\alpha}\partial^\beta\partial_\lambda A_{\beta\alpha}]$$
$$+\tfrac{8}{3}\,\partial_\mu A^{\alpha\beta}\partial_\nu A_{\alpha\beta} - 4\,\partial_\mu A^{\alpha\beta}\partial_\alpha A_{\nu\beta} - 4\,\partial_\nu A^{\alpha\beta}\,\partial_\alpha A_{\mu\beta}$$
$$+ 6\,\partial^\alpha A_\mu{}^\beta\,\partial_\alpha A_{\nu\beta} - 4\,\partial^\alpha A_\mu{}^\beta\,\partial_\beta A_{\nu\alpha} - 2\,\partial_\mu A_\nu{}^\alpha\,\partial^\beta A_{\beta\alpha}$$
$$- 2\,\partial_\nu A_\mu{}^\alpha\,\partial^\beta A_{\beta\alpha} - 2\,A^{\alpha\beta}\,(\partial_\alpha\partial_\mu A_{\nu\beta} + \partial_\alpha\partial_\nu A_{\mu\beta})$$
$$+\tfrac{5}{3}\,A^{\alpha\beta}\,\partial_\mu\partial_\nu A_{\alpha\beta} - 2\,A_\mu{}^\beta\,\partial^\alpha\,\partial_\nu A_{\alpha\beta} - 2\,A_\nu{}^\beta\,\partial^\alpha\partial_\mu A_{\alpha\beta}$$
$$- 3\,A_\mu{}^\alpha\,\partial^2 A_{\alpha\nu} - 3\,A_\nu{}^\alpha\partial^2 A_{\alpha\mu} - 2\,A_\mu{}^\beta\partial^\alpha\partial_\beta A_{\nu\alpha}$$
$$- 2\,A_\nu{}^\beta\,\partial^\alpha\,\partial_\beta A_{\mu\alpha}\ .$$

5. VACUUM SOLUTIONS AND VACUUM STATE

In discussing vacuum solutions of the conformally symmetric theories, the conformally covariant energy momentum tensor $\Theta_{\mu\nu}$ plays a fundamental role. Since $\Theta_{\mu\nu}$ has the properties of eq. (23), the general form of $\Theta_{\mu\nu}$ can be expressed as [10] [17] [18]

$$(37)\qquad \Theta_{\mu\nu} = (4\,\tau_\mu\,\tau_\nu - g_{\mu\nu}\,\tau_\lambda\,\tau^\lambda)\,\Theta(\tau)\ .$$

Here we take $\tau = \frac{1}{2}\,cx^2 + dx + e$ as a general parameter and $\tau_\lambda = \frac{\partial\tau}{\partial x^\lambda}$. Conservation of $\Theta_{\mu\nu}$ implies that

$$(38)\qquad \tau_\lambda\,\tau^\lambda\,\dot{\Theta} + 6\,c\,\Theta = 0\ ,$$

where dot refers to differentiation with respect to the parameter τ. It is easy to see that there are only two solutions of eq. (38)

$$(39)\qquad \Theta(\tau) = 0 \qquad\qquad \text{instantonlike}$$

$$(40)\qquad \Theta(\tau) = \frac{\text{const}}{(\tau_\lambda\tau^\lambda)^3} \qquad\qquad \text{meronlike}\ .$$

In both cases, we have $E = \int d^3x\,\Theta_{00} = 0$ (for eq. (40), $\tau_0 \neq 0$), i.e. they correspond to vacuum solutions. Many conformally symmetric theories are known to have these types of solutions.

The fact that $E = 0$ implies that the classical vacuum solutions can be the candidate for the vacuum in the quantum world. Instead of the naive vacuum state $|0\rangle$, a new vacuum state $|\tilde{0}\rangle$ can be introduced, in which the vacuum expectation value of quantum field $\hat{\phi}(x)$ is [19] [20]

(41) $$\langle\tilde{0}|\hat{\phi}(x)|\tilde{0}\rangle = \phi_{c\ell}(x) .$$

$\hat{\phi}(x)$ is composed of two parts

(42) $$\hat{\phi}(x) = \phi_{c\ell}(x) + \hat{\phi}'(x) ,$$

where $\phi_{c\ell}(x)$ is the classical vacuum solutions of conformally symmetric theories, while $\hat{\phi}'(x)$ is the quantum fluctuation around it. Both $\hat{\phi}(x)$ and $\hat{\phi}'(x)$ can be quantized in the Heisenberg picture by

(43) $$\hat{\phi}(x) = \frac{1}{(2\pi)^{\frac{3}{2}}} \int \frac{d^3k}{\sqrt{2|\vec{k}|}} [\hat{a}_{\vec{k}}(t)\, e^{i\vec{k}.\vec{x}} + \hat{a}^{+}_{\vec{k}}(t) e^{-i\vec{k}.\vec{x}}]$$

(44) $$\hat{\phi}'(x) = \frac{1}{(2\pi)^{\frac{3}{2}}} \int \frac{d^3k}{\sqrt{2|\vec{k}|}} [\hat{c}_{\vec{k}}(t)\, e^{i\vec{k}.\vec{x}} + \hat{c}^{+}_{\vec{k}}(t) e^{-i\vec{k}.\vec{x}}]$$

with

(45) $$[\hat{a}_{\vec{k}}(t),\ \hat{a}^{+}_{\vec{k}'}(t)] = \delta(\vec{k} - \vec{k}')$$

(46) $$[\hat{c}_{\vec{k}}(t),\ c^{+}_{\vec{k}'}(t)] = \delta(\vec{k} - \vec{k}') ,$$

and

(47) $$\hat{a}_{\vec{k}}(t) = f_{\vec{k}}(t) + \hat{c}_{\vec{k}}(t) ,$$

where $f_{\vec{k}}(t)$ is the Fourrier components of $\phi_{c\ell}(x)$

(48) $$\phi_{c\ell}(x) = \frac{1}{(2\pi)^{\frac{3}{2}}} \int \frac{d^3k}{\sqrt{2|\vec{k}|}} [f_{\vec{k}}(t) e^{i\vec{k}.\vec{x}} + f^{*}_{\vec{k}}(t) e^{-i\vec{k}.\vec{x}}] .$$

From the definition of $|\tilde{0}\rangle$ of eq. (41), it follows that

(49) $$\hat{c}_{\vec{k}}(t)|\tilde{0}\rangle = 0$$

(50) $$\hat{a}_{\vec{k}}(t)|\tilde{0}\rangle = f_{\vec{k}}(t)|\tilde{0}\rangle , \text{ for all } \vec{k} .$$

The new vacuum $|\tilde{0}>$, being an eigenstate of the annihilation operator $\hat{a}_{\vec{k}}(t)$ with eigenvalue $f_{\vec{k}}(t)$, is a coherent state which can be expressed as [21] [22]

(51) $$|\tilde{0}> = e^{-\frac{1}{2}||f||} \; e^{(\hat{a}^{+}f)}|0> ,$$

with

(52) $$<\tilde{0}|\tilde{0}> = 1 ,$$

where

(53) $$(a^{+}f) = \int d^3k \; f_{\vec{k}}(t) \; \hat{a}^{+}_{\vec{k}}(t)$$

(54) $$||\cdot f|| = \int d^3k f^{*}_{\vec{k}}(t) \; f_{\vec{k}}(t) .$$

Eqs. (48) and (51) give the relations between classical vacuum solutions on the one hand and quantum vacuum states on the other hand. If one knew the vacuum solutions, then in principle the vacuum states could be obtained. Partial information about classical fields might yield some insight into the quantized theory. As these solutions are nonperturbative, it is hoped that they may reveal new physical configurations which cannot be reached from standard perturbation theory in quantum world.

REFERENCES

[1] G. Mack and A. Salam, Ann. Phys., 53 (1969), 174.

[2] S. Ferrara, R. Gatto and A.F. Grillo, Conformal algebra in space-time, Springer Tracts in Modern Physics, 67, New York, 1973

[3] J. Wess, Nuovo Cimento, 18 (1960), 1086.

[4] D. Gross and J. Wess, Phys. Rev., D2 (1970), 753.

[5] B.W. Xu, Jour. Phys., A14 (1981), L97.

[6] B.W. Xu, Jour. Phys., A14 (1981), L125.

[7] B.W. Xu, Jour. Phys., A15 (1982), L329.

[8] A.O. Barut and B.W. Xu, Phys. Lett., 102B (1981), 37.

[9] A.O. Barut and B.W. Xu, Phys. Rev., D23 (1981), 3076.

[10] A.O. Barut and B.W. Xu, Physica, 6D (1982), 137.

[11] B.W. Xu, Phys. Energ. Fort. Phys. Nucl., 9 (1985) 241

[12] C.J. Isham, A. Salam, and J. Strathdee, Phys. Lett., 31B (1970), 300.

[13] M. Flato, J. Simon and D. Sternheimer, Ann. Phys., 61 (1970),78.

[14] A.O. Barut and B.W. Xu, Jour. Phys., A15 (1982), L207.

[15] J. Fang, W. Heidenreich, and B.W. Xu, Jour., A16 (1983), L225.

[16] C.G. Callan, S. Coleman, and R. Jackiw, Ann. Phys., 59 (1970), 42.

[17] A. Actor, Z. Phys., C3 (1980), 353.

[18] A. Actor, Ann. Phys., 131 (1981), 269.

[19] S. Fubini, Nuovo Cimento, 34A (1976), 521

[20] V. de Alfaro and G. Furlan, Nuovo Cimento, 34A (1976), 555.

[21] S. Skagerstam, Phys. Rev., D19 (1979), 2471.

[22] G.J. Ni and Y.P. Wang, Phys. Rev., D27 (1983), 969.

THE HOLONOMY OPERATOR IN YANG-MILLS THEORY

Carlos N. Kozameh*
Universität der Bundeswehr Hamburg
Fachbereich Maschinenbau/Mathematik
P.O. Box 70 08 22
D-2000 Hamburg 70
F.R.G.

I. Introduction

Several authors [1-5] have introduced non-local variables to describe gauge field theories. Their principal motivation is to exhibit the non-local behaviour of these theories, feature that is difficult to see with local fields like the connection γ_a or curvature field F_{ab}.

A typical example of this non-local behaviour constitutes the Bohm-Aharonov effect [6], in which the outcome of the experiment is best described in terms of

$$H \equiv exp\left(i \oint_C \gamma_a \, dx^a\right) \tag{1.1}$$

where γ_a is the Maxwell connection and C is a closed, unshrinkable loop.

Another motivation for introducing non-local variables is to study global properties of Yang-Mills theory like the scattering matrix between "in" and "out" states. To study this problem one begins by using the conformal invariance of Yang-Mills equations to work in compactified Minkowski space, that is, to use a rescaled metric $g'_{ab} = \Omega^2 \eta_{ab}$ as the background geometry. The scalar field Ω and metric g'_{ab} are assumed to be smooth on a compactified space consisting of Minkowski space and two boundaries $\mathcal{I}^{\pm}$. These boundaries are hypersurfaces where $\Omega = 0$ and represent the idea of infinity along null directions [7].

The method of asymptotic quantization at null infinity [8,9] uses this framework of conformal geometry as the starting point for a quantization procedure for fields that admit a regular extension onto the boundaries $\mathcal{I}^{\pm}$. The main advantages of using this method are (a) it provides a rigorous limit to the concept of $t \to \pm\infty$ for massless fields by introducing the null boundaries $\mathcal{I}^+$ and $\mathcal{I}^-$, and (b) only the radiative part of the field is projected

* Alexander von Humbolt fellow.

onto these boundaries. Thus, one quantizes the two radiative degrees of freedom which arise naturally in this formalism. The Hilbert spaces of "in" and "out" states can then be obtained by working on the boundaries $\mathcal{I}^-$ and $\mathcal{I}^+$ respectively. Finally, to obtain the S-matrix one further needs the field equations to generate the dynamic of the system and to link the "in" and "out" states introduced before in a kinematic procedure. This is usually done by working with an intermediate local field but it could also be accomplished by introducing a non-local variable defined along null geodesics (which go from $\mathcal{I}^-$ to $\mathcal{I}^+$). Itr is the purpose of this note to suggest that the holonomy operator H of the Yang-Mills connection is a good variable to describe the effects mentioned before.

In Section II we introduce this variable H, give formulas relating the holonomy operator with the connection and curvature tensors, *i.e.*, how to obtain one in terms of the other, and write field equations for H which are equivalent to the source free Yang-Mills equations [4]. It is interesting to note that the field equations couple H to the free data A given at $\mathcal{I}^-$. That is, the free initial data at $\mathcal{I}^-$ acts as a source term for the field equations for H. This is particularly useful in a quantization procedure since this gives the link between the "in" fields (constructed out of A) with the field H at a point of Minkowski space. This feature is used in Section III to obtain the quantum holonomy operator $\underline{H}$. For simplicity we restrict the discussion in this section to the Maxwell case. Some remarks about the general case as well as the construction of the S-matrix are given at the end of the section.

II. The Holonomy Operator for Yang-Mills Theory.

In this section we want to introduce the holonomy operator associated with the Yang-Mills connection. Apart from some necessary definitions and technical details condensed together at the beginning of the section, there are two questions we want to analyze. First we would like to know the relationship between this new variable and the local fields, how to write one field in terms of the other. Second we would like to write down field equations for the holonomy operator which are equivalent to the Yang-Mills equations. By solving both problems one proves the full equivalence between the holonomy operator and the Yang-Mills field.

Since answers to these questions are presently available in the literature [4] we will only indicate the main results obtained without proofs. When needed, a rough idea of the approach taken to a given result will be presented.

a) Definitions

The Yang-Mills field is usually given as a connection γ^A_{aB} on a principal fiber bundle, with

a space-time coordinates (base manifold) and A, B fiber coordinates. For simplicity we will drop the fiber indices and think of γ_a as a matrix valued form. In the same way a vector V^A (in the fiber will be denoted by V).

Given a closed curve λ on the space-time and gauge connection γ_a, the parallel propagation of an arbitrary vector V, initially at x^a, around λ is a linear map that is described by the holonomy operator. That is, the parallely propagated vector V' (on the fiber over x^a) is related to V

$$V' - V = VH \tag{2.1}$$

where H is the holonomy operator. This operator clearly depends not only on the point x^a but also on the curve λ. Although in principle one could work on path-space [5], an infinite dimensional space, it is more convenient for an initial value formulation (see next section) to restrict ourselves to a specific set of paths chosen as follows. We first pick an arbitrary point x^a in the space-time together with its future null cone. A specific closed path is then constructed by going from x^a to $\mathcal{I}$ along an arbitrary null geodesic on the cone ℓ_x, at $\mathcal{I}$ moving an infinitesimal distance along the "cut" of $\mathcal{I}$ (the intersection of the cone with $\mathcal{I}$) and then coming back to x^a along a neighboring geodesic on the cone. The closed path so constructed is the boundary of a two-dimensional blade that will be referred to as Δ_x. This "triangle" Δ_x has a surface element $\ell^{[a}M^{b]}$ with ℓ^a the tangent vector to the null geodesics and M^a the separation vector between neighboring geodesics.

The set of all paths constructed as above form a six-dimensional space. Two dimensions are needed to specify the closed paths associated with a fixed point x^a since the intersection of the future light cone of x^a with $\mathcal{I}$ is a closed two-surface. The remaining four dimensions arise by allowing x^a to move on the space-time.

Since $\mathcal{I}$ has topology $S^2 \times \mathbf{R}$ one can assign "natural" coordinates $(u, \varsigma, \bar{\varsigma})$ to it with $-\infty < u < +\infty$ the time coordinate and $\varsigma, \bar{\varsigma}$, sterographic coordinates on the sphere. One can then use $\varsigma\bar{\varsigma}$ to label the intersection of ℓ_x with $\mathcal{I}$. Thus the six-dimensional space is coordinatized by $(x^a, \varsigma, \bar{\varsigma})$. On this six-dimensional space we introduce two types of derivatives, a space-time gradient, ∇_a and the "edth" derivative $\eth$ (and its conjugate $\bar{\eth}$) on the sphere. For a precise definition of this derivative see [10] but essentially $\eth \sim \frac{\partial}{\partial \varsigma}$.

b) Relation between H and γ_a, F_{ab}.

We would like to give an explicit relation between the holonomy operator and the local fields. To write down H in terms of γ_a or F_{ab} we need the parallel transport theorem for non-Abelian connections [11] which states

$$\oint_{\partial S} \hat{\gamma}_a \, dx^a = \int_S \hat{F}_{ab} dS^{ab}. \tag{2.2}$$

The two main differences between (2.2) and ordinary Stokes theorem are: First the surface S is constructed by a one-parameter family of curves which cover the surface. Second, the $\hat{}$ symbol denotes the restriction of the non-Abelian connection or curvature to the unique lifting of each curve [11]. By choosing S to be Δ_x we immediately obtain H in terms of $\hat{\gamma}_a$ or $\hat{F}_{ab}$. If follows from its definition that H is equal to the left side of (2.2). Thus, using $dS^{ab} = \ell^{[a}M^{b]}ds\; d\varsigma$ we obtain

$$H(x,\varsigma,\bar{\varsigma}) = \int_{s_0}^{\infty} \hat{F}_{ab}\ell^a M^b \; ds. \tag{2.3}$$

We now want to study the converse problem, that is, how to obtain the connection or curvature in terms of H. For that it is convenient to introduce the null plane coordinate system [12] $(\ell_a, n_a, m_a, \bar{m}_a)$ in terms of which the Minkowksi metric reads

$$\eta_{ab} = 2\ell_{(a}n_{b)} - 2m_{(a}\bar{m}_{b)}. \tag{2.4}$$

One can easily show that the deviation vector M^a can be written as

$$M^a = (s - s_0)m^a, \tag{2.5}$$

where s is an affine length along the geodesic ℓ_x (s_0 corresponds to x^a). Using the radon transform of (2.3) one then obtains [4]

$$\hat{\gamma}_a m^a = \ell^a \nabla_a H = DH, \tag{2.6}$$

$$\hat{F}_{ab}\ell^a m^b = D^2 H. \tag{2.7}$$

The other components of the connection are obtained by taking $\eth$ and $\bar{\eth}$ derivatives on (2.6) [4]. Equations (2.3), (2.6) and (2.7) show the equivalence between the non-local variable H and the local fields γ_a and F_{ab}.

c) The field equations for H.

If one defines the self-dual (anti-self-dual) part of F_{ab} as

$$F^{\pm}_{ab} \equiv F_{ab} \mp iF^{*}_{ab} \tag{2.8}$$

where $F^{*}_{ab} = \frac{1}{2}\epsilon_{abcd}F^{cd}$, then the source free Yang-Mills equations and the Bianchi identities for the curvature tensor can be combined together into a single equation, namely

$$\nabla_{[a}F^{\pm}_{bc]} + [F^{\pm}_{[ab}, \gamma_{c]}] = 0. \tag{2.9}$$

The idea is to impose equations for H which are equivalent to (2.9). A direct way to obtain such equations is to use (2.6) and (2.7) to reexpress (2.9) in terms of H [4]. However, for asymptotically simple Yang-Mills fields [13] one can follow another approach that has several advantages [4]. First, it couples the holonomy operator to the free radiation data at $\mathcal{I}$. Second, for the self-dual (or anti-self-dual) Yang-Mills equations one obtains linear equations for H. Third, for the full Yang-Mills equations H couples only to its complex conjugate $\bar{H}$. Finally, to write down a scattering theory one should start with asymptotically simple fields, thus this approach provides the field equations for this class of solution of Yang-Mills equations.

To obtain the field equations for H one starts by introducing a three-dimensional volume V constructed as follows. It is a pencil of null rays with starting point x^a bounded by a cap on $\mathcal{I}^-$ and the triangular regions $\Delta_x(\varsigma,\bar{\varsigma})$, $\Delta_x(\varsigma,\bar{\varsigma}+d\bar{\varsigma})$, $\Delta_x(\varsigma+d\varsigma,\bar{\varsigma})$ and $\Delta_x(\varsigma+d\varsigma,\bar{\varsigma}+d\bar{\varsigma})$. One then restricts (2.9) to the lifting of the curves ℓ_x on this pencil of rays V and integrates the ^ versions of (2.9) on this volume V. (Note that for a field that is not asymptotically simple this integral will diverge.) Finally, one uses the relations (2.3), (2.6) and (2.7) to reexpress the integrals in terms of H. A detailed derivation can be found in [4]. The final results for Maxwell, self-dual Yang-Mills and general Yang-Mills are respectively

$$\bar{\eth} H = -\eth \bar{A} \tag{2.10a}$$

$$\bar{\eth} H + [H, \bar{A}] = -\eth \bar{A} \tag{2.10b}$$

$$\bar{\eth} H + [H, \bar{A}] + J(H, \bar{H}) = -\eth \bar{A} + \int_{-\infty}^{u} [\dot{\bar{A}}, A] \tag{2.10c}$$

with $\cdot$ being $\frac{\partial}{\partial u}$, $J = \int_0^\infty [D^2 H,\ s^2 D\bar{H} - s\bar{H}] ds$ and $\bar{A}$ the restriction of the connection to $\mathcal{I}$. Note that $\bar{A}$ in (2.10b) and (2.10c) is a matrix rather than a scalar.

d) Comments

(1) The right side of (2.10) is the free data given at $\mathcal{I}^-$. That is, $A(u,\varsigma,\bar{\varsigma})$ is a complex matrix-valued function that contains all the information of the radiative part of the Yang-Mills connection. This fact is specially important in a quantization procedure since one starts with only the radiative degrees of freedom.

(2) The data A when restricted to the cut of x^a acts as a source term in the field equations for H. We recall that the cuts of x^a in Minkowski space are described by the function $u = x^a \ell_a(\varsigma,\bar{\varsigma})$, where $\ell_a(\varsigma,\bar{\varsigma})$ are the four spherical-harmonics Y_{00}, Y_{1m}, $m = 1, 0, -1$ written in terms of $\varsigma,\bar{\varsigma}$ [12]. Thus the restriction of A to the cut is given by $A(u = x^a\ell_a, \varsigma, \bar{\varsigma})$.

The idea then is to seek regular solutions of (2.10).

(3) For the Maxwell case this is not difficult since the $\bar{\eth}$ operator has a simple Green's

function K on the sphere given by

$$K(\varsigma,\eta)=\frac{1}{4\pi}\frac{(1+\varsigma\bar{\eta})}{(1+\eta\bar{\eta})(\varsigma-\eta)}. \tag{2.11}$$

Thus, the general regular solution of (2.10a) can be written as

$$\begin{aligned} H(x,\varsigma) &= \oint_{S^2} \eth' K(\varsigma,\varsigma')\bar{A}(x^a\ell'_a,\varsigma')dS' \\ &= \int_{I^-} K(x,x',\varsigma)\bar{A}(x')d^3x \end{aligned} \tag{2.12}$$

with

$$K(x,x',\varsigma)\equiv \eth' K(\varsigma,\varsigma')\delta(u-x^a\ell'_a). \tag{2.13}$$

Note that the Minkowski points x^a enter (2.13) as parameters. The Maxwell field F_{ab} can be obtained using (2.7), *i.e.*,

$$F_{ab}\ell^a m^b = D^2 H = \oint \eth'\ K\ \dot{\bar{A}}(\ell^a\ell'_a)^2 dS'' \tag{2.14}$$

where $\dot{\bar{A}}=\frac{\partial\bar{A}}{\partial u}$. Equation (2.14) is the Kirchoff formulation of Maxwell theory.

(4) By imposing a self-duality condition on the Yang-Mills field one obtains (2.10b). This is a linear equation for H whose solution will obey Huygens' principle. It will only depend on the data given on the cut $u=x^a\ell_a$. For a general Yang-Mills field the solution will not only depend on the cut but also on the part of I^- lying below the cut. This shows the non-Huygens nature of the field.

(5) H is a space-time scalar which is invariant under gauge transformation that go to the identity at null infinity.

(6) One can implement an iteration scheme of (2.10c) based on (2.10b). That is to say, one regards (2.10b) as the non-interacting field equation for a self-dual field H and the commutator between H and $\bar{H}$ in (2.10c) provides the coupling or interaction with the anti-self-dual part $\bar{H}$. The iteration scheme is then to begin with a self-dual solution $(H_0,\bar{H}_0)$ and use $J(H_0,\bar{H}_0)$ as a source term for the next order in a perturbation expansion.

(7) The proof that (2.10c) and (2.9) for regular fields are equivalent is given in [4]. It amounts to write the Yang-Mills equation in the gauge and then shows that the third D derivative of (2.10c) is identical to the standard equations.

(8) One can generalize these results for a Yang-Mills field in an asymptotically flat space-time and for the gravitational holonomy operator of an asymptotically flat space-time [14].

III. Some remarks about quantization of H.

As was mentioned before, Eqs. (2.10) couple the holonomy operator with the initial free data A. Since in asymptotic quantization procedure [8,9], one gives canonical commutation relations (c.c.r.) for the fields at $\mathcal{I}^+$ (in this case $A(u,\varsigma)$) our formulation of Yang-Mills theory seems to fit very nicely with with approach. The field equation (2.10) provide the link between the fields at $\mathcal{I}$ and the fields at an interior point x^a.

We will divide this section in three parts. First we will give a brief review of Ashtekar's method of asymptotic quantization at null infinity. We will then apply this method to our fields A and H, restricting ourselves to the Maxwell case and leaving some comments about the general case at the end of the section.

a) Quantization of the Maxwell field at null infinity.

Since Maxwell's theory is conformally invariant, one can define an asymptotically flat Maxwell field as one for which the connection γ_a and curvature F_{ab} have a finite extension to $\mathcal{I}$ [13].

Denoting by A_a the restriction of this connection to $\mathcal{I}$ and fixing a gauge by setting $A_a n^a$, the component of A_a along the generator n^a, equal to zero one can easily see that all the information about A_a is coded in the complex scalar A defined before [13]. Thus, the two radiative degrees of freedom of an asymptotically flat Maxwell field are easily picked up via this formalism.

To implement a quantization procedure for the fields at $\mathcal{I}$ we first introduce a sympletic structure

$$\Omega(A_1,A_2)=\frac{1}{4\pi}\int_{\mathcal{I}}(A_1^a L_n A_{2a}-(L_n A_1^a)A_{2a})d^3\mathcal{I} \tag{3.1}$$

where L_n is the Lie derivative with respect to the null generator n^a of $\mathcal{I}$. Next we introduce operator valued distributions $\underline{A}_a$ at $\mathcal{I}$ satisfying the following canonical commutation relations

$$[\underline{A}(A_1),\underline{A}(A_2)]=\frac{h}{i}\Omega(A_2,A_1) \tag{3.2}$$

where $\underline{A}(A_1)\equiv\Omega(A_1,\underline{A})$.

We now want to decompose the field operator $\underline{A}_a$ into its creation and annihilation parts. For that we have to split the test fields A_a into positive-frequency and negative-frequency parts. This can be done unambiguously since the integral lines of n^a provide a natural definition of a "Killing time" u. Thus, A_a can be written as (suppressing the angular coordinates)

$$A_a(u)=\int_0^\infty A_a(\omega)e^{-i\omega u}d\omega+\int_0^\infty A_a(-\omega)e^{+i\omega u}d\omega=A_a^+ + A_a^- . \tag{3.3}$$

Note that $A_a^- = \overline{A_a^+}$ since the Maxwell connection is real. Thus, the positive-frequency part determines the whole field (this will not be the case for a complex Maxwell field).

We now define the annihilation and creation operators as

$$a(A) = \underline{A}(A^+), \quad a^*(A) = \underline{A}(A^-). \tag{3.4}$$

One can easily check that the only non-trivial c.c.r. are

$$[a(A_1),\ a^*(A_2)] = h\langle A_1^+, A_2^+\rangle \equiv \frac{h}{i}\Omega(\overline{A_2^+}, A_1^+) \neq 0 \tag{3.5}$$

where we have used $\Omega(A_1^+, A_2^+) = \Omega(A_1^-, A_2^-) = 0$. Note that the norm $\langle , \rangle$ defined in (3.5) is positive definite, *i.e.*,

$$\langle A^+, A^+\rangle = \int_0^\infty \omega A(\omega)\bar{A}(\omega)d\omega > 0. \tag{3.6}$$

Hence we can use this norm together with the operators a and a^* to construct an inner product space.

b) The quantum holonomy operator.

We would like to apply the quantization procedure outlined before to our fields A and H. First we will use (3.2) to write commutation relations for $\underline{A}$ and its associated creation, annihilation operators. In the process we will obtain a natural splitting of the Hilbert space. Then we will show how to obtain the field operator at an interior point x^a in terms of the "free" operator $\underline{A}$.

If one defines the (singular) operators $\underline{A}(u,\varsigma)$ and $\underline{A}^*(u,\varsigma)$ as

$$\underline{A}_a = \underline{A}\bar{m}_a + \underline{A}^* m_a, \tag{3.7}$$

then the only non-trivial c.c.r. for $\underline{A}, \underline{A}^*$ arising from (3.6) are

$$[\underline{A}(u,\varsigma), \underline{A}^*(u'\varsigma')] = \epsilon(u-u')\delta(\varsigma-\varsigma'). \tag{3.8}$$

It is not surprising that the c.c.r. (3.8) are non-local since they are given on a null rather than on a space-time hypersurface. This is precisely what one would obtain in the commutation relations for Maxwell connections $\gamma_a(x), \gamma_b(x')$ if one writes $\Delta_F(x-x')$ for $(x-x')$ a null vector [15].

The main advantage of writing $\underline{A}_a$ as in (3.7) is that one splits the connection into its self-dual and anti-self-dual parts. That is, $\underline{A}(\underline{A}^*)$ generates a Maxwell curvature tensor

$F^+_{ab}(F^-_{ab})$ that is an eigenstate of the helicity operator with helicity $s = +1(-1)$ [17]. Thus by working with $\underline{A}$ and $\underline{A}^*$ we obtain a natural decomposition of the helicity eigenstates. To see this more explicitly we introduce an orthonormal basis A^+_α and

$a_{\pm\alpha}$ = annihilation operator of a state α with helicity $s = \pm 1$.

$a^*_{\pm\alpha}$ = creation operator of a state α with helicity $s = \pm 1$.

One can easily show from (3.5) that these operators satisfy

$$[a_{s\alpha}, a_{s'\beta}] = [a^*_{s\alpha}, a^*_{s'\beta}] = 0, [a_{s\alpha}, a^*_{s'\beta}] = \delta_{ss'}\delta_{\alpha\beta}. \tag{3.9}$$

That is, the Hilbert space consists of the direct sum of the Hilbert spaces with helicities +1 and −1.

Finally, the relation between $\underline{A}, \underline{A}^*$ and $a_{s'}, a^*_{s'}$ is given by

$$\underline{A}(A_\alpha) = a_{-\alpha} + a^*_{+\alpha}, \underline{A}^*(A_\alpha) = a_{+\alpha} + a^*_{-\alpha}. \tag{3.10}$$

One sees from (3.10) that $\underline{A}$ acting on an arbitrary state will give a positive helicity construction by creating a positive helicity particle (with a^*_+) and destroying a negative helicity particle (with a_-).

We now want to define the holonomy operator $\underline{H}$. We recall that (2.9) gives the relation between the classical H and A. Thus associated with an orthonormal basis of positive frequency A^+_α there will be a basis H^+_α defined by

$$H^+_\alpha \equiv \int_I K(x, x', \varsigma) A^+_\alpha(x') d^3x', \tag{3.11}$$

which will satisfy the field equations [4]. The quantum operators $\underline{H}, \underline{H}^*$ are then defined as

$$\underline{H} \equiv \sum_\alpha (H^+_\alpha a_{-\alpha} + \bar{H}^+_\alpha a^*_{+\alpha}), \quad \underline{H}^* \equiv \sum_\alpha (H^+_\alpha a_{+\alpha} + \bar{H}^+_\alpha a^*_{-\alpha}). \tag{3.12}$$

Note that $\underline{H} = \underline{H}(x, \varsigma)$. Although one could smear out the x dependence of $\underline{H}$ by integrating with a test function f, there would still remain the ς dependence which parameterizes the S^2 family of null plane coordinate systems [18].

In a completely equivalent way one could have taken the – version of (2.9) as our definition of $\underline{H}$. By expanding $\underline{A}$ in the orthonormal basis A^+_α one can easily show that this yields (3.12).

c) Comments

1) The construction outline before applies equally to $\mathcal{I}^+$ and $\mathcal{I}^-$. Thus one can define $\underline{A}_{\text{in}}$ and $\underline{A}_{\text{out}}$. Furthermore (3.12) gives a link between the in and out fields via

$$\underline{A}_{\text{out}} = \lim_{x \to \mathcal{I}^+} D\underline{H} = \lim_{x \to \mathcal{I}^+} \sum_{\alpha} (DH_\alpha^+ a_{-\alpha} + D\bar{H}_\alpha^+ a_{+\alpha}^*)_{\text{in}} \tag{3.13}$$

which in principle determines the S matrix of the problem [19].

2) If one explicitly carries out the calculation of the S-matrix one discovers that it is trivial. This is not, as one may simply assume, a consequence that we are dealing with source free Maxwell equations. It follows from the way we choose the appropriate A^+ to define our Hilbert space. Those A_α^+ must satisfy [9]

$$\lim_{u \to \infty} A^+(u) = \lim_{u \to -\infty} A^+(u). \tag{3.14}$$

Even if one considers interactions with sources one can show that data satisfying (3.14) produces a trivial classical scattering of charged particles [20]. Thus, to construct our Hilbert space one has to rule out very interesting free initial data, like the one who yields a class of Lienard-Wiechert solutions [16] producing non-trivial scattering. If one relaxes the finiteness of the norm condition one can produce a non-trivial S matrix for the in and out states.

3) It has been shown by Ashtekar [9] that in Q.E.D. one has to abandon the concept of a Hilbert space for the Maxwell field if the corresponding Dirac state is any other than vacuum. Although this problem cannot be treated in the context of our formalism (the massive Dirac equation cannot be made regular in a neighborhood of $\mathcal{I}$ by a conformal transformation) one could study the problem of a massless Dirac or Klein Gordon field coupled to the Maxwell field.

4) In principle the general Yang-Mills case could also be considered using the formalism outlined here for the Maxwell case. However, some technical difficulties arise in the non-Abelian case which will be analyzed in subsequent work. As mentioned before, a perturbative approach to construct the S-matrix for the general case based on self-dual and anti-self-dual decompositions could prove to be useful. The "free field" self-dual Y-M particles can be obtained out of solutions of (2.10b). Since these equations are linear there is no problem in constructing a Hilbert space for the in and the out states.

It should be interesting to compute the perturbation graphs obtained by this procedure and compare them to the Feynman graphs arising from null quantization [15].

Acknowledgements:

I wish to thank Ted Newman for useful comments on this manuscript. This research was supported in part by the National Science Foundation under Grant No. PHY82-17853, supplemented by funds from the National Aeronautics and Space Administration, at the University of California at Santa Barbara.

REFERENCES

[1]. S. Mandelstam, *Ann. of Phys.* **19**, 1 (1962); S. Mandelstam, *Phys. Rev.* **175**, 1580 (1968).

[2]. I. Bialynicki-Birula, *Bull. Acad. Pol. Sci.* **11**, 135 (1963).

[3]. T.T. Wu and C.N. Yang, *Phys. Rev. D* **12**, 3843-3845 (1975).

[4]. C.N. Kozameh and E.T. Newman, *Phys. Rev. D* **31**, 801 (1985).

[5]. O. Steinmann, *Ann. Phys.* **157**, 232 (1984).

[6]. Y. Aharonov and D. Bohm, *Phys. Rev.* **115**, 485 (1959).

[7]. R. Penrose, in Batelle Rencontres, ed. C.M. DeWitt and J.A. Wheeler, Benjamin, New York (1967).

[8]. V.P. Frolov, *Fortschr. Phys.* **26**, 455 (1978).

[9]. A. Ashtekar, Syracuse University, preprint, to appear in *Notes in Quantum Gravity*, (Bibliopolis, Naples).

[10]. E.T. Newman and R. Penrose, *J. Math. Phys.* **7**, 863 (1966).

[11]. This result has been rederived several times. See ref. 4 and references therein.

[12]. In fact this is not a tetrad but an S^2 family of null tetrads. See ref. 4 and references therein.

[13]. E.T. Newman, *Phys. Rev. D* **8**, 2901 (1978).

[14]. C.N. Kozameh and E.T. Newman, to appear in the Proceedings of the 9th International School of Cosmology and Gravitation, Erice (Italy), eds. P. Bergman and V. de Sabatta.

[15]. R.A. Neville and F. Rohrlich, *Nuovo Cimento* **1A**, 625 (1971); L. Bring, O. Lindgren, and B.E.W. Nilsson, *Nucl. Phys.* **B212**, 149 (1983); S. Mandelstam, *Nucl. Phys.* **B213**, 149 (1983).

[16]. S.L. Kent, C.N. Kozameh, and E.T. Newman, *J. Math. Phys.* **26**, 300 (1985).

[17]. I. Bialynicki-Birula *et al.* *J. Math. Phys.* **22**, 2530 (1981).

[18]. By fixing a value of ς we recover the results cited in ref. 15.

[19]. To obtain a rigorous limit one should follow the same outlines as in the LSZ formalism, *i.e.*, to average the fields with wave packets.

[20]. M. Ludvigsen, *Gen. Rel. Grav.* **13**, 7 (1981).

III. <u>CONFORMAL STRUCTURES</u>

CONFORMAL GEODESICS

B.G. SCHMIDT
MAX-PLANCK-INSTITUT FÜR PHYSIK UND ASTROPHYSIK
INSTITUT FÜR ASTROPHYSIK
KARL-SCHWARZSCHILD-STR. 1
8046 GARCHING, FRG

A conformal structure $\mathcal{C}$ on a n-dimensional manifold M is determined by the following equivalence relation between nondegenerate metrics of any signature:

(1) $$\bar{g} \sim g \Leftrightarrow \bar{g} = e^{2\sigma} g \quad , \quad \sigma : M \to R$$

In the case of Lorentz metrics, one can visualize $\mathcal{C}$ as the distribution of the null cones on M . $\mathcal{C}$ determines uniquely a reduction P(M) of L(M), the frame bundle over M . P(M) consists of all frames which are orthonormal for some metric $g \in \mathcal{C}$. The structure group of P is the direct product of the generalized orthogonal group O(p,q) and the 1-dimensional group of dilatations.

A conformal structure $\mathcal{C}$ determines further a collection $\Gamma(\mathcal{C})$ of symmetric linear connections over M , by the condition that parallel propagation does not lead out of P(M) .

The coordinate components of the difference tensor S_{ik}^{ℓ} of two connections $\nabla, \bar{\nabla}$ in $\Gamma(\mathcal{C})$ are determined by a 1-form $b=b_i dx^i$ as follows [1]:

(2) $$S_{ik}^{\ell}(b) = \delta_i^{\ \ell} b_k + \delta_k^{\ \ell} b_i - g_{ik} g^{\ell s} b_s$$

$$(g_{ik} g^{\ell s} = \bar{g}_{ik} \bar{g}^{\ell s} \text{ if } \bar{g} = e^{2\sigma} g)$$

Hence, at a fixed point $x_o \in M$, there is a 1-1-relation between elements of $\Gamma(\mathcal{C})$ and 1-forms $b \in T_{xo}(M)$.

The purpose of this note is to draw attention to the following property of a conformal structure* (n > 2):

---

*I don't know where these curves have been defined for the first time. They are mentioned in Yano's book about the Lie derivative [4].

Given a tangent vector X_{x_o} and a $\Gamma_{x_o} \in \Gamma_{x_o}(\mathcal{C})$, then there exists a unique curve $x(\lambda)$ - called a conformal geodesic - such that:

(1) $x(\lambda)$ is a geodesic for some $\Gamma \in \Gamma(\mathcal{C})$ whose Ricci tensor vanishes along $x(\lambda)$

(2) $\dot{x}(0) = X_{x_o}$. $\Gamma(x_o) = \Gamma_{x_o}$

Furthermore $\Gamma(x(\lambda))$ is unique along $x(\lambda)$.

These curves arise naturally via the "prolongations of G-structures of finite type [1,2]. More directly they can be described as follows:

Let ∇ be any connection in $\Gamma(\mathcal{C})$ and $\nabla + S(b)$ the connection we want to determine along the curve $x(\lambda)$; then if $x(\lambda)$ and $b(\lambda)$ are the solutions of the following system of ordinary differential equations:

$$(3) \qquad (\nabla_{\dot{x}}\dot{x})^i = -S^i_{k\ell}(b)\,\dot{x}^k\dot{x}^\ell$$

$$(4) \qquad (\nabla_{\dot{x}}b)_i = -L_{ik}\dot{x}^k + 1/2\; S^\ell_{ik}(b)\,\dot{x}^k b_\ell$$

where S^ℓ_{ik} is as in (2) and L_{ik} is equivalent to the Ricci tensor of ∇:

$$(5) \qquad (n-2).L_{ik} = R_{ik} - \frac{1}{2(n-1)}\, g_{ik}\, g^{rs}\, R_{rs}$$

Taking into account how the tensor L_{ik} changes if we change ∇ according to (2)

$$(6) \qquad \bar{L}_{ik} = L_{ik} + (n-2)((\nabla b)_{ik} + b_i b_k + 1/2\; b_r b_s g^{rs} g_{ik})$$

one can establish [1], that a geodesic $x(\lambda)$ of $g \in \mathcal{C}$ is a conformal geodesic, if and only if the Ricci tensor of g vanishes along $x(\lambda)$. The system (3), (4) can be interpreted to express exactly this condition.

The independence of the curve $x(\lambda)$ and $\Gamma_{x(\lambda)}$ (determined by (3), (4)) from the connection ∇ used to formulate the equations (3), (4) can be shown by using an invariant definition as in [1], or directly as follows: Let $\bar{\nabla}$ be another connection in $\Gamma(\mathcal{C})$, then there exists a 1-form c , such that

$$(7) \qquad \bar{\nabla} = \nabla + S(c)$$

Hence

$$(8) \qquad \nabla + S(b) = \bar{\nabla} + S(b-c)$$

Using (6) a calculation shows that $\bar{b} = b-c$ and $x(\lambda)$ satisfy (3), (4) with ∇ , L_{ik} replaced by $\bar{\nabla}$, $\bar{L}_{ik}$

Let $\mathcal{C}$ be a flat conformal structure ($C^a{}_{bcd} = 0$ if $n > 3$). Then the conformal geodesics are all the geodesics of flat metrics in $\mathcal{C}$.

The concept of c-geodesics can be used to generalize the notion of normal coordinate systems from metrics to conformal structures in an obvious way: pick $\Gamma_{x_o} \in \Gamma(\mathcal{C})$, then the c-geodesics for this Γ_{x_o} and the connections along on the c-geodesics determine a unique connection $\in \Gamma(\mathcal{C})$ near x_o. Hence there are near x_o as many conformal-normal-connections $\overset{n}{\Gamma}_{x_o}$ as elements of $T^*_{x_o}(M)$. Each can be used to define an exponential map exp: $T_{x_o} \to M$ as usual.

The covariant derivatives of the Riemann tensor of a metric at a point x_o determine the metric uniquely near x_o in the analytic case [3]. The corresponding property for conformal structures is, that all derivatives of the conformal tensor $C^a{}_{bcd}$ with respect to a conformal normal connection determine the conformal structure locally uniquely.

All this generalizes in the most obvious way to G-structures of finite type. If Z_i are the horizontal vector fields of the final parallelisation, then the integral curves of $\xi^i Z_i$ (ξ^i const) project onto the generalized geodesics on the base manifold M .

REFERENCES

[1] B.G. SCHMIDT: Commun. Math. Phys., 36 (1974), 73-90.

[2] S. KOBAYASHI: Transformation groups in differential geometry. Berlin-Heidelberg-New York, Springer (1972).

[3] S. KOBAYASHI, K. NOMIZU: Foundations of differential geometry, Vol. I, New York, Interscience (1969).

[4] K. YANO: The theory of Lie derivatives and its applications; North Holland Publishing Co., Amsterdam (1957).

SECOND ORDER CONFORMAL STRUCTURES

J.D. Hennig

Institut für Theoretische Physik
Technische Universität Clausthal
Clausthal, Germany F.R.

1. Introduction

The concept of stepwise *higher order prolongations* of differentiable objects such as tensor fields, connection forms, differential equations, direction fields etc. has its mathematical origin in ideas of E. Cartan, connected with the investigation of infinite dimensional Lie groups and local isomorphisms between differentiable objects /1/. Together with the rigorous formulation of the notion of 'higher order contact' within the theory of 'jets' by C. Ehresmann /2/ this concept finally led to the fibre bundle formalism of *G-structures* and their prolongations (see e.g. /3/,/4/).

It is this theory of G-structures which provides the appropriate tools for a global and coordinate free description of a wide class of geometrical structures especially used in mathematical physics, such as (pseudo-) Riemannian, Galilean, conformal and symplectic structures, or non-tensorial objects as linear connections and projective structures. In the terminology of G-structures each of these examples is of 1. or 2. order, the only orders of relevance up to now from the physical point of view.

In the present article we indicate two applications of 2. order structures to space time problems.
Starting with Lorentzian and (Lorentz-) conformal structures as standard examples of 1. order objects, we consider the following canonical extensions

- (i) prolongations to *2. order* structures,
- (ii) generalizations to 1. order *affine* structures.

As a natural combination of these two we then construct

- (iii) extensions to *2. order affine* structures.

For an application of (i) we sketch the derivation of a *Weyl geometry* from given causal and projective structures (Ehlers-Pirani-Schild axiomatic).
Extension (iii) will be used to indicate a geometrical background and program for an *O(4,2) gauge theory of gravity*, where (ii) and an affine version of Poincaré gauge theory will serve as a guiding line.

2. Conformal geometry

2.1. First and second order structures

a. Prolongation of Lorentz structures: To recall roughly the mechanism of construction for G-structures and their prolongations, consider the example of a *Lorentz metric* g on a 4-dim. manifold M, and let us define the corresponding 1. and 2. order structures:
Fix $x_0 \in M$ and denote by $CS(x_0)$ the set of all coordinate systems $\varphi : M \supset U_\varphi \longrightarrow \mathbb{R}^4$, centered at x_0 ($\varphi(x_0) = 0 \in \mathbb{R}^4$), and by CT the set of all coordinate transformations $\alpha : \mathbb{R}^4 \supset V_\alpha \longrightarrow \mathbb{R}^4$ with $\alpha(0) = 0$.
As a tensor, the coordinate representation $g_{\mu\nu}(x_0)$ of g in x_0 with respect to $\varphi \in CS(x_0)$ does not react to $\alpha \in CT$ iff the differential $\alpha^\mu_{\ ,\nu}(0) = \delta^\mu_\nu$.
Thus, the transformation rule of $g_{\mu\nu}(x_0)$ only depends on *1. order contact* equivalence classes (1-jets) $j^1(\alpha)$ of transformations $\alpha \in CT$ ($\alpha \sim \beta \;:\Longleftrightarrow\; \alpha^\mu_{\ ,\nu}(0) = \beta^\mu_{\ ,\nu}(0)$). Accordingly, $g_{\mu\nu}(x_0)$ is uniquely determined by the 1. order contact class $j^1_{x_0}(\varphi)$ of the coordinate system $\varphi \in CS(x_0)$.

We recognize the natural correspondences
(i) between 1-jets $j^1(\alpha)$ and regular 4×4-matrices $(\alpha^\mu_{\ ,\nu}(0))$,
(ii) between 1-jets $j^1_{x_0}(\varphi)$ and linear frames $e(x_0) := (\partial_\mu|_{x_0})$,
hence identify these notions. Usual composition of mappings yields the group

$$(2.1)\qquad G^1(4) := \{\, j^1(\alpha) = (\alpha^\mu_{\ ,\nu}(0)) \;/\; \alpha \in CT \,\} ,$$

canonical isomorphic to $GL(4,\mathbb{R})$, and the right action of $G^1(4)$ on the set $L_{x_0}M$ of linear frames in $T_{x_0}M$,

$$(2.2)\qquad \begin{array}{ccc} & M & \\ {\scriptstyle j^1_{x_0}(\varphi)}\nearrow & & \nwarrow{\scriptstyle j^1_{x_0}(\varphi)\circ j^1(\alpha) := j^1_{x_0}(\varphi\circ\alpha)} \\ \mathbb{R}^4 & \xleftarrow[\;j^1(\alpha)\;]{} & \mathbb{R}^4 \end{array} ,$$

which means basis transformations in $T_{x_0}M$.

To define the (1. order) G-structure determined by g, consider the subset $L_{g,x_0}M$ of all $j^1_{x_0}(\varphi) \in L_{x_0}M$, which give $g_{\mu\nu}(x_0)$ in the 'standard form' $\eta_{\mu\nu} = \mathrm{diag}(-1,1,1,1)$. Obviously, the right action of $G^1(4)$ then reduces to an action of $O(3,1) \subset G^1(4)$ on $L_{g,x_0}M$. Finally, performing the whole construction for each $x_0 \in M$, we get the usual sub*bundle* L_gM *of g-frames* in the bundle LM of all linear frames; L_gM is the (1. order) *G-structure* belonging to g , with structure group $G = O(3,1)$.

The advantage of this 'jet'-description of LM is to be seen in the possibility to generalize immediately to higher order structures:
We introduce *2. order contact classes* (2-jets) $j^2(\alpha)$ of transformations $\alpha \in CT$ ($\alpha \overset{2}{\sim} \beta :\Longleftrightarrow \alpha^\mu{}_{,\nu}(0) = \beta^\mu{}_{,\nu}(0) \wedge \alpha^\mu{}_{,\nu\rho}(0) = \beta^\mu{}_{,\nu\rho}(0)$) and get the group

$$(2.3) \qquad G^2(4) := \{ j^2(\alpha) = (\alpha^\mu{}_{,\nu}(0), \alpha^\mu{}_{,\nu\rho}(0)) \,/\, \alpha \in CT \} \quad ,$$

with multiplication law (chain rule for partial differentiation)

$$(2.4) \qquad (\alpha^\mu_\nu, \alpha^\mu_{\nu\rho})(\beta^\mu_\nu, \beta^\mu_{\nu\rho}) = (\alpha^\mu_\sigma \beta^\sigma_\nu, \alpha^\mu_\sigma \beta^\sigma_{\nu\rho} + \alpha^\mu_{\sigma\tau} \beta^\sigma_\nu \beta^\tau_\rho) \quad ,$$

$\alpha^\mu_\nu := \alpha^\mu{}_{,\nu}(0)$, $\alpha^\mu_{\nu\rho} := \alpha^\mu{}_{,\nu\rho}(0)$. Hence $G^2(4) = G^1(4) \otimes A^2$ with abelian invariant subgroup $A^2 := \{(\delta^\mu_\nu, \alpha^\mu_{\nu\rho})\}$.

Similarly, we define the 2. order classes $j^2_{x_0}(\varphi)$ for $\varphi \in CS(x_0)$, i.e. *2. order frames* $\overset{2}{e}(x_0)$ at x_0 , and get a natural right action of $G^2(4)$ on the fibre $L^2_{x_0}M$ of these frames (cf. 2.2).

Returning to the local representation of g in x_0 , we remember from pseudo-Riemannian geometry the following facts:

(i) The existence of 'normal' coordinate systems $\varphi \in CS(x_0)$, such that $g_{\mu\nu}(x_0) = \eta_{\mu\nu}$ and also $g_{\mu\nu,\rho}(0) = 0$; i.e., the development of g in x_0 up to 1. order (which corresponds to 2. order in the φ's) is given in a 'standard form'.

(ii) Two normal $\varphi_1, \varphi_2 \in CS(x_0)$ always fulfil $(\alpha^\mu{}_{,\nu}(0)) \in O(3,1)$ and $\alpha^\mu{}_{,\nu\rho}(0) = 0$, where $\alpha := \varphi_1 \circ \varphi_2^{-1} \in CT$ in a neighborhood of 0 .

Translated into the jet-picture, this means:
There is a distinguished subset $L^2_{g,x_0}M \subset L^2_{x_0}M$ of those 2. order frames, which yield the standard 1. order development for g in x_0 . Moreover, the $G^2(4)$ action on $L^2_{x_0}M$ reduces to an action of $O(3,1) \otimes \{0\} \subset G^2(4)$ on $L^2_{g,x_0}M$.

Gluing together all fibres over M , we end up with the *prolongation* L^2M of LM , i.e. the *bundle of 2. order frames* over M and its sub*bundle* L^2_gM *of 2. order g-frames*, the prolongation of L_gM (cf. diag.(2.5)). We call an arbitrary $O(3,1)$ subbundle $\xi \subset L^2M$ *prolonged*, if there is a Lorentz metric g on M such that $\xi = L^2_gM$.

b. Prolongation of conformal structures: In complete analogy with the preceding constructions it is possible to treat a *conformal structure*, i.e. an equivalence class $[g]$ of Lorentz metrics, where $g' \overset{c}{\sim} g \;:\Longleftrightarrow\; g'(x) = \Omega(x)g(x)$ for positive Ω :

The usual coordinate representation of $[g]$ is given via the tensor density $[g]_{\mu\nu}(x) := -g_{\mu\nu}(x)\,(\det g_{\rho\sigma}(x))^{-1}$. Hence, in the discussion of Lorentz metrics, we have consequently to replace $g_{\mu\nu}$ by $[g]_{\mu\nu}$.

Then, in the results, replace

on the 1. order level:	$L_g M \subset LM$	by	$L_{[g]}M \subset LM$ and
	$O(3,1)$	by	$CO(3,1) := O(3,1) \circledS D$,
on the 2. order level:	$L^2_g M \subset L^2 M$	by	$L^2_{[g]}M \subset L^2 M$ and
	$O(1,3) \circledS \{0\}$	by	$CO(3,1) \circledS K^4$,

where $0 \in A^2$, and K^4 denotes the 4-dim. subgroup of A^2 given by $\{(\delta^\mu_\nu, \alpha^\mu_{\nu\rho} = \rho_\nu \delta^\mu_\rho + \delta^\mu_\nu \rho_\rho - \eta_{\nu\rho}\rho^\mu) \;/\; \rho_\sigma \text{ arbitrary}\}$. The occurrence of the dilatations D in the 1. order structure group is due to one degree of freedom of the conformal factor Ω in each $x \in M$, the additional occurrence of K^4 in the 2. order group due to 4 degrees of freedom in the gradient of Ω in x .

As a subset of LM the G-structure $L_{[g]}M$ of $[g]$ simply is the union of all $L_{g'}M$, $g' \in [g]$. Similarly for the prolongation $L^2_{[g]}M$ of $L_{[g]}M$, the *bundle of 2. order [g]-frames*.

Summarizing all structures and using the obvious natural projections from 2. to 1. order bundles we have

$$
\begin{array}{ccccc}
L^2_g M & \hookrightarrow & L^2_{[g]}M & \hookrightarrow & L^2 M \\
\updownarrow\!\!\!/ & & \downarrow & & \downarrow \\
L_g M & \hookrightarrow & L_{[g]}M & \hookrightarrow & LM
\end{array}
\quad \text{with structure groups} \quad
\begin{array}{ccccc}
O(3,1)\circledS\{0\} & \hookrightarrow & CO(3,1)\circledS K^4 & \hookrightarrow & G^2(4) \\
\updownarrow & & \downarrow & & \downarrow \\
O(3,1) & \hookrightarrow & CO(3,1) & \hookrightarrow & G^1(4)
\end{array}
\tag{2.5}
$$

Observe the different dimensionality between $L^2_{[g]}M$ and $L_{[g]}M$ in contrast to the case of $L^2_g M$ and $L_g M$, which after all is a consequence of algebraic properties of the 1. order structure groups $CO(3,1)$ and $O(3,1)$, respectively /4/ . Moreover, each $CO(3,1)$ subbundle in $L^2 M$ is 'prolonged', i.e. coincides with $L^2_{[g]}M$ for suitable $[g]$.

c. Symmetric connections and projective structures: From the foregoing it becomes clear, that a *symmetric linear connection* Γ is a 2. order object because of the 2. order transformation property of its coordinate representation $\Gamma^\rho_{\mu\nu}(x_0)$. Following the same lines of reasoning as in the metric case, we look for all 2. order frames $j^2_{x_0}(\varphi) \in L^2 M$, which yield a 'standard form' for Γ , i.e. $\Gamma^\rho_{\mu\nu}(x_0) = 0$. This determines $j^2_{x_0}(\varphi)$ up to linear transformations. Thus, we get the (2. order) G-structure $L^2_\Gamma M$, with $G = GL(4,\mathbb{R}) \circledS \{0\} \subset G^2(4)$.

As a simple implication of this construction we recognize: Given a Lorentz metric g and its Levi-Civita connection Γ_g, i.e. $g_{\mu\nu,\rho}(x_0) = 0 \iff {}_g\Gamma^{\rho}_{\mu\nu}(x_0) = 0$; then, in frame notation, this means for an arbitrary symmetric connection Γ

(2.6) $$\Gamma = \Gamma_g \iff L^2_g M \subset L^2_\Gamma M \quad .$$

In case g is physically interpreted as a space time metric, the elements of $L^2_g M$ are precisely the orthonormal local *inertial* coordinate systems.

A *projective structure* on M is an equivalence class $[\Gamma]$ of symmetric, linear connections which have the same autoparallels (as unparametrized curves). We get $\Gamma' \in [\Gamma]$ iff $\Gamma'^{\rho}_{\mu\nu}(x) = \Gamma^{\rho}_{\mu\nu}(x) - (\tau_\mu(x)\delta^\rho_\nu + \delta^\rho_\mu \tau_\nu(x))$ for suitable τ_μ. Similarly as for $L^2_{[g]}M = \bigcup L^2_{g'}M$, $g' \in [g]$, $[\Gamma]$ determines (and is determined by) the 2. order G-structure $L^2_{[\Gamma]}M = \bigcup L^2_{\Gamma'}M$, $\Gamma' \in [\Gamma]$, with structure group $GL(4,\mathbb{R}) \circledS P^4 \subset G^2(4)$, where $P^4 := \{(\delta^\mu_\nu, \tau_\mu\delta^\rho_\nu + \delta^\rho_\mu\tau_\nu) \,/\, \tau_\sigma \text{ arbitrary}\} \subset A^2$. Obviously,

(2.7) $$\Gamma' \in [\Gamma] \iff L^2_{\Gamma'}M \subset L^2_{[\Gamma]}M \quad .$$

<u>d. Weyl-structures:</u> There are various compatibility conditions between g , [g] on the one hand and Γ , $[\Gamma]$ on the other, which have a clear geometrical interpretation and may be of physical interest.
As a first example, consider a conformal structure [g] and a symmetric connection Γ such that Γ-parallel-transport preserves the [g]-light-cone-structure, i.e. Γ reduces to the subbundle $L_{[g]}M \subset LM$. We then call [g] and Γ *Weyl compatible* and the pair $([g],\Gamma)$ a *Weyl structure*. The 2. order formulation of this condition is straightforward

(2.8) $$[g]\,,\,\Gamma \text{ Weyl compatible} \iff L^2_{[g]}M \mathbin{\hat\cap} L^2_\Gamma M \neq \emptyset \quad .$$

(We write $A \mathbin{\hat\cap} B \neq \emptyset$ for principal subbundles $A , B \subset L^2M$ iff $A \cap B \neq \emptyset$ in each fibre of L^2M . Then $A \cap B$ is a principal subbundle, too, with intersection structure group; in (2.8) this group is CO(3,1) .)

The physical relevance of these notions (cf. sectión 2.2.) is due to the following well known geometrical fact: Γ is the Levi-Civita connection of a suitable Lorentz metric $g' \in [g]$ iff Γ-parallel transport of a (timelike) vector from $x \in M$ to $x' \in M$ along different paths always yields vectors of equal lengths. In this case g' is determined uniquely up to a constant scaling factor.

As an integrability condition this can be understood easily in 1. order bundle terms: Given a Weyl structure $([g],\Gamma)$, consider the reduction ω^Γ of Γ to $L_{[g]}M$ and its projection w^Γ (the classical *Weyl connection*) on the D bundle

(2.9) $$L_{[g]}M \,/\, O(3,1) \;=:\; W_{[g]}M \quad ,$$

with covariant derivative D^{Γ} . The integrability condition then is $D^{\Gamma}w^{\Gamma} = 0$. On a simply connected M this is equivalent to the existence of global w^{Γ}-compatible sections, i.e. to the existence of a Γ-compatible $O(3,1)$ bundle in $L_{[g]}M$ (given up to the action of an element $d \in D \subset CO(3,1)$).

Also, the prolongation property of an arbitrary $O(3,1)$ subbundle $\xi \subset L^2M$ can be interpreted within this context: Knowing ξ is equivalent to knowing

(i) its 1. order projection, the $O(3,1)$ bundle $\underline{\xi} \subset LM$ (i.e. a metric g_{ξ}) and

(ii) the $CO(3,1)$ bundle $\xi \cdot D \subset L^2M$ (or the corresponding Weyl structure ($[g_{\xi}]$, Γ_{ξ}) , hence the Weyl connection $w^{\Gamma_{\xi}}$ on $W_{[g_{\xi}]}M$).

As a section σ_{ξ} in $W_{[g_{\xi}]}M$, g_{ξ} determines an $\mathbb{R}_+$ valued equivariant function f_{ξ} on $W_{[g_{\xi}]}M$ ($f_{\xi}(\sigma_{\xi}(M)) = 1$). We get

(2.10) $$\xi \text{ prolonged} \iff D^{\Gamma_{\xi}} f_{\xi} = 0 \iff \Gamma_{\xi} \text{ reduces to } \underline{\xi} \quad .$$

2.2. Application: Weyl geometry

The conditions in (2.6),(2.7),(2.8) already indicated possibilities to express relations between geometrical (space time) properties in simple 2. order terms. We will proceed along these lines and reproduce a central step in the axiomatics of Ehlers, Pirani and Schild /5,6,7/, i.e. the derivation of a unique Weyl structure from a given conformal structure $[g]$ (interpreted as the causal or light cone structure of space time) and a projective structure $[\Gamma]$ (correlated with the unparametrized world lines of freely falling massive point particles):

Assuming the independent structures $[g]$, $[\Gamma]$ on M to be known, the authors of /5/ impose the following, physically motivated compatibility condition between $[g]$ and $[\Gamma]$: There is a neighborhood U_x for each $x \in M$ such that in U_x each (connected) unparametrized null $[g]$-geodesic (photon world line) through x can be arbitrarily approximated by suitable (connected) $[g]$-timelike $[\Gamma]$-autoparallels (massive particle world lines) through x .

This *null path compatibility* of $[g]$ and $[\Gamma]$ turns out to be a condition of 2. order contact. Hence we translate it into the L^2M-picture and get /7/

(2.11) $$[g]\,,\,[\Gamma] \text{ null path compatible} \iff L^2_{[g]}M \cap L^2_{[\Gamma]}M \neq \emptyset \quad .$$

The existence of a *Weyl structure* then is guaranteed by

Proposition Given null path compatible $[g]$, $[\Gamma]$.
Then there is exactly one linear connection $\Gamma' \in [\Gamma]$, such that $[g]$, Γ' constitute a Weyl structure.

For, the structure group of $L^2_{[g]}M \cap L^2_{[r]}M =: \xi$ in (2.11) is given by $CO(3,1) \circledS K^4 \cap GL(4,\mathbb{R}) \circledS P^4 = CO(3,1)$. The action of $GL(4,\mathbb{R}) \otimes \{0\} \subset G^2(4)$ on ξ then yields a $GL(4,\mathbb{R})$ subbundle in $L^2_{[r]}M$, i.e. a connection $\Gamma' \in [\Gamma]$ fulfilling the condition in (2.8). Uniqueness is shown similarly. □

Moreover, reversing the lines of reasoning leading to (2.10), if we assume the existence of an equivariant $\mathbb{R}_+$ valued function f on $W_{[g]}M$ such that $D^{\Gamma'}f = 0$, then Γ' in Proposition reduces to a *Lorentz metric* $g' \in [g]$, uniquely determined up to a scaling factor.
The example f = mass function (conformal weight -1) has been discussed in detail in /8/.

3. Affine conformal geometry

3.1. First and second order affine structures

In a sence, opposite to the prolongations from 1. to 2. order structures, the 'affine extensions' are a combination of 1. order objects with '0. order' transformations (translations). These extensions exist and are canonical for all 1. order G-structures. Applying standard techniques /4/, they can be generalized to the cases of 2. order Lorentz and conformal structures, i.e. to L^2_gM and $L^2_{[g]}M$:

a. Frame bundles: Consider first the bundle $AM = \cup A_xM$ of affine frames (p,e) , $p \in TM$, $e \in LM$, on M with natural right action of the affine group $GA(4,\mathbb{R}) = T^4 \circledS GL(4,\mathbb{R})$ [1)] and embedding $LM \hookrightarrow AM$, $e \longmapsto (0,e)$. Each Lorentz structure then induces a sub*bundle* $A_gM \subset AM$ *of affine g-frames*; similarly for a conformal structure $[g]$, which yields the sub*bundle* $A_{[g]}M = \cup A_{g'}M$, $g' \in [g]$, *of affine [g]-frames.*

The bundle $LM \overset{\alpha}{\times} GA(4,\mathbb{R})$, associated with LM through the natural left action of $GL(4,\mathbb{R})$ on $GA(4,\mathbb{R})$ and also interpreted as a principal $GA(4,\mathbb{R})$ bundle, is canonical isomorphic with AM . Correspondingly for A_gM and $A_{[g]}M$. We use the analogous constructions to define for given g and $[g]$

(i) the $T^4 \circledS O(3,1)$ *bundle of affine 2. order g-frames*
$A^2_gM := L^2_gM \overset{\alpha}{\times} (T^4 \circledS O(3,1))$, and

(ii) the $O(4,2)$ *bundle of affine 2. order [g]-frames*
$A^2_{[g]}M := L^2_{[g]}M \overset{\alpha}{\times} O(4,2)$
(where $\{1\} \otimes K^4 \subset CO(3,1) \circledS K^4$ is identical with the 4-dim. abelian subgroup of special conformal transformations in $O(4,2)$).

---

1) We adopt the notations of /9/; in particular, from now on we distinguish between holonomic (μ , ν ,..) and anholonomic indices (i,j,..). Also, given a G bundle P and a G action ϕ on F , we denote by $P \overset{\alpha}{\times} F$ the bundle ϕ -associated with P .

With obvious embeddings on the 2. order level, (2.5) then extends to [1)]

(3.1)

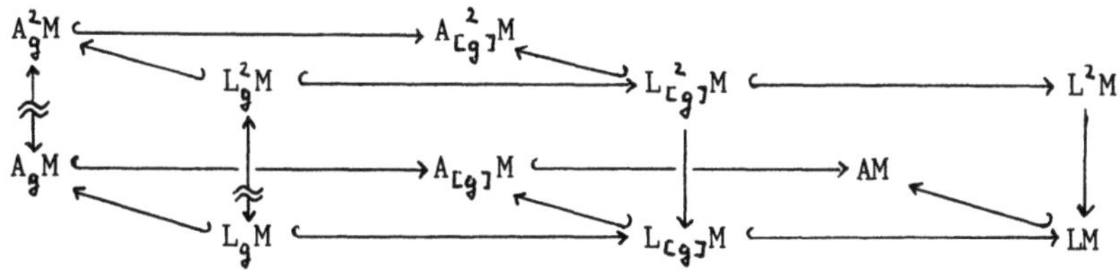

(3.2)

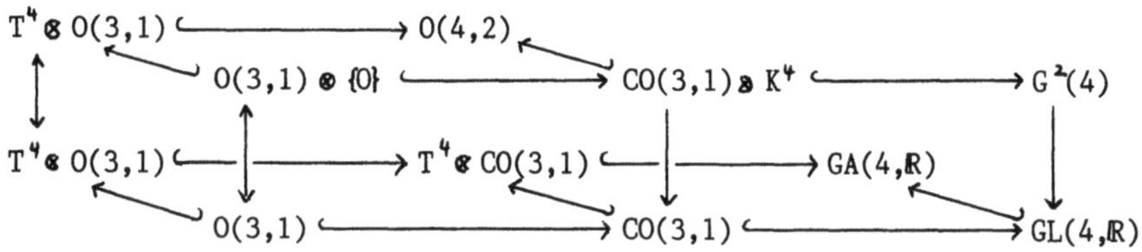

Observe: (i) Given M , g , $[g]$, all structures and mappings are canonical.
(ii) There is no natural 2. order structure for AM .
(iii) There is no projection $A^2_{[g]}M \longrightarrow A_{[g]}M$.

We use the adjective 'affine' for A^2_gM and $A^2_{[g]}M$ according to the following consideration:

A^2_gM : Take the standard action φ of $T^4 \circledS O(3,1)$ on *Minkowski space* M_o and the (vector) bundle $T^2_gM := A^2_gM \overset{\varphi}{\times} M_o$ φ-associated with A^2_gM . Then on each fibre of T^2_gM a flat Lorentz metric g_x is induced, i.e. $T^2_{g,x}M \approx M_o$. The elements of the fibre $A^2_{g,x}M \subset A^2_gM$ over x now can be interpreted as 2. order g_x-frames on the Lorentz manifold $T^2_{g,x}M$. Since $A^2_gM \approx A_gM$, T^2_gM is isomorphic with TM as an (affine) vector bundle.

$A^2_{[g]}M$: Similarly, take *conformal space* $\overline{M}_o := O(4,2)/CO(3,1) \circledS K^4$ (double covering of compactified Minkowski space $\approx S^1 \times S^3$) together with its usual conformal structure $[h]$ induced by the 'linear' isotropy subgroups ($\approx CO(3,1)$) of $O(4,2)$ [2)] . Denote by $\overline{\varphi}$ the $O(4,2)$ action on $\overline{M}_o$ and define the $\overline{\varphi}$-associated bundle $T^2_{[g]}M := A^2_{[g]}M \overset{\overline{\varphi}}{\times} \overline{M}_o$. Then each fibre $T^2_{[g],x}M$ carries an induced conformal structure $[g]_x$ such that $T^2_{[g],x}M \approx \overline{M}_o$. The elements of $A^2_{[g],x}M$ then are to be interpreted as 2. order $[g]_x$-frames on $T^2_{[g],x}M$. $T^2_{[g]}M$ is bundle isomorphic with a double covering of the compactification of a (Lorentz) tangent bundle TM .

---

1) We use a standard identification of $O(4,2)$ and its subgroups as isometry groups in $\mathbb{R}^6$ (metric $(g_{ab}) = \mathrm{diag}(-1,1,1,1,1,-1)$), i.e.
$O(4,1)$ is given by its fixpoint $e_6 = (0,0,0,0,0,1)$,
$O(3,2)$ " " $e_5 = (0,0,0,0,1,0)$,
$O(3,1)$ " fixpoints e_5 and e_6 ,
T " invariant plane $y^5 - y^6 = 1$ (fixpoint $e_5 + e_6$) ,
K " " $y^5 + y^6 = 1$ (" $e_5 - e_6$) ,
D " " $y^1 = y^2 = y^3 = y^4 = 0$.

2) I.e., by the differential in $T_x\overline{M}$ of the isotropy subgroup in each $x \in \overline{M}_o$.

b. Connections and Cartan connections: In diag. (3.1) we distinguish between (affine) 'A-bundles' and the canonically embedded 'L-bundles'. Choose now a connection form ω^A on one of these A-bundles (whose structure Lie algebra shall be denoted by $\mathfrak{t}^4 + \mathfrak{g}$). The $\mathfrak{t}^4+\mathfrak{g}$-valued restriction ω^L of ω^A to the corresponding L-bundle (structure algebra $\mathfrak{g}$) is called a *Cartan connection* /4/ and $\omega^{\mathfrak{g}}$ in the decomposition $\omega^L = \omega^{\mathfrak{t}} + \omega^{\mathfrak{g}}$ the *vertical part* of ω^L.

In the cases of $A_g M$, $A_{[g]}M$ and $A_g^2 M$ we find $\mathfrak{t}^4$ as an invariant subalgebra of $\mathfrak{t}^4 + \mathfrak{g}$. As a consequence, $\omega^{\mathfrak{g}}$ constitutes a connection form on the L-bundle. On the other hand, for $A_{[g]}^2 M$ ($\mathfrak{t}^4+\mathfrak{g} = \mathfrak{t}^4 +(\mathfrak{c}\,\mathfrak{o}(1,3)+\mathfrak{h}^4) = \mathfrak{o}(4,2)$), the subalgebra $\mathfrak{t}^4$ is not invariant, hence $\omega^{\mathfrak{g}}$ in general not a connection form on $L_{[g]}^2 M$.
This difference between connections and the vertical part of Cartan connections will be of importance in the following.

3.2. Application: Conformal gauge theory

Using the standard building blocks of classical Yang Mills theories (YMT), we discuss some geometrical 2. order aspects of an O(4,2) gauge theory of gravity (see also /10,11/), which should be understood as a 'conformal prolongation' of an affine version of Poincaré gauge theory (PGT); cf. /9/ for the Poincaré case.

The motivation for a concept like this may be associated with the extensions $L_g M \overset{\alpha}{\dashrightarrow} L_{[g]}M \overset{\beta}{\dashrightarrow} L_{[g]}^2 M \overset{\gamma}{\dashrightarrow} A_{[g]}^2 M$ in diag. (3.1), where

(α) reflects the attempt to derive space time properties such as (pseudo) Riemannian or Riemann-Cartan geometry from more 'primitive' structure elements. Following section 2.2., we consider conformal structures to be natural candidates for those elements,

(β) describes an 'unfolding' of the internal structure of conformal spaces, which is maximal in the prolongation sence: there are no higher than 2. order prolongations for a general conformal manifold (and which is trivial in the Lorentz case: equal structure groups for $L_g M$ and $L_g^2 M$),

(γ) corresponds to the necessity to incorporate translations, if one aims at an interpretation of the tetrad fields as potentials in YM sence, i.e. as connection coefficients.

a. Model spaces: To crystallize some characteristic geometrical features of conformal gauge theory, let us analyse shortly the group theoretical origin of classical gauge formalisms.

Usual constructions are based, at least implicitly, on a Lie group $S = E \otimes I$ which carries the information of all 'global' symmetries of the theory, I denoting the internal and E the external (space time) part of the group. A 'model' for space time then is given as a homogeneous space of E. Concerning the gauge scheme, the central structure element of the 'global' theory is the $\mathfrak{s}$ valued left invariant *canonical 1-form* Θ_S on S ($\Theta_S(X) = X$, $X \in \mathfrak{s} = T_{\mathbf{1}}S$ = Lie algebra of S).

We specialize first to the case E = Poicaré group, i.e. $S = (T^4 \otimes O(3,1)) \otimes I$, and indicate the immediate consequences of the existence of Θ_S .
Obviously, S has the structure of an $O(3,1) \otimes I$ bundle over Minkowski space $M_0 = T^4 = S/O(3,1)\otimes I$ (with metric g_0 on M_0 induced by the 'linear' isotropy group of S in $0 \in M_0$). Treating separately internal and external symmetries, we obtain the I bundle $V_i := T^4 \otimes I$ and the O(3,1) bundle $V_e := T^4 \otimes O(3,1)$ over $M_0 = T^4$; they serve as *model bundles* on which the (internal and external, resp.) *global* theory is formulated and which also induces the *local* structure of the general theory (see 3.2.b. for the conformal case).

The reductions $\Theta_{T^4 \otimes I}$ and $\Theta_{T^4 \otimes O(3,1)}$ of Θ_S to these model bundles induce flat connection forms: the *standard connection* ω^{i} (= i part of $\Theta_{T^4 \otimes I}$) on V_i , which yields the horizontal fibration of V_i into 'constant' sections $\sigma_i(M_0) = a \cdot T^4 \subset V_i$, $a \in I$, and the *Levi-Civita connection* $\omega^{o(3,1)}$ (= $o(3,1)$ part of $\Theta_{T^4 \otimes O(3,1)}$) on V_e giving the horizontal fibration of V_e into (holonomic) sections $\sigma_e(M_0) = b \cdot T^4 \subset V_e = L_{g_0} M_0$, $b \in O(3,1)$, the g_0-orthonormal or 'inertial' coordinate systems on M_0 .

In bundle terms, the central notion of *global gauge transformations* on V_i (V_e) then is given by those bundle automorphisms α which leave the canonical structure ω^{i} ($\omega^{o(3,1)}$) invariant. The group picture describes α via *left* multiplication in the group $V_i = T^4 \otimes I$ ($V_e = T^4 \otimes O(3,1)$).

The 'global' formalism now proceeds with the definition of globally gauge invariant free (matter) Lagrangians L_m on M_0 . For this, the 'coordinate systems' σ_i and σ_e on V_i , V_e are used. The *covariant* (i.e. section independent) expressions for L_m then are obtained via the replacements $\partial_\mu \longrightarrow D^{i}_\mu$ in the internal case (applying ω^{i}), and $\partial_\mu \longrightarrow D^{o(3,1)}_\mu$, $\delta^{i}_\mu \longrightarrow e^{i}_\mu$ in the external case (using $\omega^{o(3,1)}$ and the $\mathfrak{t}^4$ part (e^{i}_μ) of $\Theta_{T^4 \otimes O(3,1)}$). The occurrence of (e^{i}_μ) , the 'vierbein fields', in the second case is due to the non trivial product structure in $V_e = T^4 \otimes O(3,1)$, i.e. to the coupling of space time 'T -indices' μ with 'O(3,1)-indices' i .

We recapitulated these facts mainly to stress the following: The transformation properties of ω^{i} and $\omega^{o(3,1)}$ as *connection forms* stem from the *special* group structures of V_i and V_e (invariance of $\mathfrak{t}^4$ under $\mathrm{ad}_{V_i} I$ and $\mathrm{ad}_{V_e} O(3,1)$, resp.).

A characteristic difference arises, if we consider *conformal theory*, i.e. $E = O(4,2)$. We regard $O(4,2) =: V$ as a principal (model) bundle over conformal space $\overline{M}_0 = O(4,2)/CO(3,1) \circledS K^4$ (observe: $V = L^2_{[h]}\overline{M}_0$ as a $CO(3,1) \circledS K^4$ bundle). To find the natural coordinate systems φ in $\overline{M}_0$ or, equivalently, to find the (holonomic) sections σ in $L^2_{[h]}\overline{M}_0$ corresponding to σ_i and σ_e above, choose an arbitrary frame $\overset{2}{e} \in L^2_{[h]}\overline{M}_0$, located at $x \in \overline{M}_0$. Then there is a unique conformal embedding $\varphi_{\overset{2}{e}} : M_0 \longrightarrow \overline{M}_0$, $\varphi(0) = x$, of Minkowski space, such that $\overset{2}{e} = j^2_x(\varphi)$ as a 2-jet. Consequently, for each $\overset{2}{e} \in V$ there exists a unique (non global) holonomic section $\sigma_{\overset{2}{e}}$ through $\overset{2}{e}$. These *[h]-compatible coordinate systems* (and sections) obviously are the conformal generalizations of inertial coordinates in M_0 and are to be used for the coordinate dependent formulation of global conformal theory.
To pass to the covariant description, similarly as for V_e, one has to utilize
- (i) the $\mathfrak{co}(3,1) \oplus \mathfrak{h}^4$ part $\omega^{\mathfrak{co}(3,1)+\mathfrak{h}^4} = (\Theta^i_j, \tau, \tau^i)$ in $\mathfrak{o}(4,2) = \mathfrak{t}^4 + \mathfrak{o}(3,1) + \mathfrak{d} + \mathfrak{h}^4$ of the canonical 1-form $\Theta_{\mathfrak{o}(4,2)} = (\Theta^i, \Theta^i_j, \tau, \tau^i)$ (its kernel in $T_{\overset{2}{e}}V$ being tangent to $\sigma_{\overset{2}{e}}$ for each $\overset{2}{e} \in V$), and
- (ii) the translational part (Θ^i).

The crucial point, contrasting the cases of $V_i = T^4 \circledS I$ and $V_e = T^4 \circledS O(3,1)$, then is : $\omega^{\mathfrak{co}(3,1)+\mathfrak{h}^4}$ is the *vertical part of a Cartan connection* on V ; it fails to have the transformation property of a connection ($[\mathfrak{co}(3,1)+\mathfrak{h}^4, \mathfrak{t}^4] \not\subset \mathfrak{t}^4$). Consequently, in a conformal gauge theory modelled on $V = L^2_{[h]}\overline{M}_0$, one is confronted with a conflict: Either, one insists on the notion of global gauge invariance and hence has to replace connections by (vertical parts of) Cartan connections, unlike the usual YM scheme; or, one has to treat $L^2_{[h]}\overline{M}_0$ as an abstract $CO(3,1) \circledS K^4$ bundle and thus looses the geometrical background of the theory. Since we represent the geometrical viewpoint here, we reject the second possibility. Moreover, the first alternative turns out to be equivalent to the following concept:

To save both, geometry and YM analogy, the *affine extension* of $L_{[h]}\overline{M}_0$ to the bundle $A^2_{[h]}\overline{M}_0$ with structure group $O(4,2)$ seems natural. Then, $\Theta_{\mathfrak{o}(4,2)}$ extends to an $O(4,2)$ valued 1-form $\hat{\Theta}_{\mathfrak{o}(4,2)}$ on $A^2_{[h]}\overline{M}_0$, which
- is invariant under global gauge transformations (automorphisms of $A^2_{[h]}\overline{M}_0$ induced by holonomic automorphisms of $L^2_{[h]}\overline{M}_0$), and
- has the transformation character of a connection form.

As a consequence of this discussion, let us choose $A^2_{[h]}\bar{M}_0$, together with the natural (flat) connection form $\hat{\Theta}_{O(4,2)}$, as a *global model for conformal theory.*

b. Background structure: In a next step, the 'topological generalization', we replace $(\bar{M}_0, [h])$ by a conformal manifold $(M, [g])$ locally isomorphic to $(\bar{M}_0, [h])$ (which means flatness of $[g]$). Thus, the rigid *geometrical background* of our gauge approach finally is given by the O(4,2) bundle $A^2_{[g]}M$ ($\overset{loc}{\approx} A^2_{[h]}\bar{M}_0$ in the holonomic sence). This implies that all conformally invariant objects on $A^2_{[h]}\bar{M}_0$, as e.g. $\hat{\Theta}_{O(4,2)}$ or invariant Lagrangians, have well defined natural counterparts on $A^2_{[g]}M$. In particular, we find a unique (flat) connection $\hat{\Theta}$ on $A^2_{[g]}M$,which is invariant under all local conformal mappings of M (and which, together with a suitable Lagrangian, may be used for the covariant formulation of a 'free theory').

c. Anholonomic equivalence principle: According to the usual YM scheme we apply the 'generalized equivalence principle' (minimal coupling procedure) to introduce 'dynamical geometry' on $A^2_{[g]}M$. This means (cf. /9/) to pass from flat background $\hat{\Theta}$ to an arbitrary anholonomic [1] connection form $\hat{\omega} = (\hat{\omega}^i, \hat{\omega}^i_j, \hat{\varsigma}, \hat{\varsigma}^i)$ on $A^2_{[g]}M$ and to extend the Lagrangian formalism to the field $\hat{\omega}$. The 'potential' $\hat{\omega}$ and two symmetry breaking fields ϕ , σ_0 on $A^2_{[g]}M$ then are used to indicate an example for the derivation of a Riemann-Cartan structure on M :

d. Symmetry breaking: We introduce the notion of a *W-field* as an arbitrary $\mathbb{R}^6$-valued function ϕ on $A^2_{[g]}M$, equivariant with respect to the standard O(4,2) action on $\mathbb{R}^6$. Equivalently, ϕ is a global section in $A^2_{[g]}M \overset{a}{\times} \mathbb{R}^6$. In addition, we choose on $\mathbb{R}^6$ an O(4,2) *invariant potential* $U(y) := \alpha (y)^4 + \beta (y)^2$, $(y)^2 := y^a y_a := y^a g_{ab} y^b$, $\alpha > 0$; hence $U \circ \phi$ may be seen as a function on M [2]. Eventually, ϕ is assumed to yield minimal $(U \circ \phi)(x)$ for each $x \in M$.
Different values for β in U then imply:
$\beta > 0$ and minimal U fix $(y)^2 =: k < 0$ and for each $x \in M$ a ϕ orbit $O_{k,x}$ in $\mathbb{R}^6$ which meets $y_0 := \sqrt{-k}\, e_6$ (with de Sitter isotropy subgroup $H_0 := O(4,1)$).
$\beta < 0$ gives $(y)^2 =: k' > 0$ and $y_0 := \sqrt{k'}\, e_5$ (anti de Sitter isotropy group $H_0 := O(3,2)$).
$\beta = 0$ yields $(y)^2 = 0$ and $y_0 := \lambda (e_5 + e_6)$, $\lambda > 0$ (Poincaré isotropy group $H_0 := T^4 \otimes O(3,1)$).
Accordingly, we get an isotropy subbundle $\xi^0_\phi \subset A^2_{[g]}M$ with 10-dim. structure group $H_0 \subset O(4,2)$.

1) $\hat{\omega}$ is holonomic iff $\hat{\omega} = \hat{\Theta}$; i.e., similarly as in the case of PGT, one has to disregard the naturality of $\hat{\Theta}$ when applying the generalized equivalence principle.

2) which, of course, must be multiplied by a suitable horizontal invariant 4-form on $A^2_{[g]}M$ if to be considered as part of a Lagrangian.

The second symmetry breaking is motivated by a well known analogy in PGT /9/. We apply the *zero section* in $T^2_{[g]}M$ or the corresponding $\bar{M}_0$ valued equivariant function σ_0 on $A^2_{[g]}M$ (cf. sect. 3.1.a.) to remove the 'affine' degrees of freedom; i.e. we reduce to $L^2_{[g]}M \subset A^2_{[g]}M$. $H_c \cap CO(3,1) = O(3,1)$ in each case for β then implies: The W-field ϕ , together with the canonical section σ_0 , determines the *reduction to an O(3,1) bundle*

$$(3.3) \qquad \xi_\phi := \overset{\circ}{\xi}_\phi \cap L^2_{[g]}M$$

in $L^2_{[g]}M$. Thus, following sect. 2.1.d., we obtain a *Weyl structure* $([g], \Gamma_{\xi_\phi}) \mathrel{\hat{=}} \xi_\phi \cdot D \subset L^2_{[g]}M$ and a *Lorentz metric* $[g] \ni g_{\xi_\phi} \mathrel{\hat{=}} \underline{\xi}_\phi \subset L_{[g]}M$ (or the corresponding $\mathbb{R}_+$-function f_{ξ_ϕ} on $W_{[g]}M$). Metric compatibility between both structures means $D^{\Gamma_{\xi_\phi}} f_{\xi_\phi} = 0$ or ξ_ϕ to be prolonged. The different bundles $\underline{\xi}_\phi$ and ξ_ϕ (prolonged or not) may be thought responsible for the transformation rules of different (i.e. 1. and 2. order) types of physical fields, which come into consideration in a complete Lagrangian formalism on $L^2_{[g]}M$ or $A^2_{[g]}M$.

<u>e. Riemann-Cartan structure:</u> The reduction to $\xi_\phi \subset L^2_{[g]}M$ generates various fields originating from the connection form $\hat{\omega} = (\hat{\omega}^i, \hat{\omega}^i_j, \hat{\varsigma}, \hat{\varsigma}^i)$ on $A^2_{[g]}M$ (sect. 3.2.c.). Because of the complete reducibility of Ad O(3,1) in O(4,2) , the restriction of $\hat{\omega}^i_j$ to ξ_ϕ is a connection form, while the restrictions of $\hat{\omega}^i$, $\hat{\varsigma}$, $\hat{\varsigma}^i$ are tensorial forms on ξ_ϕ . In particular, considering the 1. order projections ω^i_j and ω^i on $L_{g'}M$, we return to the framework of PGT. Correspondingly, we assume ω^i to be regular and interprete it as a *deformation potential*, which determines a strong bundle automorphism $\tau : LM \longrightarrow LM$. Thus, finally, we arrive at a *Riemann-Cartan structure* $L_{\tilde{g}}M := \tau(L_{g'}M)$, $\tilde{\omega}^i_j := \tau_* \omega^i_j$, i.e., we get a Lorentz structure $\tilde{g}$ and a $\tilde{g}$-compatible (non symmetric) connection form $\tilde{\omega}^i_j$.

$(0 := [\mathbb{1}] \in O(4,2)/CO(3,1) \otimes K^4 = \bar{M}_0)$

(3.4)

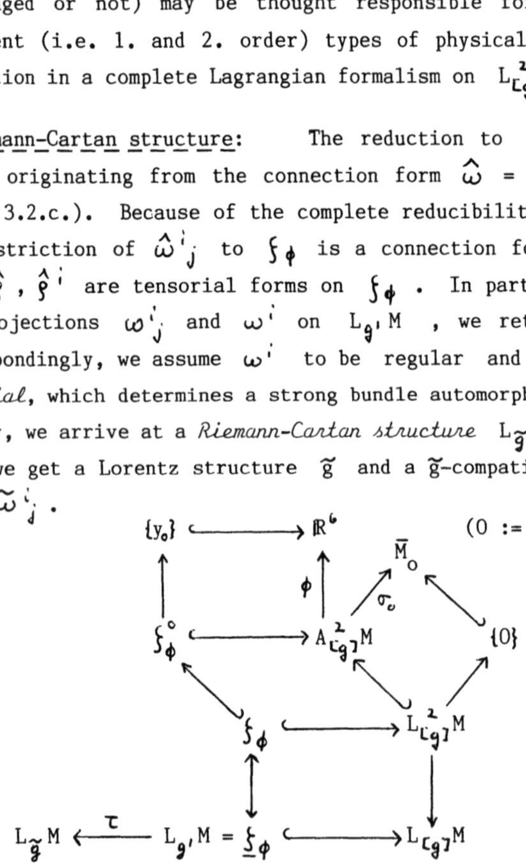

<u>f. Remarks:</u> The derivation of a Riemann-Cartan structure on M was sketched along the lines $A^2_{[g]}M \dashrightarrow \overset{\circ}{\xi}_\phi \dashrightarrow \xi_\phi \dashrightarrow$. Equivalently, one could start with a Cartan connection on $L^2_{[g]}M$ and perform the reduction to ξ_ϕ via a restricted function ϕ on $L^2_{[g]}M$. Again one observes the occurrence of a second type of connection form (Γ_{ξ_ϕ}) which originates from the interplay between bundles of different order and which gives the prolongation criterion for ξ_ϕ .

The analogy between the present concept and the affine version of PGT is stronger than we could indicate here. E.g., the idea of a 'shifting field' which connects the notions of 'diffeomorphism covariance' and 'translational gauge invariance' in PGT /9/ can be given a precise meaning in conformal theory, too. This again reflects our interpretation that, although the affine formalism on $A^2_{[g]}M$ only seems to be an extension of an equivalent $L^2_{[g]}M$ formalism, it may, however, give relevant insights into the geometry of conformal gauge theories (concerning YM analogy, incorporation of Cartan connections, the role of tetrad fields, generalization of equivalence principle, etc.).

References

/1/ Cartan, E., Ann.Sci.Ec.Norm.Sup. 21, 153 (1904); 22, 219 (1905); 25, 57 (1908)

/2/ Ehresmann, C., Comptes Rendues 233, 598, 777, 1081, (1951); 243, 1028, 1424, (1952)

/3/ Singer, I.M., Sternberg, S., J.Anal.Math. 15, 1 (1965)

/4/ Kobayashi, S., 'Transformation Groups in Differential Geometry', Springer, New York (1972)

/5/ Ehlers, J., Pirani, F.A.E., Schild, A., in 'General Relativity', Papers in Honour of J.L. Synge (ed. L. O'Raifeartaigh), Oxford (1972);

/6/ Coleman, R.A., Korte,H., J.Math.Phys. 22, 2598 (1981)

/7/ Hennig, J.D., in Lect.Notes in Phys. 139, Springer, Berlin (1981) (Proceedings, Clausthal 1978);
'G-Structures and Space Time Geometry II' (Clausthal 1978, unpublished)

/8/ Audretsch, J., Gähler, F., Straumann, N., Comm.Math.Phys. 95, 41 (1981)

/9/ Hennig, J.D., in Lect.Notes in Math. 1139, Springer, Berlin (1985) (Proceedings, Clausthal 1983)

/10/ Harnad, J.P., Pettitt, R.B., J.Math.Phys. 17, 1827 (1976)

/11/ Jadczyk, A.Z., in Lect.Notes in Math. 836, Springer, Berlin (1980) (Proceedings, Aix-en-Provence and Salamanca 1979)

THE CONFORMAL STRUCTURE OF EINSTEIN'S FIELD EQUATIONS

Helmut Friedrich
Universität der Bundeswehr Hamburg
Fachbereich Maschinenbau
P.O. Box 70 08 22
D-2000 Hamburg 70
F.R.G.

1. Introduction

Conformal techniques have played a role in the investigation of the structure and existence of solutions of Einstein's field equations in two quite different ways. They seem to have made their first appearance in this context in the analysis of the constraint equations which are implied by Einstein's equations on a space-like hypersurface. The method of solving the constrained equations by conformal methods has been extensively analysed and a discussion of its significance and a list of the relevant literature can be found in [3]. In this article I shall rather be concerned with the possibility to obtain information on the existence of global solutions of Einstein's field equations satisfying certain asymptotic conditions by exploiting the conformal structure of the full field equations.

In the early sixties various authors, among them notably Sachs [16, 17] and Bondi et al [2] were concerned with the question of what the behaviour of gravitational fields far away from their generating sources (which were thought of being confined to a spatially compact region) should be. In particular the problem whether and how notions like "incoming and outgoing radiation" etc. could be given a precise meaning for such space-times played a guiding role. On the basis of an analysis of a certain type of formal expansion of solutions of Einstein's field equations, into which entered a lot of physical intuition and knowledge about certain exact solutions and the behaviour of the solutions of the linearized equations, these authors were able to present a rather coherent picture of the situation they wanted to model. Shortly after it was shown by Penrose [14] that the complete picture could be derived in a very elegant way for space-times satisfying a few conditions on their global conformal structure.

In his characterization of a space-time $(\tilde{M},\tilde{g})$, with "physical metric" $\tilde{g}$ on the "physical manifold" $\tilde{M}$, as a space-time representing the field of a bounded gravitating source, it is required that the field be such that

- there exists a 3-dimensional manifold $\mathcal{J}$, which can be attached to $\tilde{M}$ in such a way that the manifold structure on $\tilde{M}$ extends smoothly to $M = \tilde{M} \cup \mathcal{J}$ to make the latter a smooth manifold with boundary $\mathcal{J}$,

- there exists a smooth function Ω on M with $\Omega > 0$ on $\tilde{M}$ and $\Omega \equiv 0$ but $d\Omega \neq 0$ on $\mathcal{J}$ such that the "non-physical metric"

$$g = \Omega^2 \tilde{g}$$

  extends to a smooth Lorentz metric on M.

A few further conditions have to be added here to obtain a satisfactory picture. For this and a detailed discussion of the meaning of the requirements above the reader may consult the literature [15, 11, 12]. Space-times with the properties indicated above are called "asymptotically simple". It turns out that the null geodesics for $\tilde{g}$, which as point sets coincide with the null geodesics of g on $\tilde{M}$, that approach a point on the surface $\mathcal{J}$ do this only by attaining infinite value of their affine parameter. Thus $\mathcal{J}$ represents a piece of infinity, called conformal infinity. Further information on the causal structure of conformal infinity can only be obtained by taking into account the field equations satisfied by the metric $\tilde{g}$. It is of importance here that the requirement, that the physical field allows a conformal extension as indicated above, incorporates a "fall-off condition" for the physical field $\tilde{g}$. The structures like Ω and $\mathcal{J}$, associated with the conformal structure of the field, do not belong to general relativity from the beginning. Einstein's field equations (with cosmological constant Λ)

$$\tilde{R}_{\mu\nu}[\tilde{g}] = \Lambda \cdot \tilde{g}_{\mu\nu} \qquad (1)$$

are designed to determine an <u>isometry</u> class of solutions from given initial data and are therefore not conformally invariant. Formally this is seen by replacing $\tilde{g}$ in (1) by $\Omega^{-2}g$ and expressing everything in terms of quantities derived from g and Ω. Then (1) is represented by the equivalent "conformal field equations"

$$R_{\mu\nu}[g] = \Omega^{-2}g_{\mu\nu}(\Lambda+3\nabla_\sigma\Omega\nabla^\sigma\Omega) - \Omega^{-1}(2\nabla_\mu\nabla_\nu\Omega+g_{\mu\nu}\nabla_\sigma\nabla^\sigma\Omega) \qquad (2)$$

which of course loose their meaning where Ω vanishes. Thus it is far from obvious that the fall-off conditions given above should be in harmony with the propagative properties of Einstein's field equations. On the other hand the conformal description of the asymptotic behaviour of the field was satisfied by various exact solutions of the field equations, it fits surprisingly smooth together with the type of formal expansion studied by Bondi, Sachs and others, and finally it allows to define physical notions like "radiation field" etc. in such an elegant and natural way that it is hard to believe that there should be something wrong with it.

2. The Regular Conformal Field Equations

To understand whether the characterization of the asymptotic behaviour of the fields in terms of their global conformal structure is adequate for solutions of Einstein's equations and why this should be so, I investigated the "conformal structure" of Einstein's equations, i.e. the structure of the equations (2), where the conformal factor is arbitrary and even allowed to vanish at some points. At the center of the analysis was the problem to specify, if possible, those properties of the conformal field equations which will enable one to say something about the set of solutions which possess the described asymptotic behaviour. The most obvious problem with equation (2) is the fact that the right hand terms become singular where Ω goes to zero. Multiplying by Ω^2 does not help since the principle part of the differential operator which acts on the metric g will then vanish where Ω vanishes. This difficulty is resolved by the following result [4]:

Regularity Theorem 1: *The conformal field equations (2) for the non-physical metric g and the conformal factor* Ω *can be represented by a system of first order quasilinear partial differential equations, the "regular conformal field equations", which is regular for all values of the conformal factor* Ω.

The regular conformal field equations constitute a set of field equations for the components of an orthonormal frame, for the connection coefficients, for Ω, $d\Omega$, $s = \frac{1}{4} \nabla_\mu \nabla^\mu \Omega$, for the traceless part of the Ricci-tensor and for the rescaled Weyl tensor. This set of equations, which comprises the trace-free part of (2), now read as an equation for $d\Omega$, can be derived from (2). That a regular system is obtained depends essentially on two facts. The vacuum Bianchi identities, read as an equation for the Weyl tensor, transform under conformal rescalings into a regular equation for the rescaled Weyl tensor. Moreover, from equation (2) can be derived an integrability condition which may be read as a field equation for s. By incorporating these equations into the regular system one is able to show:

Any solution of Einstein's field equations (1) supplies a solution of the regular conformal field equations. Conversely, any solution of the regular conformal field equations on a manifold M provides on the open submanifold $\tilde{M}$ of M where Ω is positive a solution of Einstein's field equations. The cosmological constant is obtained as a constant of integration, which may be fixed on a suitable initial surface.

Thus the regular conformal field equations generalize Einstein's equations slightly since they are defined and regular even where the conformal factor Ω vanishes.

Since the regular system is built up by using various integrability conditions, it is highly overdetermined and does at first sight resemble none of the systems to which the theory of partial differential equations would apply. To derive statements about the set of solutions of the regular system the following rèsult is important [4, 6].

Reduction Theorem 2: *Initial value problems for the regular conformal field equations can be reduced to initial value problems for symmetric hyperbolic systems of "reduced conformal field equations".*

In [6] it has been shown that in the regular conformal field equations can be singled out certain functions called "gauge source functions", which on the one hand can be given arbitrary functional form by a suitable choice of conformal factor, coordinate system and frame field, and which on the other hand determine the choice of gauge uniquely, if the gauge has been fixed on a suitable initial surface. If the gauge source functions are isolated in the field equations and considered as given, one finds that the regular conformal field equations imply a symmetric hyperbolic system of propagation equations for all unknowns occurring in the regular conformal field equations. It can be shown that a solution of these "reduced equations" for data which solve the constraint equations on a suitable initial surface provides in fact a solution of the complete system of regular conformal field equations. Since symmetric hyperbolic systems [10] are well understood and very general existence theorems for solutions of such systems are available [13], the regularity theorem and the reduction theorem may be used as starting point for deriving various existence theorems for asymptotically simple solutions of Einstein's field equations. The fact that the choice of gauge can be controlled in the reduced equations by the freedom in choosing the gauge source functions allow to adopt the form of the equations suitably to various interesting situations [7, 9].

Finally it may be pointed out that the results indicated above extend to the case of the coupled Yang-Mills-Einstein equations.

3. Asymptotically Simple Solutions of Einstein's Equations.

If one wants to use the results discussed above to derive existence theorems about asymptotically simple solutions of Einstein's equations, the kind of PDE problem which one has to deal with will depend on the sign of the cosmological constant, since this determines the causal structure of conformal infinity [14].

The case of negative cosmological constant, in which conformal infinity represents a timelike hypersurface has not been worked out yet. However, it may be pointed out here that the regular conformal field equations may be used here to prove the existence of asymptotically flat solutions by investigating a mixed problem where data are given on a spacelike hypersurface, which is thought as being a hypersurface in the physical space-time, and on a timelike surface which represents conformal infinity.

In the case of positive cosmological constant conformal infinity is space-like. The simply connected, geodesically complete, conformally flat standard example for this situation is provided by de-Sitter space-time, where conformal infinity consists of two components, which represent past respectively future null and timelike infinity. There are two results which suggest that being asymptotically simple or weakly asymptotically simple [15] (i.e. possessing "patches of conformal infinity" in the past and/or in the future) is a rather general property of solutions of Einstein's equations with positive cosmological constant [8, 9].

Theorem 3: *Let S be an orientable compact 3-dimensional manifold with Riemannian metric h and let d be a symmetric, covariant tensor field of valence two with vanishing divergence on (S,h). If these structures are sufficiently smooth then there exists a unique past asymptotically simple solution of Einstein's field equations (1) with* $\Lambda > 0$, *a given constant, which is diffeomorphic to* $S^3 \times \mathbb{R}$ *and such that past conformal infinity is diffeomorphic to S, the metric implied on past conformal infinity is conformal to h and the field d is up to a certain factor identified with a certain component of the rescaled Weyl tensor on past conformal infinity.*

This result assures the existence of a huge class of semi-global solutions of Einstein's equations with the desired behaviour at past null and timelike infinity. In fact the freedom to prescribe data on past conformal infinity is essentially the same as in the standard Cauchy problem for Einstein's field equations. However, the constraint equations implied by the regular conformal field equations on past conformal infinity are rather simple and it is easy to provide initial data sets for the reduced field equations.

If the initial data set S, h, d and Λ is given in such a generality as above, it will of course be difficult to gain information on the late time behaviour of the solutions. Therefore one may consider data which are "near" to the corresponding data for de-Sitter space-time. In that case S is diffeomorphic to S^3, the standard 2-sphere, the tensorfield $d = d_0$ vanishes identically on S and, after a suitable choice of the conformal factor, the metric $h = h_0$ coincides with the standard metric on the unit 3-sphere, and $\Lambda = \Lambda_0 = 3$.

Theorem 4: *There exists a neighbourhood of the de-Sitter data h_0, d_0 on S^3 such that for data h, d in this neighbourhood and for Λ sufficiently near to Λ_0 the solution of Einstein's field equations, whose existence has been asserted in theorem 3, is asymptotically simple in the past as well as in the future.*

Of course one has to (and can) make precise what the meaning of "near" and "neighbourhood" should be. Beside the fact that in the theorem is shown the existence of global solutions of Einstein's equations, two aspects of the result are of interest here. On the one hand it shows that the propagative properties of Einstein's equations are such that even fairly general data evolve into a space-time which satisfies the conditions of asymptotic simplicity. On the other hand it demonstrates that the conformal properties of the field equations which are indicated in theorems 1 and 2 may be used to prove semiglobal and in fact global existence theorems.

The most interesting case is of course that of Einstein's field equations with vanishing cosmological constant, where conformal infinity is lightlike. The standard example is provided by Minkowski space-time, where conformal infinity consists of two components, past and future null infinity, which may be thought of as set of past respectively future endpoints of null geodesics. The conformal extension can be constructed in this particular case in such a way that it contains three additional points i^-, i^+, i^0, representing in this order past and future timelike and spatial infinity. Past (future) null infinity is then the future (past) null cone of $i^-(i^+)$ and both hypersurfaces meet at i^o to form the null cone at i^o. In the conformal extension of more general space-times, in particular if sources are present, these additional points are not necessarily obtained.

Various initial value problems for the regular conformal field equations may be investigated in the ease of vanishing cosmological constant. The *"pure radiation problem"*. Here certain "free data" are prescribed on a cone whose vertex is thought of representing past timelike infinity while the cone itself is thought of representing past null infinity of the space-time to be determined from the data. Furthermore a certain completeness condition on the generators of the cone is required. A solution of this problem will represent pure gravitational radiation which comes in from infinity and interacts non-linearly with itself [7].

Theorem 5: *The "pure radiation problem" can be reduced to a characteristic initial value problem for a symmetric hyperbolic system. A solution of the pure radiation problem is determined uniquely by the free initial data (the "radiation field").*

Though this result gives evidence that the pure radiation problem is natural for Einstein's field equations, the technical performance of the existence proof is considerably aggravated by the non-smoothness of the initial surface at the vertex.

The *"hyperboloidal initial value problem"*. Here initial data are prescribed on a spacelike hypersurface with compact boundary, where the latter is thought of as the intersection of this hypersurface with (past or future) null infinity. Initial data pertaining to this situation will be called "hyperboloidal initial data". The name comes from the fact that surfaces of this type are provided by the spacelike unit hyperbolic in Minkowski space-time. In this particular case the corresponding data for the regular conformal field equations will be called "Minkowskian hyperboloidal initial data". One has the following result [5, 9].

Theorem 6: *For sufficiently smooth hyperboloidal initial data there exists a (up to questions of extensibility) unique solution of the regular conformal field equations. It provides a solution of Einstein's field equations $Ric(\tilde{g}) = 0$ which possesses a smooth "piece of conformal infinity". For hyperboloidal initial data which are sufficiently near to the Minkowskian hyperboloidal initial data the corresponding solution of the regular conformal vacuum field equations has a (past resp.*

future) Cauchy horizon whose generators converge to a point (i^+ resp. i^-) which represents (past resp. future) timelike infinity for the physical space-time, while the Cauchy horizon represents (past resp. future) null infinity.

While in the first part of the theorem no assumptions are made on the largeness of the initial data and consequently no information is given about the behaviour of the solutions far away from the initial surface, in the second part it is seen that in situations which do not deviate too much from the flat standard situation, the structure of (past resp. future) null infinity resembles that of Minkowski space-time.

The *"standard Cauchy problem"*. This corresponds to the "usual" Cauchy problem for Einstein's equations $\mathrm{Ric}(\tilde{g}) = 0$, where data are given on a spacelike hypersurface which is thought of as a Cauchy surface for the solution space-time.

It is of course possible to reproduce the "usual" existence theorem by using the regular conformal field equations. The really interesting problem, however, is, whether it is possible to specify a sufficiently general class of initial data, for which it can be shown that the solution space-time will possess a smooth past as well as future null infinity. The analysis is particularly complicated by the fact that in the interesting cases spatial infinity cannot be represented by a regular point in the conformal extension [1]. Whatever the final answer to this problem will be, I think the results obtained so far show clearly that the analysis of the conformal structure of the field equations is of considerable use in obtaining information about the global behaviour of their solutions.

The diligent reader will have noticed that the space restrictions led in this article to a style of presentation which saved him from all the gory details. For these he is asked to consult the quoted literature.

References

[1] Ashtekar, A.: Asymptotic structure of the gravitational field at spatial infinity. In: General relativity and gravitation, Vol. 2, Held, A. (ed.) New York: Plenum 1980.

[2] Bondi, H., von der Burg, M.G.J., Metzner, A.W.K.: Gravitational waves in general relativity. VII Waves from axi-symmetric isolated systems. Proc. Roy. Soc. Lond. A 269, 21-52 (1962).

[3] Choquet-Bruhat, Y., York, J.W.: The Cauchy problem. In: General relativity and gravitation, Vol. 1, Held, A. (ed.) New York: Plenum 1980.

[4] Friedrich, H.: The asymptotic characteristic initial value problem for Einstein's vacuum field equations as an initial value problem for a first order quasilinear symmetric hyperbolic system. Proc. Roy. Soc. Lond A 378, 401-421 (1981).

[5] Friedrich, H.: Cauchy problems for the conformal vacuum field equations in general relativity. Commun. Math. Phys. 91, 445-472 (1983).

[6] Friedrich, H.: On the hyperbolicity of Einstein's and other gauge field equations. Commun. Math. Phys. 100, 525-543 (1985).

[7] Friedrich, H.: On purely radiative space-times. Commun. Math. Phys. In print.

[8] Friedrich, H.: Existence and structure of past asymptotically simple solutions of Einstein's field equations with positive cosmological constant. Preprint, Hamburg 1985.

[9] Friedrich, H.: On the existence of global and semiglobal solutions of Einstein's field equations with smooth asymptotic structure. In preparation.

[10] Friedrichs, K.O.: Symmetric hyperbolic linear differential equations. Commun. Pure Appl. Math. 8, 345-392 (1954)

[11] Geroch, R.: Asymptotic structure of space-time. In: Esposito, F.R., Witten, L. (eds.) Asymptotic structure of space-time. New York: Plenum 1976.

[12] Hawking, S.W., Ellis, G.F.R.: The large scale structure of space-time. Cambridge: University Press 1973.

[13] Kato, T.: The Cauchy problem for quasi-linear symmetric hyperbolic systems. Arch. Ration. Mech. Anal. 58, 181-205 (1975).

[14] Penrose, R.: Zero rest-mass fields including gravitation: asymptotic behaviour. Proc. Roy. Soc. Lond. A 284, 159-203 (1965)

[15] Penrose, R.: Structure of space-time. In: DeWitt, C.M., Wheeler, J.A. (eds.), Battelle Rencontres. New York: Benjamin 1968.

[16] Sachs, R.K.: Gravitational waves in general relativity. VI The outgoing radiation condition. Proc. Roy. Soc. Lond. A 264, 309-338 (1961).

[17] Sachs, R.K.: Gravitational waves in general relativity. VII Waves in asymptotically flat space-time. Proc. Roy. Soc. Lond. A 270 103-126 (1962).

NONRELATIVISTIC CONFORMAL SYMMETRIES
AND
BARGMANN STRUCTURES

C. DUVAL
CPT-CNRS(*) and Université d'Aix-Marseille II
Luminy, Case 907
13288 MARSEILLE Cedex 9, FRANCE.

§1. INTRODUCTION.

▪ According to the principle of Galileian relativity, the "laws of physics" are "invariant" under the **Galilei group**

(1.1) $G_{10} := (SO(3) \times \mathbb{R}) \ltimes (\mathbb{R}^3 \times \mathbb{R}^3)$.

In special relativity, one simply replaces the Galilei group by the Poincaré group in the fundamental statements. Now some relativistic theories are actually characterized by a larger symmetry, namely the **conformal invariance** under O(4,2) - or SU(2,2). They are in fact associated to masslessness (Maxwell and Yang-Mills fields, classical massless spinning particles [43] or twistors). It is therefore tempting to address the question of a **nonrelativistic version of conformal symmetry.**

In 1972, Hagen [20] and Niederer [36] independently showed that the maximal kinematical invariance group of the free Schrödinger equation is actually larger than the Galilei group. That group has since been called the **Schrödinger group**

(1.2) $Sch_{12} := (SO(3) \times Sl(2,\mathbb{R})) \ltimes (\mathbb{R}^3 \times \mathbb{R}^3)$.

See e.g. [2,3,6,7,9-12,16,21,23,37-40] (non exhaustive bibliography). It has soon been realized that Sch_{12} has no direct relationship with O(4,2) [2,12,21,38]. The striking feature is that the Schrödinger group is indeed specific of **massive systems** (e.g. the Schrödinger field, classical (spinning) particles, etc...) rather than massless ones (e.g.

(*) Laboratoire propre du CNRS

the mass zero Galilei coadjoint orbits interpreted as symplectic models for classical "photons" [43,16]). On the other hand, the plain 15-dimensional contraction $\lim_{c\to\infty} O(4,2)$ seems to have nothing to do with symmetries of classical particle/field theory (see however [2] for a contraction procedure that involves a mass rescaling).

▪ Let us briefly sketch the historical introduction [36] of the maximal kinematical invariance group of the free Schrödinger equation

(1.3) $\quad S\Psi = 0; \qquad S := (2m)^{-1}\Delta + i\,\partial_t.$

Look for those local diffeomorphisms of spacetime $\mathbb{R}^4$, $a : x \mapsto x^*$ such that the transformed wave function Ψ^*

(1.4) $\quad \Psi^*(x^*) := F_a(x)\,\Psi(x)$

(where F_a is some complex-valued function needed in the context of unitary irreducible projective representations) is again a solution of (1.3). Infinitesimally, that amounts to finding the differential operators

(1.5) $\quad X := \mathbf{A}^K(x)\,\partial_K + B(x)\,\partial_t + C(x); \quad x = (\mathbf{x}^K,t); \; K = 1,2,3,$

that satisfy

(1.6) $\quad [\,S\,,X\,] = i\,\lambda\,S$

for some (real) function λ depending on X. The general solution of (1.5,6) is given by

(1.7) $\quad \mathbf{A}^K = \boldsymbol{\omega}^K{}_L\,\mathbf{x}^L + \boldsymbol{\beta}^K t + \boldsymbol{\gamma}^K + \alpha\, t\,\mathbf{x}^K + \chi\,\mathbf{x}^K$

(1.8) $\quad B \;=\; \alpha t^2 + 2\chi\, t + \varepsilon$

(1.9) $\quad C \;=\; im(-\boldsymbol{\beta}_K\,\mathbf{x}^K - \alpha/2\,\mathbf{x}_K\,\mathbf{x}^K + \theta) + 3/2(\alpha t + \chi)$

with $\boldsymbol{\omega} \in so(3)$ { $\boldsymbol{\omega}_{(KL)} = 0$ }, $\boldsymbol{\beta}$, $\boldsymbol{\gamma} \in \mathbb{R}^3$, α, χ, $\theta \in \mathbb{R}$ and $\lambda = \partial_t B$. Galilei transformations are generated by ($\boldsymbol{\omega}$, $\boldsymbol{\beta}$, $\boldsymbol{\gamma}$, ε) and χ (resp. α) generate dilatations (resp. inversions or expansions). They altogether span the Lie algebra sch_{12}, while θ gives rise to a central extension m_sch_{13} of sch_{12}. In contradistiction with the relativistic conformal group, time is dilated twice as much as space [20] ! Nevertheless, the occurence of inversions suggests a plausible analogy with conformal transformations of some "metric" space whose relationship with spacetime still remains to be clarified.

▪ The purpose of this article is to decipher in elementary terms the geometry underlying Sch_{12} (§3) and to discuss some of its aspects in symplectic mechanics (§4). To that end, we introduce in §5 a certain 5_dimensional extension of spacetime called a **Bargmann structure** (much in the spirit of Kaluza-Klein theory) that accomodates U(1)-central extensions of Sch_{12} arising in prequantized classical particle mechanics [43,16] and nonrelativistic quantum mechanics (§6).

Let us finally mention without further details miscellaneous applications of nonrelativistic conformal symmetries.

▪ By using the techniques developed in §5,6 it has been shown in [8] that the **harmonic oscillator** is "Bargmann-conformally related" to the free particle. The Feynman-Souriau kernel [44] for the harmonic oscillator has been recovered explicitely together with the correct Maslov index (see [37,26,41] for an alternative approach). These results extend to the time-dependent case (5.24), e.g. **Newtonian cosmology** [39].

▪ The **"virial group"** of canonical similitudes (dilatations) of the presymplectic evolution space of a test particle in a Coulomb potential shows up as a subgroup of Chr_{13} defined in (3.15). See [16].

▪ In integrating the Schrödinger equation for a **charge-Dirac_magnetic_monopole** system, Jackiw [25] emphasized the crucial $SO(3) \times SO(2,1)\uparrow$ symmetry of the problem (§4). See also [24] for an account on geometric quantization.

▪ The conformal invariance of nonrelativistic **spin wave equations** has been first elucidated in [20]. We revisit in §6 the spin_1/2 **Levy-Leblond equation** [32] in the light of our 5-dimensional setting [29,30].

▪ New **soliton** solutions of the nonlinear Schrödinger equation have been discovered [11] with the help of conformal symmetry properties.

▪ One of the major issues is the rising of a new geometry, the **chronoprojective geometry** [6,7] of classical spacetime viewed as a reduction of the O(5,2) conformal geometry [39].

Acknowledgements. We thank M. Perrin and J. Elhadad for a careful reading of the manuscript.

§2. GALILEI AND NEWTON-CARTAN SPACETIMES.

▪ Galilei spacetime. The (proper) Galilei group G_{10} is defined as the multiplicative group of all 5 x 5 matrices

$$(2.1) \qquad g = \begin{pmatrix} \mathbf{R} & \mathbf{v} & \mathbf{r} \\ \mathbf{0} & 1 & t \\ \mathbf{0} & 0 & 1 \end{pmatrix} \qquad (\mathbf{R} \in SO(3),\, \mathbf{v},\, \mathbf{r} \in \mathbb{R}^3,\, t \in \mathbb{R}).$$

The homogeneous Galilei subgroup $H_6 := SO(3) \underline{\times} \mathbb{R}^3$ is spanned by

$$(2.2) \qquad e = \begin{pmatrix} \mathbf{R} & \mathbf{v} \\ \mathbf{0} & 1 \end{pmatrix}.$$

Spacetime $G_{10}/H_6 \simeq \mathbb{R}^4$ is therefore parametrized by $(\mathbf{r},t)$ and $G_{10} \to \mathbb{R}^4$ is a principal H_6–fibre bundle, the bundle of Galilei frames of $\mathbb{R}^4$ [28]. Letting $e =: (e_1\, e_2\, e_3\, e_4)$, the H_6–invariant function of G_{10}

$$(2.3) \qquad \gamma := e_A \otimes e_B\, \delta^{AB} \qquad (A, B = 1,2,3)$$

descends to $\mathbb{R}^4$ as a symmetric 2-contravariant tensor with signature

$$(2.4) \qquad \mathrm{sign}(\gamma) = (+\,+\,+\,0).$$

It endows each fibre $\mathbb{R}^4{}_t := \mathbb{R}^3 \times \{t\}$ (space at time t) with a canonical Euclidian metric. The Maurer-Cartan 1-form $\Theta := g^{-1}dg$ of G_{10} (see (2.1))

$$(2.5) \qquad \Theta = \begin{pmatrix} \omega & \theta \\ 0 & 0 \end{pmatrix}$$

where

$$(2.6) \qquad \omega = \begin{pmatrix} \mathbf{R}^{-1}d\mathbf{R} & \mathbf{R}^{-1}d\mathbf{v} \\ \mathbf{0} & 0 \end{pmatrix}; \qquad \theta = \begin{pmatrix} \mathbf{R}^{-1}(d\mathbf{r} - \mathbf{v}\, dt) \\ dt \end{pmatrix}$$

is a privileged affine Cartan connection on our frame-bundle. Since we are in the reductive case, ω is a (flat) principal connection on $G_{10} \to \mathbb{R}^4$. If ∇ denotes the associated covariant derivative, we have

$$(2.7) \qquad \nabla \gamma = 0; \qquad \mathrm{Torsion}\, \nabla = 0.$$

From (2.6), it is clear that the time-component θ^4 of the soldering 1_form θ descends to $\mathbb{R}^4$ as the Galilei clock

(2.8) $\tau = dt$,

(2.9) $\ker \gamma = \mathbb{R}\,\tau$,

(2.10) $\nabla\tau = 0$.

Note that $\theta^1 \wedge \theta^2 \wedge \theta^3 \wedge \theta^4$ defines a canonical volume element on spacetime. We call $(\mathbb{R}^4, \gamma, \tau, \nabla)$, briefly $(\mathbb{R}^{3;1}, \nabla)$ the **flat affine Galilei spacetime**.

▪ Newton-Cartan spacetime. [13,46,47,22,28,15,19,34,35] We extend these definitions to the curved case to describe Newtonian gravitation in a purely covariant manner. Spacetime M is now assumed to be a smooth connected 4_dimensional manifold endowed with a (proper) **Galilei structure** (γ, τ)

(2.11) $\gamma = \gamma^{\alpha\beta}\, \partial_\alpha \otimes \partial_\beta\,, \qquad \gamma^{[\alpha\beta]} = 0,$

(2.12) $\tau = \tau_\alpha\, dx^\alpha, \qquad d\tau = 0, \qquad \ker \gamma = \mathbb{R}\,\tau,$

and a compatible symmetric linear connection ∇ (2.7,10), a **Galilei connection**. The clock τ (locally) defines the **absolute time-axis** $T := M/\ker(\tau)$. Contrary to general relativity, Galilei connections are not uniquely determined by the given Galilei structure. The degeneracy of the "metric" (γ, τ) forces us to treat the gravitational field ∇ rather independently (see below). Part of the ambiguity in Galilei connections is gauged out via the nontrivial constraint (curlfreeness of the Newtonian field)

(2.13) $R_\alpha{}^\beta{}_\gamma{}^\delta = R_\gamma{}^\delta{}_\alpha{}^\beta \qquad (R_\alpha{}^\beta{}_\gamma{}^\delta := R_{\alpha\sigma\gamma}{}^\delta\, \gamma^{\sigma\beta}).$

The number of independent components of the curvature R is therefore the same as in general relativity [28]. We call such connections **Newtonian connections** and $(M, \gamma, \tau, \nabla)$ a **Newton-Cartan spacetime**.

Let us finally recall the covariant expression of Newton (-Cartan) field equations

(2.14) $\mathrm{Ric} = (4\pi G\rho + \Lambda)\, \tau \otimes \tau$

where Ric denotes the Ricci tensor and G Newton's constant. The only sources of gravitation are mass density ρ and possibly the cosmological constant Λ.

§3. GALILEI AND SCHRÖDINGER SPACETIME AUTOMORPHISMS.

▪ In characterizing the automorphisms of a Newton-Cartan structure (M,γ,τ,∇) we would start with the nonrelativistic "isometries"

(3.1) $\quad \mathrm{Cor}(M,\gamma,\tau) := \{ a \in \mathrm{Diff}(M);\ a_* \gamma = \gamma;\ a^* \tau = \tau \}$.

Unfortunately, the "Coriolis" transformations form an infinite dimensional diffeological group [45]. If we then define [46]

(3.2) $\quad \mathrm{Gal}(M,\gamma,\tau,\nabla) := \mathrm{Cor}(M,\gamma,\tau) \cap \mathrm{Aff}(M,\nabla)$,

where $\mathrm{Aff}(M,\nabla)$ is the group of affine diffeomorphisms of (M,∇), then

(3.3) $\quad \mathrm{Gal}(\mathbb{R}^3;1,\nabla)_0 \simeq G_{10}$.

The neutral component of the group of automorphisms(*) of the flat Galilei structure is thus isomorphic to the Galilei group (2.1).

▪ If interested in preserving only the direction of the Galilei structure, we may define [21] the (pseudo)group of local "Leibnitz" diffeomorphisms

(3.4) $\quad \mathrm{Leib}(M,\gamma,\tau) := \{a \in \mathrm{Diff}_{loc}(M);\ a_* \gamma = \Omega^{-2}\gamma,\ \Omega \in C^\infty(M,\mathbb{R}^*)\}$

that automatically preserve the direction of τ : $(a^*\tau)\wedge\tau = 0$ - see (2.12). We again end up with an infinite-dimensional object of little physical interest. Since we wish to extend Newtonian affinities, the simplest idea that comes to mind is to look for projective Leibnitz transformations (that preserve the geodetic structure i.e. the geometry of free fall of (M,∇)). We thus define **"Chronoprojective" transformations** of a Newton-Cartan structure as

(3.5) $\quad \mathrm{Chr}(M,\gamma,\tau,\nabla) := \mathrm{Leib}(M,\gamma,\tau) \cap \mathrm{Proj}(M,\nabla)$.

Infinitesimally(**)

(3.6) $\quad \mathrm{chr}(M,\gamma,\tau,\nabla) := \{ X \in \Gamma(TM);\ L_X\gamma = \lambda\gamma;\ L_X\Gamma = \mathbf{1}\otimes\varphi + \varphi\otimes\mathbf{1};$

$\lambda \in C^\infty(M,\mathbb{R});\ \varphi \in \Gamma(T^*M) \}$.

It is then easy to prove [16] that

(3.7) $\quad L_X\tau = \mu\,\tau$

(*) i.e. the orientation preserving Galilei automorphisms.

(**) $\Gamma(TM) \equiv$ vector fields on M.

(3.8) $\lambda + \mu = \text{const.}$

(3.9) $2\,\varphi = d\mu = \mu'\,\tau \qquad (\mu' := d\mu/dt).$

In the case of $(\mathbb{R}^3;1,\nabla)$ we seek those vector fields $X = X^\alpha \partial_\alpha$ $(\alpha,\beta,\gamma = 1,\ldots,4)$ such that

(3.10) $L_X\, \gamma^{\alpha\beta} = -2\,\gamma^{\rho(\alpha}\,\partial_\rho X^{\beta)} = \lambda\,\gamma^{\alpha\beta} \quad (\gamma^{\alpha\beta} = \delta^\alpha{}_A\,\delta^\beta{}_B\,\delta^{AB};\ A, B = 1,2,3)$

(3.11) $L_X\,\tau_\alpha = \tau_\beta\,\partial_\alpha X^\beta = \mu\,\tau_\alpha \qquad (\tau_\alpha = \delta_\alpha{}^4)$

(3.12) $L_X\,\Gamma_{\alpha\beta}{}^\gamma = \partial_\alpha\partial_\beta X^\gamma = \mu'\,\delta^\gamma{}_{(\alpha}\tau_{\beta)}.$

The general solution of (3.10-12) is given by

(3.13) $X^A = \boldsymbol{\omega}^A{}_B\,\mathbf{x}^B + \boldsymbol{\beta}^A t + \boldsymbol{\gamma}^A + \alpha\, t\,\mathbf{x}^A + \chi\,\mathbf{x}^A$

(3.14) $X^4 = \alpha t^2 + \eta t + \varepsilon$

with $\boldsymbol{\omega} \in so(3)$; $\boldsymbol{\beta}$, $\boldsymbol{\gamma} \in \mathbb{R}^3$; $\alpha, \chi, \eta \in \mathbb{R}$ (compare (1.7,8)); $\mu = 2\alpha t + \eta$ and $\lambda + \mu = \eta - 2\chi$. Thus

(3.15) $chr_{13} := chr(\mathbb{R}^3;1,\nabla) \simeq (so(3) \times gl(2,\mathbb{R}))\ \underline{\times}\ (\mathbb{R}^3 \times \mathbb{R}^3)$

contains sch_{12} as a subalgebra (see (1.2)), also $sch_{12} = [chr_{13}, chr_{13}]$. Since chr_{13} acts projectively on the real line (time axis), we have called it the chronoprojective Lie algebra. We might as well have defined the **special chronoprojective (Schrödinger) Lie algebra** ($\lambda + \mu = 0$ in (3.8)) as

(3.16) $sch(M,\gamma,\tau,\nabla) := \{\, X \in \Gamma(TM);\ L_X\gamma = -\mu\,\gamma;\ L_X\tau = \mu\,\tau;$

$L_X\Gamma = \mathbf{1} \otimes \varphi + \varphi \otimes \mathbf{1};\ \mu \in C^\infty(M,\mathbb{R});\ \varphi \in \Gamma(T^*M)\,\}.$

That characterization of Newton-Cartan "conformal" automorphisms has however a cryptic geometrical status we will elucidate in §5.

■ Let us discuss some <u>global</u> aspects of the Schrödinger group Sch_{12} that may be consistently defined (up to a covering) as the subgroup of $Gl(5,\mathbb{R})$ [36]

(3.17) $$Sch_{12} := \left\{ \begin{pmatrix} \mathbf{A} & \mathbf{b} & \mathbf{c} \\ \mathbf{0} & d & e \\ \mathbf{0} & f & g \end{pmatrix};\ \mathbf{A} \in SO(3);\ \mathbf{b},\mathbf{c} \in \mathbb{R}^3;\ d,e,f,g \in \mathbb{R};\ dg - ef = 1 \right\}$$

acting projectively on $\mathbb{R}^4 \times \{1\}$ (hint: (3.5))

(3.18) $$\begin{pmatrix} \mathbf{r} \\ t \\ 1 \end{pmatrix} \mapsto \begin{pmatrix} \mathbf{A}\mathbf{r} + \mathbf{b}t + \mathbf{c} \\ dt + e \\ ft + g \end{pmatrix} \sim \begin{pmatrix} (\mathbf{A}\mathbf{r} + \mathbf{b}t + \mathbf{c})/(ft + g) \\ (dt + e)/(ft + g) \\ 1 \end{pmatrix}.$$

Let us then find out a Newton-Cartan manifold $(M, \gamma, \tau, \nabla)$ that would be a homogeneous space of Sch_{12} to be considered as the **typical Schrödinger spacetime** (locally $\mathbb{R}^{3;1}$). If Sch_8 denotes the stabilizer of the origin in $\mathbb{R}^4$ ($\mathbf{c}=0$, e=0 in (3.17,18)), then

(3.19) $M := Sch_{12} / Sch_8 \simeq (\mathbb{R}^3 \times S^1) / \mathbb{Z}_2$.

is a rank-3 vector bundle over $\mathbb{P}_1(\mathbb{R})$ [16]. Only time is compactified.

(fig.1)

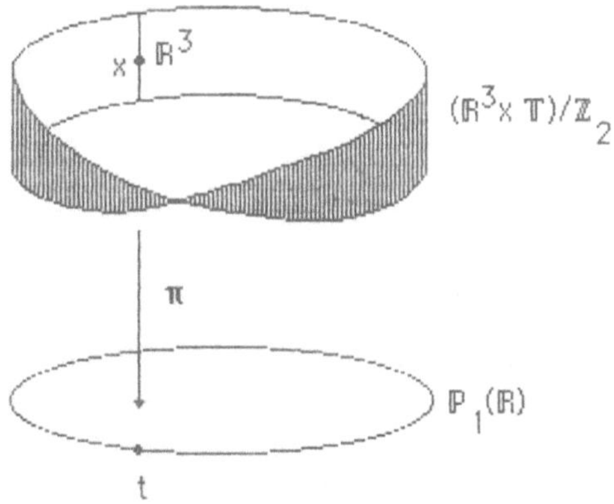

The Newton-Cartan structure on M is canonically defined by the principal H'_6-fibre bundle $Sch_{10} \to M \simeq Sch_{10}/H'_6$(*). Also $M \simeq \mathbb{P}_4(\mathbb{R}) \setminus \mathbb{P}_2(\mathbb{R})$ where $\mathbb{P}_2(\mathbb{R}) \simeq ((\mathbb{R}^3 \setminus \{0\}) \times \{0\}) / \mathbb{R}^*$.

Because H'_6 and $\mathbb{R}^4$ are in a reductive position in Sch_{10}, the Maurer-Cartan 1-form of Sch_{10}

(3.20)
$$\begin{pmatrix} \boldsymbol{\omega} & \boldsymbol{\omega}_4 & \boldsymbol{\theta} \\ \mathbf{0} & 0 & \theta^4 \\ \mathbf{0} & -\theta^4 & 0 \end{pmatrix}$$

serves to define the Schrödinger clock

(3.21) $\tau = \theta^4$,

and the Schrödinger connection

(3.22)
$$\omega = \begin{pmatrix} \boldsymbol{\omega} & \boldsymbol{\omega}_4 \\ \mathbf{0} & 0 \end{pmatrix}$$

where $\boldsymbol{\omega} = \mathbf{A}^{-1} d\mathbf{A}$; $\boldsymbol{\omega}_4 = \mathbf{A}^{-1}(d\mathbf{b} + \mathbf{c}\tau)$. The "space" metric is

(*) $Sch_{10} := (SO(3) \times SO(2)) \underline{\times} (\mathbb{R}^3 \times \mathbb{R}^3)$, $H'_6 := Sch_{10} \cap Sch_8 \simeq (SO(3) \times \mathbb{Z}_2) \underline{\times} \mathbb{R}^3$).

(3.23) $\gamma = \delta^{AB}\, \partial/\partial c^A \otimes \partial/\partial c^B$.

Schrödinger spacetime (3.19) is nonflat ! Its curvature 2-form is given by

(3.24) $\Omega^A{}_B = 0\,;\quad \Omega^A{}_4 = \theta^A \wedge \theta^4$ (A,B = 1,2,3),

whence

(3.25) $\mathrm{Ric} = 3\,\tau \otimes \tau$

and $(M, \gamma, \tau, \nabla)$ is a topologically nontrivial solution of Newton's field equations (2.14) corresponding to a vacuum with unit reduced cosmological constant ($\Lambda/3 = 1$).

▪ It is shown in [7,8,16] that the nonrelativistic **chronoprojective Weyl curvature** is

(3.26) $C_{\alpha\beta\gamma}{}^{\lambda} := R_{\alpha\beta\gamma}{}^{\lambda} - 2/3\, \delta^{\lambda}{}_{[\alpha} R_{\beta]\gamma}$

together with the consistency relation associated with the Ricci tensor

(3.27) $R_{\alpha\beta} = f\, \tau_\alpha \tau_\beta$ $(f \in C^\infty(M, \mathbb{R}))$.

We claim that Newton's field equations (2.14) can thus be interpreted as a necessary condition for a Schrödinger (chronoprojective) structure to exist on a Newton-Cartan spacetime. No general relativistic analogue ! In addition to Schrödinger spacetime (3.19), other examples of **chronoprojectively flat** manifolds ($C = 0$) are discussed in [7,8,16] (e.g. the homogeneous and isotropic **Newtonian cosmological spacetime**).

§4. GROUP COHOMOLOGY AND SCHRÖDINGER INVARIANT DYNAMICAL SYSTEMS.

▪ Let (M, σ) be a (pre)symplectic manifold with a canonical action of some Lie group G (i.e. $a^*\sigma = \sigma$, all $a \in G$). If $(g, [\ ,\ ])$ is the Lie algebra of G and if $\underline{Z}$ denotes the standard vector field of M associated with $Z \in g$, then <u>locally</u>

(4.1) $\sigma(\,\underline{Z}\,) = -\, d(\mu.Z)$ for some $\mu \in C^\infty(M, g^*)$.

In the <u>global</u> case μ is called the **momentum mapping** [43] of (M, σ, G) - strong (pre)symplectic action - and there exits a function $\theta : G \to g^*$

(4.2) $\theta(a) := \mu(a(x)) - ad^*(a).\mu(x)$ $(a \in G;\ x \in M)$

where ad^* denotes the coadjoint representation of G on g^*. We then have

(4.3) $\delta\theta(a, a') := ad^*(a).\theta(a') + \theta(a) - \theta(aa') = 0$ (all $a, a' \in G$),

(4.4) $\sigma(\underline{Z}, \underline{Z}') = \mu.[Z, Z'] + f(Z)(Z')$ (all $Z, Z' \in g$),

(4.5) $f := D(\theta)(e) \in \Lambda^2 g^*$.

Now δ gives rise to the **symplectic cohomology** denoted by $H^1(G,g^*)$ where 1-cocycles $Z^1(G,g^*)$ are defined by (4.3,5) and coboundaries by

(4.6) $[\theta \in B^1(G,g^*)] \Leftrightarrow [\ \theta(a) = \delta\mu(a) := ad^*(a).\mu - \mu;\ \mu \in g^*\ ].$

▪ **Galilei symplectic cohomology**. Denote any vector in g_{10} by

(4.7) $$Z = \begin{pmatrix} j(\boldsymbol{\omega}) & \boldsymbol{\beta} & \boldsymbol{\gamma} \\ \mathbf{0} & 0 & \varepsilon \\ \mathbf{0} & 0 & 0 \end{pmatrix} \qquad (\boldsymbol{\omega}, \boldsymbol{\beta}, \boldsymbol{\gamma} \in \mathbb{R}^3, \varepsilon \in \mathbb{R})$$

where $j(\boldsymbol{\omega})(\mathbf{v}) := \boldsymbol{\omega} \times \mathbf{v}$. The pairing between g_{10} and $g^*_{10} \ni \mu := \{\mathbf{l}, \mathbf{q}, \mathbf{p}, E\}$ is defined by

(4.8) $\mu.Z := < \mathbf{l}, \boldsymbol{\omega} > - < \mathbf{q}, \boldsymbol{\beta} > + < \mathbf{p}, \boldsymbol{\gamma} > - E\,\varepsilon.$

The symplectic cohomology of the Galilei group G_{10} is 1-dimensional : every non-trivial cocycle θ is of the form (see (2.1))

(4.9) $\theta = m\,\theta_1\,;\ \theta_1(a) = \{\mathbf{r} \times \mathbf{v}, \mathbf{r} - \mathbf{v}t, \mathbf{v}, v^2/2\}\quad (a \in G_{10}, m \in \mathbb{R}),$

the coefficient m which labels a class in $H^1(G_{10},g^*_{10})$ will be later on interpreted as the **mass**. The derivative (4.5) of θ at 1 is

(4.10) $f(Z,Z') = m\,(< \boldsymbol{\beta}, \boldsymbol{\gamma}' > - < \boldsymbol{\beta}', \boldsymbol{\gamma} >).$

▪ **Schrödinger symplectic cohomology**. The Lie algebra sch_{12} (3.17) is spanned by the 5 x 5 matrices

(4.11) $$Z = \begin{pmatrix} j(\boldsymbol{\omega}) & \boldsymbol{\beta} & \boldsymbol{\gamma} \\ \mathbf{0} & \chi & \varepsilon \\ \mathbf{0} & -\alpha & -\chi \end{pmatrix} \qquad (\boldsymbol{\omega}, \boldsymbol{\beta}, \boldsymbol{\gamma} \in \mathbb{R}^3, \alpha, \chi, \varepsilon \in \mathbb{R}).$$

The pairing between sch_{12} and $sch^*_{12} \ni \mu := \{\mathbf{l}, \mathbf{q}, \mathbf{p}, E, K, D\}$ is defined by

(4.12) $\mu.Z := < \mathbf{l}, \boldsymbol{\omega} > - < \mathbf{q}, \boldsymbol{\beta} > + < \mathbf{p}, \boldsymbol{\gamma} > - E\,\varepsilon - K\,\alpha + D\,\chi.$

with the physical interpretation : $\mathbf{l}$:= angular momentum, $\mathbf{q}$:= centre of mass, $\mathbf{p}$:= linear momentum, E:= energy, K:= inversion momentum, D:= dilatation momentum. A tedious calculation [16] then shows that

(4.13) $\dim(H^1(Sch_{12}, sch^*_{12})) = 1$

just as in the Galilei case and

(4.14) $\theta \in H^1(Sch_{12}, sch^*_{12}) \Leftrightarrow [\theta = m\,\theta_1\,;\ m \in \mathbb{R}],$

(4.15) $\theta_1(a) = \{\mathbf{c \times b},\ \mathbf{cd - be},\ \mathbf{bg - cf},\ \|\mathbf{bg - cf}\|^2/2, \|\mathbf{cd - be}\|^2/2, <\mathbf{bg - cf}, \mathbf{cd - be}>\}$

(all $a \in Sch_{12}$). Note that the derivative f of θ at 1 is again given by (4.10,11).

- The barycentric decomposition [43].

Let (M,σ) be a connected symplectic manifold with a strong symplectic action of some Lie group G. Suppose that G' be a closed abelian invariant subgroup of G with Lie algebra $g' \subset g$. If θ defines a nontrivial class in $H^1(G,g^*)$, the induced 2-form $\sigma' := f|g'$ depends on θ only and G acts symplectically on (g',σ'). If μ' is the induced momentum mapping of G', then μ' is a submersion $M \to g'^*$ if $\ker(\sigma') = \{0\}$ - i.e. if (g',σ') is a symplectic vector space as will be assumed from now on. Then (M,σ) is symplectomorphic to the **direct symplectic product** $(g',\sigma') \times (M'',\sigma'')$ where $\iota: M'' := \{x \in M; \mu'(x) = 0\} \to M$ is an embedding and $\sigma'' := \iota^*\sigma$. If $\theta' := \theta|g'$, then $G'' := (\theta')^{-1}(\{0\})$ acts canonically on (M'',σ'').

In the Galilei case, $g' = \mathbb{R}^3 \times \mathbb{R}^3$ (with the symplectic 2-form σ' given by (4.10,7)) represents the space of **centre of mass motions** of a dynamical system of total **mass** $m > 0$; M" is then interpreted as the space of orbital motions with dynamical group $G'' = SO(3) \times \mathbb{R}$. Elementary massive galilean dynamical systems are associated with coadjoint orbits of G", namely $(S^2, s\,\mathrm{surf}) \times \{E_0\}$ where $s \in \mathbb{R}^+$ is the **spin** and $E_0 \in \mathbb{R}$ the **internal energy.**

In the Schrödinger case, the situation is almost the same as before. If the total mass m of a Sch_{12}-invariant dynamical system (M,σ) is nonzero, it defines a class in $H^1(Sch_{12}, sch^*_{12})$ and M splits up into the direct symplectic product of $(g' = \mathbb{R}^6, \sigma')$ - centre of mass motions - where σ' is given by (4.10,11) and some symplectic manifold (M'',σ'') representing orbital motions. Again, massive elementary Schrödinger dynamical systems are characterized by the fact that M" be a coadjoint orbit of the semi-simple Lie group $G'' = SO(3) \times SL(2,\mathbb{R})$, symplectomorphic to $(S^2, s\,\mathrm{surf}) \times (H^2, c\,\mathrm{surf})$ where the Casimir invariant s is still interpreted as the spin and H^2 is a sheet of a certain hyperboloid in $\mathbb{R}^{2,1}$. Unfortunately enough, neither the Casimir number c, nor the internal phase space H^2 seem to retain a clearcut physical interpretation. This point has already been emphasized by Perroud [40] in the context of representation theory. If we insist on the additional **Schrödinger invariance** of Galilei massive elementary dynamical systems, then necessarily

(4.16) $\quad E_0 = 0,$

corresponding to the trivial $SL(2,\mathbb{R})$ orbit $c = 0$. **No internal energy** ! The overall additive constant in the definition of galileian energy is killed by assuming Schrödinger invariance.

▪ An example : the Dirac charge-monopole system.

One can describe the classical motions of a charge-monopole system $\{q,q^*\}$ interacting via the magnetic field $\mathbf{B} := qq^* \mathbf{r}\, r^{-3}$ ($\mathbf{r} := \mathbf{r}_1 - \mathbf{r}_2$, $r := \|\mathbf{r}\|$) by the presymplectic 2-form of $[\mathbb{R}^6 \times \{(\mathbb{R}^3 \setminus \{0\}) \times \mathbb{R}^3\}] \times \mathbb{R}$

(4.17) $\quad \sigma := \sigma_{bar} + \sigma_{orb}$

(4.18) $\quad \sigma_{bar} := M < d\mathbf{V}_{\wedge}(d\mathbf{R} - \mathbf{V}\, dt)>$

(4.19) $\quad \sigma_{orb} := m <d\mathbf{v}_{\wedge}(d\mathbf{r} - \mathbf{v}\, dt)> + qq^*\, \mathrm{surf}$

with $\mathbf{R} := (m_1\mathbf{r}_1 + m_2\mathbf{r}_2)/M$; $M := m_1 + m_2$; $m := m_1 m_2 /M$; "surf" denotes the canonical surface element of the unit sphere $S^2 \subset \mathbb{R}^3$, i.e.

(4.20) $\quad \mathrm{surf} = 1/2\, r^{-3} < \mathbf{r}\,,\, d\mathbf{r} \times d\mathbf{r} >.$

The foliation $\ker(\sigma)$ then yields the familiar equations of motion. It has been shown [16] that the infinitesimal action of Sch_{12} ((3.13,14) with $\eta = 2\,\chi$) can be lifted to the evolution space $\mathbb{R}^3 \times \mathbb{R}^3 \times \mathbb{R} \ni (\mathbf{R}, \mathbf{V}, t)$ according to

(4.21) $\quad Z := (\omega^A{}_B\, R^B + \beta^A t + \gamma^A + \alpha t R^A + \chi R^A)\, \partial/\partial R^A + (\alpha t^2 + 2\chi t + \varepsilon)\partial/\partial t$

$$+ (\omega^A{}_B\, V^B + \beta^A + \alpha(R^A - V^A t) - \chi V^A)\, \partial/\partial V^A$$

in such a way that

(4.22) $\quad L_Z\, \sigma_{bar} = 0,$

i.e. that Z be an infinitesimal (pre)symplectomorphism of the barycentric evolution space (free particle). If we then look for those vector fields Z that Lie-transport the orbital presymplectic structure σ_{orb}, we end up with the nontrivial symmetry $so(3) \times sl(2,\mathbb{R})$ ($\beta = \gamma = 0$) whose momentum mapping (4.1) reads (compare [25])

(4.23) $\quad \mathbf{l} = m\, \mathbf{r} \times \mathbf{v} - qq^*\, \mathbf{r}/r$

(4.24) $\quad E = m\, \|\mathbf{v}\|^2/2$

(4.25) $\quad D = m <\mathbf{v}, \mathbf{r} - \mathbf{v}t>$

(4.26) $\quad K = m\, \|\mathbf{r} - \mathbf{v}t\|^2/2.$

§5. BARGMANN STRUCTURES AND RELATED CONFORMAL AUTOMORPHISMS.

We deal here with a new setting that incorporates Newton-Cartan structures and allows for a geometrically transparent definition of "conformal" nonrelativistic symmetries.

▪ A **Bargmann manifold** [18] is a principal $(\mathbb{R},+)$ bundle $\pi:\tilde{M}\to M$ over a 4-dimensional smooth connected manifold M (spacetime) such that

(5.1) $\tilde{M}$ is endowed with a Lorentz metric g of signature $(++++-)$,

the group generator ξ satisfies

(5.2) $g(\xi,\xi)=0; \qquad \tilde{\nabla}\xi=0$

where $\tilde{\nabla}$ denotes the Levi-Cività connection of $(\tilde{M},g)$. The difference with standard Kaluza-Klein theory is threefold : the principal fibration is assumed to be <u>null</u> rather than space-like, <u>non-compact</u> (no elementary "mass" for the time being) and <u>parallel</u> rather than merely isometric.

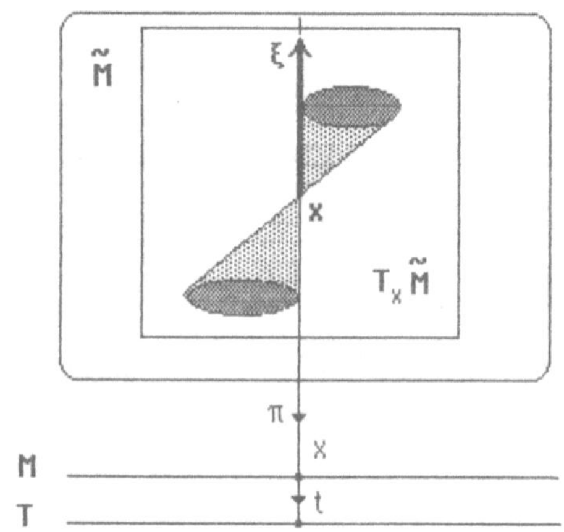

(fig.2)

The 1-form $\tilde{\tau} := g(\xi)$ is basic and closed, hence

(5.3) $\tilde{\tau}=\pi^*\tau\,; \qquad d\tau=0.$

Since ξ is in particular an isometry, the 2-contravariant symmetric tensor

(5.4) $\gamma := \pi_* g^{-1}$

descends to M and (γ,τ) satisfies (2.11,12). Algebraic inspection shows that the signature of γ is as in (2.4), hence (M,γ,τ) is a <u>Galilei structure</u>. This point of view has independently been espoused in [48].

Given two infinitesimal automorphisms $\tilde{X}, \tilde{Y}$ of $\pi : \tilde{M} \to M$(*),

(5.5) $\qquad \nabla_X Y := \pi_* \tilde{\nabla}_{\tilde{X}} \tilde{Y}$

is a well defined vector field of M, depending only on the projections X, Y of $\tilde{X}, \tilde{Y}$. Moreover ∇ is a Galilei connection, in fact a Newtonian connection on (M, γ, τ) - hint: the curvature $\tilde{R}$ of $\tilde{\nabla}$ trivially satisfies (2.13) where indices are now raised by g^{-1}. We have thus associated to our Bargmann structure a **unique Newton-Cartan structure**.

▪ **Bargmann automorphisms** consist in those isometries of $(\tilde{M}, g)$ that are also automorphisms of the principal bundle $\pi : \tilde{M} \to M$, i.e.

(5.6) $\qquad \mathrm{Barg}(\tilde{M}, g, \xi) := \mathrm{Isom}(\tilde{M}, g) \cap \mathrm{Aut}(\tilde{M}, \xi).$

Now $\mathrm{Aut}(\tilde{M}, \xi)$ extends Diff(M) and because isometric vertical automorphisms reduce to $(\mathbb{R},+)$, $\mathrm{Barg}(\tilde{M}, g, \xi)$ is clearly a $(\mathbb{R},+)$ central extension of $\mathrm{Gal}(M, \gamma, \tau, \nabla)$ (see (3.2)) as expressed by the exact sequence

(5.7) $\qquad 1 \to \mathbb{R} \to \mathrm{Barg}(\tilde{M}, g, \xi) \to \mathrm{Gal}(M, \gamma, \tau, \nabla) \to 1.$

▪ Consider $\tilde{M} = \mathbb{R}^5$ with

(5.8) $$g = \begin{pmatrix} \mathbf{1} & \mathbf{0} & \mathbf{0} \\ \mathbf{0} & 0 & 1 \\ \mathbf{0} & 1 & 0 \end{pmatrix} \qquad \xi = \begin{pmatrix} \mathbf{0} \\ 0 \\ 1 \end{pmatrix}.$$

Then $(\mathbb{R}^{4,1}, \xi) := (\mathbb{R}^5, g, \xi)$ is a Bargmann structure - **the flat Bargmann structure** - that induces the flat Newton-Cartan structure and

(5.9) $\qquad \mathrm{Barg}(\mathbb{R}^{4,1}, \xi)_0 =: B_{11}$

is the subgroup of the affine de Sitter group $SO(4,1)^\uparrow \ltimes \mathbb{R}^5$ that preserves ξ, i.e. the group of all 6x6 matrices of the form [31,33]

(5.10) $$\begin{pmatrix} \mathbf{R} & \mathbf{v} & \mathbf{0} & \mathbf{r} \\ \mathbf{0} & 1 & 0 & t \\ -{}^t\mathbf{vR} & -v^2/2 & 1 & s \\ \mathbf{0} & 0 & 0 & 1 \end{pmatrix}$$

(*) $\tilde{X} \in \mathrm{aut}(\tilde{M}, \xi)$ iff $\{ \tilde{X} \in \Gamma(T\tilde{M}), [\tilde{X}, \xi] = 0\}$.

where $\mathbf{R} \in SO(3)$; $\mathbf{v}$, $\mathbf{r} \in \mathbb{R}^3$; t, $s \in \mathbb{R}$. In view of (5.7,10) the **Bargmann group** $B_{11} \simeq H_6 \rtimes \mathbb{R}^5$ is a central nontrivial $(\mathbb{R},+)$ extension of the Galilei group G_{10}. The introduction of $U(1)$_extensions labelled by the mass [1] will be explained later on.

▪ In generalizing the notion of Bargmann automorphisms stricto sensu to the case of **conformal Bargmann automorphisms**, we define

(5.11) $\quad \mathrm{Conf}(\tilde{M},g,\xi) := \mathrm{Conf}(\tilde{M},g) \cap \mathrm{Aut}(\tilde{M},\xi)$

with $\mathrm{Conf}(\tilde{M},g) := \{\, a \in \mathrm{Diff}_{loc}(\tilde{M});\ a^*g = \Omega^2 g;\ \Omega \in C^\infty(\tilde{M},\mathbb{R}^*)\,\}$, infinitesimally

(5.12) $\quad \mathrm{conf}(\tilde{M},g,\xi) := \{\tilde{X} \in \Gamma(T\tilde{M}); [\xi,\tilde{X}] = 0; L_{\tilde{X}}\, g = \lambda g; \lambda \in C^\infty(\tilde{M},\mathbb{R})\}$.

By using (5.3–5) and (3.16), we find that

(5.13) $\quad \tilde{X} \in \mathrm{conf}(\tilde{M},g,\xi) \Rightarrow X := \pi_* \tilde{X} \in \mathrm{sch}(M,\gamma,\tau,\nabla)$,

and by repeating the preceding argument we get the exact sequence

(5.14) $\quad 0 \to \mathbb{R} \to \mathrm{conf}(\tilde{M},g,\xi) \to \mathrm{sch}(M,\gamma,\tau,\nabla) \to 0$,

that insures that infinitesimal **conformal Bargmann automorphisms centrally extend special chronoprojective automorphisms** (or Schrödinger transformations) of the underlying Newton-Cartan structure.

In the <u>flat case</u> (5.8), a simple calculation yields

$$\begin{aligned}(5.15) \quad \tilde{X} \in \mathrm{conf}(\mathbb{R}^{4,1},\xi) \Leftrightarrow \tilde{X} = {} & (\boldsymbol{\omega}^A{}_B\, \mathbf{x}^B + \boldsymbol{\beta}^A t + \boldsymbol{\gamma}^A + \alpha t \mathbf{x}^A + \chi \mathbf{x}^A)\partial/\partial \mathbf{x}^A \\ & + (\alpha t^2 + 2\chi t + \varepsilon)\partial/\partial t \\ & + (-\boldsymbol{\beta}_A \mathbf{x}^A - \alpha/2\, \mathbf{x}_A \mathbf{x}^A + \theta)\partial/\partial x^5\end{aligned}$$

with $\boldsymbol{\omega} \in so(3)$; $\boldsymbol{\beta}$, $\boldsymbol{\gamma} \in \mathbb{R}^3$; $\alpha, \chi, \theta \in \mathbb{R}$. Conspicuously, Schrödinger transformations (5.13) are most readily introduced in terms of Bargmann structures (compare (3.16)) whose infinitesimal conformal automorphisms (a **subalgebra of o(5,2)** - rather than o(4,2)!) canonically define a $(\mathbb{R},+)$ central extension of the Schrödinger Lie algebra.

▪ The <u>conformal structure</u> of $(\tilde{M},g)$ is completely determined by the <u>Weyl tensor</u> $\tilde{C}$

(5.16) $\quad \tilde{C}_{ijk}{}^l := \tilde{R}_{ijk}{}^l - 2/3\{\, \delta^l{}_{[i}\tilde{R}_{j]k} + g^{lm}\tilde{R}_{m[i}g_{j]k}\,\} + 1/6\,\tilde{R}\,\delta^l{}_{[i}g_{j]k}$

$(i,j,k,l = 1,\ldots,5)$ which conveys all (traceless) information on the curvature $\tilde{R}$ of $(\tilde{M},g)$. On a Bargmann manifold $(\tilde{M},g,\xi)$ the identity

(5.17) $\quad \tilde{R}_{ijk}{}^l \xi^k = 0.$

holds (see (5.2)). The quantities $\tilde{C}_{ijk}{}^l$ (5.16) and

(5.18) $\tilde{C}_{ij}{}^{l} := \tilde{C}_{ijk}{}^{l}\xi^{k}$

are thus invariant under the substitution

(5.19) $g \mapsto g^{*} := \Omega^{2} g; \qquad \xi \mapsto \xi^{*} = \xi$

entering definition (5.11). Yet, to consistently define the **Bargmann-Weyl tensor** $\tilde{C}$ of $(\tilde{M}, g, \xi)$ with all symmetries characteristic of the Levi-Cività curvature (in particular (5.17)) we have to put $\tilde{C}_{ij}{}^{l} := 0$ in (5.18). Hence

(5.20) $\tilde{C}_{ijk}{}^{l} := \tilde{R}_{ijk}{}^{l} - 2/3\{\delta^{l}{}_{[i}\tilde{R}_{j]k} + g^{lm}\tilde{R}_{m[i}g_{j]k}\},$

(5.21) $\tilde{R}^{ij} = \pi^{*}(f)\,\xi^{i}\xi^{j}, \qquad (f \in C^{\infty}(M,\mathbb{R})).$

Note that $\tilde{C}$ (5.20) does project on spacetime M as the chronoprojective Weyl curvature C (3.26,27). Details concerning 2nd-order chronoprojective Cartan structures can be found in [39].

- Example. Newtonian potentials $V(\mathbf{r},t)$ associated with conformally flat Bargmann manifolds $\tilde{M} \simeq \mathbb{R}^{5} \ni (\mathbf{r} = (\mathbf{x}^{A}), t, s)$,

(5.22) $g = \delta_{AB}\, d\mathbf{x}^{A} \otimes d\mathbf{x}^{B} + dt \otimes ds + ds \otimes dt - 2V(\mathbf{r},t)\, dt \otimes dt,$

(5.23) $\xi = \partial/\partial s,$

satisfy $\partial_{A}\partial_{B}V = 1/3\, \Delta V \delta_{AB}$, hence are of the general form

(5.24) $V(\mathbf{r},t) = a(t)\mathbf{r}^{2}/2 + \langle \mathbf{b}(t), \mathbf{r} \rangle + c(t).$

§6. THE SCHRÖDINGER AND LEVY-LEBLOND EQUATIONS.

- The Schrödinger equation (1.3) can be written on $(\mathbb{R}^{4,1}, \xi)$ as

(6.1) $\Delta\tilde{\Psi} + 2\partial_{4}\partial_{5}\tilde{\Psi} = 0; \qquad \partial_{5}\tilde{\Psi} = im/\hbar\, \tilde{\Psi},$

where $\tilde{\Psi} : \mathbb{R}^{4,1} \to \mathbb{C}$ is thus related to the usual Schrödinger wave function $\Psi : \mathbb{R}^{3;1} \to \mathbb{C}$ by $\tilde{\Psi}(x, x^{5}) = \exp(imx^{5}/\hbar)\,\Psi(x)$. Our claim is that (6.1) reads intrinsically [18,48]

(6.2) $\square\tilde{\Psi} = 0; \qquad \xi(\tilde{\Psi}) = im/\hbar\, \tilde{\Psi}.$

Gravitational (minimal) coupling is then straightforward : the Schrödinger equation on a curved Bargmann manifold $(\tilde{M}, g, \xi)$ is exactly the same as in (6.2), $\square$ being the Laplace-Beltrami operator of $(\tilde{M}, g)$. A quantum spinless test particle is described by a **harmonic function** $\tilde{\Psi} : \tilde{M} \to \mathbb{C}$ such that $s^{*}\tilde{\Psi} = \underline{m}(s)\tilde{\Psi}$ (all $s \in (\mathbb{R},+)$) where

(6.3) $\underline{m}(s) := e^{ims/\hbar}$

is a **character of the structural group** of $\pi:\tilde{M} \to M$ **that defines the mass m** of the particle. The corresponding spacetime wave function Ψ (a section of the associated line bundle $\tilde{M} \times_{\underline{m}} \mathbb{C}$) satisfies the **Schrödinger-Kuchar** equation [27,17,18] on the induced Newton-Cartan manifold(*).

The **Schrödinger equation** (6.2) is **Bargmann conformally invariant**. More precisely, let $\Omega \in C^\infty(\tilde{M}, \mathbb{R}^*)$ and put as in (5.19)

(6.4) $g^* := \Omega^2 g, \qquad \xi^* := \xi.$

Clearly $d\Omega \wedge \tilde{\tau} = 0$, i.e. Ω is a function of time only and if $\tilde{R}$ is the scalar curvature of $(\tilde{M}, g)$ we readily get $\tilde{R}^* = \Omega^{-2}\tilde{R} = 0$ since the field equations (2.14) $\widetilde{Ric} = (4\pi G\rho + \Lambda)\tilde{\tau} \otimes \tilde{\tau}$ are assumed to hold. Letting

(6.5) $\tilde{\Psi}^* := \Omega^{-3/2}\tilde{\Psi},$

we find the scale-invariance relationship

(6.6) $\Delta_{g^*}\tilde{\Psi}^* = \Omega^{7/2}\Delta_g\tilde{\Psi} = 0; \qquad \xi^*(\tilde{\Psi}^*) = im/\hbar\, \tilde{\Psi}^*.$

The **mass is invariant** against the substitution (6.4,5). Furthermore, if $a \in \mathrm{Conf}(\tilde{M}, g, \xi)$ (5.11) then by (6.5,6), $(a^{-1})^*(\Omega^{-3/2}\tilde{\Psi})$ is again a solution of (6.2) and $\mathrm{Conf}(\tilde{M}, g, \xi)$ acts on the set of solutions of the Schrödinger equation according to

(6.7) $\tilde{\Psi} \mapsto a_*(\Omega^{-3/2}\tilde{\Psi}).$

The associated infinitesimal action (see (5.12))

(*) In the setting of <u>geometric quantization</u>, one would consider wave functions as <u>half-densities</u> $\Psi_t^{\#}$ (locally $\Psi_t |\det \mathbf{g}|^{1/4}$) of a (riemannian) configuration space $(\mathbf{Q}, \mathbf{g})$. Quantization of the geodesic flow on a curved configuration space leads to the Schrödinger equation $iL_{\partial/\partial t}\Psi^{\#} = -\hbar/2m\, \boldsymbol{\Delta}^{\#}\Psi^{\#}$ $(= -\hbar/2m(\boldsymbol{\Delta} - \mathbf{R}/6)\Psi |\det \mathbf{g}|^{1/4}$, $\mathbf{R}$ is the scalar curvature of $(\mathbf{Q}, \mathbf{g}))$ (e.g. [42]). If $(\tilde{M}, g, \xi)$ is the Bargmann manifold $\tilde{M} := \mathbf{Q} \times \mathbb{R}^2$, $g = \mathbf{g} + dt \otimes ds + ds \otimes dt$, $\xi = \partial/\partial s$, then the preceding Schrödinger equation can be cast into the form : $\Delta^{\#}\Psi^{\#} = 0$, $L_\xi \tilde{\Psi}^{\#} = im/\hbar\, \tilde{\Psi}^{\#}$ where $\Delta^{\#}$ denotes the Laplacian of half-densities of $(\tilde{M}, g)$ (hint: $\tilde{R} = \mathbf{R}$; $\det(g) = \det(\mathbf{g})$). If Newton - Cartan field equations (2.14) hold, $\mathbf{Q}$ is Ricci flat, we get rid of the extra term $\tilde{R}/6$ and thus recover the equations (6.2).

(6.8) $(\tilde{X},\lambda)(\tilde{\Psi}) = \tilde{X}(\tilde{\Psi}) + 3/4\,\lambda\tilde{\Psi}$

agrees in the flat case with Niederer's (1.7-9) - cf. (5.15) and $\lambda = 2/5\,\mathrm{div}(\tilde{X})$. In view of (6.3), $\mathrm{Conf}(\tilde{M}, g, \xi)$ "acts" on the solutions of the Schrödinger equation via the quotient

(6.9) $m\_\mathrm{Conf}(\tilde{M}, g, \xi) := \mathrm{Conf}(\tilde{M}, g, \xi)/\ker(\underline{m})$.

In the case of $(\mathbb{R}^{4,1}, \xi)$, we recover the nontrivial central $U(1)$ $(\simeq \mathbb{R}/\ker(\underline{m}))$ extension m_Sch_{13} of Sch_{12}.

▪ <u>The Levy-Leblond equation</u> [29,32] for a spin 1/2 particle is strictly equivalent to the Schrödinger-Pauli equation in the free case. It admits a Bargmann invariant formulation hidden in the original derivation of the "square root" of the Schrödinger equation.

The Clifford algebra over a 5-dimensional space admits an irreducible 4-dimensional complex representation. In the case of $\mathbb{R}^{4,1}$ with Lorentz metric (g_{ab}) given by (5.8), we may choose the γ-matrices generators as

(6.10) $\gamma^A := \begin{pmatrix} \sigma_A & 0 \\ 0 & -\sigma_A \end{pmatrix}$; $\gamma^4 := \begin{pmatrix} 0 & 0 \\ 1 & 0 \end{pmatrix}$; $\gamma^5 := \begin{pmatrix} 0 & -2i \\ 0 & 0 \end{pmatrix}$

where the σ_A's $(A = 1,2,3)$ denote the Pauli matrices. We have $(a, b = 1,\dots,5)$

(6.11) $\gamma^{(a}\gamma^{b)} = g^{ab}$.

The elements $\tilde{A}$ of the Clifford algebra (generated by $1_{\mathbb{C}^{2,2}}$ and the γ's (6.10)) such that $\tilde{A}^{-1}\gamma^a\tilde{A} = A^a{}_b\,\gamma^b$ $(A \in SL(5,\mathbb{R}))$ form the $\mathbb{Z}_2$-covering group $\mathrm{Spin}(\mathbb{R}^{4,1})$ of the de Sitter group $SO(4,1)\uparrow$. The subgroup $\mathrm{Spin}(\mathbb{R}^{4,1},\xi)$ that leaves $\gamma_5\,(=\gamma_a\xi^a)$ invariant is isomorphic to $SU(2) \times \mathbb{R}^3$, the universal covering of the homogeneous Galilei group H_6 (2.2) (see also [4,5] for earlier related results). The representation spinor-space $\mathbb{C}^{2,2}$ is furthermore endowed with a canonical quaternionic structure.

In our formalism, the **Levy-Leblond equation** simply reads [30]

(6.12) $\gamma^a\partial_a\tilde{\Psi} = 0;\quad \xi^a\partial_a\tilde{\Psi} = im/\hbar\,\tilde{\Psi}$.

Thus $\tilde{\Psi}(x, x^5) = \exp(imx^5/\hbar)\,\Psi(x)$; $\Psi = (\Psi', \Psi'')$ is a $\mathbb{C}^{2,2}$-valued function of $\mathbb{R}^{3;1}$ such that Ψ', Ψ'' both satisfy the free Schrödinger-Pauli equation. It is worth noticing that (6.12) is well adapted for the geometric prescription of (minimal) gravitational coupling (cf. the C.O.W. experiment [14] on neutron interferometry in the gravitational field). To that end, assume the existence of a spin structure $Spin(\tilde{M},g,\xi)$, i.e. a $SU(2)\times\mathbb{R}^3$ principal bundle that equivariantly $\mathbb{Z}_2$-covers the bundle of Galilei frames $(a, b, j = 1,...,5)$

(6.13) $\quad H_6(\tilde{M},g,\xi) := \{ (e_a = e^j{}_a \partial_j)_{x\in\tilde{M}},\ g(e_a, e_b) = g_{ab},\ e_5 = \xi \}.$

The local spinor-tensors are then defined by

(6.14) $\quad \gamma^j := \gamma^a e^j{}_a.$

Given a spinor field $\tilde{\Psi} \in \Gamma(Spin(\tilde{M},g,\xi)\times_{(SU(2)\times\mathbb{R}^3)} \mathbb{C}^{2,2})$, its covariant derivative is defined via a section of $H_6(\tilde{M},g,\xi)$ by

(6.15) $\quad \tilde{\nabla}_j \tilde{\Psi} := \partial_j\tilde{\Psi} + \tilde{\Gamma}_j\tilde{\Psi}; \qquad \tilde{\Gamma}_j := 1/4\, \tilde{\nabla}_j e^k{}_a\, \gamma_k\, \gamma^a.$

The **covariant Levy-Leblond equation** is then

(6.16) $\quad D\tilde{\Psi} := \gamma^j \tilde{\nabla}_j \tilde{\Psi} = 0; \qquad L_\xi\tilde{\Psi} = \xi^j\tilde{\nabla}_j \tilde{\Psi} = im/\hbar\, \tilde{\Psi}.$

By "squaring" it, we recover the Schrödinger equation (6.2) modulo a term involving the scalar curvature $\tilde{R}$ that vanishes once Newton's field equations are taken into account. It turns out that the **Levy-Leblond equation** (6.16) is also **Bargmann conformally invariant** since it transforms under the substitution

(6.17) $\quad g^* = \Omega^2 g; \qquad \xi^* = \xi; \qquad \tilde{\Psi}^* = \Omega^{-2}\, \tilde{\Psi}$

according to

(6.18) $\quad D^*\tilde{\Psi}^* = \Omega^{-3}\, D\tilde{\Psi} = 0; \qquad L_{\xi^*}\tilde{\Psi}^* = im/\hbar\, \tilde{\Psi}^*.$

REFERENCES

[1] **BARGMANN, V.** Ann. Math. 59 1 (1954).
[2] **BARUT, A.O.** Helv. Phys. Acta, 46 496 (1973).
[3] **BOYER, C.P. & PENAFIEL, M.** Nuov. Cim. 31B 195 (1976).
[4] **BROOKE, J.A.** Ph.D. Thesis, University of Alberta 1980 (unpublished).
[5] **BROOKE, J.A.** J. Math. Phys. 19 952 (1978); 21 617 (1980).
[6] **BURDET, G., DUVAL, C. & PERRIN, M.** J. Math. Phys. 24 1752 (1983).
[7] **BURDET, G., DUVAL, C. & PERRIN, M.** Publ. R.I.M.S. 19 813(1983).
[8] **BURDET, G., DUVAL, C. & PERRIN, M.** Lett. Math. Phys. 10 255 (1985).
[9] **BURDET, G. & PERRIN, M.** Lett. Nuov. Cim. 4 651 (1972).
[10] **BURDET, G. & PERRIN, M.** J. Math. Phys. 26 292 (1985).
[11] **BURDET, G. & PERRIN, M.** "Exact Solutions of Nonlinear Schrödinger Equation in Time Dependent Parabolic Density Profiles", Preprint Marseille CPT-85/P.1770 (to appear in Lett. Math. Phys).
[12] **BURDET, G., PERRIN, M. & SORBA, P.** Comm. Math. Phys. 34 123 (1973).
[13] **CARTAN, E.** Ann. Scient. Ec. Norm. Sup. 40 325 (1923); 41 1 (1924).
[14] **COLLELA, R., OVERHAUSER, A.W. & WERNER, S.A.** Phys. Rev. Lett. 34 1472 (1975).
[15] **DIXON, W.G.** Comm. Math. Phys. 45 167 (1975).
[16] **DUVAL, C.** "Quelques procédures géométriques en dynamique des particules", Thèse, Université d'Aix-Marseille 2, 1982 (unpublished).
[17] **DUVAL, C. & KÜNZLE, H.P.** Gen. Rel. Gravit. 16 333 (1984).
[18] **DUVAL, C., BURDET, G., KÜNZLE, H.P. & PERRIN, M.** Phys. Rev. D 31 1841 (1985).
[19] **EHLERS, J.** "Über den Newtonschen Grenzwert der Einsteinischen Gravitations theorie", in Grundlagenprobleme der modernen Physik, p. 65, edited by NITSCH, J. et al., Bibliographisches Institut Mannheim, 1981.
[20] **HAGEN, C.R.** Phys. Rev. D 5 377 (1972).
[21] **HAVAS, P. & PLEBANSKI, J.** J. Math. Phys. 19 482 (1978).
[22] **HAVAS, P.** Rev. Mod. Phys. 36 938 (1964).
[23] **HAVAS, P.** Helv. Phys. Acta 45 802 (1972).
[24] **HORVATHY, P.** "Etude géométrique du monopole magnétique", Thèse de 3ème cycle, Université de Provence, 1983; J. Geom. Phys 1 39 (1984).
[25] **JACKIW, R.** Ann. of Phys. 129 183 (1980).
[26] **JAKUBIEC, A. & KIJOWSKI, J.** Rep. Math. Phys. 10 21 (1976).
[27] **KUCHAR, K.** Phys. Rev. D 22 1285 (1980).
[28] **KÜNZLE, H.P.** Ann. Inst. H. Poincaré 17 A 337 (1972) and references therein.
[29] **KÜNZLE, H.P. & DUVAL, C.** Ann. Inst. H. Poincaré 41 363 (1984).
[30] **KÜNZLE, H.P. & DUVAL, C.** "Relativistic and Nonrelativistic Classical Field Theory on Fivedimensional Spacetime", Preprint CPT-85/P.1795, Marseille.
[31] **LEVY-LEBLOND, J.M.** "Galilei Group and Galileian Invariance", in Group Theory and its Applications, p. 221, edited by LOEBL, E.M., Vol 2 (Academic, New York, 1971).
[32] **LEVY-LEBLOND, J.M.** Comm. Math. Phys. 6 286 (1967).
[33] **LEVY-LEBLOND, J.M.** Riv. Nuovo Cimento 4 99 (1974).
[34] **MÜLLER-HOISSEN, F.** Dissertation Göttingen, 1983 (unpublished).
[35] **MÜLLER-HOISSEN, F.** Ann. Inst. H. Poincaré 40 21 (1984).
[36] **NIEDERER, U.** Helv. Phys. Acta 45 802 (1972).
[37] **NIEDERER, U.** Helv. Phys. Acta, 46 191 (1973).

[38] **NIEDERER, U.** Helv. Phys. Acta, 47 119 (1974).
[39] **PERRIN, M., BURDET, G. & DUVAL, C.** "Chronoprojective Invariance of the Fivedimensional Schrödinger Formalism", Preprint CPT-84/P.1668, Marseille (to appear in Class. & Quant Grav.).
[40] **PERROUD, M.** Helv. Phys. Acta 50 233 (1977).
[41] **RAY, J.** Phys. Rev. A 26 729 (1982).
[42] **SNIATICKI, J.** Geometric Quantization and Quantum Mechanics, Springer 1980.
[43] **SOURIAU, J.M.** Structure des systèmes dynamiques (Dunod, Paris, 1970).
[44] **SOURIAU, J.M.** "Construction explicite de l'indice de Maslov", in Proceedings of the 4th International Colloquium on Group Theoretical Methods in Physics, Nijmegen 1975. Lecture Notes in Physics, Vol. 50, p. 117, Springer 1976.
[45] **SOURIAU, J.M.** "Mécanique classique et géométrie symplectique", Preprint CPT-84/PE.1695, Marseille.
[46] **TRAUTMAN, A.** C.R. Acad. Scien. PARIS 257 617 (1963).
[47] **TRAUTMAN, A.** "Comparison of Newtonian and Relativistic theories of space-time", in Perspectives in Geometry and Relativity, p. 413, edited by HOFFMANN, B., Indiana University Press, Bloomington 1964.
[48] **TULCZYJEW, W.M.** "An intrinsic Formulation of Nonrelativistic Analytical Mechanics & Wave Mechanics" Preprint IFM-Torino 1985 (to appear in J. Geom. Phys).

IV. CONFORMAL SPINORS

WAVE EQUATIONS FOR CONFORMAL MULTISPINORS*

Miguel Lorente
Max Planck Institut, D-8130 Starnberg, West Germany
and
Facultad de Ciencias Físicas, Universidad Complutense
28040 Madrid, Spain

I. INTRODUCTION

In 1936 Dirac wrote a famous paper[1] in which he derived a conformal invariant wave equation for massless systems of spin 0, 1/2 and 1, in a similar fashion to the relativistic wave equation of massive particles of spin 1/2.

The method used by Dirac is extended to massless particles of arbitrary spin. This generalization to multispinor is equivalent to that used by Bargmann and Wigner in his equations of relativistic particles of arbitrary spin.

The free states obeying the conformal wave equations can be used to describe the interactions of the fundamental entities, proposed by Weizsäcker (urs), giving raise to superselection rules as it has been done by Castell[2] and Heidenreich[6].

II. SPINOR REPRESENTATION OF THE CONFORMAL GROUP

For the generators of the fundamental representation of the conformal group, Dirac[1] uses the operators:

$$\beta_a = (\gamma_1, \gamma_2, \gamma_3, \gamma_4, \gamma_5, i)$$
$$\gamma_a = (\gamma_1, \gamma_2, \gamma_3, \gamma_4, \gamma_5, -i)$$

where the β's and γ's satisfy $\beta_a\gamma_b + \beta_b\gamma_a = 2\eta_{ab}$, $\eta_{ab} = \mathrm{diag}(+, +, +, -, -, +)$ being the metric tensor. The generators for the fundamental representation are then

$$J_{ab} = \frac{1}{i}(x_a\partial_b - x_b\partial_a) + \frac{1}{2i}\beta_a\gamma_b \equiv M_{ab} + S_{ab}$$

The generators of the contragradient representation are obtained by interchanging the β's and the γ's, namely ,

$$J_{ab} = \frac{1}{i}(x_a\partial_b - x_b\partial_a) + \frac{1}{2i}\gamma_a\beta_b$$

Both representations have the spin content (1/2,0) and (0,1/2) respectively. These representations act on some spinor fields of one index ψ_α and $\psi_{\dot\alpha}$ respectively.

Multispinor field transform under the direct product of the fundamental and contragradient representations (2j and 2k times, respectively). They transform under the infinitesimal generators in the following way[4]:

$$\delta\psi_{\alpha_1\alpha_2\ldots\alpha_{2j}} = \varepsilon^{ab}\frac{1}{2i}\sum_{i=1}^{2j}(\beta_a\gamma_b)_{\alpha_i}^{\alpha'_i}\psi_{\alpha'_1\alpha'_2\ldots\alpha'_{2j}}$$

$$\delta\psi_{\dot\alpha_1\dot\alpha_2\ldots\dot\alpha_{2k}} = \varepsilon^{ab}\frac{1}{2i}\sum_{i=1}^{2k}(\gamma_a\beta_b)_{\dot\alpha'_i}^{\dot\alpha'_i}\psi_{\dot\alpha_1\dot\alpha_2\ldots\dot\alpha_{2k}}$$

where the multispinors are traceless and totally symmetric in the α_i and $\dot\alpha_i$. They correspond to the representations (j+1, j, 0) and (k+1, 0, k) in the Yao[3] classification. All these representations are defined on the light cone ($x^2=0$) and therefore, using homogeneous coordinates on a 4-dimensional manifold, when restricted to the Poincaré group, they become massless particles of arbitrary spin[4].

The wave equation will be worked out with the help of the second order Casimir operator, i.e.

$$C_2 = \frac{1}{2}J_{ab}J^{ab} = \frac{1}{2}M_{ab}M^{ab} + M_{ab}S^{ab} + \frac{1}{2}S_{ab}S^{ab}$$

Each term of the Casimir operator commutes with the others and with the operator $x^a\partial_a$. Therefore, they must have a common set of eigenfunctions, which are at the same time, the carrier space where the representation acts. Let us calculate the three parts of the Casimir operator.

$$x^a\partial_a \equiv x\,\partial$$

$$\frac{1}{2}M_{ab}M^{ab} \equiv \frac{1}{2}M^2 = -x^2\partial^2 + (x\partial)^2 + 4(x\partial)$$

$$M_{ab}S^{ab} \equiv MS = -(\beta x)(\gamma\partial) + (x\partial) \quad , \quad (j = \tfrac{1}{2},\ k = 0)$$

$$= -(\gamma x)(\beta\partial) + (x\partial) \quad , \quad (j = 0,\ k = \tfrac{1}{2})$$

$$= -\sum_{r=1}^{2j}(\beta x)_{(r)}(\gamma\partial)_{(r)} + 2j(x\partial), \quad (j,\ k = 0)$$

$$= -\sum_{r=1}^{2k}(\gamma x)_{(r)}(\gamma\partial)_{(r)} + 2k(x\partial), \quad (j = 0,\ k)$$

where the index (r) means the direct product of the unit matrix with the matrix β^a, γ^a in the place (r). Besides

$$\frac{1}{2} S_{ab}S^{ab} = -\frac{1}{8}\left(\sum_{r=1}^{2j}(\beta_a\gamma_b)_{(r)}\right)\left(\sum_{i=1}^{2j}(\beta^a\gamma^b)_{(r)}\right) \qquad (j,\ k=0)$$

$$= -\frac{1}{8}\left(\sum_{r=1}^{2k}(\gamma_a\beta_b)_{(r)}\right)\left(\sum_{i=1}^{2k}(\gamma^a\beta^b)_{(r)}\right) \qquad (k,\ j=0)$$

The eigenvalues of the last operator can be calculated with the help of the alternative expressions[3]

$$S_{ab}S^{ab} = S^2 = 2(\vec{J}^2 + \vec{K}^2) + R_o(R_o+4) - 2(P_-P_+ + Q_-Q_+ + S_-S_+ + T_-T_+)$$

and the particular representation of the generators

$$\vec{J} = \frac{1}{2}\begin{pmatrix} \vec{\sigma} & 0 \\ 0 & 0 \end{pmatrix}, \quad \vec{K} = \frac{1}{2}\begin{pmatrix} 0 & 0 \\ 0 & \vec{\sigma} \end{pmatrix}, \quad R_o = \frac{1}{2}\begin{pmatrix} 1 & 0 \\ 0 & -1 \end{pmatrix}$$

$$P_+ = \frac{i}{2}\left[\begin{pmatrix} 0 & 1 \\ 0 & 0 \end{pmatrix} + \begin{pmatrix} 0 & \sigma_3 \\ 0 & 0 \end{pmatrix}\right], \quad P_- = \frac{i}{2}\left[\begin{pmatrix} 0 & 0 \\ 1 & 0 \end{pmatrix} + \begin{pmatrix} 0 & 0 \\ \sigma_3 & 0 \end{pmatrix}\right]$$

$$Q_+ = \frac{i}{2}\left[\begin{pmatrix} 0 & 1 \\ 0 & 0 \end{pmatrix} - \begin{pmatrix} 0 & \sigma_3 \\ 0 & 0 \end{pmatrix}\right], \quad Q_- = \frac{i}{2}\left[\begin{pmatrix} 0 & 0 \\ 1 & 0 \end{pmatrix} - \begin{pmatrix} 0 & 0 \\ \sigma_3 & 0 \end{pmatrix}\right]$$

$$S_+ = \frac{i}{2}\left[\begin{pmatrix} 0 & \sigma_1 \\ 0 & 0 \end{pmatrix} + i\begin{pmatrix} 0 & \sigma_2 \\ 0 & 0 \end{pmatrix}\right], \quad S_- = \frac{i}{2}\left[\begin{pmatrix} 0 & 0 \\ \sigma_1 & 0 \end{pmatrix} - i\begin{pmatrix} 0 & 0 \\ \sigma_2 & 0 \end{pmatrix}\right]$$

$$T_+ = \frac{i}{2}\left[\begin{pmatrix} 0 & \sigma_1 \\ 0 & 0 \end{pmatrix} - i\begin{pmatrix} 0 & \sigma_2 \\ 0 & 0 \end{pmatrix}\right], \quad T_- = \frac{i}{2}\left[\begin{pmatrix} 0 & 0 \\ \sigma_1 & 0 \end{pmatrix} + i\begin{pmatrix} 0 & 0 \\ \sigma_2 & 0 \end{pmatrix}\right]$$

Applying S^2 to the highest state of some irreducible representation (j+1, j, 0) we get

$$S^2 = 2j(j+1) + j(j+4)$$

and similarly for the representation (k+1, 0, k)

$$S^2 = 2k(k+1) + k(k+4)$$

Finally the operator $(x\partial)$ has the eigenvalue n, i.e. the homogeneity degree. Collecting all the eigenvalues of the operators M^2, MS, S^2 and $(x\partial)$, we get for the Casimir operator:

$$C_2 = n^2 + 4n + m + 3j(j+2) = 3(j^2-1)$$

and similar expresion for the representation (k+1, 0, k),

$$C_2 = n^2 + 4n + m + 3k(k+2) = 3(k^2-1) \quad .$$

III. WAVE EQUATION FOR THE SPINOR FIELD OF j = 1/2

Following Dirac[1], we take the eigenvalue equations for the operator MS as the wave equation.

$$\{-(\beta x)(\gamma\partial) + (x\partial)\}\psi = m\psi$$
$$(x\partial)\psi = n\psi$$

therefore $(\beta x)(\gamma\partial)\psi = (n-m)\psi$ where we have omitted the spinorial index. We have two cases:

$n \neq m$. Multiplying by $(\gamma\partial)$ from the left and using (A 4) we get

$$(\gamma\partial)(\beta x)(\gamma\partial)\psi = [6 + 2(x\partial) - (\gamma\partial)(\beta\partial)](\gamma\partial)\psi = (n-m)(\gamma\partial)\psi$$

In the light cone $(\beta\partial)(\gamma\partial)\psi = \partial^2\psi = 0$, therefore

$$(4 + 2n)(\gamma\partial)\psi = (n-m)(\gamma\partial)\psi.$$

Since $(\gamma\partial)\psi = 0$ implies the trivial solution $\psi = 0$, we are left with

$$4 + 2n = n-m \qquad \text{or} \qquad m = -n - 4 \quad .$$

This equation together with Casimir C_2 gives $n = -1$, and $m = -3$, respectively. As Dirac[1] proved, the solution can be written

$$\psi = (\beta x)\chi$$

where χ is an arbitrary scalar function with homogeneity degree $n' = -2$.

$n = m$. From the Casimir operator, we obtain $n = m = -2, -3$. In this case, Dirac proved[1] that there is a gauge freedom, i.e.

$$\psi \to \psi' = \psi + (\beta x)\chi$$

In order that the wave equation should be satisfied we must have

$$(\beta x)(\gamma\partial)(\beta x)\chi = (6 + 2n')(\beta x)\chi - x^2(\beta\partial)\chi = 0 \ , \quad (x\partial)\chi = n'\chi$$

If $n'=-3$, we can define the function χ out of the light cone and multiplying from the left by (γx) we get $(\gamma x)(\beta\partial)\chi = 0$ which is the wave equation for the field ($j=0$, $k=1/2$). As Heidenreich[6] has pointed out, the last equation corresponds to some irreducible representation (5/2, 0, 1/2) which is contained in the indecomposable representation (3/2, 1/2, 0). If we use the representation (3/2,0,1/2) we can carry out the same arguments, with the only interchange of β's and γ's. In this case, the gauge freedom gives the indecomposable representation $D_I(3/2,0,1/2)$ containing the irreducible representation $D(5/2,1/2,0)$, as it can be seen in case b) of Appendix B, with $n=1$.

IV. WAVE EQUATIONS FOR SPINOR FIELDS OF $j = 1$

The eigenvalue equation for the operator MS gives

$$\{-(\beta x)_1(\gamma\partial)_1 - (\beta x)_2(\gamma\partial)_2 + 2(x\partial)\}\psi = m\psi \qquad \text{or}$$

$$\{(\beta x)_1(\gamma\partial)_1 + (\beta x)_2(\gamma\partial)_2\}\psi = (2n-m)\psi \quad .$$

As before we have two cases:

$2n \neq m$. Applying $(\gamma\partial)_1\ (\gamma\partial)_2$ from the left we get

$$2(6 + 2(n-1))(\gamma\partial)_1(\gamma\partial)_2\psi = (2n-m)(\gamma\partial)_1(\gamma\partial)_2\psi$$

Then $m = 2(-n - 4)$ and from the Casimir operator $n=-1$, and $m=-6$. In this case, the solution can be written as

$$\psi = (\beta x)_1(\beta x)_2\ \chi \qquad \text{with} \qquad (x\partial)\chi = (n-2)\chi$$

where χ is an arbitrary scalar function of degree $n' = -3$.

$2n = m$. From the Casimir operator, we get $n=-3$, $m = -6$. We have a gauge freedom

$$\psi \rightarrow \psi' = \psi + (\beta x)_1\chi \quad , \qquad (x\partial)\chi = (n-1)\chi$$

The wave equation imposes

$$\{(\beta x)_1\ [-(\gamma x)_1(\beta\partial)_1 + 6 + 2n'] + (\beta x)_2(\gamma\partial)_2(\rho x)_1\}\chi = 0$$

Multiplying by $(\gamma x)_1$ on the left, we get

$$[-x^2(\gamma x)_1(\beta\partial)_1 + 6 + 2n' + 2 + x^2(\beta x)_2(\gamma\partial)_2]\chi = 0$$

If $n'=-4$, it can be proved from the Casimir operator that this equation corresponds to the representation (3,1/2,1/2) which is an irreducible representation contained in the indecomposable representation (2,1,0), as it can be seen in case b) of Appendix B, with n=2.

V. WAVE EQUATION FOR MULTISPINOR FIELDS OF ARBITRARY SPIN

The eigenvalue equation for the operator MS gives

$$\sum_{i=1}^{2j}(\beta x)_i(\gamma\partial)_i\psi = (2jn-m)\psi \qquad \text{for} \quad \text{repr. } (j+1,j,0)$$

with two cases:

$2jn \neq m$. Applying $\prod_i(\gamma\partial)_i$ from the left, we get

$$2j(6 + 2(n-1)\prod_i(\gamma\partial)_i\psi = (2jn-m)\prod_i(\gamma\partial)_i\psi \quad .$$

Therefore $m = 2j(-n-4)$ hence $n = 2j - 3$, $m = -2j(2j + 1)$

$2jn = m$. From the Casimir operator, we get

$$n = -(2j + 1) \quad , \quad m = -2j(2j + 1)$$

Again we have a gauge freedom: $\psi \rightarrow \psi' = \psi + (\beta x)_1\chi$. The wave equation imposes the following conditions on the function χ :

$$[-x^2(\gamma x)_1(\beta\partial)_1 + 6 + 2n' + 2(2j-1) + x^2 \sum_{i\neq 1} (\beta x)_i(\gamma\partial)_i]\chi = 0 \quad .$$

If $n' = n-1 = -2j-2$, χ satisfies the wave equation corresponding to the irreducible representation $D(j+2, j-\frac{1}{2}, \frac{1}{2})$ contained in the indecomposable representation $D_I(j+1,j,0)$, as it can be seen from case b) with $n = 2j$ in Appendix B.

VI. WAVE EQUATIONS INVARIANT UNDER SO(4,2)

The simplest nontrivial representation of SO(4,2) including reflexions is 8-dimensional[4,5]. It is built up with the help of the matrices L_a $(a = 1,2,\dots,8)$. Satisfying $\{L_a, L_b\} = 2\eta_{ab}$. In the Dirac basis they are represented by

$$L_\mu = \gamma_\mu \times \sigma_3 \quad , \quad L_5 = i\, 1\!\!1 \times \sigma_2 \quad , \quad L_6 = -\, 1\!\!1 \times \sigma_1$$

The wave equation now reads for $m = n = -2$

$$L_a L_b M^{ab}\, \psi = -2\psi$$

If we introduce $L_7 \equiv -iL_0L_1L_2L_3L_5L_6$, which satisfies $\{L_7, L_a\} = 0$, it is easy to prove that $\phi_\pm = 1/2(1 \pm L_7)\psi$ is also a solution of the wave equation, and $\phi_\pm$ is an eigenfunction for L_7, with eigenvalues ± 1, respectively.

Applying the operator $(1\!\!1 - L_5L_6)$ to the wave equation, in a reference system in which the coordinates take the values $x_a = (0,0,0,0,1,1)$, one easily obtains $(1 - L_5L_6)\psi = 0$. But $L_5L_6 \equiv \Delta$ is the intrinsic dilatation operator, therefore $\Delta\psi = \pm\psi$. Writting L_7 and L_5L_6 in Dirac basis we obtain only two independent components for the wave function.

If we take the indecomposable representation (5/2,0,1/2) acting on the wave function, ψ, with degree of homogeneity $n = -3$, we can single out the invariant subspace, using the appropiate projection operator, as in Ref. 4.

APPENDIX A. SOME USEFUL RELATIONS

A1. $(\beta x)(\gamma x) = (\gamma x)(\beta x) = x^2$

A2. $$\begin{cases} (x\partial)(\beta x) = (\beta x)(x\partial) + (\beta x) \\ (x\partial)(\gamma x) = (\gamma x)(x\partial) + (\gamma x) \end{cases}$$

A3 $\begin{cases} (\beta\partial)(x\partial) = (x\partial)(\beta\partial) + (\beta\partial) \\ (\gamma\partial)(x\partial) = (x\partial)(\gamma\partial) + (\gamma\partial) \end{cases}$

A4 $\begin{cases} (\beta x)(\gamma\partial) + (\beta\partial)(\gamma x) = 6 + 2(x\partial) \\ (\gamma x)(\beta\partial) + (\gamma\partial)(\beta x) = 6 + 2(x\partial) \end{cases}$

A5 $\begin{cases} [(\beta x)(\gamma\partial)]^2 = -x^2\partial^2 + [4 + 2(x\partial)] \; (\beta x)(\gamma\partial) \\ [(\gamma x)(\beta\partial)]^2 = -x^2\partial^2 + [4 + 2(x\partial)] \; (\gamma x)(\beta\partial) \end{cases}$

APPENDIX B. INDECOMPOSABLE REPRESENTATION OF SU(2,2) DEFINED ON THE ENVELOPPING ALGEBRA

Let $\mu_1 = (0,0,1,-1)$, $\mu_2 = (1,-1,0,0)$ be the positive compact roots and $\alpha_1 = (0,1,0,-1)$, $\alpha_2 = (1,0,-1,0)$, $\alpha_3 = (1,0,0,-1)$, $\beta =$ $= (0,1,-1,0)$ the positive non-compact roots of the Algebra A_3. Let us defined the corresponding root vectors

$$E_{\mu_1} , E_{\mu_2} , E_{\alpha_1} , E_{\alpha_2} , E_{\alpha_3} , E_\beta \quad .$$

A basis for the envelopping algebra U of all the vectors associated with negative roots, can be taken as

$$\Omega_- = \{1\!\!1 , E^{\ell}_{-\alpha_1} E^m_{-\alpha_2} E^n_{-\alpha_3} E^r_{-\beta} E^s_{-\mu_1} E^t_{-\mu_2}\}$$

where ℓ, m, n, r, s, t are non negative integers, and $1\!\!1$ corresponds to all the exponents equal to zero. A representation for the algebra su(2,2) has been given in Ref. 7, provided we make the following identification

$$E_{-\mu_1} \leftrightarrow f_{43} , \quad E_{-\mu_2} \leftrightarrow f_{21} , \quad E_{-\alpha_1} \leftrightarrow if_{42}$$

$$E_{-\alpha_2} \leftrightarrow if_{31} , \quad E_{-\alpha_3} \leftrightarrow if_{41} , \quad E_{-\beta} \leftrightarrow if_{32}$$

A Verma module ρ_Λ of su(2,2) is defined by the conditions

$$\begin{cases} H_i \, 1\!\!1 = \Lambda_i \, 1\!\!1 \quad , \quad i = 1,2,3,4 \qquad \Lambda_i \in \mathbb{C} \\ E_\alpha \, 1\!\!1 = 0 \quad \text{for all positive roots} \quad \alpha \quad . \end{cases}$$

Verma submodules ρ_M are defined by $\Omega_- Y$ where Y is the extremal vector corresponding to the highest weight M, given by the conditions

$$\begin{cases} H_i Y = M_i Y \quad , \quad i = 1,2,3,4 \qquad M_i \in \mathbb{C} \\ E_\alpha Y = 0 \quad \text{for all the positive roots} \quad \alpha \quad . \end{cases}$$

The extremal vectors of su(2,2) have been given in Ref. 7.

In the case that $\rho_M \subset \rho_\Lambda$, the representation ρ_Λ is called indecomposable (it contains an invariant subspace). The quotiens of ρ_Λ with respect to all ρ_M becomes irreducible.

We can defined a bilinear form[8] in the envelopping algebra U by the following operations:

$$(x,y) \equiv \Lambda(\gamma(x^\sigma y)) \quad , \quad x,y \in U$$

where x^σ means conjugation, namely,

$$E_\alpha^\sigma = E_{-\alpha} \, , \quad E_{-\alpha}^\sigma = E_\alpha \quad , \quad H_\alpha^\sigma = H_\alpha \quad ,$$

$\gamma(x)$, the projection into the subspace Ω_o generated by Cartan generators H_α

$$\gamma(x) = \begin{cases} x & \text{if} \quad x \in \Omega_o \\ 0 \, , & \text{otherwise} \end{cases}$$

and $\Lambda(H_\alpha)$ is a linear form in the dual space of the Cartan subalgebra, i.e.

$$\Lambda(H_\alpha) = 2\,\frac{(\Lambda,\alpha)}{(\alpha,\alpha)} \quad ,$$

Λ being the highest weight of ρ_Λ.

The bilinear form becomes a scalar product when it is non-degenerate positive definite namely $(x,x) \geq 0$ and $(x,x) = 0$ iff $x = 0$.

All the extremal vectors Y satisfy $(Y,Y)=0$. We call a proper maximal ideal I_o of ρ_Λ, a subspace generated by all the vectors of null norm.

With the help of the bilinear form we can define infinitesimally unitary representations for the algebra su(2,2), by the conditions[8]:

i/ The components of the highest weight Λ are real,

ii/ the scalar product is a non-degenerate definite positive bilinear form. This is true iff

$$\Lambda(H_\mu) \geq 0 \quad \text{for all compact roots} \quad \mu_1, \mu_2$$
$$\Lambda(H_\alpha) + R(H_\alpha) \leq 0 \text{ for all non-compact roots } \alpha_1,\alpha_2,\alpha_3,\beta.$$

In Ref. 8 we have calculate the highest weight and extremal vectors of the unitary representations of su(2,2). A basis for these representations with envelopping algebra U, is given, by the quotient space of the Verma module ρ_Λ with respect to the maximal proper ideal I_o. This ideal is generated by all extremal vectors quoted below.

case a) $\Lambda = (\frac{-m-2}{4}, \frac{-m-2}{4}, \frac{3m+2}{4}, \frac{-m+2}{4})$, $m = 1,2,...$

Extremal vectors: $Y = E^{m+1}_{-\mu_1}$, $M = \Lambda-(m+1)\mu_1$

$Y = E_{-\mu_2}$, $M = \Lambda- \mu_2$

$Y = E_{-\beta}E_{-\mu_1}-mE_{\alpha_1}$, $M = \Lambda- \alpha_1$

case b) $\Lambda = (\frac{n-2}{4}, \frac{-3n-2}{4}, \frac{n+2}{4}, \frac{n+2}{4})$, $n = 1,2,3,...$

Extremal vectors: $Y = E_{-\mu_1}$, $M = \Lambda - \mu_1$

$Y = E^{n+1}_{-\mu_2}$, $M = \Lambda - (n+1)\mu_2$

$Y = E_{-\beta}E_{-\mu_2} + nE_{-\alpha_2}$, $M = \Lambda - \alpha_2$

case c) $\Lambda = (\frac{n-m}{4} -1, \frac{-3n-m}{4} -1, \frac{n+3m}{4} +1, \frac{n-m}{4} +1)$, $n,m = 1,2,3...$

Extremal vectors: $Y = E^{m+1}_{-\mu_1}$, $M = \Lambda - (m+1)\mu_1$

$Y = E^{n+1}_{-\mu_2}$, $M = \Lambda - (n+1)\mu_2$

$Y = E_{-\beta}E_{-\mu_1}E_{-\mu_2}-mE_{-\alpha_1}E_{-\mu_2}+nE_{-\alpha_2}E_{-\mu_1}-mnE_{-\alpha_3}$, $M = \Lambda - \alpha_3$

Finally we can calculate the restriction of su(2,2) to the maximal compact subgroup SU(2)xSU(2)xU(1) given by its generators

$J_{\pm} = E_{\pm\mu_2}$, $J_3 = \frac{1}{2}(H_1-H_2)$, $K_{\pm} = E_{\pm\mu_1}$, $K_3 = \frac{1}{2}(H_1-H_2)$

$R_o = \frac{1}{2}(H_1 + H_2 - H_3 - H_4)$

Defining the eigenvalues of J_3, K_3, R_o by j,k,d, respectively we have the following correspondence with the components of the highest weight Λ:

$j = \frac{1}{2}(\Lambda_1 - \Lambda_2)$, $k = \frac{1}{2}(\Lambda_3 - \Lambda_4)$, $d = \frac{1}{2}(\Lambda_1+\Lambda_2-\Lambda_3-\Lambda_4)$

With this identification we obtain the standard classification[9] of all the unitary representations of su(2,2) in Ω_- (equivalent to negative energy), defined by the numbers (j,k,d),

case a) $(0, \frac{m}{2}, -\frac{m}{2} -1)$: mass = 0 , helicity = $\frac{m}{2}$

case b) $(\frac{n}{2}, 0, -\frac{n}{2} -1)$: mass = 0 , helicity = $\frac{n}{2}$

case c) $(\frac{n}{2}, \frac{m}{2}, -\frac{n}{2} - \frac{m}{2} -2)$: mass $\neq 0$, spin = $\frac{n+m}{2}$

case d) $(0, \frac{m}{2}, < -\frac{m}{2} -1)$: mass $\neq 0$, spin = $\frac{m}{2}$

case e) $(\frac{n}{2}, 0, < -\frac{n}{2} - 1)$: mass $\neq 0$, spin $= \frac{n}{2}$

case f) $(\frac{n}{2}, \frac{m}{2}, < -\frac{m}{2} - \frac{n}{2} - 2)$: mass $\neq 0$, spin $= \left|\frac{n-m}{2}\right|, \ldots, \frac{n+m}{2}$

REFERENCES

* Work supported in part by the Max Planck Gesellschaft

1. P.M.A. DIRAC, Wave equations in conformal space, Ann. of Math. 37, 429 (1936)

2. L. CASTELL, DRISCHNER, C.F. VON WEIZSÄCKER, Quantum theory and the structure of space and time, Vol. 1-4

3. T. YAO, Unitary irreducible representations of SU(2,2), Jour. Math. Phys. 8, 1931 (1967); 9, 1615 (1968); 12, 315 (1971)

4. G. MACK & A. SALAM, Finite-component field representations of the conformal group, Ann. of Phys. 53, 174 (1969)

5. BUDINI & P. FURLAM & R. RACZKA, Weyl and conformal covariant field theories, Il Nuovo Cimento, 52, 191, (1979)

6. W. HEIDENREICH, Die dynamischen gruppen $SO_0(3,2)$ und $SO_0(4,2)$ als Raum-Zeit-Gruppen von elementarteilchen, Doctoral Thesis. Technischen Universität Munchen 1981

7. M. LORENTE & B. GRUBER, Construction of extremal vectors for Verma submodules of Verma modules, Journ. Math. Phys. 25, 1674 (1984)

8. M. LORENTE & B. GRUBER, Bilinear forms on envelopping algebras (preprint)

9. G. MACK, All unitary ray representations of the conformal group SU(2,2) with positive energy, Com. Math. Phys. 55, 1 (1977)

GLOBAL CONFORMAL TRANSFORMATIONS OF SPINOR FIELDS

P. Budinich, L. Dąbrowski* and H.R. Petry**
Scuola Internazionale Superiore di Studi Avanzati,
Strada Costiera 11, 34014 Trieste, Italy

ABSTRACT

Conformal transformations of spinor fields are discussed from the global point of view. The two inequivalent spinor structures exist on the minimal conformal compactification M of the Minkowski space time. They are interachanged by the space and space-time inversions. It is suggested that Dirac spinor fields should be coupled to a gauge potential in order to get a nontrivial unitary representation of the conformal group in the space of solutions of massless Dirac equation on M.

1. INTRODUCTION

There are several good reasons for considering the conformal symmetry in physics [c.f.1-5]. From the differential-geometric point of view, conformal transformations in the Minkowski space-time $R^{1,3}$ exhibit undesirable singularities either on a null cone, or on a null plane. A way out is to pass over to a bigger space, which we assume to be a minimal one, so that only a lower-dimensional part is added to $R^{1,3}$ (known as the 'light cone at infinity'). Obtained in such a way M, the minimal conformal compactification of Minkowski space-time, has a nontrivial global topology. In this paper we are interested in spinor fields on M, and, more generally, in global conformal transformations of spinor fields.

There are two interesting models for M. In the first one, where M is the projective null cone in $R^{2,4}$, a pair of 4-component spinors arises locally from the 8-component spinor of SO(2,4). Extending them to the whole M seems to be inconsistent [19]. In this paper we work with the second realization of M, as an underlying manifold of the unitary group U(2). The description of spinor structures on U(2) is straightforward.

After introducing the general setting for global conformal transformations of Dirac spinor fields and implementing the conformal group in a unitary way, we describe the conformal transformations of spinors on U(2) and interprete two kinds of spinor

* On leave of absence from Institute of Theoretical Physics, University of Wrocław, Poland.
** Permanent address: Institut für Theoretische Kernphysik, Universität Bonn, Nussallee 14-16, D-5300, West Germany.

fields in terms of asymptotic conditions in Minkowski space-time.

The content of the paper is as follows. In §2 we recall the way in which one introduces spinor fields in curved space-times and formulate a general setting for lifting conformal transformations to spinors. We describe the two inequivalent spinor structures on U(2) in §3. We characterize in §4 the form which the two related kinds of spinors assume after pull back to Minkowski space-time. In §5 we present the action of the full conformal group on spinor fields on U(2). We obtain the spectrum of the Dirac operator in §6. Finally, §7 contains a discussion of our results.

2. CONFORMAL TRANSFORMATIONS OF SPINOR FIELDS: GENERAL SETTING

Let M be a connected, four-dimensional, Lorentzian space-time equipped with a metric tensor g. Denote by D the Dirac representation of the group Pin(1,3) in C^4, where Pin(1,3) is the nontrivial double covering of the full Lorentz group O(1,3)

$$0 \to Z_2 \to \mathrm{Pin}(1,3) \overset{\rho}{\to} O(1,3) \to 0 \qquad (2.1)$$

and ρ is the vector (twisted adjoint, c.f. [6]) representation. <u>Dirac spinor fields</u> are defined as D-equivariant

$$\Psi(ph) = D^{-1}(h)\,\Psi\ , \qquad (2.2)$$

functions Ψ from some principal Pin(1,3)-bundle S into C^4. Here ph denotes the right translation of $p \in S$ by $h \in \mathrm{Pin}(1,3)$. (The equivalent definition is in terms of sections of the associated bundle $S \times_D C^4$). One requires the usual relation between spinors and tensors: the tangent bundle TM of M has to be isomorphic to the vector bundle associated to S with the vector representation ρ of Pin(1,3). This is equivalent to the existence of <u>spinor structure</u> S, η on M, i.e. prolongation of the bundle F of orthonormal frames to the structure group Pin(1,3), described by the commutative diagram

$$\begin{array}{ccc} S & \leftarrow & S \times \mathrm{Pin}(1,3) \\ \eta \downarrow & & \downarrow \eta \times \rho \\ F & \leftarrow & F \times O(1,3) \end{array} \qquad (2.3)$$

where the horizontal arrows denote right actions. If the structure group O(1,3) of F reduces to a subgroup G, the structure group Pin(1,3) of S reduces in the natural way to H, the double cover of G. For instance, $G = SO_o(1,3)$ and $H = \mathrm{Spin}_o(1,3) \simeq SL(2,C)$ if M is space and time oriented. The topological conditions for M to admit a spinor structure are well known [7,8], c.f. also [10] as well as the possibility of weakening the obstructions by coupling spinors to an external gauge potential (generalized spinor structure, or spin^c-structure in the case of gauge group U(1) [6,16,17,18].

Assuming that a spinor structure exists on M, it is, generally, not unique. Two spinor structures S,η and S',η' are <u>equivalent</u> if, and only if, there exists a (strong, or based) bundle isomorphism $\beta: S' \to S$, which intertwines η and η': $\eta' = \eta \circ \beta$. The equivalence classes of spinor structures are in a bijective correspondence with elements of the first cohomology group $H^1(M,Z_2)$ of M with coefficients in Z_2 [9,10] or, equivalently, with distinct homomorphisms from $\pi_1(M)$, the fundamental group of M, into Z_2 [11]. Apart from the possible inequivalence of S and S' as principal bundles, two spinor structures S,η and S,η' may be inequivalent due to different morphisms η and η'. In this case the two distinct covariant derivatives ∇ and ∇' can be defined. They are both related to the Levi-Civita connection on F, pulled back to S either by η, or by η' respectively (we identify the Lie algebras of Pin(1,3) and O(1,3) by means of $T\rho$, the derivative of ρ). Quite often one can perform a compensating Pin(1,3)-gauge transformation in a dense submanifold of M, and work with the same local expression for the covariant derivative ∇ . Then, spinor fields related to S,η' obey antisymmetric boundary conditions along all noncontractible loops in M which represent homotopy classes mapped into $-1 \in Z_2$ by the homomorphism corresponding to spin structure S,η'. The possible physical relevance has been discussed in the literature [12,13,14,15].

Now we formulate general results [20] on lifting conformal maps to morphisms of spin structures, which are needed in the sequel (c.f. [21] for the connected component of the group of isometries). Let $f: M' \to M$ be a conformal map with a positive conformal factor $\Omega: M \to R$, i.e. $f^*g = \Omega^2 g'$, where g' and g are metric tensors respectively on M' and M. Denote by $\hat{f}: F' \to F$ the natural morphism between bundles of orthonormal frames over M' and M induced by the rescaled derivative $\Omega^{-1}Tf$ of f.

<u>Proposition</u>.

For any spinor structure S,η over M there exists exactly one spinor structure S',η' over M' such that a given conformal map $f: M' \to M$ and the induced bundle morphism $\hat{f}: F' \to F$ lifts to a bundle morphism $\tilde{f}: S' \to S$ making the diagram (2.4) commute

$$\begin{array}{ccc} S' & \xrightarrow{\tilde{f}} & S \\ \eta \downarrow & & \downarrow \eta \\ F' & \xrightarrow{\hat{f}} & F \\ \downarrow & & \downarrow \\ M' & \xrightarrow{f} & M \end{array} \qquad (2.4)$$

Lifting $\tilde{f}$ is unique up to a sign. For $M' = M$ and $f \in Conf_o(M)$, the connected component of conformal group of M, the assignment $f \to \tilde{f}$ preserves the composition rule (up to a sign). This yields a representation either of $Conf_o(M)$, or at most of its double cover, in the space of spinor fields by

$$\Psi'(p') = \Omega^{3/2}\,\Psi(f(p')). \qquad (2.5)$$

The scaling degree 3/2 is fixed by requiring the invariance (up to factor) of the massless Dirac operator $\not{D}$

$$(\not{D}\Psi)' = \Omega^{-1}\not{D}\,\Psi.$$

A conformally invariant scalar product can be defined in the space of solutions of the massless Dirac equation on a Lorentzian manifold with a global space-like hypersurface. After passing to equivalence classes with respect to the induced norm one can unitarily implement $Conf_o(M)$ in the resulting Hilbert space.

3. TWO INEQUIVALENT SPINOR STRUCTURES ON M=U(2)

Let the points of the Minkowski space-time $R^{1,3}$ be represented by real linear combinations $x = x^m\sigma_m$ of antihermitian Pauli matrices σ_m, m=0,1,2,3; $\sigma_0=\begin{pmatrix} i & 0\\ 0 & i\end{pmatrix}$, $\sigma_1=\begin{pmatrix} 0 & i\\ i & 0\end{pmatrix}$, $\sigma_2=\begin{pmatrix} 0 & 1\\ -1 & 0\end{pmatrix}$, $\sigma_3=\begin{pmatrix} i & 0\\ 0 & -i\end{pmatrix}$. Let

$$\Sigma_m(x) = \frac{\partial}{\partial\tau}\Big|_o (x+\tau\sigma_m) \qquad (3.1)$$

be the standard orthonormal frame in $R^{1,3}$. Consider the underlying manifold of the unitary group U(2) together with the pseudoriemannian metric tensor, defined by the global right invariant frame $\sigma^R(u) \equiv \{(\sigma_m)^R(u)\}$, or, equivalently by the left invariant one $\sigma^L(u) \equiv \{(\sigma_m)^L(u)\}$, where to any antihermitian 2x2 matrix y we associate the two vector fields on U(2) by

$$y^R(u) = \frac{\partial}{\partial\tau}\Big|_o[(\exp\,\tau y)u] \qquad \text{and} \qquad y^L(u) = \frac{\partial}{\partial\tau}\Big|_o[u(\exp\,\tau y)] \qquad (3.2)$$

The Cayley map

$$f : R^{1,3} \to U(2)\,, \qquad x \to u = (1+x)(1-x)^{-1} \qquad (3.3)$$

is a dense conformal embedding of $R^{1,3}$, with the conformal factor

$$\Omega^2(x) = 4\det^{-1}(1-x^2)\,. \qquad (3.4)$$

From the diffeomorphism (not group isomorphism)

$$U(1) \times SU(2) \ni \left(z,\begin{pmatrix} a & b\\ c & d\end{pmatrix}\right) \longrightarrow \begin{pmatrix} za & b\\ zc & d\end{pmatrix} \in U(2)$$

it follows that M=U(2), the minimal conformal compactification of the Minkowski space-time,has the topology $S^1 x S^3$. Being parallelizable with $\pi_1(M) = Z_2$,M admits exactly two inequivalent spinor structures. They are given by the same (trivial) bundle

$S = M \times Pin(1,3)$ and two distinct morphisms η^R and η^L defined by

$$\eta^R(\tilde{\sigma}(u)) = \sigma^R(u) \tag{3.5a}$$

and

$$\eta^L(\tilde{\sigma}(u)) = \sigma^L(u) , \tag{3.5b}$$

where $\tilde{\sigma}(u)$ is some global section of S. The automorphism $\beta: S \to S$ intertwining η^R and η^L does not exist because the frames $\sigma^R(u)$ and $\sigma^L(u)$ are related by the nontrivial, u-dependent SO(3)-rotation $\rho(u)$

$$(\sigma_m)^R(u) = (\sigma_m)^L \rho(u) = (u^+ \sigma_m u)^L(u) . \tag{3.6}$$

Note that the phase of u is annihilated in Eq. (3.6), $Spin(3) = SU(2) \subset SL(2,C) = Spin(1,3)$ and $\rho \circ \ell$ is a noncontractible loop in SO(3), where $\ell: \tau \to \exp[\tau/2(\sigma_o + \sigma_3)]$ generates $\pi_1(M)$.

With respect to the global section $\tilde{\sigma}$ of S the Dirac spinor fields, related to two spinor structures, are just ordinary functions from M to C^4. One can however perform locally a SU(2)-gauge transformation $h(u) = u \det^{-½}(u)$ which covers $\rho(u)$ given by (3.6) and work with the same frame $\sigma_R(u)$ in both cases. With respect to $\sigma_R(u)$ the second kind of spinors can be thought of as functions defined on the submanifold $\det(1+u) \neq 0$ obeying antisymmetric boundary conditions along loops homotopic to $\ell(\tau)$. Then, the local expressions for connections (and covariant derivatives) are equal

$$\tilde{\sigma}^* \eta_R^* \Gamma = (\tilde{\sigma} h)^* \eta_L^* \Gamma . \tag{3.7}$$

4. THE FORM OF TWO SPINOR STRUCTURES IN $R^{1,3}$

In order to find how the two inequivalent spinor structures are manifested in $R^{1,3}$ we apply the results of §2.

Let $F' = R^{1,3} \times O(1,3)$ be the bundle of orthonormal frames in $R^{1,3}$ trivialized by the Cartesian frame $\Sigma(x)$ given by (3.1). The obvious spinor structure in $R^{1,3}$ is (S',η'), where $S' = R^{1,3} \times Pin(1,3)$ is trivialized by $\tilde{\Sigma}(x)$ and $\eta': S' \to F'$ is given by $\eta'(\tilde{\Sigma}(x)) = \Sigma(x)$. For the Cayley map f (3.3) the morphism $\hat{f}$ is given by

$$\hat{f}(\sigma'_m(x)) = \det^{½}(1-x^2)[(1-x)^{-1}\sigma_m(1+x)^{-1}]_R = \det^{½}(1-x^2)[(1+x)^{-1}\sigma_m(1-x)^{-1}]_L .$$

Therefore the morphisms $\tilde{f}_R: S' \to S$, $\tilde{f}_L: S' \to S$ assume the form

$$\tilde{f}_R(\tilde{\Sigma}(x)) = \pm \tilde{\sigma}_R(u(x)) S_R(x) \tag{4.1a}$$

and

$$\tilde{f}_L(\tilde{\Sigma}(x)) = \pm \tilde{\sigma}_L(u(x))\, S_L(x) \tag{4.1b}$$

where

$$S_R(x) = (1-x)^{-1}\det^{1/2}(1-x) \tag{4.2a}$$

and

$$S_L(x) = (1+x)^{-1}\det^{1/2}(1+x) \tag{4.2b}$$

are SL(2,C)-rotations.

Now we characterize the two kinds of Dirac spinor fields in the language of $R^{1,3}$. The behaviour on paths mapped by f into loops homotopic to the generator of $\pi_1(M)$ is important. Since all such loops meet the surface det(1+u)=0 we need a description of its points in terms of objects in $R^{1,3}$. Consider the family of straight lines in $R^{1,3}$, passing through all points $w \in R^{1,3}$ in all directions $v \neq 0$. Any such a line $Y_\tau = w+\tau v$, $\tau \in R$, asymptotically approaches in both directions the point $u_{w,v}$ of the 'light cone at infinity'

$$\lim_{\tau\to\infty} u(Y_\tau) = \lim_{\tau\to\infty} u(Y_{+\tau}) = \begin{cases} -1 & \text{if } \langle v,v\rangle \neq 0 \\ (-\langle v,w\rangle + v^j\sigma_j)(\langle v,w\rangle + iv^o)^{-1} & \text{if } \langle v,v\rangle = 0 \end{cases},$$

where $\langle w,v\rangle = w^i v^i - w^o v^o$. Also any $u \in U(2)$, such that det(1+u) = 0, can be obtained in this manner. Next, by inserting (4.1a,b) and (3.4) into (2.1) we obtain the local components of transformed spinor fields in $R^{1,3}$

$$\psi'_R(x) = \Omega^{3/2}(x) D^{-1}(S_R(x))\ \psi(u(x)),$$

$$\psi'_L(x) = \Omega^{3/2}(x) D^{-1}(S_L(x))\ \psi(u(x)).$$

The asymptotic overall fall off is

$$\lim_{\tau\to\infty} \Omega^{3/2}(Y_\tau) = \begin{cases} \tau^{-3}|\langle v,v\rangle|^{-3/2} & \langle v,v\rangle \neq 0 \\ \sqrt{2}/4\ \tau^{-3/2}(|v|^2+\langle v,w\rangle^2)^{-3/4} & \langle v,v\rangle = 0 \end{cases}.$$

The SL(2,C)-transformations (4.2a,b) asymptotically behave as

$$\lim_{\tau\to\infty} S_R(Y_\tau) = \begin{cases} \tilde{v}\langle v,v\rangle^{-1/2} & \text{if } \langle v,v\rangle \neq 0 \\ (2\tau)^{-1/2}(1+\tilde{w}+\tau\tilde{v})(\langle v,w\rangle + iv^o)^{-1/2} & \text{if } \langle v,v\rangle = 0 \end{cases} \tag{4.3a}$$

$$\lim_{\tau\to\infty} S_L(Y_\tau) = \begin{cases} \tilde{v}\langle v,v\rangle^{-1/2} & \text{if } \langle v,v\rangle \neq 0 \\ (2\tau)^{-1/2}(1+\tilde{w}+\tau\tilde{v})(\langle v,w\rangle + iv^o)^{-1/2} & \text{if } \langle v,v\rangle = 0 \end{cases} \tag{4.3b}$$

where $\underset{\sim}{v} = (v^o\sigma_o - v^i\sigma_i)$ and we choose a square root in a continuous way; the only ambiguity corresponds to ± in (4.1a,b). The matrix elements of (4.3a,b) are frame dependent and divergent in general. However, in any frame there is a crucial distinction between spinor fields coming from different spin structures in M:

$$\lim_{\tau\to\infty} S_R(Y_\tau)\, S_R(Y_{-\tau}) = \lim_{\tau\to\infty} S_L(Y_\tau) S_L(Y_{-\tau}) \times \begin{cases} +1 & \text{if } \quad \langle v,v\rangle \neq 0 \\ -1 & \phantom{\text{if }} \quad \langle v,v\rangle = 0 \end{cases} .$$

Therefore the arising two kinds of spinor fields in $R^{1,3}$ are twisted in such a way that they become asymptotically antisymmetric with respect to one another in all isotropic directions and symmetric in all nonisotropic directions in $R^{1,3}$.

5. CONFORMAL TRANSFORMATIONS IN M

The local action in $R^{1,3}$ of the connected component of the conformal group can be realized in terms of the four-fold covering group $SU(2,2) = \{T \in GL(4,C),\ \det T=1,\ THT^+ = H\}$, where $H=\begin{pmatrix} 0 & 1 \\ 1 & 0 \end{pmatrix}$. With translation $x \to x+t$, $t^+ = -t$; Lorentz rotation $x \to \rho(S)x$, $S \in SL(2,C)$; dilatation $x \to e^{2\lambda}x$, $\lambda \in R$ and special conformal transformation

$$x \to (x-k\langle x,x\rangle)(1-\langle k,x\rangle+\langle k,k\rangle\langle x,x\rangle)^{-1}$$

there are associated the following SU(2,2)-matrices T

$$\begin{pmatrix} 1 & t \\ 0 & 1 \end{pmatrix},\ \begin{pmatrix} S & 0 \\ 0 & (S^+)^{-1} \end{pmatrix},\ \begin{pmatrix} e^{\lambda} & 0 \\ 0 & e^{-\lambda} \end{pmatrix} \text{ and } \begin{pmatrix} 1 & 0 \\ k & 1 \end{pmatrix}$$

respectively. The action of $T = \begin{pmatrix} a, & b \\ c, & d \end{pmatrix}$ is

$$x \to (ax+b)(cx+d)^{-1} .$$

The corresponding action in M=U(2) is

$$u \to (Au+B)(Cu+D)^{-1} ,$$

where $A = a+b+c+d$, $B = d-a+b-c$, $C = d-a-b+c$ and $D = a-b-c+d$. Spinor fields associated with a fixed spinor structure transform according to (2.5). It is not so for discrete transformations P,T and PT

$$P : x \to \underset{\sim}{x} = \sigma_2 x^T \sigma_2^{-1},\quad T : x \to \underset{\sim}{x} = \sigma_2 \bar{x}\, \sigma_2^{-1},\quad PT : x \to -x,$$

which generate discontinuous components of $Conf(R^{1,3})$. On M=U(2) they become

$$P : u \to \sigma_2 u^T \sigma_2^{-1},\quad T : u \to \sigma_2 \bar{u}\, \sigma_2^{-1},\quad PT : u \to u^{-1} .$$

Since P and PT interchange the global frames σ_L and σ_R on M, their lifts to the morphisms $\tilde{P}$ and $\widetilde{PT}$ necessarily interchange the two inequivalent spinor structures on M. The time reversion T preserves both spinor structures.

6. SPECTRUM OF THE DIRAC OPERATOR ON M

We first compute the spectrum of the Dirac operator $\not{D} = i\eta^{mn}\gamma_m(\sigma_n^L+\Gamma_n^L)$ related to the spinor structure S,η_L. The (constant) matrices $\gamma_m = \begin{pmatrix} 0, & \sigma_m \\ \tilde{\sigma}_m, & 0 \end{pmatrix}$ obey

$$\{\gamma_m,\gamma_n\}_+ = \eta_{mn} = \mathrm{diag}(-,+,+,+) \quad \text{and} \quad \Gamma_m^L = (\sigma^L)^*\Gamma(\sigma_m^L)$$

is the spinor connection on M obtained from the Levi-Civita connection Γ. The conditions to be torsion-free and to preserve the metric, i.e. $[\Gamma_m^L,\gamma_n]-[\Gamma_n^L,\gamma_m]=\gamma([\sigma_m^L,\sigma_n^L])$ and Γ_m^L is a linear combination of skew bilinears in γ_m's, uniquely determine Γ_m^L: $\Gamma_o^L=0$, $\Gamma_j^L = \tfrac{1}{2}\begin{pmatrix} \sigma_j & 0 \\ 0 & \sigma_j \end{pmatrix}$, $j=1,2,3$. It can be seen by analogy with the spin-orbit coupling in quantum mechanics that the possible spectrum of $\not{D}$ is

$$k - \chi[\varepsilon(2\ell+1)+\tfrac{1}{2}], \tag{6.1}$$

where $k \in Z$, $2\ell \in Z$ and $\ell \geq 0$, $\varepsilon = \pm 1$ for $\ell \geq \tfrac{1}{2}$ and $\varepsilon = +1$ for $\ell = 0$, and $\chi = \pm 1$ is the chirality. The corresponding eigenfunctions

$$\Psi_{k\ell\varepsilon\chi}(\phi,A) = e^{ik\phi}\gamma_o(1+\chi\gamma_5)\times\sum_{p=-\ell}^{\ell} D^{(\ell)}_{p,q}(A)\left[(\ell,p,\tfrac{1}{2},\tfrac{1}{2}|\ell,\tfrac{1}{2},\ell+\tfrac{1}{2}\varepsilon,p)\begin{pmatrix}1\\0\\1\\0\end{pmatrix}+(\ell,p,\tfrac{1}{2},-\tfrac{1}{2}|\ell,\tfrac{1}{2},\ell+\tfrac{1}{2}\varepsilon,p)\begin{pmatrix}0\\1\\0\\1\end{pmatrix}\right],$$

where $u = e^{i\phi}A$, $D^{(\ell)}(A)$ is the representation of $A \in SU(2)$ with spin ℓ and $(|)$ denote the Clebsch-Gordan coefficients (degeneracy is partially labelled by the subscript q on the r.h.s.) are well defined on $U(2) = (U(1)\times SU(2))/Z_2$ if:

k is even and ℓ is integer, or, k is odd and ℓ is integer plus half. (6.2)

The spectrum of $\not{D}$ for the other spinor structure (S,η_R) can be similarly obtained by using the vierbein σ_m^R and $\Gamma_m^R = -\Gamma_m^L$. The allowed values for k and ℓ are exactly complementary to (6.2) in (6.1). This could also be seen by adopting the alternative picture c.f. §2. Then, $\Psi_{k\ell\varepsilon\chi}$ differ by sign at (ϕ,A) and $(-\phi,-A)$ (antisymmetric boundary conditions in $M=U(2)$).

None of the eigenvalues (6.1) is zero, and the natural unitary representation of $\mathrm{Conf}_o(R^{1,3})$ in the space of solutions of $\not{D}\psi=0$ on $M = U(2)$ is trivial.

7. CONCLUDING REMARKS

The postulate of a global smooth action of the conformal group leads to a conformal compactification M of Minkowski space-time, or to its universal (open) covering. Assumed to be minimal, M can be interpreted in terms of asymptotic conditions for fields in $R^{1,3}$, and can be used to study the conformally covariant Yang-Mills systems, also coupled to fermions [c.f. 22,23]. Being multiply connected, M admits two inequivalent spinor structures. They are both on the same footing and they are interchanged by parity and/or total reversions. To implement these fundamental reversions and to have standard properties of Dirac spinor fields one should consider both spinor structures at the same time and linearly combine the associated spinor fields.

The two massless, free Dirac operators associated with two spinor structures on M are intertwined by P and PT. Both of them have no zero eigenvalues. To obtain a nontrivial unitary representation of the connected component of $\mathrm{Conf}(R^{1,3})$ in the space of solutions of $\not{D}\psi=0$ a possibility is to minimally couple spinors to an additional gauge potential on M. In the simplest U(1)-case a candidate is $A=\pm(n+\frac{1}{4})\det^{-1}(u)d(\det(u))$ for $n\in Z$, where the zero eigenfunctions are of definite chirality (opposite for different spinor structures). The work is in progress on replacing spinors in the external potential A by a coupled system with a true dynamical field undergoing also the conformal transformations. In the framework of complex geometry and minimal conformal compactification M^c of complexified Minkowski space-time it is natural to consider a $\mathrm{Spin}^c=(\mathrm{Spin}\times U(1))/Z_2$ structure. On M, the real slice of M^c, there are also exactly two inequivalent Spin^c structures, which are interchanged by P and PT [24].

To get solutions of $\not{D}\psi=0$ on M (we do not consider coverings of M) another possibility is to introduce the torsion in M. The 'parallelizing' torsion on M can be eliminated by allowing the conformal factor Ω to be complex and by properly rotating vierbeins in M into the complexified directions of M^c [25]. The effects are equivalent to introducing the U(1)-gauge potential which couples to spinors by local phase rotations.

ACKNOWLEDGEMENTS

Two of the authors (P.B. and L.D.) are indebted to Prof. Roger Penrose for a discussion we had after completing this work, and for sending us a chapter on spinor structures of [24] prior to publication.

REFERENCES

[1] Penrose R., in Battelle Recontres 1967, Eds. C.M. De Witt and J.A. Wheeler, Benjamin N.Y. 1968.

[2] Segal I.E., Mathematical Cosmology and Extragalactic Astronomy, Acad. Press N.Y. 1976.

[3] Todorov I.T., Mintchev M.C. and Petkova V.B., Conformal Invariance in Quantum Field Theory, Scuola Normale Superiore, Pisa, 1978.

[4] Patashinskii A.Z. and Pokrovskii V.L., Fluctuation theory of phase transitions, Pergamon, Oxford 1979.

[5] Belavin A., Polyakov A.M., and Zamolodchikov A.B., Nucl. Phys.B241, 333, 1984.

[6] Atiyah M.F., Bott R. and Shapiro A., Topology 3 Suppl. 1, 3, 1964.
[7] Karoubi M., Ann. Ec. Norm. 4, 161, 1968.
[8] Haefliger A., C.R. Acad. Sci. Paris 243, 558, 1956.
[9] Milnor J., Enseignement Math. 9, 198, 1963.
[10] Greub W. and Petry H.R., in Lect. Notes in Math. 676, Springer Verlag, 1978.
[11] Whiston G.S., Int. J. Theor. Phys. 12, 225, 1975.
[12] Isham C.J., Proc. R. Soc. Lond. A364, 591, 1978.
[13] Petry H.R., J. Math. Phys. 20, 231, 1979.
[14] Ford L.H., Phys. Rev. D21, 933, 1980.
[15] Goncharov Y.P., Phys. Lett. B119, 403, 1982.
[16] Hawking S. and Pope C., Phys. Lett. B73, 42, 1978.
[17] Forger M. and Hess H., Commun. Math. Phys. 64, 269, 1979.
[18] Avis S.J. and Isham C.J., Commun. Math. Phys. 72, 103, 1980.
[19] Dąbrowski, L., Nuovo Cim. A80, 177, 1984.
[20] ______, Ph.D. Thesis, 1984 unpublished.
[21] Chichilnisky G., Trans. An. Math. Soc. 172, 307, 1972.
[22] Harnad J., Shnider S. and Vinet L., J. Math. Phys. 20, 931, 1979.
[23] Legare M., J. Math. Phys. 24, 1219, 1983.
[24] Penrose R. and Rindler W., Spinors and Space-time, Vol. II. Cambridge University Press, to be published.
[25] Todorov I.T., this volume.

PURE SPINORS FOR CONFORMAL EXTENSIONS OF SPACE-TIME

Paolo Budinich

International School for Advanced Studies (I.S.A.S.)

Strada Costiera, 11 - 34014 Trieste - Italy

1. Introduction

E. Cartan [1] based his definition of "simple" spinors, later on named "pure", on their equivalence with maximal null planes in euclidean complex spaces $C^{2\nu-1}$ and $C^{2\nu}$. This equivalence implies a bijective correspondence, named in a previous paper with A. Trautman [2] "the Cartan map", between pure spinor directions and maximal totally null planes. To the transitive action of Pin and Spin groups on pure spinors there corresponds the transitive action of the corresponding orthogonal groups on maximal totally null planes building up invariant manifolds (sometimes named quadric Grassmannians diffeomorphic to coset spaces [3]) laying on the projective null-quadric of the euclidean complex space.

For $\nu \geq 4$ pure spinor -directions build up invariant manifolds: non linear subsets (pure spinors-components are subject to quadratic constraints) in linear spinor-spaces, in bijective Cartan correspondence with invariant manifolds in euclidean spaces. For $\nu \leq 3$ instead, pure spinors fill the whole spinor- space (no constraints) and spinor-directions are bijectively mapped on invariant manifolds on projective quadrics of the corresponding euclidean spaces.

The Cartan conception of pure spinors assigns then to 2- and 4- component spinors ($\nu \leq 3$) the important and exceptional role of linearizing non linear projective manifolds (quadric Grassmannians) of complex 3-, 4-, 5-, and 6-dimensional complex euclidean spaces. Spinors of higher dimensional spaces (≥ 7) with 8, 16, ... components instead, if simple or pure, are fundamentally non linear in a similar way as the tensor-manifolds (quadric Grassmannians) Cartan-equivalent to their directions. Linearity in nature too seems to be the exception rather than the rule; and this suggests the conjecture that, if spinors play a fundamental role in the laws of elementary phaenomena, they should be rather conceived as simple or pure spinors rather than vectors in linear spaces.

The study of pure spinors has been extended by C. Chevalley [4] to the equivalent case of real neutral spaces $\mathbb{R}^{\nu,\nu-1}$, and $\mathbb{R}^{\nu,\nu}$ (we indicate with $\mathbb{R}^{p,q}$ a pseudo-euclidean real space with p-space and q-time directions).

Of physical interest are specially $\mathbb{R}^{2\nu-2,1}$ and $\mathbb{R}^{2\nu-1,1}$ real spaces (for G.U.T.) and possibly also, and perhaps more naturally, $\mathbb{R}^{\nu+1,\nu-2}$ and $\mathbb{R}^{\nu+1,\nu-1}$ real spaces, conformal extensions of $\mathbb{R}^{4,1}$ and of Minkowski spaces-time $\mathbb{R}^{3,1}$ (for $\nu > 3$ and for $\nu > 2$ respectively).

In this short note we will try to extend the concept and definitions of pure spinors to the latter real spaces, and to draw some consequences of possible physical interest.

2. Pure $\mathbb{R}^{\nu+1,\nu-1}$- and $\mathbb{R}^{\nu+1,\nu-2}$-spinors

We recall that pure $\mathbb{R}^{\nu,\nu}$ and $\mathbb{R}^{\nu,\nu-1}$-spinors [4][5][8] admit a transitive action of Spin and Pin groups respectively and their directions are isomorphic to tensor-manifolds (sets of totally null ν-and $(\nu-1)$-planes) diffeomorphic to $SO(\nu)$. And of their $2^{\nu-1}$ components ($2^{\nu-1}-1$ for directions) only $\binom{\nu}{2}$ are then independent (all of them for $\nu \leq 3$).

Restricting for the moment to $\mathbb{R}^{\nu,\nu}$, if $G_1,..G_{2\nu}$ are the generators of the corresponding Clifford algebra $Cl(\nu,\nu) = \mathbb{R}(2^\nu)$, a $\mathbb{R}^{\nu,\nu}$-spinor ξ_+ or ξ_- is pure iff

$$\xi_\pm >< \xi_\pm = r_\pm^{j_1..j_\nu} G_{j_1..j_\nu} \tfrac{1}{2}(1 \pm \Gamma_{2\nu+1}) \in Cl(\nu,\nu) \simeq \mathbb{R}(2^\nu) \qquad (1)$$

where $r_\pm^{j_1..j_\nu}$ are the (real) components of totally skew $\mathbb{R}^{\nu,\nu}$ ν-tensors, $\Gamma_{2\nu+1} = G_1 G_2 ... G_{2\nu}$ may be considered as representing a unit (space)-vector orthogonal to $\mathbb{R}^{\nu,\nu}$ and $G_{j_1..j_\nu}$ is the totally antisymmetrized product of $G_1, .., G_{j_\nu}$. Furthermore we define

$$< \xi = \xi^T C$$

where C is defined by:

$$C\, G_j^T = (-1)^\nu G_j C \qquad j = 1,...2\nu.$$

Eq.(1) implies the constraint equations, bilinear in pure spinor components (in number of 1, 10, 66, for $\nu = 4,5,6$ respectively):

$$< \xi_\pm G_{j_1..j_p} (1 \pm \Gamma_{2\nu+1}) \xi_\pm > = 0, \qquad \text{for } p < \nu \qquad (2)$$

and viceversa (2) implies (1). Both leave only $\binom{\nu}{2}$ of the $2^{\nu-1}-1$ spinor direction components independent (*). Eq. (2) and then (1) are identically satisfied for $\nu \leq 3$. The above may be extended to $\mathbb{R}^{\nu,\nu-1}$ spinors ξ ; ξ is pure iff

$$\xi > < \xi = r^{j_1 \cdot\cdot j_{(\nu-1)}} G_{j_1 \cdot\cdot j_{(\nu-1)}} \qquad j \leq 2\nu-1 \tag{1'}$$

with the same meaning, mutatis mutandis (Spin being substituted by Pin group), as in $\mathbb{R}^{\nu,\nu}$-case ($Cl(\nu,\nu-1) = \mathbb{R}(2^{\nu-1}) \oplus \mathbb{R}(2^{\nu-1})$). Constraint equations are:

$$< \xi \; G_{j_1 \cdot\cdot j_p} \; \xi > = 0 \qquad \text{for } p < \nu - 1$$

in the same number as in eq.(2).
Let now ζ represent an $\mathbb{R}^{\nu+1,\nu-1}$-spinor and $\Gamma_1, \Gamma_2 .. \Gamma_{2\nu}$ represent the generators of $Cl(\nu+1,\nu-1)$ Clifford algebras.
Since

$$Cl(\nu+1,\nu-1) \simeq \mathbb{R}(2^{\nu}) \simeq Cl(\nu,\nu), \tag{3}$$

we may expect to be able to define (see ref.[6,8] for further possible definitions of $\mathbb{R}^{\nu+1,\nu-1}$ pure spinors) $\mathbb{R}^{\nu+1,\nu-1}$-pure spinors ζ with the same properties as $\mathbb{R}^{\nu,\nu}$-ones ξ ; that is, transitivity of both Spin and orthogonal groups; Cartan map; dependence on $\binom{\nu}{2}$ parameters of both spinor directions and tensor-manifolds. The only substantial difference will be that while in the neutral case the elements of the manifolds may be identified with totally null ν-planes, in $\mathbb{R}^{\nu+1,\nu-1}$ case they will only contain totally null (ν-1)-planes (they are flags). With this aim in mind we will then define a $\mathbb{R}^{\nu+1,\nu-1}$-spinor as pure if it may be bijectively mapped to a corresponding $\mathbb{R}^{\nu,\nu}$-pure spinor.
We have then

<u>Theorem</u> <u>The $\mathbb{R}^{\nu+1,\nu-1}$-spinor ζ_+ or ζ_- is pure if and only if</u>

$$\zeta_\pm > < \zeta_\pm = r_\pm^{j_1 \cdot\cdot j_\nu} \Gamma_{j_1 \cdot\cdot j_\nu} \tfrac{1}{2}(1 \pm \Gamma_{2\nu+1}) \in Cl(\nu+1,\nu-1) \simeq \mathbb{R}(2^{\nu}) \tag{4}$$

<u>where $r_\pm^{j_1 \cdot\cdot j_\nu}$ are the real components of an $\mathbb{R}^{\nu+1,\nu-1}$-semi ν-vector.</u>

<u>Proof</u> We have (for $\nu > 2$)

(*) Eq.(1) and (2) imply the Cartan map : $\xi_\pm$ may be multiplied by an arbitrary factor and then spinors are substituted by spinor directions; correspondingly, the space ΛV of p-vectors in $\mathbb{R}^{\nu,\nu}$, linear space of $Cl(\nu,\nu)$, is substituted by the corresponding projective space $\mathbb{P}\Lambda V$.

$$Cl(\nu,\nu) = Cl(2,2) \otimes \mathbb{R}(2)_{(1)} \otimes .. \otimes \mathbb{R}(2)_{(\nu-2)} \tag{5}$$

correspondingly we may decompose the $\mathbb{R}^{\nu,\nu}$ spinor ξ in

$$\xi = \phi_1 \oplus \phi_2 \oplus .. \oplus \phi_{2^{(\nu-2)}} \tag{6}$$

where each ϕ_j is a 4-component $\mathbb{R}^{2,2}$-spinor, which, in turn, may be decomposed in the direct sum of two pure 2-components Weyl spinors ϕ^+ and ϕ^-

$$\phi_j = \phi_j^+ \oplus \phi_j^-$$

Let g_μ be the generator of $Cl(2,2) \simeq \mathbb{R}(4)$ and let $\gamma_5 = g_1 g_2 g_3 g_4$ represent a unit space vector orthogonal to $\mathbb{R}^{2,2}$ $(\gamma_5^2 = 1;\ \{\gamma_5,\gamma_\mu\}_+ = 0)$ then for each $\phi_j^\pm$ the Cartan map may be represented by

$$\phi^\pm > < \phi^\pm = r^\pm_{\mu\nu}\, g^{\mu\nu}\, \tfrac{1}{2}(1 \pm \gamma_5) \in Cl(2,2) \simeq \mathbb{R}(4) \tag{7}$$

where $r^\pm_{\mu\nu}$ represent semi 2-vector (real) components in $R^{2,2}$.

Since $Cl(2,2) \simeq \mathbb{R}(4) \simeq Cl(3,1)$ we may rearrange the Clifford algebra elements in (7) in such a way to obtain a Cartan map for $R^{3,1}$ Weyl spinors ψ. In fact, a possible easy choice is to interchange g_2 and γ_5 alone $(g_2^2 = -1;\ \gamma_5^2 = +1)$ and then obviously $\{\gamma_1, \gamma_2, \gamma_3, \gamma_4\} = \{g_1, \gamma_5, g_3, g_4\}$ generate $Cl(3,1) \simeq R(4)$ and $\gamma_1\gamma_2\gamma_2\gamma_4 = g_2$. One obtains (*) then from (7)

$$\psi^\pm > < \psi^\pm = f^\pm_{\mu\nu}\, \gamma^{\mu\nu}\, \tfrac{1}{2}(1 \pm \gamma_5') \tag{7'}$$

where $f^\pm_{\mu\nu}$ now represent the components of the selfdual and antiselfdual e.m. tensor. $\gamma'_5 = i\gamma_1\gamma_2\gamma_3\gamma_4$ represents a space unit vector orthogonal to $R^{3,1}$ ($= ig_2$ with the above choice). Obviously one may go back from (7') to (7): as an example taking as imaginary unit $\Sigma = \gamma_1\gamma_2\gamma_3\gamma_4\gamma_5'$ and then assuming $g_2 = \Sigma\gamma_2$ which, with the remaining generators of Cl(3,1), gives the generators of Cl.(2.2).

(*) One may also start from customary Cl(3,1) generators $\gamma_i = \sigma_1 \otimes \sigma_i$, $\gamma_4 = i\sigma_2 \otimes 1$ where σ_i are Pauli matrices. Then Cl(2,2) generators g_μ are obtained by taking $\Sigma = \gamma_1\gamma_2\gamma_3\gamma_4\gamma_5$ as imaginary unit and assuming $g_1 = \gamma_1 = \sigma_1 \otimes \sigma_1$; $g_2 = \Sigma\gamma_2 = \sigma_1 \otimes i\sigma_2$; $g_3 = \gamma_3 = \sigma_1 \otimes \sigma_3$; $g_4 = \gamma_4 = i\,\sigma_2 \otimes 1$ with this choice $\gamma_5 = \sigma_3 \otimes 1$. From these another useful set of Cl(3,1) <u>real</u> generators are obtained:

$$\gamma_1 = g_1 g_4 = \sigma_3 \otimes \sigma_1;\ \gamma_2 = g_3 g_4 = \sigma_3 \otimes \sigma_3;\ \gamma_3 = g_4\gamma_5 = \sigma_1 \otimes 1;\ \gamma_4 = g_4 = i\sigma_2 \otimes 1.$$

From (7') and the analogous of (5)

$$Cl(\nu+1, \nu-1) = Cl(3,1) \otimes \underset{(1)}{\mathbb{R}(2)} \otimes ..\underset{(\nu-2)}{\mathbb{R}(2)} \qquad (5')$$

and we will have that the $\mathbb{R}^{\nu+1,\nu-1}$ spinor ζ splits in

$$\zeta = \psi_1 \oplus \psi_2 \oplus..\oplus \psi_{2(\nu-2)} \qquad (6')$$

each ψ_j being an $\mathbb{R}^{3,1}$ spinor. But then if ξ satisfies (1), ζ satisfies (4) and the analogous of (2)

$$< \zeta_\pm \; \Gamma_{j_1..j_p} \tfrac{1}{2}(1 \pm \Gamma_{2\nu+1})\xi_\pm > \; = 0 \qquad \text{for } p < \nu \qquad (2')$$

by which the number of independent parameters of ζ-directions will be $\binom{\nu}{2}$. Furthermore the Spin group acts transitively on $\zeta_\pm$- pure as the corresponding orthogonal group on the $\mathbb{R}^{\nu+1,\nu-1}$ ν-tensor.

Remark 1 The Theorem may be easily extended also for $\mathbb{R}^{\nu+1,\nu-2}$ spinor ζ which will be pure if and only if

$$\zeta >< \zeta = z^{j_1..j_{\nu-1}} \Gamma_{j_1..j_{(\nu-1)}} \qquad j_\tau \leq 2\nu -1 \qquad (8)$$

where the tensor components $z^{j_1..j_{\nu-1}}$ may now be complex since $Cl(\nu+1,\nu-2) = C(2^{\nu-1})$.

Remark 2 It may be shown that the complex character of (8) may be completely assigned to the generators of the Clifford algebras (we need both $\Gamma_1\Gamma_2.. \Gamma_{2\nu-1}$ and $\Gamma_{2\nu} = G_1G_2...G_{2\nu-1}$ in $Cl(\nu+1, \nu-2)$; and their product gives an imaginary unit) in such a way that the tensor components in (8) may be taken real; they represent the direct sum of two (ν-1)-semi vectors plus their intersection, building up a flag (projective) isomorphic to the direction of ζ .

3. Projective spinor-spaces and projective quadrics for ν= 2,3.

Let us go back to $\mathbb{R}^{\nu,\nu}$-pure spinors. The totally null planes Cartan-corresponding to their direction lay on the projective quadric $x^2 = t^2$ (where $x^2 = x_1^2 +.. + x_\nu^2$ and the same for t^2) diffeomorphic, in general, to

$$\frac{S_{\nu-1} \times S_{\nu-1}}{Z_2}$$

Two important concepts are then introduced by the Cartan-map that of projective geometry both in spinor and pseudoeuclidean spaces, and the compact, topologically non trivial, feature of projective quadrics where the manifolds diffeomorphic to spinor directions lay. Due to the substantial equivalence of neutral spaces-spinors with those of the corresponding conformal extensions of space-time, implicit in the equalities $Cl(\nu,\nu) \simeq \mathbb{R}(2^{\nu}) \simeq Cl(\nu+1,\nu-1)$, the same feature implied by the Cartan map will be relevant for space-time and its conformal extensions $\mathbb{R}^{\nu+1,\nu-1}$, for which the projective $x^2 = t^2$ quadric will be, in general, diffeomorphic to (containing the null component of the manifold):

$$x^2 = t^2 \simeq \frac{S_{\nu} \times S_{\nu-2}}{Z_2}$$

(They may however present topological features different from the neutral case above). We wish to stress here that these properties of projective geometry and compactness are characteristic of the Cartan map even before the non linear properties of the set of pure spinors matters (at $\nu \geq 4$). Therefore, in order to examine their relevance, we may, taking advantage of the equalities (5) and (5'), recall the elementary case of $\mathbb{R}^{2,2}$ for which the maximal quadric Grassmannian is [3]

$$\Phi_2^{(+)}(\mathbb{R}^{2,2}) \simeq \frac{SO(2) \otimes SO(2)}{SO(2)} \simeq SO(2) \tag{9}$$

and compare it with the equivalent familiar case of two component Weyl spinors in $\mathbb{R}^{3,1}$. With obvious notations, we find from (7')

$$\psi^{\pm}><\psi^{\pm} = \sigma^{jh} f_{jk} \pm \sigma^{j} E_{j} = (iH_j \pm E_j)\,\sigma^{j} = z_j^{\pm}\,\sigma^{j}$$

or, (dropping for economy $\pm$ superscripts) for $\psi = \begin{vmatrix} \phi_o \\ \phi_1 \end{vmatrix}$

$$\begin{vmatrix} \phi_o\phi_1 & -\phi_o^2 \\ \phi_1^2 & -\phi_o\phi_1 \end{vmatrix} = \begin{vmatrix} z_3 & z_1 - iz_2 \\ z_1 + iz_2 & -z_3 \end{vmatrix} \tag{10}$$

where σ^j are Pauli matrices and we have adopted the pseudoscalar $\Sigma_3 = \sigma_1\sigma_2\sigma_3$ as imaginary unit i($\vec{H}$ is an axial $\mathbb{R}^3$-vector), by which $\psi^{\pm}><\psi^{\pm}$ element of Cl(3,1) is identified as element $(\overset{(1)}{V} + \overset{(2)}{V})$ (first equality) of Cl(3,0) and then of $Cl(C^3)$ $(z_j^- = -\bar{z}_j^+)$. In this last context the isotropy $z^2 = 0$ is expressed by the equation characterizing e.m. plane waves ($z^{\pm} = iH \pm E$):

$$E^2 - H^2 = 0 = E \cdot H \tag{11}$$

The $\mathbb{R}^{3,1}$ isotropic fourvector laying on the $\mathbb{R}^{3,1}$ projective null quadric may be easily obtained as bilinear ψ^+, ψ^--component polynomia (intersection of the manifolds corresponding to ψ^+ and ψ^-); to this end it is enough to take eq.(8) with $\nu=3$ which contains then a term $K_\mu \gamma^\mu \gamma^5$ (with the property (*) $K^\mu f^\pm_{\mu\nu} = 0$) such that

$$K_\mu = \langle \psi^+ \gamma_\mu \gamma_5 \psi^- \rangle = \{ \langle \phi \sigma_i \dot{\phi} \rangle ; \langle \phi \dot{\phi} \rangle \}$$

and satisfying the isotropy condition $K_\mu K^\mu = K^2 - K_o^2 = 0$. We have indicated with ϕ and $\dot{\phi}$ the two non zero components of ψ^+ and ψ^- respectively. K_μ components are generally complex; real ones may be obtained by substituting ϕ with its conjugate $\phi^c = -i\sigma_2 \phi$ and we obtain $K_\mu = X_\mu$ real given by

$$\{K_i, K_o\} = \{ \langle \phi^c \sigma_i \dot{\phi} \rangle ; \langle \phi^c \dot{\phi} \rangle \} = \{ X, Y, Z, T \} =$$
$$= \{ (|\phi_o|^2 - |\phi_1|^2), (\phi_o \bar{\phi}_1 + \bar{\phi}_o \phi_1), i(\phi_1 \bar{\phi}_o - \bar{\phi}_1 \phi_o), (|\phi_o|^2 + |\phi_1|^2) \} \tag{12}$$

which may be taken to represent a $\mathbb{R}^{3,1}$ light-vector. The corresponding projective quadric : $X^2 + Y^2 + Z^2 = T^2$ may be represented by

$$x_1^2 + x_2^2 + x_2^2 = 1$$

If we consider the $\mathbb{R}^{3,1}$-spinor $\psi = \begin{vmatrix} \phi_o \\ \phi_1 \end{vmatrix}$ we may choose a particular affine chart to represent its direction: $Z = \phi_1 / \phi_o$. The Cartan map may be then obtained (**) by considering Z as complex coordinates of the Argand plane, on which the Riemann sphere S_2 is stereographically projected. We have $Z = \rho e^{i\phi}$ and then locally up to a factor ρ, ϕ_1/ϕ_o represents an element of SO(2) as foreseen from (9) from which we also see that SO(2) is also a stability group of spinor ψ-direction. This can also be seen directly by considering the standard spinor $\psi_{st} = \begin{vmatrix} 1 \\ o \end{vmatrix}$ and a stability group is represented by SO(2) rotations of S_2 (about the z-axis) leaving the North Pole N fixed. S_2 above represents the conformal compactification of Argand plane, to which then the line at infinity $\{\infty\}$ (point) must be added (corresponding to N), where spinor directions are represented. To SO(3,1) transformations in $\mathbb{R}^{3,1}$ there correspond SL(2,C) spinor transformations and, due to the conformal nature of the stereographic projection, holomorphic mappings (i.e. complex-analytic) of z in the plane.

(*) K_μ, $f_{\mu\nu}$ build up then a flag Cartan-equivalent to the $\mathbb{R}^{4,1}$, 4-component spinor $\psi = \psi^+ \oplus \psi^-$.

(**) The Cartan-map may be also obtained directly from (10) and (11) precisely $E^2=1=H^2$. In the text we show instead the possible connection with the already classical way of introducing space-time spinors by R. Penrose [7].

These may form a subgroup in an infinite dimensional Lie group. An exhaustive presentation of all this may be found in Vol. I of ref. [7].

Let us now consider the $\mathbb{R}^{2,2}$-spinor $\phi^{\pm}$. We may reapeat the same procedure as before and obtain from (7)

$$\phi^{\pm} >< \phi^{\pm} = \tau^{jk} r^{\pm}_{jk} \pm \tau^{j} E_{j} = (H_{j} \pm E_{j})\, \tau^{j} = D^{\pm}_{j}\, \tau^{j} \qquad (10')$$

or (we drop for economy the ± superscript).

$$\begin{vmatrix} \phi_o \phi_1 & -\phi_o^{2} \\ \phi_1^{2} & -\phi_o\phi_1 \end{vmatrix} = \begin{vmatrix} D_3 & D_1 - D_2 \\ D_1 + D_2 & -D_3 \end{vmatrix}$$

where we have adopted $\tau^1 = \sigma^1$; $\tau^2 = -i\,\sigma^2$, $\tau^3 = \sigma^3$; and now we obtain instead of one equation of isotropy $Z^2 = 0$ (and its complex conjugate) two independent ones for D^+ and D^-; each of the form

$$D_1^{2} - D_2^{2} + D_3^{2} = 0;$$

(they are equivalent to (11) however with metric $g_{jk} = (+1, -1, +1)$.

For projective coordinates they correspond to a torus (take the affine chart $(D_2)^2 = 1$).

But we may also again compute $K_\mu = <\phi^+ g_\mu \gamma_5\, \phi^- >$ now real $= \{(\phi_o^+ \phi_o^- - \phi_1^+ \phi_1^-), - (\phi_o^+ \phi_o^- + \phi_1^+ \phi_1^-), - (\phi_1^+ \phi_o^- + \phi_o^+ \phi_1^-), (\phi_o^+ \phi_1^- - \phi_1^+ \phi_o^-)\}$ satisfying

$$K_1^2 - K_2^2 + K_3^2 - K_4^2 = 0$$

and the projective quadric may then be identified with a torus $k_1^2 + k_3^2 = 1 = k_2^2 + k_4^2$.

We could now repeat the procedure as in the previous case, the torus substituting the Riemann sphere, and, due to the conformal properties of stereographic projection SO(2,2) will reduce to the conformal group in compactified $\mathbb{R}^{1,1}$.

Let us adopt an isotropic basis

$$\theta_1 = \tfrac{1}{2}(g_1 + g_2); \qquad \eta_1 = \tfrac{1}{2}(g_1 - g_2)$$

$$\theta_2 = \tfrac{1}{2}(g_3 + g_4); \qquad \eta_2 = \tfrac{1}{2}(g_3 - g_4)$$

$$\{\theta_i, \theta_j\}_+ = 0 = \{\eta_i, \eta_j\}_+; \quad \{\theta_i, \eta_j\}_+ = \delta_{ij}$$

then the standard spinor $\phi_{st} = \begin{vmatrix}1\\0\end{vmatrix}$ may be represented by the minimal left Cl(2,2) ideal $\theta_2\theta_1$ (image of the corresponding maximal isotropic plane in Cl(2,2)); and the action of the spin group is

$$\exp\,[\alpha^{\pm}\theta_2\theta_1 + \beta^{\pm}\eta_2\eta_1 + \gamma^{\pm}]\ \theta_2\theta_1 = e^{\gamma^{\pm}}(1+\beta^{\pm}\eta_2\eta_1)\ \theta_2\theta_1 \,.$$

It is seen that if we wish to generate real spinors $\gamma^{\pm}$, $\beta^{\pm}$ are real and in the corresponding chart ϕ_1/ϕ_o will be real

$$\phi_1^{\pm}/\,\phi_o^{\pm} = \beta^{\pm}$$

we have the dependence of spinor direction on two real independent numbers, which may be represented by hyperbolic functions corresponding to the fact that in this case the group is not SO(2) but SO(1,1). SO(1,1) plus dilatations ($e^{\alpha^{\pm}\theta_2\theta_1}$, $e^{\gamma^{\pm}}$) is also, locally, a stability group of $\phi_{st} = \theta_2\theta_1$.(*) Globally instead it will be the conformal group in $R^{1,1}$.Also in this case we will have SL(2,R) corresponding to SO(2,2) and after stereographic projection from the torus to the plane to two independent Lie groups of transofrmations instead of one complex analytic (and its complex conjugate) apt ot be embedded in an infinite Lie group.It appears that the compact SO(2) belongs to complex spinors; while non compact SO(1,1) to real ones, both appearing as stability groups and as group diffeormorphic to spinor directions in $\mathbb{R}^{3,1}$ and $\mathbb{R}^{2,2}$ spinors(**). The above may also be applied to a 4-component $\mathbb{R}^{3,2}$ spinor direction (in a particular affine chart)

$$\xi = \begin{vmatrix}1\\ r_1\\ r_2\\ r_3\end{vmatrix}$$

generated by the action on the standard spinor $\theta_2\theta_1$

$$[1 + r_1\eta_1\eta_2]\ [1 + r_2\ \eta_2\ \tfrac{1}{2}(1+\gamma_5)]\ 1+r_3\eta_1\tfrac{1}{2}(1+\gamma_5)]\ \theta_2\theta_1$$

the stability group of $\theta_2\theta_1$ is the six parameter groups with generators:

(*) In both cases to $\phi_{st} = \begin{vmatrix}1\\0\end{vmatrix}$ there corresponds the $E^2 - H^2 = 0$ where $\{E_1, E_2, E_3\} = \{\tfrac{1}{2},0,0\}$; $\{H_1, H_2, H_3\} = \{0, -\tfrac{1}{2}, 0\}$.

(**) See note on page 4 ; taking the real generators γ_μ of Cl(3,1) we could generate real spinors.

$$e^{\gamma}\,[1-a_1\,\theta_2\theta_1]\,[1+a_2\,\gamma_5\theta_1]\,[1+a_3\gamma_5\theta_2]\,[1+a_4\tfrac{1}{2}(1-\gamma_5)\,\eta_1]$$

$$[1+a_5\tfrac{1}{2}(1-\gamma_5)\eta_2]\;\theta_2\theta \;=\; \theta_2\theta_1$$

4. The general case.Outlook.

For conformal extension of spce-time $\mathbb{R}^{\nu+1,\nu-1}$ spinor directions are diffeomorphic (if pure for $\nu\geq 4$) to Grassmannians whose quadric part lay in compact projective quadrics of the general form

$$Q_{\nu+1,\nu-1} = \frac{S_{\nu} \times S_{\nu-2}}{Z_2}$$

which contains the conformal compactification of the lower dimensional $\mathbb{R}^{\nu,\nu-2}$; precisely $Q_{\nu+1,\nu-1}$ consists of $\mathbb{R}^{\nu,\nu-2}$, plus its light-cone at infinity plus its $(2\nu-4)$dimensional projective light-cone. This in turn contains the conformal compactification of $\mathbb{R}^{\nu-1,\nu-3}$and so on, and the essential appearance of only light-cones is justified the conformal embeddings implementing massless physical systems. One may envisage in this "onion" structure of light-cones each embedded in a higher dimensional one a possible instrument of dimensional reduction. But perhaps the most interesting feature that results from this analysis is the possibility that projective spinor spaces and (for $\nu\geq 4$) pure spinor sets are diffeomorphic, if real, to compact projective manifolds in corresponding pseudoeuclidean spaces building up coset spaces possibly diffeomorphic to Lie groups of physical interest. This fact and the observation that for real (Majorana) spinors the stability groups seem to have the feature of conformal groups in $\mathbb{R}^{1,1}$ may suggest a natural origin of the role of this group in physics.

Another consequence of the adoption of the Cartan map is that the non-trivial topological structure of compact manifolds diffeomorphic to projective spinor manifolds may induce to take those manifolds rather than the pseudoeuclidean spaces as the basis of spinor structures. This, considering also the projective features of these spaces, could have far reaching consequences in corresponding field theories, and could induce to take seriously the existence of more non-equivalent spinor structures in topologically non-trivial manifolds which seem naturally to arise from the conformal extension of space-time.

These problems, the problem of internal symmetry, its compactness and its breaking, naturally arising from simplicity or purity, will be the object of further studies.

Acknowledgements

The author is pleased to thank Professor A. Trautman for illuminating discussions as well as Professors R. Raczka, R.W. Tucker, P. Furlan and L. Dabrowski for very interesting conversations.

References

[1] E. Cartan, Léçons sur la Théorie des Spineurs, Hermann, Paris (1938).

[2] P. Budinich, A. Trautman, Remarks on pure spinors, ISAS Preprint 87/85/EP.

[3] I.R. Portheous, Topological Geometry, Cambridge Univ. Press, Cambridge (1981).

[4] C. Chevalley, The Algebraic Theory of Spinors, New York (1954).

[5] P. Budinich, Pure spinors and quadric Grassmannians, ISAS Int/Rep. 37/85/EP.

[6] I.M. Benn and R.W. Tucker, Pure spinors and real Clifford algebras, Univ. of Lancaster preprint (1984).

[7] R. Penrose, W. Rindler, Spinors and space-time, Cambridge Univ. Press, Cambridge, (1984).

[8] P. Budinich and L. Dabrowski, Lett. Math. Phys. 10, 7, 1985.

Complex Clifford Analysis over the Lie Ball

by

John Ryan

School of Mathematics, University of Bristol, University Walk,

Bristol BS8 1TW, Britain.

Introduction:

In 1904 A.C. Dixon published a paper [7] in which a study was made of a differential equation, which today would be recognised as the time independent, massless Dirac equation. The equation is viewed in [7] as a natural generalization of the classical Cauchy-Riemann equations, and the function theoretic properties of its solutions are studied, via a generalized Cauchy integral formula. During the 1930's Fueter published a number of papers (e.g. [8,9]) in which the quaternion algebra is used to study an analogue of this equation over R^4. More recently Delanghe [5], Iftimie [12], Delanghe - Brackx [6], Brackx - Delanghe - Sommen [3], amongst others, have used real Clifford algebras to study properties of solutions to a homogeneous Dirac equation defined over R^n. This analysis is referred to as Clifford analysis [3]. Applications of this analysis, within mathematical physics, have been developed by a number of authors (e.g. [4,10,13,21]). In particular, in [13] Imaeda, while investigating Maxwell's equations, extends Fueter's quaternionic analysis to $\mathbb{C}^4 (\cong \mathbb{C}(2)$ - the algebra of 2 × 2 complex matrices).

In a series of recent papers (e.g. [18,19,20]) the author has used results, and ideas developed in [13], together with complex Clifford algebras, to develop a function theory for solutions to a Dirac equation defined over $\mathbb{C}^n$, where n is even. In this paper we restrict this holomorphic function theory to a special domain in $\mathbb{C}^n$. This domain is called the Lie ball [14], and it is Cartan's classical domain of type 4 (see [11]). We use a Runge approximation theorem to construct a holomorphic function which satisfies the equation introduced here, and is defined on the Lie ball, but which cannot be extended holomorphically beyond any point of its boundary. In a later work we shall use the methods employed here to obtain similar results for more general classes of domains than those considered here. In conclusion, we use the generalized Cauchy integral formula employed here to construct non-analytic continuous extensions to bounded holomorphic solutions to the Dirac equation, on the Lie ball. We characterize these continuous functions by means of integrals over real (n - 2) dimensional submanifolds, with boundary, of a null cone in $\mathbb{C}^n$. These integrals are closely related to the integral representations of solutions to the wave equation in space-time of even dimensions given by Riesz in [17], and to the formulae described by Penrose in [15].

Preliminaries:

In this section we develop the necessary algebraic and analytic background required for the rest of the paper.

In [2] and [16, chapter 13] it is shown that from the space R^n, with orthonormal basis $\{e_j\}_{j=1}^n$ it is possible to construct a 2^n dimensional, associative algebra A_n, with identity 1, and with $R^n \subseteq A_n$. Moreover the elements $\{e_j\}_{j=1}^n$ satisfy the relation $e_ie_j + e_je_i = 2\delta_{ij}$, where δ_{ij} is the Kroneker delta. This algebra is an example of a real Clifford algebra. By taking the tensor product of this algebra with the complex numbers we obtain the 2^n dimensional complex Clifford algebra, $A_n(\mathbb{C})$. The complex subspace spanned by the vectors $\{e_j\}_{j=1}^n$ is identified with $\mathbb{C}^n$ by the mapping $e_j \mapsto (0,\dots,0,1,0,\dots,0)$, where the unit appears in the jth place. A vector $z_1e_1+\dots+z_ne_n$ in $\mathbb{C}^n$ is denoted by $\underline{z}$. The null cone $\{\underline{z} \in \mathbb{C}^n : \underline{z}^2 = 0\}$ is denoted by $N(\underline{0})$, and for each point $\underline{z}_1 \in \mathbb{C}^n$, the null cone $\{\underline{z} \in \mathbb{C}^n : (\underline{z}-\underline{z}_1)^2 = 0\}$ is denoted by $N(\underline{z}_1)$.

Suppose that $D^n(R)$ is the disc, of radius R, lying in R^n, and centred at the origin, then we denote the component of $\mathbb{C}^n\backslash X$, where $X = \bigcup_{\underline{z} \in \partial D^n(R)} N(\underline{z})$, containing the interior of $D^n(R)$ by $\widetilde{B}^n(R)$. It may be deduced that

$$\widetilde{B}^n(R) = \{\underline{z} \in \mathbb{C}^n : ||\underline{z}||^2 + (||\underline{z}||^4 - |z_1^2+\ldots+z_n^2|^2)^{\frac{1}{2}} < R^2\} ,$$

where $||\underline{z}||$ is the Euclidean norm on $\mathbb{C}^n$. The domain $\widetilde{B}^n(R)$ is called a <u>Lie ball</u>, [14].

<u>Definition 1</u>: Suppose that U is a domain in $\mathbb{C}^n$, and that $f : U \to A_n(\mathbb{C})$ is a holomorphic function, which satisfies the equation

$$\sum_{j=1}^{n} e_j \frac{\partial f}{\partial z_j}(\underline{z}) = 0 \tag{1}$$

for each $\underline{z} \in U$. Then $f(\underline{z})$ is called a <u>left regular function</u> [18].

A similar definition may be given for right regular functions. Equation (1) may be seen to be a generalization of the Dirac equation studied in [3] and elsewhere.

From now on we shall assume that the integer n is even. As a special case of a theorem given in [18] we have:

<u>Theorem 1</u>: Suppose that $f : \widetilde{B}^n(R) \to A_n(\mathbb{C})$ is a left regular function then for each r with $0 < r < R$, and each $\underline{z}_0 \in \widetilde{B}^n(r)$ we have

$$f(\underline{z}_0) = \frac{1}{w_n} \int_{\partial D^n(r)} G(\underline{z} - \underline{z}_0) D\underline{z} f(\underline{z}),$$

where w_n is the surface area of the unit sphere in R^n, $G(\underline{z} - \underline{z}_0) = (\underline{z} - \underline{z}_0)\{\underline{z} - \underline{z}_0\}^{-n}$ and

$$D\underline{z} = \sum_{j=1}^{n} e_j (-1)^j dz_1 \wedge \ldots \wedge dz_{j-1} \wedge dz_{j+1} \wedge \ldots \wedge dz_n . \qquad \square$$

From [20] we have the following Runge approximation theorem:

<u>Theorem 2</u>: Suppose that V is a contractable subdomain of $D^n(r)$ and that U(V) is the component of $\mathbb{C}^n - Y$, where $Y = \bigcup_{z \in \partial\overline{V}} N(\underline{z})$, containing V. Suppose also that $f : U(V) \to A_n(\mathbb{C})$ is a bounded left regular function. Then, for each $R > r$, and each $\varepsilon > 0$, there is a left regular function

$$g_{\varepsilon,R} : \widetilde{B}^n(R) \to A_n(\mathbb{C})$$

such that

$$\sup_{\underline{z} \in U(V)} ||g_{\varepsilon,R}(\underline{z}) - f(\underline{z})|| < \varepsilon ,$$

where $|| \;\; ||$ denotes the Euclidean norm on $A_n(\mathbb{C})$. □

Left Regular Functions on $\widetilde{B}^n(R)$:

Definition 2: We denote the right $A_n(\mathbb{C})$ module of left regular functions defined over the Lie ball of radius R, by $\Omega(\widetilde{B}^n(R),A_n(\mathbb{C}))$.

Theorem 3: For each real, positive number R, there exists a function $f \in \Omega(\widetilde{B}^n(R),A_n(\mathbb{C}))$ which may not be holomorphically continued beyond any point of the boundary of $\widetilde{B}^n(R)$.

Proof: Suppose that the sequence $\{\underline{z}_o\}_{i=o}^{\infty}$ is a dense subset of $\partial D^n(R)$. Then, for each $R_1 > R$ there is a sequence $\{V_j\}_{j=1}^{\infty}$ where each V_j is a subdomain of $D^n(R_1)$ with the following properties:

1. each domain V_j is contractable to a point, within R^n
2. there is a disc $D^n(r_j)$ contained in V_j where $r_j < r_{j+1} < R$, and $\lim_{j\to\infty} r_j = R$
3. each $\overline{V}_j$ contains the points $\underline{x}_o,\ldots,\underline{x}_{j-1}$ in its interior, but it does not contain the point $\underline{x}_j$

and

4. for j the open set $\bigcap_{i=1}^{j} V_j$ is connected, and contractable to a point, within R^n.

The domains $\{V_j\}$ may be constructed by considering suitable homotopy retracts of the domain $\mathring{D}^n(R_1)$ within R^n.

Now consider the function

$$f(\underline{z}) = G(\underline{z}-\underline{x}_o) + \sum_{j=1}^{\infty} G(\underline{z}-\underline{x}_j) - g_j(\underline{z}) ,$$

where $\underline{z} \in \widetilde{B}^n(R)$, and $g_j : \widetilde{B}^n(R_1) \to A_n(\mathbb{C})$ is a left regular function, such that

$$\sup_{\underline{z} \in U(V_j)} ||G(\underline{z}-\underline{x}_j) - g_j(\underline{z})|| \le \frac{\varepsilon}{j^2}$$

for some $\varepsilon > 0$. It now follows from property 4 of the sets V_j that for each $k \geq 1$ and for each $\underline{z} \in N(\underline{x}_k) \cap (\widetilde{B}^n(R) \bigcup_{j \neq k} (N(\underline{x}_j) \cap (\widetilde{B}^n(R)))$, where $(\widetilde{B}^n(R))$ denotes the closure of the set $\widetilde{B}^n(R)$, there is a continuous function

$$\lambda_{\underline{z}} : (0,1) \to U(V_k) \cap \widetilde{B}^n(R)$$

such that $\lim_{t \to 1} \lambda_{\underline{z}}(b) = \underline{z}$, and the function $f_k(\underline{z}') = f(\underline{z}') - G(\underline{z}' - \underline{x}_k)$ is bounded on the set $\lambda_{\underline{z}}((0,1))$.

Consequently, the limit as t tends to 1 of $f(\lambda_{\underline{z}}(t))$ is not finite. Moreover, it is straightforward to deduce that the set of all such $\underline{z}$'s is dense in $\partial\, (\widetilde{B}^n(R))$. The result follows. □

We now proceed to consider boundary problems associated to left regular functions defined on closed neighbourhoods of the Lie ball, $\widetilde{B}^n(R)$.

Theorem 4: Suppose that $U \subseteq \mathbb{C}^n$ is a domain containing $\widetilde{B}^n(R)$ and $g : U \to A_n(\mathbb{C})$ is a left regular function. Then there exists a continuous function

$$\Gamma_g : \mathbb{C}^n \backslash (\partial D^n(R) \cup Y) \to A_n(\mathbb{C}) ,$$

where $Y = \{\underline{z} \in N(\underline{x}_i) \cap N(\underline{x}_j)$ for some $\underline{x}_i, \underline{x}_j \in \partial D^n(R)\}$, such that

i $$\Gamma_g\big|_{\widetilde{B}^n(R)} = g$$

and

ii $$\Gamma_g(\underline{z}) = 0$$

for each $\underline{z}$ with $N(\underline{z}) \cap \widetilde{B}^n(R) = \phi$.

Proof: Suppose that $\underline{z}_0 \in \mathbb{C}^n/R^n$ and $N(\underline{z}_0) \cap D^n(R) \neq \phi$, $N(z_0) \cap \partial D^n(R) \neq \phi$ then it may be deduced that the set $X(\underline{z}_0) = N(\underline{z}_0) \cap D^n(R)$ is a $(n-2)$ dimensional manifold, with boundary, and that this manifold is a submanifold of an $(n-2)$ dimensional sphere.

If $K(\underline{z}_0,\varepsilon)$, $\subseteq \partial D^n(R)$, is a closed neighbourhood of $X(\underline{z}_0) \cap \partial D^n(R)$ of volume ε, for some suitable $\varepsilon \in R^+$, and such that $\partial K(\underline{z}_0,\varepsilon)$ is homeomorphic to $S' \times X(\underline{z}_0)$, then we have from Stokes' theorem that

$$\frac{1}{w_n}\int_{\partial D^n(R)} G(\underline{z}-\underline{z}_o)D\underline{z}g(\underline{z}) = \frac{1}{w_n}\int_{K(\underline{z}_o,\varepsilon)\cup T(\underline{z}_o,\varepsilon)} G(\underline{z}-\underline{z}_o)D\underline{z}g(\underline{z}) ,$$

where $T(\underline{z}_o,\varepsilon)$ is an $(n-1)$ dimensional submanifold of $\widetilde{B}^n(R)$, with boundary, and satisfying the conditions:

1 $$\partial T(\underline{z}_o,\varepsilon) = \partial K(\underline{z}_o,\varepsilon)$$

2 $$T(\underline{z}_o,\varepsilon) \cap X(\underline{z}_o) = \phi$$

and

3 $T(\underline{z}_o,\varepsilon)$ is an S^1 fibration of an $(n-2)$ dimensional submanifold, $K'(\underline{z}_o,\varepsilon)$ of $K(\underline{z}_o,\varepsilon)$.

Moreover, we have that for each point $\underline{x} \in K'(\underline{z}_o,\varepsilon)$ the fibre S^1 lies in the plane passing through $\underline{x}$, and spanned by the vectors $\underline{x}$-Re $\underline{z}_o$ and i Im $\underline{z}_o$, where

$$\text{Re }\underline{z} = x_1e_1+\ldots+x_ne_n$$

$$\text{Im }\underline{z} = iy_1e_1+\ldots+iy_ne_n$$

with $\underline{z}_o = (x_1+iy_1)e_1+\ldots+(x_n+iy_n)e_n$.

We may now introduce the following homotopy:

$$H : T(\underline{z}_o,\varepsilon)\times[0,1] \to \widetilde{B}^n(R) : H(\underline{x}+r\cos\theta(\underline{x}-\text{Re }\underline{z}_o)+ir\sin\theta\,\text{Im}\,\underline{z}_o,t)$$

$$= \underline{x}+r\{\cos\theta(\underline{x}-\text{Re }\underline{z}_o)\psi_\varepsilon(\underline{x})\cos\tfrac{\pi}{2}t+\text{Im }\underline{z}_o\,\psi_\varepsilon(\underline{x})\sin\tfrac{\pi}{2}t+i\,\text{Im }\underline{z}_o\sin\theta\} ,$$

where r is the radius of the fibre S^1, θ is a parameterization of this circle, $t \in [0,1]$, and $\psi_\varepsilon : R^n \to R$ is a C^∞ function with compact support $D^n(R-\varepsilon)$, and

$$\psi_\varepsilon\Big|_{D^n(R-2\varepsilon)} = 1 .$$

As $H(T\underline{z}_o,\varepsilon),[0,1]) \cap N(\underline{z}_o) = \phi$ we have from Stokes' theorem that

$$\frac{1}{w_n}\int_{K(\underline{z}_o,\varepsilon)\cup T(\underline{z}_o,\varepsilon)} G(\underline{z}-\underline{z}_o)D\underline{z}g(\underline{z}) = \frac{1}{w_n}\int_{K(\underline{z}_o,\varepsilon)\cup H(T(\underline{z}_o,\varepsilon),1)} G(\underline{z}-\underline{z}_o)D\underline{z}g(\underline{z}) .$$

We now have

$$\frac{1}{w_n}\int_{H(T'(\underline{z}_o,\varepsilon),1)} G(\underline{z}-\underline{z}_o)D\underline{z}g(z) = \frac{1}{w_n}\int_{H(T'(\underline{z}_o,\varepsilon),1)} \left\{\frac{\underline{y}}{(\|y\|^2-z^2)^{\frac{n}{2}}} - \frac{z\underline{\psi}}{(\|y\|^2-z^2)^{\frac{n}{2}}}\right\}D\underline{z}g(\underline{z}), \qquad (2)$$

where $T'(\underline{z}_o,\varepsilon)$ is the restriction of $T(z_o,\varepsilon)$ to the fibering of $D^n(R-2\varepsilon)\cap K(\underline{z}_o)$, $\underline{y} = \underline{x}-\text{Re }\underline{z}_o$, $\underline{\psi} = \text{Im }\underline{z}_o\ (\|i\text{ Im }\underline{z}_o\|)^{-1}$ and $z = (-1 + \cos\theta + i\sin\theta)\ \|i\text{ Im }\underline{z}_o\|$.

On placing $k = {}^n/2$ we now have from the residue theorem of one variable complex analysis [1] that the right hand side of equation (2) evaluates to

$$\frac{2\pi i}{w_n}\int_{D^n(R-2\varepsilon)\cap K(\underline{z}_o)} \sum_{p=o}^{k}\binom{k}{p}\frac{1}{(2\|y\|)}k\left\{\underline{y}\text{ Res}\left(\frac{g_{p,y}(\underline{z})}{\underline{z}_1^{\,p}}\right) - \underline{\psi}\text{ Res}\left(\frac{g_{p,y}(\underline{z})}{z_1^{\,p-1}}\right)\right\}\underline{n}(\underline{x})dK(\underline{z}_o),$$

where $z_1 = z + \|i\text{ Im }\underline{z}_o\|$, $\underline{z} = \underline{x}+z_1\text{ Im }\underline{z}_o$, $\text{Res}\left|\frac{g_{p,y}(\underline{z})}{z_1^{\,p}}\right|$ denotes the residue of $(\|y\|+z)^{p-k}\,g(\underline{z})z_1^{-p}$, $\underline{n}(x)$ is a unit vector in R^n normal to the surface $K(\underline{z}_o)$ at $\underline{x}$, and orthogonal to the real vector $i\text{ Im }\underline{z}_o$, also $dK(\underline{z}_o)$ is the Lebesgue measure on the $(n-2)$ dimensional manifold $K(z_o)$.

It may now be deduced that

$$\lim_{\varepsilon\to o}\frac{1}{w_n}\int_{K(\underline{z}_o,\varepsilon)\cup T(\underline{z}_o,\varepsilon)} G(\underline{z}-\underline{z}_o)D\underline{z}g(z) = \frac{2\pi i}{w_n}\int_{K(\underline{z}_o)} \sum_{p=o}^{k}\binom{k}{p}\frac{1}{(2\|y\|)}k\left\{y\text{ Res}\left|\frac{g_{p,y}(\underline{z})}{z_1^{\,p}}\right| - \underline{\psi}\text{ Res}\left|\frac{g_{p,y}(z)}{z_1^{\,p-1}}\right|\right\}\underline{n}(\underline{x})dK(\underline{z}_o).$$

It may be observed that for each sequence $\{\underline{z}_i\}_{i=1}^{\infty}\subseteq ¢^n\backslash R^n$, with limit $\underline{z}\in ¢^n\backslash R^n$ we have that

$$\lim_{i\to\infty} K(\underline{z}_i) = K(\underline{z}).$$

Consequently we have that the function

$\Gamma_g : ¢^n\backslash(R^n\cup Y) \to A_n(¢)$

$$: \Gamma_g(\underline{z}_o) = \frac{2\pi i}{w_n}\int_{K(z_o)}\sum_{p=o}^{k}\binom{k}{p}\frac{1}{(2\|y\|)}k\left\{\underline{y}\text{ Res}\left(\frac{g_{p,y}(\underline{z})}{z_1^{\,p}}\right) - \underline{\psi}\text{ Res}\left(\frac{g_{p,y}(\underline{z})}{z_1^{\,p-1}}\right)\right\}\underline{n}(\underline{x})dK(\underline{z}_o)$$

is a continous function.

On placing $\Gamma_g(\underline{x}) = 0$ for each $\underline{x}\in R^n\backslash D^n(R)$ and $\Gamma_g(x) = g(x)$ for each $x\in \mathring{D}^n(R)$ we may continuously extend the function Γ_g to $¢^n\backslash(\partial D^n(R)\cup Y)$.

In the cases where $\underline{z}_0 \in \tilde{B}^n(R)/\mathcal{B}^n(R)$ the residue integrals given in the proof of theorem 4 correspond to integrals for left regular functions given in [4], [13], and [21].

References:

[1] L.V. Ahlfors, 'Complex Analysis', 2nd edition, McGraw-Hill, 1966.

[2] M.F. Atiyah, R. Bott and A. Shapiro, 'Clifford modules' Topology 3, 1965, 3-38.

[3] F. Brackx, R. Delanghe and F. Sommen, 'Clifford Analysis' Research Notes in Mathematics, No.76, Pitman, 1982.

[4] J. Bures and V. Soucek, 'Generalized hypercomplex analysis and its integral formulae' to appear in Complex Variables: Theory and Application.

[5] R. Delanghe, 'On regular analytic functions with values in a Clifford algebra' Mathematische Annalen 185, 1970, 91-111.

[6] R. Delanghe and F. Brackx, 'Hypercomplex function theory and Hilbert modules with reproducing kernel' Proceedings of the London Mathematical Society 37, 1978, 545-576.

[7] A.C. Dixon, 'On the Newtonian potential' Quarterly Journal of Mathematics 35, 1904, 283-296.

[8] R. Fueter, 'Die Funktionentheorie der Differentialgleichungen $\Delta u = 0$ and $\Delta\Delta u = 0$ mit Vier Reallen Variablen' Commentarii Mathematici Helvetici, 7, 1934, 307-330.

[9] R. Fueter, 'Über die Analytische Darstellung der Regulären Funktionen einer Quarternionenvariablen' Commentarii Mathematici Helvetici 8, 1935, 371-378.

[10] F. Gursey and H.C. Tze, 'Complex and quaternionic analyticity in chiral and gauge theories I' Annals of Physics 128, 1980, 29-130.

[11] K. Hua, 'Harmonic Analysis of Functions of Several Complex Variables in Classical Domains' Translations of Math. Monographs, 6, AMS, Providence, Rhode Island, 1963.

[12] V. Iftimie, 'Fonctions hypercomplexes' Bull. Math. Soc. Sci. Math. R.S. Roumanie 57, 1965, 279-332.

[13] K. Imaeda, 'A new formulation of electromagnetism' Nuovo Cimento 32B, 1976, 138-162.

[14] M. Morimoto, 'Analytic functionals on the Lie sphere' Tokyo Journal of Mathematics 3, 1980, 1-35.

[15] R. Penrose, 'Null hypersurface initial data for classical fields of arbitrary spin and for general relativity' General Relativity and Gravitation 12, 1980, 225-265.

[16] I. Porteous, 'Topological Geometry' Van Nostrand Company, 1969.

[17] M. Riesz, 'A geometric solution of the wave equation in space-time of even dimension' Communications on Pure and Applied Mathematics XIII, 1960, 329-351.

[18] J. Ryan, 'Special functions and relations in complex Clifford analysis' Complex Variables: Theory and Application 2, 1983, 177-198.

[19] J. Ryan, 'Duality in complex Clifford analysis' Journal of Functional Analysis 61, 1985, 117-135.

[20] J. Ryan, 'Runge approximation theorems in complex Clifford analysis, together with some of their applications' to appear.

[21] V. Soucek, 'Complex quaternionic analysis applied to spin $\frac{1}{2}$ massless fields' Complex Variables: Theory and Application 1, 1983, 327-346.

V. LIE GROUPS, -ALGEBRAS AND SUPERALGEBRAS

PLANCHEREL THEOREM FOR THE UNIVERSAL COVER OF THE CONFORMAL GROUP

Rebecca A. Herb*

Department of Mathematics
University of Maryland
College Park, Maryland 20742

Joseph A. Wolf†

Department of Mathematics
University of California
Berkeley, California 94720

§0. INTRODUCTION

Ever since the discovery of conformal invariance of Maxwell's equations, conformal groups and conformal structures have had an important role in mathematical physics. For the most part, this role has been confined to consequences of the geometric action of the conformal group and to the use of certain unitary representations. Now there is a fairly explicit theory of harmonic analysis on the conformal group. It seems likely that this theory will be of some physical importance, e.g. in partial wave analysis.

When a group G acts by geometric symmetries on a space X, it also acts on various spaces of functions on X. These functions are better understood by taking the symmetries into account. That is, of course, the basic idea in Fourier analysis. It has also been exploited in the use of spherical harmonics, where $G = SO(3)$ and $X = S^2$, and in the application of other sorts of special functions.

Now the machinery is available for the case where G is the conformal group or one of its coverings, and X is either the space G itself or is a symmetric homogeneous space of G.

In this article we describe some of those developments, first sketching the general theory and then describing the case of the simply connected covering group $\widetilde{SU}(2,2)$ of the conformal group.

*Research partially supported by National Science Foundation grant DMS-84-01374.
†Research partially supported by National Science Foundation grant DMS-85-13467.

We assume that the reader is well acquainted with Mackey theory but not so well acquainted with Harish-Chandra theory.

The results in §§1 and 2 have been known for some time. The Plancherel formula in §3 was worked out by Harish-Chandra in the 1960's (published somewhat later in [1], [2] and [3]) for semisimple groups with finite center. We recently developed another approach ([4],[5]) which allows infinite center, as in the group $\widetilde{SU}(2,2)$. In §4 we work out the constants to obtain an explicit formula for $\widetilde{SU}(2,2)$.

§1. TEMPERED REPRESENTATIONS OF SEMISIMPLE GROUPS

We describe the representations involved in the Plancherel formula for a semisimple group. To do this for a class of semisimple groups, one must ensure that certain subgroups belong to the same class. Our class consists of the reductive Lie groups G that have a closed normal abelian subgroup Z such that

(1.1) $\quad Z$ centralizes the identity component G^0 and G/ZG^0 is finite

and

(1.2) $\quad$ If $x \in G$ then $\mathrm{Ad}(x)$ is an inner automorphism of $\mathfrak{g}_{\mathbb{C}}$.

Here "reductive" means that the Lie algebra $\mathfrak{g}$ of G is (commutative)$\oplus$(semisimple).

If $\pi \in \hat{G}$, the set of equivalence classes of irreducible unitary representations of G, then π has three types of characters. The *central character* ζ_π is the scalar valued function on the center Z_G that is given by $\pi(z) = \zeta_\pi(z)\cdot I$ where I is the identity on the representation space $\mathcal{H}(\pi)$. The *infinitesimal character* χ_π is the map on the center $\mathcal{Z}(\mathfrak{g})$ of the enveloping algebra $\mathcal{U}(\mathfrak{g})$ given by $d\pi(D) = \chi_\pi(D)\cdot I$ for every Casimir operator D. We view it as a homomorphism $\mathcal{Z}(\mathfrak{g}) \to \mathbb{C}$ of associative algebras. The *character* or *distribution character* $\Theta(\pi)$ is the Schwartz distribution on G given by

$$(1.3) \qquad \Theta(\pi:f) = \operatorname{trace} \pi(f) \quad \text{for} \quad f \in C_c^\infty(G)$$

where $\pi(f) = \int_G f(x)\pi(x)dx$. The equivalence class of π is specified by $\Theta(\pi)$. The differential equations

$$(1.4) \qquad D\Theta(\pi) = \chi_\pi(D)\cdot\Theta(\pi) \quad \text{for} \quad D \in \mathcal{Z}(\mathfrak{g})$$

lead to the information that $\Theta(\pi)$ is integration against a locally L^1 function $T(\pi)$,

$$(1.5) \qquad \Theta(\pi:f) = \int_G T(\pi:x)f(x)dx \quad ,$$

which is real analytic on a certain dense open subset ("regular set") G' in G.

We may suppose $Z \cap G^0 = Z_{G^0}$. Then $\pi \in \hat{G}$ belongs to the *relative discrete series* if its coefficients

$$(1.6) \qquad \phi_{u,v}(x) = \langle u, \pi(x)v\rangle_{\mathcal{H}(\pi)} \quad , \qquad u,v \in \mathcal{H}(\pi) \; ,$$

are L^2 modulo Z. The representations we will use will be constructed from relative discrete series representations.

Choose a Cartan involution θ of G. In other words, θ is an automorphism of G, $\theta^2 = 1$, θ is the identity on $Z_G(G^0)$, and the fixed point set K of θ satisfies: $K/Z_G(G^0)$ is a maximal compact subgroup of $G/Z_G(G^0)$.

Let $\mathfrak{h}$ be a Cartan subalgebra of $\mathfrak{g}$, i.e. a maximal diagonalizable (over $\mathbf{C}$) subalgebra. Then $H = \{x \in G: Ad(x)\xi = \xi \text{ for all } \xi \in \mathfrak{h}\}$ is the corresponding Cartan subgroup. One can find $x \in G^0$ such that $Ad(x)\mathfrak{h}$ and xHx^{-1} are θ-stable.

If G has relative discrete series representations, then K contains a Cartan subgroup of G. Conversely, let $T \subset K$ be a Cartan subgroup of G, $\mathfrak{t}$ its Lie algebra, and $\Phi^+ = \Phi^+(\mathfrak{g},\mathfrak{t})$ a system of positive roots. Let $\rho = \frac{1}{2}\Sigma_{\Phi^+}\alpha$ and set

$$(1.7) \qquad \Lambda' = \lambda \in i\mathfrak{t}^*: \begin{cases} \langle\lambda,\alpha\rangle \neq 0 \text{ for all } \alpha \in \Phi^+ \text{ and} \\ e^{\lambda-\rho} \text{ is well defined on } T^0 . \end{cases}$$

If $\lambda \in \Lambda'$ then there is a relative discrete series representation π_λ^0 of G^0 such that

$$(1.8) \qquad T(\pi_\lambda^0 : x) = (\text{constant})\Delta(x)^{-1} \sum_{w \in W^0} \det(w)\, e^{w\lambda}(x)$$

where $x \in T^0 \cap G'$, $\Delta(x) = \prod_{\alpha\in\Phi^+}(e^{\alpha/2} - e^{-\alpha/2})(x)$, and W^0 is the Weyl group of (G^0,T^0). The central character of π_λ^0 is $e^{\lambda-\rho}|_{Z_{G^0}}$. If $\chi \in Z_G(G^0)^\wedge$ agrees with π_λ^0 on Z_{G^0}, then

$$(1.9) \qquad \chi \otimes \pi_\lambda^0 \in (G^\dagger)^\wedge \; , \qquad G^\dagger = Z_G(G^0)G^0$$

is well defined and is a relative discrete series representation of $G^\dagger$. Finally, the relative discrete series of G consists of the

$$(1.10) \qquad \pi_{\chi,\lambda} = \text{Ind}_{G^\dagger}^G(\chi \otimes \pi_\lambda^0) \; , \qquad \lambda \in \Lambda' \; , \quad \chi \in Z_G(G^0)^\wedge$$

where π^0_λ and χ agree on Z_{G^0}.

Now let H be any θ-stable Cartan subgroup of G. Then $\mathfrak{h} = \mathfrak{t}_H \oplus \mathfrak{a}_H$, ± 1 eigenspaces of $\theta|_{\mathfrak{h}}$, and $H = T_H \times A_H$ where $T_H = H \cap K$ and $A_H = \exp_G(\mathfrak{a}_H)$. Then the centralizer $Z_G(A_H) = M_H \times A_H$ where $\theta M_H = M_H$ and T_H is a Cartan subgroup of M_H. Let $\Phi^+_{\mathfrak{a}_H} = \Phi^+(\mathfrak{g},\mathfrak{a}_H)$ be a system of positive $\mathfrak{a}_H$-roots, $\mathfrak{n}_H = \sum_{\alpha \in \Phi^+_{\mathfrak{a}_H}} \mathfrak{g}_\alpha$, N_H the corresponding analytic subgroup of G, and $P_H = M_H A_H N_H$ the associated "cuspidal parabolic" subgroup of G. If $\nu \in \Lambda'_{M_H}$, η^0_ν is the corresponding relative discrete series representation of M^0_H, and for $\chi \in Z_{M_H}(M^0_H)^\wedge$ consistent with η^0_ν then $\eta_{\chi,\nu} = \mathrm{Ind}^{M_H}_{M^\dagger_H}(\chi \otimes \eta^0_\nu)$ is the corresponding relative discrete series representation of M_H. If $\sigma \in \mathfrak{a}^*_H$ now $\eta_{\chi,\nu} \otimes e^{i\sigma} \in (M_H A_H)^\wedge$ extends to P_H, trivial on N_H, and we have the unitary representation

$$\pi_{\chi,\nu,\sigma} = \mathrm{Ind}^G_{P_H}(\eta_{\chi,\nu} \otimes e^{i\sigma}) \quad . \tag{1.11}$$

It does not depend on choice of $\Phi^+_{\mathfrak{a}_H}$. The representations (1.11) of G constitute the H-*series*. If $A_H = \{1\}$ that is the *relative discrete series*. If A_H is maximal it is the *principal series*. Given ν and χ, $\pi_{\chi,\nu,\sigma}$ is irreducible for almost all σ. The irreducible constituents of representations $\pi_{\chi,\nu,\sigma}$, H variable, are the *tempered* representations of G.

When we are dealing with several Cartan subgroups we will write

(1.12a) $\pi(H:\chi:\nu:\sigma)$ for the H-series representation $\pi_{\chi,\nu,\sigma}$,

(1.12b) $\Theta(H:\chi:\nu:\sigma)$ for the distribution character of $\pi_{\chi,\nu,\sigma}$, and

(1.12c) $\Theta(H:\chi:\nu:\sigma:f)$ for the trace of $\pi_{\chi,\nu,\sigma}(f)$, $f \in C^\infty_c(G)$.

§2. TEMPERED REPRESENTATIONS OF $\widetilde{SU}(2,2)$

The conformal group is usually realized as the identity component $SO(2,4)$ of the orthogonal group of the real bilinear form $-x_1y_1 - x_2y_2 + x_3y_3 + \dots + x_6y_6$ on $\mathbb{R}^{2,4}$. The space of light-like lines in $\mathbb{R}^{2,4}$ is the conformal completion of Minkowsky space $\mathbb{R}^{1,3}$, and that gives the action of $SO(2,4)$ there. We will find it much more convenient to use the complex form

$$SU(2,2) = \{x \in GL(4;\mathbb{C}) : xJx^* = J \text{ and } \det x = 1\} \tag{2.1}$$

of the double cover of $SO(2,4)$, $J = \begin{pmatrix} -1 & & & \\ & -1 & & \\ & & 1 & \\ & & & 1 \end{pmatrix}$, for the linear group. The two to one correspondence is

(2.2) $SU(2,2) \to SO(2,4)$ by Λ^2 (vector representation) ,

antisymmetrization of the usual representation of SU(2,2) on $\mathbb{C}^4$.

Write G for the universal covering group $\widetilde{SU}(2,2)$ of SU(2,2), $\psi: G \to SU(2,2)$ for the covering, and $Z_1 = \mathrm{Ker}(\psi)$. Since SU(2,2) has center of order 4, Z_1 has index 4 in $Z = Z_G$.

The maximal compactly embedded subgroup $K = [K,K] \times Z_K^0$ where $[K,K] \cong SU(2) \times SU(2)$ maps one-to-one to SU(2,2), and where $Z_K^0 \cong \mathbb{R}$. Let $\mathfrak{t}$ denote the compactly embedded Cartan subalgebra of $\mathfrak{g}$ given by

$$\text{(2.3)} \qquad \mathfrak{t} = \{\mathrm{diag}(i\theta_1, i\theta_2, i\theta_3, i\theta_4): \ \theta_j \text{ real}, \ \Sigma\theta_j = 0\} .$$

Then $\mathfrak{t} = (\mathfrak{t} \cap [\mathfrak{k},\mathfrak{k}]) \oplus \mathfrak{z}_K$ with

$$\text{(2.4)} \qquad \begin{aligned} \mathfrak{t} \cap [\mathfrak{k},\mathfrak{k}] &= \{\mathrm{diag}(i\theta_1, -i\theta_1, i\theta_2, -i\theta_2): \ \theta_j \text{ real}\} \\ \mathfrak{z}_K &= \{\mathrm{diag}(i\theta, i\theta, -i\theta, -i\theta): \ \theta \text{ real}\} . \end{aligned}$$

This gives us a parameterization of the corresponding Cartan subgroup T of G:

$$\text{(2.5a)} \qquad T = \{t(\theta_1,\theta_2)z_u: \ 0 \leq \theta_1 < 2\pi, \ 0 \leq \theta_2 < 2\pi, \ -\infty < u < \infty\}$$

where

$$\text{(2.5b)} \qquad t(\theta_1,\theta_2) = \exp_G \mathrm{diag}(i\theta_1, -i\theta_1, i\theta_2, -i\theta_2) \quad \text{and}$$

$$\text{(2.5c)} \qquad z_u = \exp_G \mathrm{diag}(iu, iu, -iu, -iu) .$$

Notice that ψ sends $t(\theta_1,\theta_2)z_u$ to $\mathrm{diag}\left(e^{i(\theta_1+u)}, e^{i(-\theta_1+u)}, e^{i(\theta_2-u)}, e^{i(-\theta_2-u)}\right)$, so $Z = Z_G = \psi^{-1}(\{\pm I, \pm iI\})$ is given by

$$\text{(2.6)} \qquad \{t(0,0)z_{k\pi}, \ t(\pi,\pi)z_{k\pi}, \ t(0,\pi)z_{(k+\frac{1}{2})\pi}, \ t(\pi,0)z_{(k+\frac{1}{2})\pi} : \ k \in \mathbb{Z}\} .$$

Let $\varepsilon_j: \mathrm{diag}(a_1,a_2,a_3,a_4) \mapsto a_j$ as usual. Then $\Phi^+(\mathfrak{g},\mathfrak{t}) = \{\varepsilon_i - \varepsilon_j: \ 1 \leq i < j \leq 4\}$. The compact roots are $\{\varepsilon_1-\varepsilon_2, \ \varepsilon_3-\varepsilon_4\}$. The only sets of strongly orthogonal (sums and differences are not roots) noncompact positive roots are, up to K-conjugacy,

$$\text{(2.7)} \qquad \emptyset , \quad \{\varepsilon_1-\varepsilon_3\} \quad \text{and} \quad \{\varepsilon_1-\varepsilon_3, \ \varepsilon_2-\varepsilon_4\} .$$

It follows from general theory that $\mathfrak{g}$ has exactly three conjugacy classes of Cartan subalgebras: $\mathfrak{t}$, $\mathfrak{j}$ and $\mathfrak{h}$, where $\mathfrak{t}$ is given by (2.3) and

$$(2.8) \qquad \mathfrak{j} = \left\{ \begin{pmatrix} i\theta_1 & 0 & v & 0 \\ 0 & i\theta_2 & 0 & 0 \\ v & 0 & i\theta_1 & 0 \\ 0 & 0 & 0 & i\theta_4 \end{pmatrix} : v, \theta_j \text{ real}, \ 2\theta_1 + \theta_2 + \theta_4 = 0 \right\},$$

$$(2.9) \qquad \mathfrak{h} = \left\{ \begin{pmatrix} i\theta_1 & 0 & v_1 & 0 \\ 0 & i\theta_2 & 0 & v_2 \\ v_1 & 0 & i\theta_1 & 0 \\ 0 & v_2 & 0 & i\theta_2 \end{pmatrix} : v_j, \theta_j \text{ real}, \ \theta_1 + \theta_2 = 0 \right\} .$$

Consider the "intermediate" Cartan subgroup J corresponding to $\mathfrak{j}$. First,

$$(2.10a) \qquad \mathfrak{a}_J = \left\{ \begin{pmatrix} 0 & 0 & v & 0 \\ 0 & 0 & 0 & 0 \\ v & 0 & 0 & 0 \\ 0 & 0 & 0 & 0 \end{pmatrix} \right\} \quad \text{and} \quad \mathfrak{t}_J = (\mathfrak{t}_J \cap [\mathfrak{k},\mathfrak{k}]) + (\mathfrak{t}_J \cap [\mathfrak{m}_J, \mathfrak{m}_J])$$

where

$$(2.10b) \qquad \mathfrak{t}_J \cap [\mathfrak{k},\mathfrak{k}] = \left\{ \begin{pmatrix} i\theta & & & \\ & -i\theta & & \\ & & i\theta & \\ & & & -i\theta \end{pmatrix} \right\}, \quad \mathfrak{t}_J \cap [\mathfrak{m}_J, \mathfrak{m}_J] = \left\{ \begin{pmatrix} 0 & & & \\ & i\theta & & \\ & & 0 & \\ & & & -i\theta \end{pmatrix} \right\}$$

Here $T_J^0 = \{t(\theta-u, \theta+u)z_u\}$ contains Z_G. As $T_J = Z(\mathfrak{a}_J)T_J^0$ where $Z(\mathfrak{a}_J)$ is generated by Z_G and $\gamma_{\varepsilon_1-\varepsilon_3} = \exp_G \operatorname{diag}(i\pi, 0, -i\pi, 0) = t(\frac{\pi}{2}, -\frac{\pi}{2})z_{\pi/2} \in T_J^0$ we have

$$(2.11a) \qquad T_J = \{t(\theta-u, \theta+u)z_u\}, \text{ connected} .$$

Similarly $M_J = Z(\mathfrak{a}_H)M_J^0$ where $Z(\mathfrak{a}_H)$ is generated by Z_G, $\gamma_{\varepsilon_1-\varepsilon_3}$, and $\gamma_{\varepsilon_2-\varepsilon_4} = t(-\frac{\pi}{2}, \frac{\pi}{2})z_{\pi/2} \in T_J^0$, so

$$(2.11b) \qquad M_J \cong \widetilde{SU}(1,1) \times S^1 , \text{ connected} .$$

Now consider the maximally split Cartan subgroup H corresponding to $\mathfrak{h}$. Here

$$(2.12a) \qquad \mathfrak{a}_H = \left\{ \begin{pmatrix} 0 & 0 & v_1 & 0 \\ 0 & 0 & 0 & v_2 \\ v_1 & 0 & 0 & 0 \\ 0 & v_2 & 0 & 0 \end{pmatrix} \right\} \quad \text{and} \quad \mathfrak{t}_H = \left\{ \begin{pmatrix} i\theta & & & \\ & -i\theta & & \\ & & i\theta & \\ & & & -i\theta \end{pmatrix} \right\} \subset [\mathfrak{k},\mathfrak{k}] .$$

$T_H = Z(\mathfrak{a}_H)T_H^0$ with $Z(\mathfrak{a}_H)$ generated by $\gamma_{\varepsilon_1-\varepsilon_3}$ and $\gamma_{\varepsilon_2-\varepsilon_4}$ as above, and $T_H^0 = \{t(\theta,\theta): 0 \leq \theta < 2\pi\}$. Compute

$$\gamma_{\varepsilon_2-\varepsilon_4} \in \gamma_{\varepsilon_1-\varepsilon_3}\cdot T_H^0$$

$$t(0,0)z_{k\pi},\ t(\pi,\pi)z_{k\pi} \in (\gamma_{\varepsilon_1-\varepsilon_3})^{2k}\cdot T_H^0$$

$$t(0,\pi)z_{(k+\frac{1}{2})\pi},\ t(\pi,0)z_{(k+\frac{1}{2})\pi} \in (\gamma_{\varepsilon_1-\varepsilon_3})^{2k+1}\cdot T_H^0$$

to see that

(2.12b) $$T_H = \bigcup_{n=-\infty}^{\infty} (\gamma_{\varepsilon_1-\varepsilon_3})^n T_H^0 = M_H \cong \mathbb{Z}\times T_H^0 \quad .$$

Now we have the Cartan subgroups and the associated cuspidal parabolic subgroups. So we can parametrize the tempered series.

The space Λ' of (1.7) for the Cartan subgroup T of G is

(2.13a) $$\Lambda_T' = \{\lambda_{n,m,h} = \tfrac{n}{2}(\varepsilon_1-\varepsilon_2)+\tfrac{m}{2}(\varepsilon_3-\varepsilon_4)+\tfrac{h}{4}(\varepsilon_1+\varepsilon_2-\varepsilon_3-\varepsilon_4),\ n,m \text{ integers} \neq 0,\ h \text{ real},\ n\pm m\pm h \neq 0\} \quad .$$

Here notice that

(2.13b) $$\exp\lambda_{n,m,h}: t(\theta_1,\theta_2)z_u \mapsto e^{in\theta_1}e^{im\theta_2}e^{ihu} \quad .$$

Since G is connected, its relative discrete series consists of the

(2.14a) $$\pi(T\colon n\colon m\colon h) = \pi_{\lambda_{n,m,h}} \quad \text{where} \begin{cases} n,m \text{ integers} \neq 0 \\ h \text{ real} \\ n\pm m\pm h \neq 0 \quad . \end{cases}$$

The Weyl group $W(G,T)$ is generated by reflections in compact roots $\varepsilon_1-\varepsilon_2$ and $\varepsilon_3-\varepsilon_4$. Thus

(2.14b) $$\pi(T\colon n\colon m\colon h) = \pi(T\colon n'\colon m'\colon h') \Leftrightarrow n=\pm n',\ m=\pm m',\ h=h' \quad .$$

The space Λ' of (1.7) for the Cartan subgroup T_J of M_J is

(2.15a) $$\Lambda_J' = \{\lambda_{n,h} = \tfrac{n}{4}(\varepsilon_1-\varepsilon_2+\varepsilon_3-\varepsilon_4)+\tfrac{h}{2}(\varepsilon_2-\varepsilon_4)\colon n \text{ integer},\ h\neq 0\} \quad .$$

Here

(2.15b) $$\exp\lambda_{n,h}: t(\theta-u,\ \theta+u)z_u \longmapsto e^{in\theta}e^{2ihu} \quad .$$

Since M_J is connected, its relative discrete series consists of the

(2.16a) $$\eta(T_J\colon n\colon h) = \eta_{\lambda_{n,h}} \quad \text{where } n \text{ integer},\ h\neq 0 \ .$$

Parameterize $\mathfrak{a}_J^*$ by $\sigma_s\begin{pmatrix} 0 & 0 & v & 0 \\ 0 & 0 & 0 & 0 \\ v & 0 & 0 & 0 \\ 0 & 0 & 0 & 0 \end{pmatrix} = sv$; so σ_s comes by Cayley transform from $\frac{s}{2}(\varepsilon_1-\varepsilon_3) \in i\mathfrak{t}^*$. Now the J-series of G consists of the

$$\text{(2.16b)} \qquad \pi(J{:}\ n{:}\ h{:}\ s) = \mathrm{Ind}_{M_J A_J N_J}^{G}(\eta(T_J{:}\ n{:}\ h) \otimes e^{i\sigma_s})$$

for $n \in \mathbb{Z}$, h and s real, $h \neq 0$. The Weyl group W(G,J) is generated by reflection in the real root, which is 1 on $\mathfrak{t}_J$ and -1 on $\mathfrak{a}_J$, so

$$\text{(2.16c)} \qquad \pi(J{:}\ n{:}\ h{:}\ s) = \pi(J{:}\ n'{:}\ h'{:}\ s') \Longleftrightarrow n=n',\ h=h',\ s=\pm s' \ .$$

The space Λ' of (1.7) for the Cartan subgroup T_H of M_H is

$$\text{(2.17a)} \qquad \Lambda_H' = \{\lambda_n = \tfrac{n}{4}(\varepsilon_1-\varepsilon_2+\varepsilon_3-\varepsilon_4)\colon\ n \text{ integer}\} \ .$$

Here

$$\text{(2.17b)} \qquad \exp \lambda_n\colon\ t(\theta,\theta) \longrightarrow e^{in\theta} \ .$$

Now the relative discrete series of $M_H = \langle \gamma_{\varepsilon_1-\varepsilon_3} \rangle \times T_H^0$ consists of the unitary characters

$$\text{(2.18a)} \qquad \eta(T_H{:}\ n{:}\ h)\colon\ (\gamma_{\varepsilon_1-\varepsilon_3})^m\, t(\theta,\theta) \mapsto e^{\pi imh}\, e^{in\theta} \ ,$$

n integer and $0 \leqq h < 2$. Parameterize $\mathfrak{a}_H^*$ by $\sigma_{s,t}\begin{pmatrix} 0 & 0 & v_1 & 0 \\ 0 & 0 & 0 & v_2 \\ v_1 & 0 & 0 & 0 \\ 0 & v_2 & 0 & 0 \end{pmatrix} = sv_1 + tv_2$

Then the H-series of G (which is the principal series) consists of the

$$\text{(2.18b)} \qquad \pi(H{:}\ n{:}\ h{:}\ s{:}\ t) = \mathrm{Ind}_{M_H A_H N_H}^{G}(\eta(T_H{:}\ n{:}\ h) \otimes e^{i\sigma_{s,t}})$$

for n integer, $0 \leqq h < 2$, s and t real. The Weyl group W(G,H) is generated by conjugation by

$$\begin{pmatrix} i & & & \\ & 1 & & \\ & & -i & \\ & & & 1 \end{pmatrix}, \quad \begin{pmatrix} 1 & & & \\ & i & & \\ & & 1 & \\ & & & -i \end{pmatrix} \quad \text{and} \quad \begin{pmatrix} 0 & i & 0 & 0 \\ i & 0 & 0 & 0 \\ 0 & 0 & 0 & i \\ 0 & 0 & i & 0 \end{pmatrix} .$$

The first two are trivial on $\mathfrak{t}$, hence on $T_H \subset T$; the first sends $\sigma_{s,t} \to \sigma_{-s,t}$ and the second sends $\sigma_{s,t} \to \sigma_{s,-t}$. The third, call it w, is -1 on $\mathfrak{t}_H$, interchanges $\gamma_{\varepsilon_1-\varepsilon_3}$ and $\gamma_{\varepsilon_2-\varepsilon_4}$, and sends $\sigma_{s,t} \to \sigma_{t,s}$. Compute
w: $\gamma^m_{\varepsilon_1-\varepsilon_3} t(\theta,\theta) \to \gamma^m_{\varepsilon_2-\varepsilon_4} t(-\theta,-\theta) = \gamma^m_{\varepsilon_1-\varepsilon_3} t(m\pi-\theta,\ m\pi-\theta)$. We conclude that

$$\text{(2.18c)} \qquad \pi(H{:}n{:}h{:}s{:}t) = \pi(H{:}n'{:}h'{:}s'{:}t') \Longleftrightarrow \text{either } (n',h',s',t') = (n,h,\pm s,\pm t)$$
$$\text{or } (n',h',s',t') = (-n,h+n,\pm t,\pm s) \ .$$

§3. PLANCHEREL THEOREM FOR SEMISIMPLE GROUPS

We describe the Plancherel formula for the class of reductive Lie groups specified in (1.1) and (1.2). Here we enlarge Z if necessary so that $Z \cap G^0 = Z_{G^0}$ -- just replace Z by ZZ_{G^0}. Let $\mathrm{Car}(G)$ denote a set of representatives of the conjugacy classes of Cartan subgroups of G, chosen so that $\theta H = H$ for all $H \in \mathrm{Car}(G)$. The Plancherel formula says that, if $f \in C_c^\infty(G)$, then

$$(3.1) \qquad f(x) = c_G \sum_{H \in \mathrm{Car}(G)} c_{H\cap G^0}^{-1} \int_{\chi \in Z_{M_H}(M_H)^\wedge} \deg(\chi) \times \sum_{\substack{\nu \in \Lambda'_H \\ \nu,\chi \text{ agree}}} \int_{\sigma \in \mathfrak{a}_H^*} \Theta(H:\chi:\nu:\sigma: r_x f) \times \left| \prod_{\alpha \in \Phi^+(\mathfrak{g},\mathfrak{h})} \langle \nu + i\sigma, \alpha\rangle \prod_{\beta \in \Phi_{\mathbb{R}}^+(\mathfrak{g},\mathfrak{h})} \bar{p}_\beta(\chi:\sigma) \right| d\sigma d\chi$$

where $r_x f$ is the right translate of f by x, $(r_x f)(y) = f(yx)$. In this section we explain the ingredients of (3.1).

First, for the formula to make any sense at all, we must normalize Haar measures on the groups over which we integrate.

Let $G_1 = ZG^0/Z$, let θ_1 denote the Cartan involution derived from θ, let K_1 denote the fixed point set of θ_1, and let B_1 be a fundamental (as compact as possible) Cartan subgroup of G_1. Warning: this notation differs slightly from that of [4], and the following normalizations of Haar measures are streamlined over the ones in [4], because we do not need certain auxiliary groups for the final formula (3.1).

Write $\langle\ ,\ \rangle$ for the Killing form on $\mathfrak{g}_1$ and $(\ ,\ )$ for the associated positive definite form, $(\xi,\eta) = -\langle \xi, \theta_1\eta\rangle$. Split $B_1 = T_1 \times A_1$ as in §1. Then T_1 is a torus; give it Haar measure of total mass $1/|\pi_1(G_{1\mathbb{C}})|$, where $G_{1\mathbb{C}}$ is the complexification of G_1 and $\pi_1(G_{1\mathbb{C}})$ is its fundamental group. The exponential map is a diffeomorphism from $\mathfrak{a}_1$ to A_1; give A_1 the Haar measure corresponding to the $(\ ,\ )$-euclidean structure of $\mathfrak{a}_1$. Now B_1 carries the product Haar measure.

Let $\mathfrak{g}_1'$ denote the regular subset of $\mathfrak{g}_1$. It consists of all elements of $\mathfrak{g}_1$ whose centralizers are Cartan subalgebras. The subset $\mathfrak{e}$, all elements of $\mathfrak{g}_1$ whose centralizers are conjugate to $\mathfrak{b}_1$, is open in $\mathfrak{g}_1$ and inherits a measure from the $(\ ,\ )$-euclidean structure there. Define a G_1-invariant measure on G_1/B_1 by

$$(3.2a) \qquad \int_{\mathfrak{e}} f(\xi)d\xi = \int_{G_1/B_1} \left\{ \int_{\mathfrak{b}_1} \left| \prod_{\alpha\in\Phi^+} \alpha(\xi)\right|^2 \cdot f(\mathrm{Ad}(x)\xi)d\xi \right\} d(xB_1)$$

where $\Phi^+ = \Phi^+(\mathfrak{g}_1,\mathfrak{b}_1)$ is a positive root system and where $f \in C_c^\infty(\mathfrak{e})$. Normalize

Haar measure on G_1 by

$$\int_{G_1} f(x)dx = \int_{G_1/B_1} \left\{ \int_{B_1} f(xb)db \right\} d(xB_1) \tag{3.2b}$$

for $f \in C_c^\infty(G_1)$. Now we have normalized Haar measure $d(xZ)$ on $ZG^0/Z = G_1$.

Fix a Haar measure on $Z_G(G^0)$. If it is compact use the invariant measure of total mass 1. If it is discrete use counting measure. Then

$$\int_{Z_G(G^0)} f(z)dz = \sum_{z_0 \in Z_G(G^0)/Z} \int_Z f(z_0 z)dz \tag{3.3a}$$

specifies Haar measure on Z, and

$$\int_{ZG^0} f(x)dx = \int_{ZG^0/Z} \left\{ \int_Z f(xz)dz \right\} d(xZ) \tag{3.3b}$$

defines Haar measure on ZG^0. At last, we have Haar measure on G defined by

$$\int_G f(x)dx = \sum_{y \in G/ZG^0} \int_{ZG^0} f(yx)dx \quad . \tag{3.3c}$$

It is independent of choice of Z.

Now that Haar measure on G is normalized, the operators

$$\pi(H:\chi:\nu:\sigma:r_\chi f) = \int_G f(yx)\pi(H:\chi:\nu:\sigma:y)dy \tag{3.4a}$$

are specified for $f \in C_c^\infty(G)$, and it makes sense to talk about their traces

$$\Theta(H:\chi:\nu:\sigma:r_\chi f) = \text{trace } \pi(H:\chi:\nu:\sigma:r_\chi f) \quad . \tag{3.4b}$$

Those traces are the basic ingredient in the Plancherel formula (3.1).

Next, we look at the measures $d\chi$ on the $Z_{M_H}(M_H^0)^\wedge$ that occur in (3.1). Given our choice of Haar measure on $Z_G(G^0)$, we fixed Haar measure on Z by (3.3a), and that normalizes Haar measure on $\hat{Z}$ by

$$f(x) = |Z_G(G^0)/Z| \int_{\hat{Z}} f_\zeta(x)d\zeta \tag{3.5a}$$

where, for $f \in C_c^\infty(G)$ and $\zeta \in \hat{Z}$ we denote

$$f_\zeta(x) = \int_Z f(xz)\zeta(z)dz \quad . \tag{3.5b}$$

We now normalize $d\chi$ by

(3.6)
$$\int_{Z_{M_H}(M_H^0)^\wedge} \phi(\chi)\deg(\chi)d\chi = \int_{\hat{Z}} \sum_{\chi \in Z_{M_H}(M_H^0)^\wedge_\zeta} \phi(\chi)\deg(\chi)d\zeta$$

where $Z_{M_H}(M_H^0)^\wedge_\zeta = \{\chi \in Z_{M_H}(M_H^0)^\wedge : \chi|_Z \text{ has } \zeta \text{ as summand}\}$. This is equivalent to the normalization in [4, Lemma 6.12].

The measure $d\sigma$ on $\mathfrak{a}_H^*$ in (3.1) is also normalized through the abelian Fourier transform. If $f \in C_c^\infty(\mathfrak{a}_H)$ then the (,)-euclidean structure on $\mathfrak{g}_1 = [\mathfrak{g},\mathfrak{g}] \supset \mathfrak{a}_H$ specifies $\hat{f}: \mathfrak{a}_H^* \to \mathbb{C}$ by

(3.7a)
$$\hat{f}(\sigma) = \int_{\mathfrak{a}_H} f(\xi)\, e^{i\sigma(\xi)}\, d\xi \quad ,$$

and we normalize $d\sigma$ by

(3.7b)
$$f(\xi) = (2\pi)^{-\dim \mathfrak{a}_H} \int_{\mathfrak{a}_H^*} \hat{f}(\sigma)\, e^{-i\sigma(\xi)}\, d\sigma \quad .$$

The constant c_G in (3.1) is given by

(3.8)
$$c_G = |\pi_1(G_{1\mathbb{C}})| \cdot \frac{W(G^0,\, B\cap G^0)}{|G/Z_G(G^0)G^0| \cdot (2\pi)^{r+p}}$$

where B is a fundamental Cartan subgroup of G (e.g. the inverse image of $B_1 \subset G_1 = ZG^0/Z$), where $W(G^0, B\cap G^0)$ is the Weyl group

$$\{x \in G^0 : \mathrm{Ad}(x)\mathfrak{b} = \mathfrak{b}\}/(B \cap G^0) \quad ,$$

where $r = |\Phi^+(\mathfrak{g},\mathfrak{b})|$, and where $p = \operatorname{rank} G - \operatorname{rank} K = \dim \mathfrak{a}_B$.

Given $H \in \mathrm{Car}(G)$, $\theta H = H$, let $\Phi_{\mathbb{R}}(\mathfrak{g},\mathfrak{h})$ denote the set of real roots in $\Phi(\mathfrak{g},\mathfrak{h})$. So $\Phi_{\mathbb{R}}(\mathfrak{g},\mathfrak{h}) = \{\alpha \in \Phi(\mathfrak{g},\mathfrak{h}) : \alpha(\mathfrak{h}) \subset \mathbb{R}, \text{ i.e. } \alpha(\mathfrak{t}_H) = 0\}$ and is a root system.

We can assume $\mathfrak{a}_B \subset \mathfrak{a}_{\mathfrak{h}}$ so $\mathfrak{h} \subset \mathfrak{m}_B + \mathfrak{a}_B$. Then $\Phi_{\mathbb{R}}(\mathfrak{m}_B + \mathfrak{a}_B, \mathfrak{h})$ is spanned by strongly orthogonal roots, hence is a direct sum of simple root systems with that property. For each simple summand there is a number that comes out of the theory of two-structures and evaluates to

(3.9)

summand	A_1	B_{2n}	B_{2n+1}	C_ℓ	D_{2n}	G_2	F_4	E_7	E_8
number	1	2^{n-1}	2^n	1	2^{n-1}	2	2	8	16

and $Q(\mathfrak{g},\mathfrak{h})$ is the product (over the simple summands) of those numbers. Let $R(\mathfrak{g},\mathfrak{h})$ denote the set of strongly orthogonal roots of $(\mathfrak{g},\mathfrak{b})$ used to define $\mathfrak{h}$ by the Cayley transform procedure. Then

(3.10)
$$c_{H\cap G^0} = |W(G^0,\, H\cap G^0)| \cdot |H\cap K^0/H\cap K^0\cap M_B^\dagger| \cdot Q(\mathfrak{g},\mathfrak{h}) \cdot \prod_{\alpha\in R(\mathfrak{g},\mathfrak{h})} \|\alpha\| \quad .$$

Given $\alpha \in \Phi_{\mathbb{R}}(\mathfrak{g},\mathfrak{h})$ we denote

$h_\alpha^* \in \mathfrak{a}_H$: element dual to $\alpha^\vee = 2\alpha/\|\alpha\|^2$,

$x_\alpha \in \mathfrak{g}_\alpha$ (α-root space): normalized by $[x_\alpha, \theta x_\alpha] = h_\alpha^*$,

$z_\alpha = x_\alpha - \theta x_\alpha$ and $\gamma_\alpha = \exp_G(\pi z_\alpha)$.

Z_{G^0} and the γ_α generate a subgroup $Z(\mathfrak{a}_H)$ of $Z_{M_H}(M_H^0)$ such that $H \cap G^0 = Z(\mathfrak{a}_H)H^0$. If $\sigma \in \mathfrak{a}_H^*$ and $\chi \in Z_{M_H}(M_H^0)^\wedge$ then

$$(3.11a) \qquad p_\alpha(\chi:\sigma) = \sinh\left(\frac{2\pi\langle\sigma,\alpha\rangle}{\langle\alpha,\alpha\rangle}\right)\left\{\cosh\left(\frac{2\pi\langle\sigma,\alpha\rangle}{\langle\alpha,\alpha\rangle}\right) I_k - \tfrac{1}{2} e^{\rho_\alpha}(\gamma_\alpha)\,[\chi(\gamma_\alpha) + \chi(\gamma_\alpha)^{-1}]\right\}^{-1}$$

is a scalar matrix, where $k = \deg(\chi)$ and ρ_α is half the sum of $\{\beta \in \Phi^+(\mathfrak{g},\mathfrak{h}) : \beta|_{\mathfrak{a}_H}$ is a multiple of $\alpha\}$. The factor $\bar{p}_\alpha(\chi:\sigma)$ in (3.1) is the value of this scalar,

$$(3.11b) \qquad \bar{p}_\alpha(\chi:\sigma) = \deg(\chi)^{-1} \cdot \mathrm{trace}\; p_\alpha(\chi:\sigma) \quad .$$

This completes the description of the terms involved in the Plancherel formula (3.1).

§4. EXPLICIT PLANCHEREL FORMULA FOR $\widetilde{SU}(2,2)$

The first step is to normalize Haar measure as in §3 for $G = \widetilde{SU}(2,2)$. This comes down to the following.

(a) Note that the Killing form $\langle \xi,\eta\rangle = \mathrm{trace}(\mathrm{ad}(\xi)\mathrm{ad}(\eta))$ on $\mathfrak{g} = \mathfrak{su}(2,2)$ is given by

$$(4.1) \qquad \langle \xi,\eta\rangle = 8 \cdot \mathrm{trace}(\xi\eta) \quad .$$

This defines the euclidean structure on $\mathfrak{g}$ by

$$(4.2) \qquad (\xi,\eta) = -\langle \xi , \theta\eta\rangle = 8 \cdot \mathrm{trace}(\xi\eta^*) \quad .$$

It gives a volume element on the open subset $\mathfrak{e} = \bigcup_{x \in G} \mathrm{Ad}(x)\mathfrak{t}'$, and on $\mathfrak{t}$ itself.

(b) $\int_{\mathfrak{e}} f(\xi)d\xi = \int_{G/T}\left\{\int_{\mathfrak{t}} |\prod_{\alpha\in\Phi^+}\alpha(\xi)|^2 f(\mathrm{Ad}(x)\xi)d\xi\right\} d(xT)$ defines the G-invariant measure on G/T, where $\Phi^+ = \Phi^+(\mathfrak{g},\mathfrak{t})$ and $f \in C_c^\infty(\mathfrak{e})$.

(c) Normalize Haar measure on the compact Cartan subgroup T/Z of $G/Z = SU(2,2)/\{\pm I, \pm iI\}$ to have total volume $\tfrac{1}{4}$. Then Haar measure on G is given by

$$(4.3)\qquad \int_G f(x)dx = \int_{G/T}\left\{\int_{T/Z}\sum_{z\in Z} f(xtz)d(tZ)\right\}d(xT) \quad ,$$

where $f\in C_c^\infty(G)$.

Our Haar measure on $Z = Z_G = Z_G(G^0)$ is counting measure, and $|Z_G(G^0)/Z| = 1$ in (3.5a). So the Haar measure on $\hat{Z} = Z_G(G^0)^\wedge$ is normalized by (3.5) to total mass 1.

In (2.13), $\exp(\lambda_{n,m,h})|_Z = 1$ precisely when $h/2$ is an integer and

$$(-1)^n = (-1)^m = (-1)^{h/2} \quad ,$$

so we may view

$$\hat{Z} = \{\exp(\lambda_{0,0,h})|_Z\colon 0\leqq h<4\} \cup \{\exp(\lambda_{1,0,h})|_Z\colon 0\leqq h<4\} \quad .$$

Now

$$\int_{\hat{Z}} \phi(\zeta)d\zeta = \tfrac{1}{8}\sum_{n=0,1}\int_0^4 \phi(\exp(\lambda_{n,0,h})|_Z)dh \quad .$$

That gives us (remember: $G = M_T$),

$$(4.4)\qquad \int_{Z_G(G^0)^\wedge}\sum_{\substack{\nu\in\Lambda_T\\ \nu,\chi \text{ agree}}}\phi(\nu)d\chi = \tfrac{1}{8}\sum_{\substack{m,n\\ \text{integers}}}\int_{-\infty}^{\infty}\phi(\lambda_{n,m,h})dh \quad .$$

In (2.15), $\exp(\lambda_{n,h})|_Z = 1$ precisely when $n/2$ is an integer and $(-1)^h = (-1)^{n/2}$, so $\hat{Z} = \{\exp(\lambda_{0,h})|_Z\colon 0\leqq h<2\} \cup \{\exp(\lambda_{1,h})|_Z\colon 0\leqq h<2\}$. Now

$$\int_{\hat{Z}} \phi(\zeta)d\zeta = \tfrac{1}{4}\sum_{n=0,1}\int_0^2 \phi(\exp(\lambda_{n,h})|_Z)dh \quad .$$

That gives us

$$(4.5)\qquad \int_{Z_{M_J}(M_J^0)^\wedge}\sum_{\substack{\nu\in\Lambda_J\\ \nu,\chi \text{ agree}}}\phi(\nu)d\chi = \tfrac{1}{4}\sum_{n \text{ integer}}\int_{-\infty}^{\infty}\phi(\lambda_{n,h})dh \quad .$$

In (2.18a) express $\eta(T_H\colon n\colon h) = \chi_h\otimes\exp(\lambda_n)$. Then $(\chi_h\otimes\exp\lambda_n)|_Z = 1$ exactly when h and $n/2$ are integers with $(-1)^h = (-1)^{n/2}$, so

$$\hat{Z} = \{\chi_h\otimes\exp\lambda_0|_Z\colon 0\leqq h<2\} \cup \{\chi_h\otimes\exp\lambda_1|_Z\colon 0\leqq h<2\} \quad .$$

Using (2.12b) and (2.17), now

$$(4.6)\qquad \int_{Z_{M_H}(M_H^0)^\wedge}\sum_{\substack{\nu\in\Lambda_H\\ \nu,\chi \text{ agree}}}\phi(\nu)d\chi = \tfrac{1}{4}\sum_{n \text{ integer}}\int_0^2\phi(\chi_h\otimes\lambda_n)dh \quad .$$

Haar measure on $\mathfrak{a}_J$ is given by $\int_{\mathfrak{a}_J} f(\xi)d\xi = \int_{-\infty}^{\infty} f(r\xi_1)dr$ where $\|\xi_1\| = 1$, say $\xi_1 = \frac{1}{4}\begin{pmatrix} 0 & 0 & 1 & 0 \\ 0 & 0 & 0 & 0 \\ 1 & 0 & 0 & 0 \\ 0 & 0 & 0 & 0 \end{pmatrix}$ using (4.1). Since $\sigma_s(\xi_1) = s/4$, the normalization (3.7) becomes

$$(4.7) \qquad \int_{\mathfrak{a}_J^*} \phi(\sigma)d\sigma = \frac{1}{4}\int_{-\infty}^{\infty} \phi(\sigma_s)ds \ .$$

Similarly, Haar measure on $\mathfrak{a}_H$ is given by $\int_{\mathfrak{a}_H} f(\xi)d\xi = \int_{-\infty}^{\infty}\int_{-\infty}^{\infty} f(r_1\xi_1 + r_2\xi_2)dr_1dr_2$ where $\xi_1 = \frac{1}{4}\begin{pmatrix} 0 & 0 & 1 & 0 \\ 0 & 0 & 0 & 0 \\ 1 & 0 & 0 & 0 \\ 0 & 0 & 0 & 0 \end{pmatrix}$ and $\xi_2 = \frac{1}{4}\begin{pmatrix} 0 & 0 & 0 & 0 \\ 0 & 0 & 0 & 1 \\ 0 & 0 & 0 & 0 \\ 0 & 1 & 0 & 0 \end{pmatrix}$, so $\sigma_{s,t}(\xi_1) = s/4$ and $\sigma_{s,t}(\xi_2) = t/4$ give us

$$(4.8) \qquad \int_{\mathfrak{a}_H^*} \phi(\sigma)d\sigma = \frac{1}{16}\int_{-\infty}^{\infty}\int_{-\infty}^{\infty} \phi(\sigma_{s,t})dsdt \ .$$

As seen just before (2.14), $W(G^0, B\cap G^0) = W(G,T)$ has order 4 in (3.8). Since G is connected, $|G/Z_G(G^0)G^0| = 1$. Evidently $r=6$ and $p=0$. Finally $G_{1\mathbb{C}} = (G/Z)_{\mathbb{C}} = SL(4;\mathbb{C})/\{\pm I, \pm iI\}$ has fundamental group of order 4. So

$$(4.9) \qquad c_G = 16/(2\pi)^6 \ .$$

Note $\mathfrak{m}_T = \mathfrak{g}$. Since $\Phi_{\mathbb{R}}(\mathfrak{g},\mathfrak{t}) = \phi$, $\Phi_{\mathbb{R}}(\mathfrak{g},\mathfrak{j})$ is of type A_1, and $\Phi_{\mathbb{R}}(\mathfrak{g},\mathfrak{h})$ is of type $A_1 \times A_1$, in each case (3.9) gives $Q(\mathfrak{g},\cdot) = 1$. From (4.1), all roots $\alpha = \varepsilon_i - \varepsilon_j$ have $\|\alpha\|^2 = \|\varepsilon_i\|^2 + \|\varepsilon_j\|^2 = \frac{1}{4}$, so $\prod_{\alpha\in R(\mathfrak{g},\cdot)} \|\alpha\|$ is 1, $\frac{1}{2}$, $\frac{1}{4}$ for $\mathfrak{t}$, $\mathfrak{j}$, $\mathfrak{h}$. $W(G,T)$, $W(G,J)$ and $W(G,H)$ were seen in §2 to have respective orders 4, 2 and 8. As $G = M_T^\dagger = M_B^\dagger$, $|T\cap K^0/T\cap K^0\cap M_T^\dagger| = |J\cap K^0/J\cap K^0\cap M_T^\dagger| = |H\cap K^0/H\cap K^0\cap M_T^\dagger| = 1$. Now

$$(4.10) \qquad c_T = 4\ , \qquad c_J = 1\ , \quad \text{and} \quad c_H = 2\ .$$

$\Phi_{\mathbb{R}}(\mathfrak{g},\mathfrak{t}) = \phi$ so there are no $\bar{p}_\beta(X:\sigma)$-terms for T.

$\Phi^+_{\mathbb{R}}(\mathfrak{g},\mathfrak{j}) = \{\beta\}$, the Cayley transform of $\varepsilon_1 - \varepsilon_3$.

Compute $\frac{2\pi\langle\sigma_s,\beta\rangle}{\langle\beta,\beta\rangle} = \pi s$, $\rho_\beta(\mathrm{diag}(\pi i, 0, -\pi i, 0)) = 3\pi i$, so $e^{\rho_\beta}(\gamma_\beta) = -1$, and $(\exp\lambda_{n,h})(\gamma_\beta) = (\exp\lambda_{n,h})(t(\frac{\pi}{2}, -\frac{\pi}{2})z_{\pi/2}) = e^{i\pi n}e^{i\pi h}$. Thus

$$(4.11)\qquad \bar{p}_\beta(\exp \lambda_{n,h}: \sigma_s) = \frac{\sinh(\pi s)}{\cosh(\pi s) + (-1)^n \cos(\pi h)} .$$

$\Phi_{\mathbb{R}}(\mathfrak{g},\mathfrak{h}) = \{\beta_1,\beta_2\}$, respective Cayley transforms of $\varepsilon_1-\varepsilon_3$ and $\varepsilon_2-\varepsilon_4$.

Compute $\frac{\langle 2\pi\sigma_{s,t}, \beta_j\rangle}{\langle \beta_j,\beta_j\rangle} = \pi s$ for j=1, πt for j=2. As above, $e^{\rho_\beta}(\gamma_\beta) = -1$ for $\beta = \beta_1,\beta_2$. Also

$$\begin{aligned} &\eta(T_H\colon n\colon h)(\gamma_{\beta_1}) = e^{i\pi h} \\ \text{and}\quad &\eta(T_H\colon n\colon h)(\gamma_{\beta_2}) = e^{i\pi n}e^{i\pi h} .\end{aligned}$$

So

$$(4.12)\qquad \prod_{\beta\in\Phi^+_{\mathbb{R}}(\mathfrak{g},\mathfrak{h})} \bar{p}_\beta(\eta(T_H:n:h):\sigma_{s,t}) = \frac{\sinh(\pi s)}{\cosh(\pi s)+\cos(\pi h)}\;\frac{\sinh(\pi t)}{\cosh(\pi t)+(-1)^n\cos(\pi h)} .$$

Finally, using $\|\varepsilon_i\|^2 = 1/8$, we glance back at (2.13a) to check

$$\prod_{1\leqq i<j\leqq 4}\left\langle \frac{n}{2}(\varepsilon_1-\varepsilon_2) + \frac{m}{2}(\varepsilon_3-\varepsilon_4) + \frac{h}{4}(\varepsilon_1+\varepsilon_2-\varepsilon_3-\varepsilon_4),\ \varepsilon_i-\varepsilon_j\right\rangle$$

$$= n\tfrac{1}{2}(n-m+h)\,\tfrac{1}{2}(n+m+h)\tfrac{1}{2}(-n+m+h)\tfrac{1}{2}(-n-m+h)m\,\|\varepsilon_i\|^6 ,$$

that is,

$$(4.13)\qquad \prod_{\alpha\in\Phi^+(\mathfrak{g},\mathfrak{t})} \langle\lambda_{n,m,h},\alpha\rangle = 2^{-22}nm(n+m+h)(n+m-h)(n-m+h)(n-m-h) .$$

Similarly, using (2.15a) and the fact that σ_s comes from $\frac{s}{2}(\varepsilon_1-\varepsilon_3)$ by Cayley transform,

$$(4.14)\qquad \prod_{\alpha\in\Phi^+(\mathfrak{g},\mathfrak{j})} \langle\lambda_{n,h}+i\sigma_s,\alpha\rangle = -2^{-22}ihs|(n+h)+is|^2|(n-h)+is|^2$$

and, using (2.17a) and the fact that $\sigma_{s,t}$ comes from $\frac{1}{2}(s(\varepsilon_1-\varepsilon_3)+t(\varepsilon_2-\varepsilon_4))$ by Cayley transform,

$$(4.15)\qquad \prod_{\alpha\in\Phi^+(\mathfrak{g},\mathfrak{h})} \langle\lambda_n+i\sigma_{s,t},\alpha\rangle = 2^{-22}st|n+i(s+t)|^2|n+i(s-t)|^2 .$$

Now we are ready to put specific values into (3.1). Break the sum over $\mathrm{Car}(G) = \{T,J,H\}$ into

$$(4.16)\qquad f(x) = f_T(x) + f_J(x) + f_H(x) \quad\text{for}\quad f\in C_c^\infty(G) .$$

Then, from (3.1) and the preceding results of this section,

$$(4.17)\qquad f_T(x) = 2^{-29}\pi^{-6} \sum_{m,n=-\infty}^{\infty} \int_{-\infty}^{\infty} \Theta(T{:}\ n{:}\ m{:}\ h{:}\ r_x f) \times$$

$$\times\ |nm(n+m+h)(n+m-h)(n-m+h)(n-m-h)|\,dh\ ,$$

$$(4.18)\qquad f_J(x) = 2^{-28}\pi^{-6} \sum_{n=-\infty}^{\infty} \int_{-\infty}^{\infty}\int_{-\infty}^{\infty} \Theta(J{:}\ n{:}\ h{:}\ s{:}\ r_x f) \times$$

$$\times\ \left| hs|n+h+is|^2|n-h+is|^2 \frac{\sinh(\pi s)}{\cosh(\pi s)+(-1)^n\cos(\pi h)} \right| dhds\ ,$$

and

$$(4.19)\qquad f_H(x) = 2^{-31}\pi^{-6} \sum_{n=-\infty}^{\infty} \int_0^2 \left\{ \int_{-\infty}^{\infty}\int_{-\infty}^{\infty} \Theta(H{:}\ n{:}\ h{:}\ s{:}\ t{:}\ r_x f) \times\right.$$

$$\left.\times\ \left| st|n+i(s+t)|^2|n+i(s-t)|^2 \frac{\sinh(\pi s)}{\cosh(\pi s)+\cos(\pi h)} \frac{\sinh(\pi t)}{\cosh(\pi t)+(-1)^n\cos(\pi h)} \right| dsdt \right\} dh\ .$$

Combining these and using (2.14b), (2.16c) and (2.18c), we finally arrive at

4.20 THEOREM. *Let* G *be the universal covering of the conformal group. In the normalizations and notation described above, if* $f \in C_c^\infty(G)$ *and* $x \in G$ *then*

$$2^{27}\pi^6 f(x) =$$

$$\sum_{m,n=1}^{\infty} \int_{-\infty}^{\infty} \Theta(T{:}n{:}m{:}h{:}r_x f)\,\left|nm(n+m+h)(n+m-h)(n-m+h)(n-m-h)\right| dh$$

$$+ \sum_{n=-\infty}^{\infty} \int_{-\infty}^{\infty} \left\{ \int_0^{\infty} \Theta(J{:}n{:}h{:}s{:}r_x f) \left| hs|n+h+is|^2|n-h+is|^2 \frac{\sinh(\pi s)}{\cosh(\pi s)+(-1)^n\cos(\pi h)} \right| ds \right\} dh$$

$$+ \tfrac{1}{2} \sum_{n=-\infty}^{\infty} \int_0^2 \left\{ \int_0^{\infty}\int_0^{s} \Theta(H{:}n{:}h{:}s{:}t{:}r_x f) \left| st|n+i(s+t)|^2 \times\right.\right.$$

$$\left.\left.\times\ |n+i(s-t)|^2 \frac{\sinh(\pi s)}{\cosh(\pi s)+\cos(\pi h)} \frac{\sinh(\pi t)}{\cosh(\pi t)+(-1)^n\cos(\pi h)} \right| dtds \right\} dh$$

REFERENCES

[1] Harish-Chandra, "Harmonic analysis on real reductive groups, I," J. Funct. Anal. 19 (1975), 104-204.

[2] Harish-Chandra, "Harmonic analysis on real reductive groups, II," Invent. Math. 36 (1976), 1-55.

[3] Harish-Chandra, "Harmonic analysis on real reductive groups, III," Ann. of Math. 104 (1976), 117-201.

[4] R.A.Herb and J.A.Wolf, "The Plancherel Theorem for general semisimple groups," Invent. Math., to appear.

[5] R.A.Herb and J.A.Wolf, "Rapidly decreasing functions on general semisimple groups," Invent. Math., to appear.

HARMONIC ANALYSIS ON RANK ONE SYMMETRIC SPACES

G. van Dijk

Mathematisch Instituut, Rijksuniversiteit Leiden,
Postbus 9512, 2300 RA Leiden, The Netherlands.

0. Introduction

Some physical considerations (e.g. the study of the scattering amplitude of two particles in the t channel [8]) have lead Limič, Niederle and Raczka to the study of harmonic analysis on hyperboloids and cones. From the abstract of their 1967 paper [8] in the Journal of Mathematical Physics we quote:

"The eigenfunction expansions associated with the second order invariant operator on hyperboloids and cones are derived. The global unitary irreducible representations of the $SO_0(p,q)$ groups related to hyperboloids and cones are obtained. The decomposition of the quasi-regular representations into the irreducible ones is given and the connection with the Mautner theorem and nuclear spectral theory is discussed".

Their study is related to earlier work done by Tolar, Barut and Salam. The ideas developed by Limič, Niederle and Raczka were taken up in the seventies by such mathematicians as Molcanov [10], Strichartz [19], Rossmann [14] and in particular Faraut [1]. They started a systematic study of the harmonic analysis of rank one pseudo-Riemannian semisimple symmetric spaces. The main emphasis was on the classical isotropic spaces: the hyperboloids over the reals, the complex numbers and the quaternions. Since 1980 a group of mathematicians at Leiden University, have tried to complete the above program. Recently we succeeded in giving the decomposition of the quasi-regular representation into irreducible ones for all rank one spaces. M.T. Kosters treated the hyperboloid over the octonions [6] and obtained partial results for the space $SL(n,\mathbb{R})/GL(n-1,\mathbb{R})$ [6]. The Plancherel formula for the latter space was recently obtained by the author and Poel [22]. The non-isotropic symplectic spaces were clarified by W.A. Kosters [7]. He also has complete results now for the space $F_{4(4)}/Spin(4,5)$. Finally the author obtained the Plancherel formula for the spaces $SL(n,\mathbb{R})/GL_+(n-1,\mathbb{R})$. In this lecture we shall report on our work in a historical perspective.

1. Orthogonal groups

Denote by $O(p,q)$ the group of $(p+q)\times(p+q)$ matrices, with non-zero determinant, which leave invariant the quadratic form

$$-x_1^2 - x_2^2 - \dots - x_p^2 + x_{p+1}^2 + \dots + x_{p+q}^2 .$$

Let $SO(p,q)$ be the subgroup of $O(p,q)$ consisting of matrices of determinant equal to one. $SO_0(p,q)$ [or $SO_e(p,q)$, or $SO^{\uparrow}(p,q)$] is the connected component of the identity of $SO(p,q)$. The following special cases are well-known in physics:

$p = 1$, $q = 5$: the Euclidean conformal group

$p = 1$, $q = 4$: de Sitter group

$p = 1$, $q = 3$: Lorentz-group .

For $p = 1$, $SO_0(p,q)$ can be described more explicitly as:

$$SO_0(1,q) = \{g \in O(1,q) : \det g = 1, g_{11} \geq 1\} .$$

Let $p \geq 1$, $q \geq 1$.
Put $X = SO_0(p,q)/SO_0(p,q-1)$.
Let e_n be the basis vector in $\mathbb{R}^n$ given by $e_n = (0,0,\dots,0,1)$. The map $g \mapsto g\cdot e_n$ $(g \in SO_0(p,q))$ has stabilizer $SO_0(p,q-1)$ and hence we get a natural isomorphism of X onto one sheet of the hyperboloid

$$-x_1^2 - x_2^2 - \dots - x_p^2 + x_{p+1}^2 + \dots + x_{p+q}^2 = 1 .$$

$G = SO_0(p,q)$ acts on X by left translations. Furthermore, it is well-known that X carries a (up to a positive constant unique) positive G-invariant measure dx. So G also acts on $L^2(X)$ by unitary operators: $g\cdot f(x) = f(g^{-1}x)$ $(g \in G, x \in X, f \in L^2(X))$. Our main purpose is the decomposition of the space $L^2(X)$ into irreducible subspaces for G.
If $q = 1$, so $X = SO_0(p,1)/SO(p)$ this decomposition is well-known (cf. [20]). Note that X is a Riemannian symmetric space in this case. For $q > 1$, X is pseudo-Riemannian.

In general the decomposition of $L^2(X)$ contains a continuous and discrete part (if $q = 1$, no discrete part occurs!). We also pay attention to the question whether the decomposition is *multiplicity-free* or not. We recall the definition. A unitary representation π of G on a Hilbert space $\mathcal{H}$ is said to be (or to decompose) multiplicity-free if the commutant of $\pi(G)$ in the algebra of continuous linear operators of $\mathcal{H}$ into itself, is abelian.

2. Hyperboloids and symmetric spaces

Let J be the $(p+q) \times (p+q)$-matrix given by

$$\begin{pmatrix} 1 & & & \\ & \ddots & & \\ & & 1 & \\ & & & -1 \end{pmatrix}$$

and σ the involutive automorphism of $G = SO_0(p,q)$ defined by $\sigma(g) = JgJ$. Then $SO_0(p,q-1)$ is precisely the set of fixed points of σ:

$$SO_0(p,q-1) = \{g \in G : \sigma(g) = g\} .$$

DEFINITION. *Let* G *be a connected Lie group and* σ *an involutive automorphism of* G. *Let* G_σ *be the subgroup of fixed points of* σ *and* G_σ^0 *its connected component of the identity. Choose a closed subgroup* H *with* $G_\sigma^0 \subset H \subset G_\sigma$. *The pair* (G,H) *is called a* <u>*symmetric pair*</u> *and* $X = G/H$ *a* <u>*symmetric space*</u>.

So $(SO_0(p,q), SO_0(p,q-1))$ is a symmetric pair and the hyperboloid $X = SO_0(p,q)/SO_0(p,q-1)$ is a symmetric space. We give some more examples.

I. Let G_1 be a connected Lie group and put $G = G_1 \times G_1$. Define $\sigma(x,y) = (y,x)$. Then σ is an involution on G , $G_\sigma = \mathrm{diag}(G)$ and $X = G/G_\sigma$ is naturally isomorphic to G_1 by the map:

$$g \mapsto (g,e)G_\sigma \qquad (g \in G_1)$$

G acts on $X = G_1$ by

$$(x,y)\cdot g = xgy^{-1} \qquad (x,y,g \in G_1) .$$

Assume that G_1 admits a left- and right-invariant positive measure dx .

$L^2(X) \cong L^2(G_1)$ is decomposed into bi-G_1-invariant irreducible subspaces; this is precisely the content of the Plancherel-theorem for G_1 , provided G_1 is supposed to be a type I group.

The (abstract) *Plancherel-formula* looks like:

$$\int_{G_1} |f(x)|^2 dx = \int_{\hat{G}_1} \Theta_\pi(\tilde{f}*f)\, d\mu(\pi)$$

$$= \underbrace{\int_C \Theta_\pi(\tilde{f}*f)\, d\mu(\alpha)}_{\text{continuous part}} + \underbrace{\sum_\beta d_\beta \quad \Theta_\beta(\tilde{f}*f)}_{\text{discrete part}}$$

(f a smooth function on G_1 with compact support).

Or

$$f(e) = \int_{\hat{G}_1} \langle f, \Theta_\pi \rangle d\mu(\pi) .$$

Here Θ_π is the (distribution-)*character* of π .

<u>Special cases</u>:

$$G = \mathbb{R} \quad : f(0) = \int_{-\infty}^{\infty} \langle f, e^{2\pi i x y}\rangle dy \qquad \text{(Fourier-integral)}$$

$$G = \mathbb{R}/\mathbb{Z} : f(0) = \sum_{n \in \mathbb{Z}} \langle f, e^{2\pi i n \phi}\rangle \qquad \text{(Fourier-series)}.$$

For *semisimple* groups G_1 , Harish-Chandra has determined the Plancherel formula explicitly. Probably this will turn out to be one of the greatest efforts of the twentieth century in mathematics (see [4]). The decomposition is multiplicity-free in this case by a result of Segal and Godement [17] (G_1 non-necessarily semisimple).

II. Let G be a connected semisimple Lie group with finite centre. Here we choose now σ to be a Cartan-involution of G . $H = G_\sigma$ is *compact* and connected in this case.

Examples. $G = SO_0(p,1)$, $H = SO(p)$, σ as at the beginning of this section

$G = SO(n)$, $H = SO(n-1)$, σ as at the beginning of this section

The decomposition of $L^2(SO(n)/SO(n-1))$ is perhaps best known, since it is part of most introductory courses in harmonic analysis. The reader is referred to [15]. For $n = 2$ we are back at ordinary Fourier series theory and the decomposition is given by Parseval's theorem. For $n > 2$ one is lead to "spherical harmonics".
For G non-compact, the decomposition is more complicated. The following abstract form of a Plancherel formula was derived by Godement [3]: for all smooth functions f on G/H with compact support one has

$$f(eH) = \int_\Lambda \langle f,\phi_\lambda\rangle \, d\mu(\lambda) \ .$$

Here Λ is the parameter-set of the zonal spherical functions ϕ_λ which are positive-definite, and μ a (uniquely determined) positive Radon measure on Λ .

So the role of characters in the group case is taken over by positive-definite spherical functions on G . Each ϕ_λ is an eigenfunction of all Laplace-operators on G/H . The explicit form of μ is determined by Harish-Chandra [5]. The decomposition of $L^2(X)$ is *multiplicity-free*, which is commonly expressed in mathematics by saying that the pair (G,H) is a Gelfand pair. Furthermore only representations with a vector fixed by H in their representation-space arise in this decompostion.

We note that in both cases I and II, the Dirac-δ-distribution is decomposed into elementary (extremal) positive-definite distributions.

3. Pseudo-Riemannian rank one symmetric pairs

For simplicity of the formulation we restrict ourselves to rank one semisimple symmetric pairs in this section. Most theorems are however true for general symmetric pairs or even in a more general context.

Rank one pairs comprise not only the real hyperbolic pairs $(SO_0(p,q), SO_0(p,q-1))$. The following list is taken from [23].

G	H	G/H	
Spin(p,q+1)	Spin(p,q)	$SO_0(p,q+1)/SO_0(p,q)$	p,q ≥ 1
Spin(p,q+1)	G_σ	$SO(p,q+1)/S(O(p,q)\times O(1))$	p,q ≥ 1
SU(p,q+1)	$S(U(p,q)\times U(1))$		p,q ≥ 1
Sp(p,q+1)	$Sp(p,q)\times Sp(1)$		p,q ≥ 1
$F_{4(-20)}$	Spin(1,8)		
$SL(n+1,\mathbb{R})$	$S(GL_+(n,\mathbb{R})\times GL_+(1,\mathbb{R}))$		n ≥ 1
$SL(n+1,\mathbb{R})$	$S(GL(n,\mathbb{R})\times GL(1,\mathbb{R}))$		n ≥ 2
$Sp(n+1,\mathbb{R})$	$Sp(n,\mathbb{R})\times Sp(1,\mathbb{R})$		n ≥ 2
$F_{4(4)}$	Spin(4,5)		

A form of an (abstract) Plancherel formula for G/H , which gives some hope for actually computing it in concrete cases, was only recently found (see [22] for details).

Call $\Box$ the Laplace-Beltrami operator on $X = G/H$. A distribution T on X is said to be *spherical* if

(i) T is H-invariant

(ii) T is an eigendistribution of $\Box$:

$$\Box T = \lambda T \quad \text{for some complex number } \lambda .$$

Then one has the following theorem (see [22]).

THEOREM. *Let* S *be a "good" parametrization of the elements of* $\hat{G}$ *which allow an H-fixed distribution-vector. There exists a (non-necessarily unique) Radon measure* m *on* S *such that*

(i) $\phi(eH) = \int_S \langle T_s,\phi\rangle \, dm(s)$ *for all smooth* ϕ *with compact support on* X

(ii) *for* $s \in S$, T_s *is a spherical and extremal positive-definite distribution.*

Unicity occurs as soon as one knows that the representation of G on $L^2(X)$ is multiplicity-free.

The main ingredient of the proof of the theorem is L. Schwartz' theory of *reproducing kernels* [16] and Thomas' theory of desintegration of such kernels (see e.g. [21]).

4. Multiplicity

The following result was recently obtained by the author (see [23]): For all semi-simple symmetric rank one pairs, listed in section 3, $L^2(X)$ is multiplicity-free, *except* for the pairs

$$(Spin(1,q+1), Spin(1,q)) \qquad (q \geq 1)$$

[or $(SO_0(1,q+1, SO_0(1,q))$] .

Indeed, the decomposition of $L^2(X)$ shows multiplicity $= 2$ in the continuous part for these pairs.

In [23] we actually proved a much stronger result: any unitary representation of G which can be realized inside the space of distributions on $X = G/H$, is multiplicity-free. So in particular $L^2(X)$ is multiplicity-free. (G and H as before).

5. Explicit form of the Plancherel formula

In the following scheme we mention the rank one spaces and the authors, who found an explicit form of the Plancherel formula for these spaces.

Hyperbolic spaces (over $\mathbb{R}$, $\mathbb{C}$, $\mathbb{H}$).

over $\mathbb{R}$: $X = SO_0(p,q+1)/SO_0(p,q)$ - Molcanov [10] (1981)
$X = SO_0(p,q+1)/S(O(p,q)\times O(1))$ -
$p=1$, $q=2$: Gelfand, Graev and Vilenkin [2]
$p=1$: Shintani [18]
all $p,q \geq 1$: Limič, Niederle, Raczka [8], Strichartz [19], Rossmann [14], Molcanov [10], Faraut [1] (1979).

over $\mathbb{C}$: $X = SU(p,q+1)/S(U(p,q)\times U(1))$ -
$p = 1$: Matsumoto [9]
all $p,q \geq 1$: Faraut [1] (1979).

over $\mathbb{H}$	: $X = Sp(p,q+1)/Sp(p,q)\times Sp(1)$	- Faraut [1] (1979)
	$F_{4(-20)}/Spin(1,8)$	- M.T. Kosters [6] (1982)
	$SL(n+1,\mathbb{R})/GL(n,\mathbb{R})$	- Poel en Van Dijk [22] (1984)
		n = 2 : Molcanov [12]
	$SL(n+1,\mathbb{R})/GL_+(n,\mathbb{R})$	- Van Dijk (1984)
		n = 1 : Molcanov [11]
	$Sp(n+1,\mathbb{R})/Sp(n,\mathbb{R})\times Sp(1,\mathbb{R})$	- W.A. Kosters [7] (1985)
	$F_{4(4)}/Spin(4,5)$	- W.A. Kosters (1985) .

The idea of the construction is to start with the determination of the spectral resolution of the Laplace-Beltrami operator. The trick of Limič c.s. to reduce everything to second order ordinary differential operators by using the action of a maximal compact subgroup of G on $L^2(X)$, works only for isotropic spaces: the hyperbolic spaces over $\mathbb{R}$, $\mathbb{C}$, $\mathbb{H}$ and $F_{4(-20)}/Spin(1,8)$. In the other cases one has to choose another way to overcome the difficulties (see e.g. [22]).

To get an idea of the form of such a Plancherel formula we give it for $X = SO(p,q)/S(O(1)\times O(p-1,q))$, with $p > 1$ and q odd (see [1], Théorème 10)

$$\phi(eH) = \frac{1}{2\pi}\int_0^\infty \langle\phi,\zeta_{i\nu}\rangle \frac{d\nu}{|c(i\nu)|^2}$$

$$+ \sum_{\rho+2r+1>0} \langle\phi,\zeta_{\rho+2r+1}\rangle \; \mathrm{Residu}\left[\frac{1}{c(s)c(-s)}, \rho+2r+1\right]$$

for all smooth ϕ on X with compact support. Here $\rho = \frac{1}{2}n-1$, $n = p+q$,

$$c(s) = \frac{\Gamma(\frac{p}{2})\;\Gamma(\frac{q}{2})}{\sqrt{\pi}} \; \frac{2^{\rho-s}\;\Gamma(s)}{\Gamma((s+\rho)/2)\,\Gamma((s+p-\rho)/2)\;\Gamma((s+q-\rho)/2)}$$

References

[1] FARAUT, J., *Distributions sphériques sur les espaces hyperboliques*, J. Math. Pures Appl. 58, 369-444 (1979).

[2] GELFAND, I.M., GRAEV, M.I. and VILENKIN, N.Y., *Generalized Functions vol. 5: Integral Geometry and Representation Theory*, Academic Press, New-York-London 1966.

[3] GODEMENT, R., *Introduction aux travaux de A. Selberg*, Séminaire Bourbaki, t.9, exp. 144 (1956/57).

[4] HARISH-CHANDRA, *Collected Papers*, 4 volumes, Springer-Verlag, Berlin etc. 1983.

[5] HARISH-CHANDRA, *Spherical functions on a semisimple Lie group I,II*, Collected Papers, volume 2 (1983).

[6] KOSTERS, M.T., *Spherical distributions on rank one symmetric spaces*, Phd. Thesis, University of Leiden (1983).

[7] KOSTERS, W.A., *Harmonic analysis on symmetric spaces*, Phd. Thesis, University of Leiden (1985).

[8] LIMIČ, N., NIEDERLE, J. and RĄCZKA, R., *Eigenfunction expansions associated with the second-order invariant operator on hyperboloids and cones, III*, J. Math. Phys. 8, 1079-1093 (1967).

[9] MATSUMOTO, S., *The Plancherel formula for a pseudo-Riemannian symmetric space*, Hiroshima Math. J. 8, 181-193 (1978).

[10] MOLCANOV, V.F., *The Plancherel formula for hyperboloids*, Proceedings of the Steklov Institute of Mathematics, 2, 63-83 (1981).

[11] MOLCANOV, V.F., *Harmonic analysis on the pseudo-Riemannian symmetric spaces of the group* $SL(2,\mathbb{R})$, Math. USSR Sbornik 46, no.4, 493-505 (1983).

[12] MOLCANOV, V.F., *The Plancherel formula for the pseudo-Riemannian space* $SL(3,\mathbb{R})/GL(2,\mathbb{R})$, Sibirsk Math. J. 23, 142-151 (1982) (Russian).

[13] NIEDERLE, J., *Decomposition of discrete most degenerate representations of* $SO_0(p,q)$ *when restricted to representations of* $SO_0(p,q-1)$ *or* $SO_0(p-1,q)$, J. Math. Phys. 8, 1921-1930 (1967).

[14] ROSSMANN, W., *Analysis on real hyperbolic spaces*, J. Funct. Anal. 30, 448-477 (1978).

[15] SCHEMPP, W., DRESELER, B., *Einführung in die harmonische Analyse*, B.G. Teubner, Stuttgart, 1980.

[16] SCHWARTZ, L., Sousespaces Hilbertiens d'espaces vectoriels topologiques et noyaux associés, J. Anal. Math. 13, 115-256 (1964).

[17] SEGAL, I.E., *The two-sided regular representation of a unimodular locally compact group*, Annals of Math. 51, 293-298 (1950).

[18] SHINTANI, T., *On the decomposition of the regular representation of the Lorentz group on a hyperboloid of one sheet*, Proc. Japan Acad. 43, 1-5 (1967).

[19] STRICHARTZ, R.S., *Harmonic analysis on hyperboloids*, J. Funct. Anal. 12, 341-383 (1973).

[20] TAKAHASHI, R., *Sur les représentations unitaires des groupes de Lorentz généralisés*, Bull. Soc. Math. France 91, 289-433 (1962).

[21] THOMAS, E.G.F., *The theorem of Bochner-Schwartz-Godement for generalized Gelfand pairs*, Functional Analysis: Surveys and results III, K.D. Bierstedt and B. Fuchsteiner (eds.), Elsevier Science Publishers B.V. (North Holland) (1984).

[22] VAN DIJK, G., POEL, M., *The Plancherel formula for the pseudo-Riemannian space* $SL(n,\mathbb{R})/GL(n-1,\mathbb{R})$, preprint, Univ. of Leiden (1984), to appear in Comp. Math..

[23] VAN DIJK, G., *On a class of generalized Gelfand pairs*, Report nr. 18, Mathematical Institute, Univ. of Leiden (1985).

A spin-off from highest weight representations; Conformal covariants, in particular for $O(3,2)$.

Hans Plesner Jakobsen
Mathematics Institute, Universitetsparken 5
DK-2100 Copenhagen Ø, Denmark

0. Introduction

Let $\mathcal{D} = G/K$ be a hermitian symmetric space of the non-compact type and let $E_{\tau_i} = G \times_K V_{\tau_i}$ be holomorphic vector bundles over $\mathcal{D}$; $i=1,2$, defined by finite-dimensional representations τ_i of a maximal compact subgroup K of the group G of holomorphic transformations of $\mathcal{D}$. Denote by $\Gamma_h(E_{\tau_i})$ the space of holomorphic sections of E_{τ_i} and let U_{τ_i} denote the representation of G on $\Gamma_h(E_{\tau_i})$ obtained from left translation of G on E_{τ_i}.

Consider a differential operator

$$(0.1) \qquad D: s \in \Gamma_h(E_{\tau_1}) \to Ds \in \Gamma_h(E_{\tau_2}) .$$

Definition. D is covariant if

$$(0.2) \qquad \forall g \in G \ \forall s \in \Gamma_h(E_{\tau_1}): DU_{\tau_1}(g)s = U_{\tau_2}(g)Ds .$$

The bundles E_{τ_i} may be parallellized; then $\Gamma_h(E_{\tau_i})$ becomes the space $\mathcal{O}(V_{\tau_i})$ of V_{τ_i}-valued holomorphic functions on $\mathcal{D}$ and D becomes a matrix-valued differential operator.

We may restrict (0.2) to the Shilov boundary of $\mathcal{D}$; for appropriate realizations of $\mathcal{D}$ and for suitable choices of G, among the spaces obtained as such are n-dimensional Minkowski space as well as $U(n)$, $n=1,2,\ldots$. Secondly, the representations involved are of positive (or one-sided) energy. For these as well as more abstract reasons, one is interested in

Problem I: Determine all such (D,τ_1,τ_2), and more generally,

Problem II: i) Determine all invariant subspaces of $\mathcal{O}(V_{\tau_1})$ and identify the subquotients. ii) In particular, determine which subspaces correspond to kernels of differential operators.

Let $\mathfrak{g}$ denote the Lie algebra of G . We note here that, due to analyticity, D is covariant if and only if

$$(0.3) \qquad \forall f \in \mathcal{O}(V_{\tau_1}) \forall x \in \mathfrak{g} : DdU_{\tau_1}(x)f \subset dU_{\tau_2}(x)Df ,$$

where dU denotes the differential of the representation U . In fact, as explained in [4] or [7], the problem is completely (modulo coverings of G) equivalent, by duality, to the algebraic problem of determining all homomorphisms between certain highest weight modules (≡ vacuum vector representations) of $\mathfrak{g}$. We remark that such a homeomorphism into a vacuum vector representation is completely determined by a second vacuum in the given space. We further remark that by this transformation, even the space $\mathcal{D}$ seems to disappear from the discussion. In some sense this is true, and this implies that the results hold for a number of different realizations, but, in fact, one realization is still there; as a subset of a subspace of the complexification of $\mathfrak{g}$.

In the following sections we explain in more detail about the highest weight modules involved. Then we turn to the special case $\mathfrak{g} = so(3,2) = sp(n,\mathbb{R})$ and describe a complete solution to the classification problem. It should be noted that this is an example of a group with two root lengths. We have also recently obtained the full classification for su(2,2) , and state the result without proof. The details for su(2,2) will appear in [8] in which also a more detailed (but still far from complete) bibliography is given.

Finally, it should be remarked that our Problem II, though quite formidable in its full generality, still is only a special case of the programs of Dobrev ([3], and references therein) and of Angelopoulos ([1]). (See also these proceedings.)

1. Simple Lie algebras

A Lie algebra $\mathfrak{g}$ is called simple if it contains no ideals except 0 and itself, and such that, furthermore, $\mathfrak{g}$ is non-abelian. Then $\mathfrak{g} = [\mathfrak{g},\mathfrak{g}]$. We will assume that $\mathfrak{g}$ is complex, but usually we will have in mind that $\mathfrak{g}$ is the complexification of some specific real Lie algebra $\mathfrak{g}_{\mathbb{R}}$; $\mathfrak{g} = (\mathfrak{g}_{\mathbb{R}})^{\mathbb{C}} = \mathfrak{g}_{\mathbb{R}} \otimes_{\mathbb{R}} \mathbb{C}$.

Let $\mathfrak{h}$ be a Cartan subalgebra of $\mathfrak{g}$. Then, under the adjoint action, $\mathfrak{h}$ can be diagonalized simultaneously. Let $\alpha \in \mathfrak{h}^*$ and set

$$(1.1) \qquad \mathfrak{g}^{\alpha} = \{x \in \mathfrak{g} \mid \forall h \in \mathfrak{h}: [h,x] = \alpha(h)x\} .$$

Further set

$$(1.2) \qquad \Delta = \{\alpha \in \mathfrak{h}^* \mid g^{\alpha} \neq \{0\}\} .$$

Δ is called the set of <u>roots</u>, and $\mathfrak{g}^{\alpha}$ is the space of <u>root-vectors</u> belonging to α. It is a fact that $\forall \alpha \in \Delta: \dim_{\mathbb{C}} \mathfrak{g}^{\alpha} = 1$, and $\alpha \in \Delta \Leftrightarrow -\alpha \in \Delta$.

On $\mathfrak{g}$ there is a symmetric bilinear form B; the killing form, and the restriction of B to $\mathfrak{h}$ induces, via duality, a non-degenerate form $(\cdot,\cdot)$ on $\mathfrak{h}^*$ and hence on Δ. On the real span of Δ, this form is real and positive definite. Furthermore,

$$(1.3.) \qquad \alpha,\beta \in \Delta \Rightarrow \langle \alpha,\beta \rangle \underset{\text{def.}}{=} \frac{2(\beta,\alpha)}{(\alpha,\alpha)} \in \mathbb{Z} .$$

For $\alpha \in \Delta$ let $h_{\alpha} \in [g^{\alpha}, g^{-\alpha}] \subseteq \mathfrak{h}$ be determined by $\alpha(h_{\alpha}) = 2$. Then

$$(1.4) \qquad \langle \alpha,\beta \rangle = \beta(h_{\alpha}) .$$

Set

$$(1.5) \qquad S_{\alpha}(\beta) = \beta - \beta(h_{\alpha})\alpha .$$

Then $S_{\alpha}(\Delta) = \Delta$, and the reflexions S_{α}, $\alpha \in \Delta$, generate the so-called Weyl group of Δ.

Finally the elements in a basis Σ of Δ are called the <u>simple</u> roots. Δ decomposes according to Σ into

$$(1.6) \qquad \Delta = \Delta^{+} \cup \Delta^{-}$$

where Δ^{+} denotes the set of roots whose coordinates w.r.t. Σ all are non-negative integers, and $\Delta^{-} = -\Delta^{+}$. (A good reference to this section is [6].)

<u>2. Highest weight modules</u>

Fix a basis Σ of Δ. Set

(2.1) $\quad \mathfrak{g}^+ = \sum_{\alpha\in\Delta^+} \mathfrak{g}^\alpha \,, \quad \mathfrak{g}^- = \sum_{\alpha\in\Delta^+} \mathfrak{g}^{-\alpha}\,,$ and let $\rho = \frac{1}{2} \sum_{\alpha\in\Delta^+} \alpha$.

Then

$$\mathfrak{g} = \mathfrak{g}^- \oplus \mathfrak{h} \oplus \mathfrak{g}^+ \quad . \tag{2.2}$$

We let $\mathcal{U}(\mathfrak{g})$ denote the universal enveloping algebra of $\mathfrak{g}$. It follows from (2.2) that

$$\mathcal{U}(\mathfrak{g}) = \mathcal{U}(\mathfrak{g}^-)\mathcal{U}(\mathfrak{h})\mathcal{U}(\mathfrak{g}^+) \quad . \tag{2.3}$$

Let $\Lambda \in \mathfrak{h}^*$. The <u>Verma module</u> $M(\Lambda)$ of highest weight Λ is defined as follows:

1) $M(\Lambda)$ is a representation space of $\mathcal{U}(\mathfrak{g})$ and contains a vector v such that

i) $M(\Lambda) = \mathcal{U}(\mathfrak{g})\cdot v_\Lambda$

(2.4) ii) $\forall h \in \mathfrak{h}: h\cdot v_\Lambda = \Lambda(h)\cdot v_\Lambda$

iii) $\forall x \in \mathfrak{g}^+: x\cdot v_\Lambda = 0$.

2) $M(\Lambda)$ is maximal in this respect. ($M(\Lambda) = \mathcal{U}(\mathfrak{g}^-) \otimes v_\Lambda$ as a representation of $\mathfrak{h}$) .

More generally, a highest weight module of h.w. Λ is a module that satisfies 1) above. A special instance of this is a <u>generalized Verma module</u> $M_p(\Lambda)$ which is a quotinet of $M(\Lambda)$ and corresponds to induction from a (not necessarily minimal) parabolic p . We shall be interested in generalized Verma modules corresponding to holomorphic induction, and for that reason we only furnish the details for this case:

Assume from now on that $\mathfrak{g}$ corresponds to a hermitian symmetric space (a good reference for what follows is [5]). Let

$$\mathfrak{g}_{\mathbb{R}} = \mathfrak{k}_{\mathbb{R}} \oplus \mathfrak{p}_{\mathbb{R}} \tag{2.5}$$

be a Cartan decomposition of the underlying real Lie algebra. Then (2.2) decomposes further into

$$\mathfrak{g} = \mathfrak{p}^- \oplus \mathfrak{k}^- \oplus \mathfrak{h} \oplus \mathfrak{k}^+ \oplus \mathfrak{p}^+ \quad \text{where} \tag{2.6}$$

$$\mathfrak{k} = (\mathfrak{k}_{\mathbb{R}})^{\mathbb{C}} = \mathfrak{k}^- \oplus \mathfrak{h} \oplus \mathfrak{k}^+ \,, \quad \mathfrak{p} = \mathfrak{p}_{\mathbb{R}}^{\mathbb{C}} = \mathfrak{p}^+ \oplus \mathfrak{p}^- \,, \quad \text{and} \tag{2.7}$$

$$[\mathfrak{p}^+,\mathfrak{p}^+] \cong [\mathfrak{p}^-,\mathfrak{p}^-] = 0 \ .$$

In the present case there is a unique simple non-compact root β $(\mathfrak{g}^{\beta} \subset \mathfrak{p}^{+})$, and $\Sigma \setminus \{\beta\}$ is a set of simple roots for $\mathfrak{k}_1 = [\mathfrak{k},\mathfrak{k}]$. (We have that $\mathfrak{k} = \mathfrak{k}_1 \oplus \eta(\mathfrak{k})$ where $\eta(\mathfrak{k})$ is the 1-dimensional center of $\mathfrak{k}$; $\eta(\mathfrak{k}) = \mathbb{C}\cdot h_0$.)

Let $\mathfrak{h}_1 = \mathfrak{h} \cap \mathfrak{k}_1$ and let γ_r denote the <u>highest root</u> ($\mathfrak{g}^{\gamma_r}$ satisfies: $\mathfrak{g}^{\gamma_r} \subset \mathfrak{p}^{+}$ and $\forall x \in \mathfrak{g}^{\gamma_r}$ $\forall y \in \mathfrak{k}^{+}$: $[x,y] = 0$). Then $\Lambda \in \mathfrak{h}^*$ can be written as

(2.8) $$\Lambda = (\Lambda_0,\lambda) \quad \text{where} \quad \Lambda_0 = \Lambda|_{\mathfrak{h}_1} \quad \text{and} \quad \lambda = \Lambda(h_{\gamma_r}) \ .$$

From now on we assume:

(2.9) $$\forall \mu \in \Sigma\setminus\{\beta\}: \Lambda_0(h_\mu) \in \mathbb{N} \cup \{0\} \quad \text{and} \quad \lambda \in \mathbb{R} \ .$$

It follows that Λ_0 determines a finite dimensional representation $V(\Lambda_0)$ of $\mathfrak{k}_1$;

(2.10) $$V(\Lambda_0) = \mathcal{U}(\mathfrak{k}_1)/I$$

where I is a left ideal in $\mathcal{U}(\mathfrak{k}_1)$. If $V(\Lambda_0)$ is given together with its highest weight vector v_{Λ_0} , I is determined as $I = \{u \in \mathcal{U}(\mathfrak{k}_1) \mid u\cdot v_{\Lambda_0} = 0\}$. However, usually it is the other way around; I determines $V(\Lambda_0)$.

The generalized Verma module $M_{\mathfrak{k}}(\Lambda)$ is defined by

(2.11) $$M_{\mathfrak{k}}(\Lambda) = \mathcal{U}(\mathfrak{g}) \underset{\mathcal{U}(\mathfrak{k}\oplus\mathfrak{p}^{+})}{\otimes} V(\Lambda)$$

where $V(\Lambda_0)$ is extended to a representation $V(\Lambda)$ of $\mathcal{U}(\mathfrak{k} \oplus \mathfrak{p}^{+})$ by letting h_0 act by $\Lambda(h_0)$ and letting $\mathfrak{p}^{+}$ act as zero. Clearly,

(2.12) $$M_{\mathfrak{k}}(\Lambda) = \mathcal{U}(\mathfrak{p}^{-}) \otimes V(\Lambda)$$

as a representation of $\mathfrak{k}$. Further, if $\widetilde{A}(\Lambda) = \mathcal{U}(g)\cdot I\cdot v_{\Lambda}$ then

(2.13) $$M_{\mathfrak{k}}(\Lambda) = {}^{M(\Lambda)}/_{\widetilde{A}(\Lambda)}$$

is a realization of $M_{\mathfrak{k}}(\Lambda)$ as a quotient of $M(\Lambda)$.

<u>3. Homomorphisms between highest weight modules</u>

Let R_{Λ} and R_{Λ_1} be highest weight modules of h.w.'s Λ and Λ_1 , respectively. A homomorphism φ of R_{Λ_1} into R_Λ is a map from R_{Λ_1} to R_Λ which commutes with the representations. In particular, $\varphi(v_{\Lambda_1}) = \tilde{v}_{\Lambda_1} \in R_\Lambda$ is a vector which satisfies 1) ii) and iii) in (2.4) for Λ_1. We assume that φ is non-trivial i.e. that $\tilde{v}_{\Lambda_1} \neq 0$. Conversely, if a non-zero vector in R satisfies 1) ii) and iii), then one can clearly define a map $\bar{\varphi}: M(\Lambda_1) \to R_\Lambda$, and out of $\bar{\varphi}$ one can construct a map of any given quotient of $M(\Lambda_1)$ into an appropriate subquotient of R_Λ. It is also clear that it may happen that the induced map between a quotient of R_{Λ_1} and a quotient of R_Λ may be zero.

In particular, a map $\varphi: M(\Lambda_1) \to M(\Lambda)$ may induce the trivial map from $M_k(\Lambda_1)$ to $M_k(\Lambda)$.

In this area, the most important theorem is the BGG (Bernstein-Gelfand-Gelfand) theorem. To formulate it we need:

Definition 3.1. Let $\chi, \psi \in \mathfrak{h}^*$. A sequence of roots $\gamma_1, \ldots, \gamma_k$ is said to satisfy condition (A) for the pair (χ, ψ) if

i) $\chi = S_{\gamma_k} \cdot\cdot\cdot\cdot\cdot S_{\gamma_1}\chi$

ii) Put $\chi_0 = \psi$, and $\chi_i = S_{\gamma_i} \cdots S_{\gamma_1}\psi$. Then

$\forall i=1,\ldots,k$: $\langle \chi_{i-1}, \gamma_i \rangle \in \mathbb{N}$.

Under these circumstances, (χ, ψ) is said to satisfy $(\bar{A})$.

Theorem 3.2 (BGG; [2]). i) There is a non-zero homomorphism from $M(\Lambda_1)$ if and only if $(\Lambda_1+\rho, \Lambda+\rho)$ satisfies $(\bar{A})$. ii) If there is a homomorphism from a (sub-)quotient of $M(\Lambda_1)$ to a (sub-)quotient of $M(\Lambda)$, then $(\Lambda_1+\rho, \Lambda+\rho)$ satisfies $(\bar{A})$.

Let $\Delta_n^+ = \{\alpha \in \Delta^+ \mid \mathfrak{g}^\alpha \subset \mathfrak{p}^+\}$ be the set of non-compact positive roots. The following is proved in [7].

Corollary 3.3. If there is a non-trivial homomorphism from $M_k(\Lambda_1)$ to a (sub-)quotient of $M_k(\Lambda)$, then $(\Lambda_1+\rho, \Lambda+\rho)$ satisfies $(\bar{A})$ with a sequence $\gamma_1, \ldots, \gamma_k$ of positive non-compact roots.

4. $O(3,2) \cong Sp(2,\mathbb{R})$

The following realization of $\mathfrak{g} = \mathfrak{o}(3,2) = \mathfrak{sp}(2,\mathbb{R})$ is convenient because it displays $\mathfrak{p}^+$ and $\mathfrak{p}^-$ directly as $\mathfrak{k}$-representation spaces. Let

(4.1) $$k^+ = \left[\begin{array}{cc|cc} 0 & 1 & \multicolumn{2}{c}{0} \\ 0 & 0 & & \\ \hline \multicolumn{2}{c|}{0} & 0 & 0 \\ & & -1 & 0 \end{array}\right], \quad k^- = \left[\begin{array}{cc|cc} 0 & 0 & \multicolumn{2}{c}{0} \\ 1 & 0 & & \\ \hline \multicolumn{2}{c|}{0} & 0 & -1 \\ & & 0 & 0 \end{array}\right], \quad \text{and} \quad h_k = [k^+, k^-],$$

(4.2) $$\mathfrak{p}^- = \left\{\left[\begin{array}{cc|c} \multicolumn{2}{c|}{0} & 0 \\ \hline z_a & z_b & 0 \\ z_b & z_c & \end{array}\right] \;\middle|\; z_a, z_b, \text{ and } z_c \in \mathbb{C}\right\}, \quad \text{and}$$

(4.3) $$\mathfrak{p}^+ = \left\{\left[\begin{array}{c|cc} 0 & z_a & z_b \\ & z_b & z_c \\ \hline 0 & \multicolumn{2}{c}{0} \end{array}\right] \;\middle|\; z_a, z_b, \text{ and } z_c \in \mathbb{C}\right\}.$$

Then

(4.4) $$\mathfrak{k} = \mathbb{C}\cdot k^+ \oplus \mathbb{C}\, k^- \oplus \mathbb{C}\cdot h_k \quad \text{and} \quad \mathfrak{g} = \mathfrak{k} \oplus \mathfrak{p}^+ \oplus \mathfrak{p}^- .$$

We let z_a, z_b, and z_c denote the elements of $\mathfrak{p}^-$ corresponding to an entry 1 at the appropriate place in (4.2), and we let z_a^+, z_b^+, and z_c^+ denote the analogous elements of $\mathfrak{p}^+$ in (4.3). Also let

(4.5) $$h_a = [z_a^+, z_a], \quad h_b = [z_b^+, z_b], \quad \text{and} \quad h_c = [z_c^+, z_c].$$

The elements h_k, h_a, h_b, and h_c are all of the form $h_{\overline{\gamma}}$ for roots $\overline{\gamma}$. We denote these by μ, β, α, and γ, respectively. As before, β is the simple non-compact root, and $\gamma = \gamma_r$ is the highest root.

We identify $\mathfrak{h}$ and $\mathfrak{h}^*$ with $\mathbb{C}^2$. Let e_1 and e_2 be the usual basis vectors. Then

(4.6) $$\beta = 2e_2, \quad \alpha = e_1+e_2, \quad \gamma = 2e_1, \quad \text{and} \quad \mu = e_1-e_2.$$

Furthermore,

(4.7) $$h_a = e_2, \quad h_b = e_1+e_2, \quad h_c = e_1, \quad \text{and} \quad h_k = e_1-e_2.$$

A generalized Verma module Λ is determined by

(4.8) $$\Lambda = (\lambda, \lambda-n); \quad n \in \mathbb{N} \cup \{0\}, \quad \text{and} \quad \rho = (2,1).$$

In the following, n will be held fixed and λ will be allowed to vary. The following is obtained, by trial and error, from Corollary 3.3.

Lemma 4.1. If there is a non-trivial homomorphism $M_{\mathfrak{k}}(\Lambda_1) \to M_{\mathfrak{k}}(\Lambda)$ then the sequence of reflexions corresponding to the sequence of roots in condition (A) is, for λ in the given intervals satisfying throughout that $2\lambda \in \mathbb{Z}$; and furthermore, whenever S_γ or S_β take part, satisfying that $\lambda \in \mathbb{Z}$:

$$\begin{array}{lll} \lambda < -1 & : & \text{None} \\ -1 \le \lambda \le \frac{n}{2}-2 & : & S_\gamma \qquad\qquad (n > 1) \\ \frac{n}{2}-1 \le \lambda \le n-2 & : & S_\alpha\ ,\quad S_\gamma S_\alpha \qquad (n > 1) \\ n-2 < \lambda < n & : & S_\alpha \\ n \le \lambda & : & S_\beta\ ,\quad S_\alpha S_\beta\ ,\quad S_\alpha\ (= S_\gamma S_\alpha S_\beta) \end{array} \tag{4.9}$$

We have from (2.12) that $M_{\mathfrak{k}}(\Lambda)$ is genrated by expressions of the form $z_1 \cdot \ldots \cdot z_r \otimes v$ with $z_1,\ldots,z_r \in \mathfrak{p}^-$ and $v \in V(\Lambda)$. It is obvious that $\mathfrak{k}$ preserves the degree r of such an expression. To describe the representations of $\mathfrak{k}$ which occur in a general $M_{\mathfrak{k}}(\Lambda)$ we must first describe the $\mathfrak{k}$-representation in $\mathcal{U}(\mathfrak{p}^-)$: Let

$$\det z = z_a z_c - \frac{1}{4} z_b^{\,2} . \tag{4.10}$$

The highest weight vectors of $\mathcal{U}(\mathfrak{p})$ as a $\mathfrak{k}$-representation are then

$$z_c^{\,r} \det z^s \quad ; \qquad r,s \in \mathbb{N} \cup \{0\}. \tag{4.11}$$

This is obvious from the representation theory of $U(2)$. Observe that the $\mathfrak{k}_1$ weight of (4.11) is r and that the degree is $r+2s$. The $\otimes$-product $\mathcal{U}(\mathfrak{p}^-) \otimes V(\Lambda)$ is then easily computed either directly from the $U(2)$-theory or from [9]. Recall that $V(\Lambda)$ denotes the $\mathfrak{k}$-representation defined by (Λ_0,λ). Observe that if $\Lambda = (\lambda,\lambda-n)$ and if a $\mathfrak{k}$-irreducible subspace of $\mathfrak{k}_1$-weight $\tilde{n}$ occurs in the $\otimes$-product of degree d expressions in $\mathcal{U}(\mathfrak{p}^-)$ with $V(\Lambda)$, then if $2d > n+\tilde{n}$, the $\mathcal{U}(\mathfrak{p}^-)$ terms must all contain a factor of $\det z$.

We list here the relevant commutators

$$\begin{array}{lll} [k^+,z_a] = -z_b\ , & [k^+,z_b] = -2z_c\ , & [k^+,z_c] = 0 \\ [k^-,z_a] = 0\ , & [k^-,z_b] = -2z_a\ , & [k^-,z_c] = -z_b \\ [k^+,z_a^+] = 0\ , & [k^+,z_b^+] = 2z_a^+\ , & [k^+,z_c^+] = 2_b^+ \\ [k^+,h_c] = k^+\ , & [k^+,h_b] = 0\ , & [k^+,h_a] = -k^+ \\ [z_c^+,z_a] = 0\ , & [z_c^+,z_b] = k^-\ , & [h_k,z_c] = 2z_c \\ [z_b^+,z_a] = k^-\ , & [z_b^+,z_c] = k^+\ , & [z_a^+,z_b] = k^+ \end{array} \tag{4.12}$$

The following is then straightforward

Lemma 4.2. Inside $\mathcal{U}(\mathfrak{g})$,

$$\begin{aligned} z_a^+ \det z^s &= s \det z^{s-1} z_c (h_a + 3/2 - s) \quad \text{modulo} \quad \mathcal{U}(\mathfrak{g})\cdot k^+ \\ (4.13) \qquad z_b^+ \det z^s &= -s \det z^{s-1} (z_b/2 (h_b + 3 - 2s) - z_c k^-) \quad \text{modulo} \quad \mathcal{U}(\mathfrak{g}) k^+ \\ z_c^+ \det z^s &= s \det z^{s-1} (z_a (h_c + 3/2 - s) - z_b k^-/2) \quad \text{modulo} \quad \mathcal{U}(\mathfrak{g}) k^+ . \end{aligned}$$

Observe that the representation space in $\mathcal{U}(\mathfrak{p}^-)$ whose highest weight vector is given by (4.11), is spanned by the elements $((\operatorname{ad} k^-)^i z_c{}^r)\cdot \det z^s$ for $i=0,\ldots,r$.

Since $\mathfrak{p}^- \otimes (\overset{r}{\underset{s}{\otimes}} \mathfrak{p}^-) = \det z \overset{r-1}{\underset{s}{\otimes}} \mathfrak{p}^- \oplus \overset{r+1}{\underset{s}{\otimes}} \mathfrak{p}^-$, one can easily establish the following (it suffices to prove the first)

Lemma 4.3. Let $\varepsilon = (\beta+1)^{-1}(4\beta+2)^{-1}$. Inside $\mathcal{U}(\mathfrak{p}^-)$,

$$\begin{aligned} z_c (\operatorname{ad} k^-)^\alpha z_c^\beta &= (2\beta-\alpha+2)(2\beta-\alpha+1)\varepsilon (\operatorname{ad} k^-)^\alpha z_c^{\beta+1} + \\ &\quad 2\beta\cdot\alpha(\alpha-1)(2\beta+1)^{-1} \det z (\operatorname{ad} k^-)^{\alpha-2} z_c^{\beta-1} \\ (4.14) \qquad z_b (\operatorname{ad} k^-)^\alpha z_c^\beta &= -2(2\beta-\alpha+1)\cdot\varepsilon (\operatorname{ad} k^-)^{\alpha+1} z_c^{\beta+1} + \\ &\quad 4\alpha\beta(2\beta+1)^{-1} \det z (\operatorname{ad} k^-)^{\alpha-1} z_c^{\beta-1} \\ z_a (\operatorname{ad} k^-)^\alpha z_c^\beta &= \varepsilon (\operatorname{ad} k^-)^{\alpha+2} z_c^{\beta+1} + 4\beta(2\beta+1)^{-1} \det z (\operatorname{ad} k^-)^\alpha z_c^{\beta-1} \end{aligned}$$

For later use we observe that if a vector $v \neq 0$ in a $\mathfrak{k}$-representation space satisfies that $k^+\cdot v = 0$ and $h_k\cdot v = \tilde{n}\cdot v$ for an integer $\tilde{n} > 1$, then

$$(4.15) \qquad (z_a + \tilde{n}^{-1} z_b k^- + \tilde{n}^{-1}(\tilde{n}-1)^{-1} z_c (k^-)^2)\cdot v$$

is a highest weight vector in $\mathfrak{p}^- \otimes V$ of $\mathfrak{k}_1$-weight $\tilde{n}-2$. Likewise,

$$(4.16) \qquad (z_b + 2\tilde{n}^{-1} z_c k^-)\cdot v$$

is a highest weight vector; its weight is $\tilde{n}$, and it suffices that $\tilde{n}$ be a positive integer.

Lemma 4.4. Let $v \in (\overset{d}{\otimes} \mathfrak{p}^-) \otimes V(\Lambda)$ be a highest weight vector of weight $\tilde{n} = y+n-x$ and let $2d = x+y$. Then, if $\tilde{n} > 0$ (4.16) defines a non-zero element of $\mathcal{U}(\mathfrak{p}^-) \otimes V(\Lambda)$; and if $\tilde{n} > 1$, so does (4.15).

<u>Proof</u>. The two cases are similar, so we only consider (4.16). With no loss of generality we can assume that v does not contain a factor of $\det z$. In particular, we may assume that $x \leq n$. It follows that

$$(4.17) \quad v = (\mathrm{ad}\, k^-)^x z_c^{\frac{x+y}{2}} \cdot v \; + \text{terms from } \; \mathcal{U}(\mathfrak{p}^-) \otimes \operatorname{span}\left\{(k^-)^i v_\Lambda\right\}_{i=1}^n .$$

Thus, it suffices to consider the v_Λ-coefficient of (4.16), i.e.

$$(4.18) \qquad z_b^-(\mathrm{ad}\, k^-)^x z_c^{\frac{x+y}{2}} + 2(y+n-x)^{-1} z_c (\mathrm{ad}\, k^-)^{x+1} z_c^{\frac{x+y}{2}} .$$

This, however, is easily computed to be, with $\delta = (x+y+2)^{-1}(x+y+1)^{-1}$,

$$(4.19) \quad \begin{aligned} &2\delta(y+1)[-1+y(y+n-x)^{-1}](\mathrm{ad}\, k^-)^{x+1} z_c^{\frac{x+y}{2}+1} \\ &+ (x+y+1)^{-1} \cdot 2 \cdot x \cdot (x+y)[1+(x+1)(y+n-x)^{-1}] \det z \cdot (\mathrm{ad}\, k^-)^{x-1} z_c^{\frac{x+y}{2}-1} , \end{aligned}$$

and this is clearly always non-zero. □

Let us now turn to the problem of determining when there can be a homomorphism into $M_{\mathfrak{k}}(\Lambda)$. First of all, we proved in ([7], Proposition 1.6) that anything of the form $S_{\overline{\gamma}}$ with $\overline{\gamma}$ <u>long</u> defines a homomorphism. In case $2\lambda \notin 2\mathbb{Z}$, the same argument implies that S_α defines a homomorphism, since it is the only possible non-compact root at such λ's . In fact, it is possible to find another sequence which satisfies condition (A) for the pair $(S_\alpha(\Lambda+\rho), \Lambda+\rho)$ if and only if λ is an integer, and $\lambda \geqq n-1$. By a result due to Boe, it follows (cf. [7], Proposition 1.4) that for $\frac{n}{2} - 1 \leq \lambda < n-1$ and $2\lambda \in \mathbb{Z}$, S_α does define a homomorphism. In the remaining cases for S_α as well as for $S_\gamma S_\alpha$ and $S_\alpha S_\beta$, one is led to consider a highest weight vector q in $\mathcal{U}(\mathfrak{p}^-) \otimes V(\Lambda)$ which satisfies

$$(4.20) \quad \begin{aligned} &\text{a)} \quad \mathfrak{p}^+ q = k^+ q = 0 \\ &\text{b)} \quad q = \det z^s \, \overline{q} \quad \text{for some} \quad s \in \mathbb{N} . \end{aligned}$$

We will always assume that the s in b) is the biggest possible such. Let us further assume that the weight Λ_1 of q is $(\lambda_1, \lambda_1 - n_1)$. It then follows from (4.13) that

$$(4.21) \qquad s \det z^{s-1} z_c (\lambda_1 + 3/2 + s)\bar{q} + \det z^s z_a^+ \bar{q} = 0 .$$

Due to the fact that the ideal generated by $\det z$ is <u>prime</u>, and due to the assumption that s is biggest possible ($\bar{q}$ does not contain a factor $\det z$), it follows that

$$(4.22) \qquad \lambda_1 + 3/2 + s = 0 .$$

This equation has three interesting consequences: i) only one s is possible, ii) λ_1, and hence λ must satisfy: $2\lambda \notin 2\mathbb{Z}$, and iii) $-\lambda_1 \geqq 5/2$ $(s \geq 1)$.

Returning to $S_\gamma S_\alpha$ and $S_\alpha S_\beta$, it is easy to see (cf. the remarks following (4.11)) that if a homomorphism exists, it must be defined for a q of the form (4.20). Hence, since both exist only if $2\lambda \in 2\mathbb{Z}$, this is impossible. Finally, there can be no multiplicities for S_α due to consequences i) and iii).

This still leaves open the question whether some quotinets exist which are not defined by homomorphisms. However, Lemma 4.4 together with an easy count of multiplicities (cf. the proof for $su(2,2)$ in [8]) gives that there are no such quotients. Observe that at the situation in $sp(2,\mathbb{R})$ corresponding to the place in $su(2,2)$ where a non-homomorphic quotient exists, namely $S_\gamma S_\alpha$ with $\lambda = (n-2)/2$ (n even), the corresponding $\mathfrak{k}$-type does not belong to $\mathcal{U}(\mathfrak{p}^-) \otimes V(\Lambda)$. We can then state:

<u>Theorem 4.5</u>. For $\mathfrak{g} = sp(2,\mathbb{R})$, all quotients of $M_{\mathfrak{k}}(\Lambda)$ are defined by homomorphisms, and there are no multiplicities. For n fixed, in the language of (4.9) the full list of non-trivial homomorphisms into $M_{\mathfrak{k}}(\Lambda)$ is obtained for λ in the intervals below satisfying the requirement that $2\lambda \in \mathbb{Z}$:

$$(4.23) \qquad \begin{array}{lcll} -1 \leq \lambda \leq \frac{n}{2} - 2 & : & S_\gamma & (n \geq 2) \\ \frac{n}{2} - 1 \leq \lambda \leq n - 3/2 & : & S_\alpha & (n \geq 1) \\ n - 1/2 \leq \lambda & : & S_\beta & \text{when } \lambda \in \mathbb{Z},\ S_\alpha \text{ when } \lambda \notin \mathbb{Z} . \end{array}$$

<u>5. $\mathfrak{g} = su(2,2)$</u>

Let e_1, e_2, e_3 be the standard basis of $\mathbb{R}^3$. Then the positive non-compact roots are

(5.1) $\Delta_n^+ = \{\beta = e_1-e_2, \quad \alpha_1 = e_1-e_3, \quad \alpha_2 = e_1+e_3, \quad \text{and } \gamma=\gamma_r=e_1+e_2\}$.

The positive compact roots are

(5.2) $\Delta_c^+ = \{\mu = e_2+e_3, \quad \text{and} \quad \nu = e_2-e_3\}$.

We have that $\rho = (2,1,0)$ and the Λ's that define the $M_k(\Lambda)$'s are of the form

(5.3) $\Lambda = (r, \frac{n+m}{2}, \frac{n-m}{2})$,

with $n,m \in \mathbb{N} \cup \{0\}$. Below we assume that $n \geq m$ and write $\Lambda+\rho = (z,x,y)$. Thus, $y \geq 0$, and in the following, $z \in \mathbb{Z} + y$ throughout.

<u>Theorem 5.1.</u> [8] Let $M_k(\Lambda)$ be the generalized Verma module of highest weight $\Lambda = (z,x,y)-\rho$. Then the subspace structure is defined by homomorphisms except in the case maked by Q . There are no multiplicities:

i) $x > y+1$:

$z \leq -x$	:	None
$-x < z \leq -y-1$	:	S_γ
$-y < z \leq y$	:	S_{α_2}
$y+1 \leq z \leq x-1$	:	$S_{\alpha_1}, S_{\alpha_2}, S_{\alpha_1}S_{\alpha_2}$
$z = x$	:	None
$x+1 \leq z$	:	S_β .

ii) $x = y+1$:

$z \leq 1-x$	:	None
$1-x < z \leq x-1$	:	S_{α_2}
$z \triangleq x$	:	$S_{\alpha_1}S_{\alpha_2}$
$x+1 \leq z$	:	$S_\beta, S_{\alpha_1}S_{\alpha_2}$.

iii) $y = 0$:

$z \le -x$:	None	
$-x < z \le -1$:	S_γ	
$z = 1$:	S_{α_1}, S_{α_2}, $S_{\alpha_1}S_{\alpha_2}$, $S_\gamma S_{\alpha_1}S_{\alpha_2}$(Q)	$(x \ge 2)$
$1 < z \le x-1$:	S_{α_1},S_{α_2}, $S_{\alpha_1}S_{\alpha_2}$	
$z = x$:	$S_{\alpha_1}S_{\alpha_2}$	
$x+1 \le z$:	S_β, $S_{\alpha_1}S_{\alpha_2}$	

References

1. A. Angelopoulos, The unitary irreducible representations of $\overline{SO}_0(4,2)$, Commun. Math. Phys. 89, 41-57 (1983).

2. I.N. Bernstein, I.M. Gelfand, and G.I. Gelfand, Differential operators on the base affine space and a study of $\mathfrak{g}$-modules; in Lie Groups and Their Representations (I.M. Gelfand, Ed.) Adam Hilger, London; 1975.

3. V.K. Dobrev, Elementary representations and intertwining operators for SU(2,2). I., J. Math. Phys. 26 235-251 (1985).

4. M. Harris and H.P. Jakobsen, Singular holomorphic representations and singular modular forms, Math. Ann. 259, 227-244 (1982).

5. S. Helgason, Differential geometry, Lie groups, and symmetric spaces. Academic Press, New York, San Francisco, London; 1978.

6. J. Humphreys, Introduction to Lie algebras and representation theory. Springer Verlag, New York, Heidelberg, Berlin; 1972.

7. H.P. Jakobsen, Basic covariant differential operators on hermitian symmetric spaces, to appear in Ann. Sci. Éc. Norm. Sup.

8. H.P. Jakobsen, Conformal covariants, to appear in Publ. RIMS. Kyoto Univ.

9. G.D. James, The representation theory of the symmetric groups. Lecture Notes in Math. # 682, Springer Verlag, Berlin-Heidelberg-New York; 1978.

TENSOR CALCULUS IN ENVELOPING ALGEBRAS

E. ANGELOPOULOS
Universite de Dijon
Physique-Mathématique
BP 138
21004 - DIJON Cedex
France

ABSTRACT

The technique of reduction of tensor products $V \otimes W$ of $\underline{g}$-modules ($\underline{g}$ being a reductive complex Lie algebra), V being semi-simple finite-dimensional, by means of tensor calculus in the enveloping algebra U of $\underline{g}$ is exposed. It leads to considertations on Galois extensions of the center of U by the Weyl group of $\underline{g}$. Its use in view of the study of the unitarizability of $\underline{g}_o$-modules, $\underline{g}_o$ being a real form of $\underline{g}$ is indicated.

INTRODUCTION

The objects and techniques presented here have been used by the author as tools for the characterization of the unitary dual of some semisimple real Lie groups or families of them, among which the conformal group [10]. They have grown up slowly: at the beginning there were just tricks and shorthand notations, used to shorten lengthy calculations inside enveloping algebras. As the algebras grew bigger, the tricks grew bigger too, yielding a formalism of U-valued tensor calculus. The computational power grained by this formalism has to do with producing formulas on the reduction of the tensor product $V \otimes W$ of $\underline{g}$-modules into isotypic components, formulas which do not depend on the weights of a Cartan subalgebra on V and W. One is then lead to considerations on some rings of matrices with entries in U, homomorphic to $\mathrm{End}_{\underline{g}}(V \otimes W)$; solving the eigenvalue problem for such matrices leads to an algebraic extension of the center Z : An algebraic extension which can be used as a parametrization of Z, quite easy to manipulate for either finite or infinite dimensional $\underline{g}$-modules (which is not the case for the dominant weight formalism).

Thus, what first appeared as simple tricks related to particular algebras has been developped to a quite general formalism, which we shall outline here.

The paper's organisation is the following:
Sec. 1 is devoted to present the notations used and the motivations for this study, which concern the unitarizability problem. In Sections 2 and 3 the construction of the unitary dual of the Lorentz Lie algebra is sketched and the techniques used are discussed, to extract generalizable features which lead to tensor calculus. In Section 4 (theorem 1) the tensor formalism of the tensor product reduction is introduced; it uses spaces of intertwinning operators between $\underline{g}$-modules, denoted $\mathrm{Hom}_{\underline{g}}(\mathcal{L}(V),U)$-this is a notation often used here. Section 5 discusses why and how to extend Z, and in Theorem 2 this extension is explicitated, in a condensed form, for classical Lie algebras. Section 6 gives hints about the techniques used to obtain theorem 2, which lie upon exterior tensor calculus. Section 7 concludes with some remarks.

The talk effectively given at the Symposium contained one more example (the dual of $sl(2,\mathbb{R})$ which has been omitted, to add Sec. 6, judged more important.

I would like to thank Professors Doebner, Dobrev and Heidenreich, as well as Mrs. Illgauds for having so well organized our stay in Clausthal. Also I would like to thank Marie-Thérèse Rochet for her help to the realization of this manuscript.

1. In all what follows $\underline{g}$ will denote a reductive complex Lie algebra, U or $U(\underline{g})$ its enveloping algebra, Z or $Z(\underline{g})$ the center of U , $\underline{g}_o$ a real form of $\underline{g}$ with Cartan decomposition $\underline{g}_o = \underline{k}_o \oplus \underline{p}_o$ and, by complexification, $\underline{g} = \underline{k} \oplus \underline{p}$, $\underline{k}_o$ being the maximal compact subalgebra. A representation of $\underline{g}$ or a $\underline{g}$-module (indistinctly) will be a couple (R,W) where W is a complex vector space and R a Lie-algebra-homomorphism from $\underline{g}$ to the Lie algebra $\underline{gl}(W)$; by restriction, (R,W) is also a $\underline{g}_o$-module. The extension of R to an associative-algebra-homomorphism from U to the algebra of linear self-mappings of W, $\mathcal{L}(W)$, will be again denoted by R. When there is no risk of confusion we shall write Yf instead of R(Y)f for $Y \in U$, $f \in W$. The notation (π,V), (π',V') that is, a small greek letter in the first place, will always denote finite dimensional semisimple $\underline{g}$-modules. $(ad,\underline{g})$ will denote the adjoint; $\underline{p}$ being a $\underline{k}$-invariant subspace of $\underline{g}$, the corresponding factor of $(ad|_{\underline{k}},\underline{g})$ will be denoted $(\rho,\underline{p})$.

Tensor products of $\underline{g}$-modules will be broadly used and denoted $(R\otimes R', W\otimes W') = (R,W) \otimes (R',W')$; the reader is supposed to be acquainted to definitions and elementary properties of them, as well to elementary tensor calculus, in particular the Einstein summation convention, that is $F_A{}^A = \Sigma_{A\in I,\ B\in I,\ A=B}\ F_B{}^A$ for every monomial expression F, the summation range of A being some fixed finite set I. When a metric tensor is available, no distinction of upper and lower indices will be made, that is the Feynman summation convention will be used (this concerns sections 2 and 3).

The unitarizability of a $\underline{g}_o$-module (R,W) is closely related to the reduction of tensor products of $\underline{k}$-modules, $\underline{k} \oplus \underline{p}$ being a Cartan decomposition of $\underline{g}$, more precisely to the reduction of $(\rho,p)\otimes(\pi,V)$ for arbitrary V. Indeed, assume that W is $\underline{k}$-finite, that is, every isotypic $\underline{k}$-invariant subspace of W has finite multiplicity: since the restriction of R to $\underline{k}_o$ is unitarizable, one can express W as some completion of $\dot{W} = \oplus_i (\oplus_\alpha V_{i,\alpha})$, with $V_{i,\alpha}$ and $V_{i,\beta}$ equivalent simple $\underline{k}$-modules, $V_{i,\alpha}$ and $V_{j,\beta}$ nonequivalent for $j\neq i$; $\dot{W}$ has a $\underline{k}_o$-invariant prehilbert structure (which is not unique: multiplying by a positive number the restriction of the scalar product to some self-dual $V_{i,\alpha}$ does not affect $\underline{k}_o$-invariance) and there is no loss of generality in identifying W to $\dot{W}$ (take, e.g. $W = R(U)f$ with $f\in V_{i,\alpha}$ that is, W monogeneous), evacuating thus any topological considerations.

It is clear that R is unitarizable iff, for every i, α, for every $f\in V_{i,\alpha}$ one has

$$(1)\qquad (R(Y)f|R(Y)f) = (f|R(Y^*Y)f) > 0$$

for every $Y\in\underline{g}$, the involution $Y \to Y^*$ being the antilinear extension to $\underline{g}$ of the principal Lie algebra antiautomorphism of $\underline{g}_o$: this is the transcription of $\underline{g}_o$-invariance of the scalar product, already satisfied by $Y\in\underline{k}_o$; if, moreover, W is simple, the scalar product is unique up to a factor, and it is positive definite iff R is unitarizable.

It should be immediately pointed out that a direct approach of the unitarizability problem is outside the main stream of recent research in this topic, most probably because the positivity of generalized matrix elements of the form (1) seems quite hard to establish, noncommutative calculations inside enveloping algebras becoming quickly repelling. However, it fits for small algebras, like $\underline{sl}(2,\mathbb{R})$ [1], [2] and $\underline{so}(3,1)$ [3], and the use of tensor calculs smoothens the way for more general cases [4].

The general scheme of the direct approach consists on observing that the linear span of $R(Y)$ for $Y \in \underline{p}$, $f \in V = V_{i,\alpha}$ is homomorphic to $\underline{p} \otimes V$ as a $\underline{k}$-module, hence the $\underline{k}$-content of W consists of a lattice of points which can be granted with some (total or partial) order, every point of the lattice corresponding to a $\underline{k}$-isotypic component. One can then look for necessary conditions, in terms of algebraic relations between $R(Z(\underline{g}))$ and $R(Z(\underline{k}))$, or, equivalently, geometric relations between $R(Z(\underline{g}))$ and $\underline{k}$-lattice, by taking elements of $Z(\underline{g})$ which can be expressed as linear combinations of Y^*Y's for $Y \in U(\underline{g})$; and for sufficient conditions by examining for which isotypic components V one can write

$$(2) \qquad (f|f) = \Sigma_i (f_i|f_i) \ , \quad f_i \in V_i$$

with $V_i < V$ with respect to the lattice's order for every occurring i , so that explicite checking of positivity at the remaining points is sufficient. To work out these topics one has to dispose of adequate tools so that the required calculations inside enveloping algebras can be carried out.

2. We shall begin by exposing a classical example, the Lorentz Lie algebra $\underline{g}_o = \underline{so}(3,1) = \underline{k}_o \oplus \underline{p}_o$ with $\underline{k}_o = \underline{so}(3)$; for sake of brevity there will be no calculations, but sufficiently enough intermediate results so that the reader who desires to check may easily do so.

Let $\{J_i, K_j\}_{i,j} = 1,2,3$ be a basis of $\underline{g}_o$, $\{J_i\}$ a basis of $\underline{k}_o$, with commutation relations:

$$(3) \qquad [J_i, J_j] = - [K_i, K_j] = \varepsilon_{ijk} J_k$$

$$[J_i, K_j] = [K_i, J_j] = \varepsilon_{ijk} K_k$$

ε being a completely skew-symmetric tensor with $|\varepsilon_{ijk}| = 1$ or 0. The principal antiautomorphism is given by

$$(4) \qquad (J_i)^* = - J_i \qquad (K_i)^* = - K_i$$

We shall introduce 3-vector formalism to get rid of indices i, so that the casimir element w of $\underline{k}$ is given by

(5) $\quad w = \vec{J}^*.\vec{J} = -\vec{J}.\vec{J} = -J_i J_i = w^*$

and the spanning elements of $Z(\underline{g})$ by

(6) $\quad a = \vec{J}^*.\vec{J} - \vec{K}^*.\vec{K} = \vec{K}.\vec{K} - \vec{J}.\vec{J} = a^*; \quad b = \vec{J}.\vec{K} = b^*$

One can define another 3-vector element in $U(\underline{g}_o)$:

(7) $\quad \vec{K} \wedge \vec{J} : (\vec{K} \wedge \vec{J})_i = \varepsilon_{ijk} K_j J_k$

satisfying

(8) $\quad \vec{K}.(\vec{K} \wedge \vec{J}) = \varepsilon_{ijk} 1/2[K_i,K_j]J_k = w; \quad \vec{J}.(\vec{K} \wedge \vec{J}) = (\vec{K} \wedge \vec{J}).\vec{J} = b$

Moreover, the following relations hold in U:

(9) $\quad 1/2[w,\vec{K}] = \vec{K} - \vec{K} \wedge \vec{J}$

(10) $\quad 1/2[w,\vec{K} \wedge \vec{J}] = -\vec{K}w + \vec{J}b = -\mathrm{ad}(1/2w).(\mathrm{ad}(1/2w)-1)(K)$

Consider now simple $\underline{g}$-modules (R,W) such that

(11) $\quad W = \oplus_{2j\in N} W_j; \quad W_j = \mathrm{Ker}(R(w)-(j^2+j)1), \quad \dim W_j < \infty$

that is W_j is the isotypic component of the 2j+1-dimensional simple $\underline{k}$-module (π_j,V_j), j being the maximal eigenvalue of $\pi_j(J_3.\sqrt{-1})$. Since $R(w)$ is diagonalizable and its eigenvalues determine the isotypic component, equations (9) and (10), together with $[w,\vec{J}] = 0$ provide a system of equations in $R(U) \subset \mathscr{L}(W)$, which enables to write:

(12) $\quad R(\vec{K}) = \vec{K}^+ + \vec{K}^o + \vec{K}^- ; \quad K_i^\alpha W_j \subset W_{j+\alpha 1}$

defining thus <u>transition operators</u> from each $\underline{k}$-component to another one, or, equivalently, reducing the tensor product $\underline{p} \otimes W_j$. The operators $\vec{K}^\alpha$ are defined componentwise, that is by their action on each W_j:

(13) $\quad R(1/2[w,\vec{K}])\Big|_{W_j} = \vec{K}^+.(j+1) - \vec{K}.(j)$

(14) $$R(1/2[w,[1/2w,\vec{K}]])\Big|_{W_j} = \vec{K}^{+}.(j+1)^2 + \vec{K}^{-}.j^2$$

Writing $\vec{K},\vec{J}$ instead of $R(\vec{K})$, $R(\vec{J})$, one gets:

(15a) $$\vec{K}^{+}.(2j+1).(j+1) = \vec{K}.(j+1)^2 - \vec{K} \wedge \vec{J}.(j+1) + \vec{J}.b$$

(15b) $$\vec{K}^{-}.(2j+1).j = \vec{K}.j^2 + \vec{K} \wedge \vec{J}.j + \vec{J}.b$$

which define completely $\vec{K}^{\pm}$ when $j(2j+1) \neq 0$; for $j = 0$ one must have $\vec{K}^{-} = \vec{K}^{0} = \vec{J} = 0$ and for $2j+1, \vec{K}^{-} = 0$, so that the transition operators are always defined.
Moreover, one obtains from (15b), using (5), (6) and (8):

(16) $$-(2j+1).j\ \vec{K}.\vec{K}^{-} = j^4 - j^2(1+a) - b^2.$$

Let us now investigate unitarizability. First of all one must have $(K_i^{\alpha})^* = - K_i^{-\alpha}$ because of $\underline{g}_0$-invariance of the scalar product. Next, one sees that $[\underline{k},\vec{K}^{\alpha}.\vec{K}^{\beta}] = 0$, so that $\vec{K}^{\alpha}.\ \vec{K}^{\beta}$ does not vanish only if $\alpha+\beta = 0$, hence $\vec{K}^{-\alpha}.\vec{K}^{\alpha} = \vec{K}.\vec{K}^{\alpha}$;
Equation (16) then yields, for every $\varphi \in W_j$:

(17) $$(j^4-j^2(R(a)+1)-R(b)^2).(\varphi|\varphi)=(2j^2+j)(\varphi|(\vec{K}^{-})^*.\vec{K}^{-}\varphi)$$
$$= (2j^2+j)\Sigma_i(K_i^{-}\varphi|K_i^{-}\varphi).$$

Now $R(a)$ and $R(b)$ must be real numbers, so let $R(b) = \lambda\mu$, $R(a) = -1+\mu^2-\lambda^2$ be a parametrization of them (the change $(\lambda,\mu) \to (-\lambda,-\mu)$ does not affect $R(Z)$) , with $\lambda,\mu \in \mathbb{C}$, so that (17) becomes:

(18) $$(j^2-\mu^2).(j^2+\lambda^2).(\varphi|\varphi) = (2j^2+j)\ \Sigma_i(K_i^{-}\varphi|K_i^{-}\varphi)$$

By induction one sees that positive-definiteness implies that $(j^2-\mu^2).(j^2+\lambda^2) > 0$ for every j such that $W_j \cap \mathrm{Ker}\ \vec{K}_{-} = \{0\}$. If j_0 is the lowest value of j , the second member must vanish, so that one must have, say, $\mu = j_0$. Since $R(b)$ is real, when $j_0 \neq 0$ λ must be real also, so that for $j_0 \neq 0$ positive-definiteness is granted for every j , and the $\underline{k}$-lattice of isotypic components is $j_0 + \mathbb{N}$. For $j_0 = 0$ the $\underline{k}$-lattice may reduce to one point, $j = 0$, (the trivial $\underline{g}$-module) obtained for $\lambda = \sqrt{-1}$, $\mu = 0$; otherwise unitarizability is equivalent to $1 + \lambda^2 > 0$ which implies $j^2 + \lambda^2 > 0$ for every

$j > j_o = 0$ and the $\underline{k}$-lattice is again $j_o + \mathbb{N}$. Notice that small imaginary values of λ still give rise to unitarizable modules: this is the complementary series.

3. This brief sketch of Naïmark's classification [3] needs some comments. First of all, this is a prototype of the scheme exposed in section 1: direct investigation of positive-definiteness of the scalar product, and classification principle involving algebraic and geometrical considerations on $R(Z)$ and the $\underline{k}$-lattice. It should immediately be pointed out that one cannot expect formulas like (17) to hold at abitrary points of the $\underline{k}$-lattice in the general case, because the commutant of $\underline{k}$ in $\underline{g}$ is not abelian in general.

A point to be stressed is that the exposition above does not use any considerations on Cartan subalgebras of $\underline{g}$ or $\underline{k}$, that is the spectrum of J_3 plays no role at all: only the spectrum of w is needed. This seems to have been overlooked by Naïmark himself in [3] so that his calculations are quite lengthy; however, he needs J_3 to construct explicitly representations on functional spaces, so that the only grief is that he has not opened wide doors for future generalizations. Indeed, this is a quite general feature, concerning not only $\underline{so}(3)$ (Racah's multiple-j-symbols are a good example of global calculus), but compact subalgebras of real Lie algebras in general: Vogan's classification of the linear dual of real semisimple groups [5] uses global considerations on $\underline{k}$-submodules.

Another point is the following: the full set of the commutation relations of $\underline{g}$ has not been used, except in order to express $\vec{K}.\vec{K}$, $\vec{K}.(\vec{K}\wedge\vec{J})$, $\vec{K}.\vec{K}^{\alpha}$ in terms of a,b,w and j. Up to then only commutation relations of type $[U(\underline{k}),U(g)]$ have been used; in particular, the only thing used to define the transition operators K_i^{α} is that they behave like 3-vectors under $\underline{k}$. Let $\{e_i\}$ be an orthonormal basis of the $\underline{k}$-module $V_1 \approx \mathbb{C}^3$: a U-valued 3-vector, say $\vec{K}$, can be defined as a homomorphism K of $\underline{k}$-modules from V_1 to U, such that $K_i = K(e_i)$, that is, $K \in \mathrm{Hom}_{\underline{k}}(V_1,U)$. One then has:

<u>Proposition:</u> Let $\underline{g}$ be any Lie algebra containing $\underline{k} = \underline{so}(3)$, and let $K \in \mathrm{Hom}_{\underline{k}}(V_1,U(\underline{g}))$, For every $\underline{g}$-module (R,W) such that W admits a direct sum decomposition as a $\underline{k}$-module into isotypic components $W_j \approx U_j \otimes V_j$ (U_j being a $\underline{k}$-trivial $\underline{k}$-module, labelling multiplicity, and V_j the $(2j+1)$-dimensional simple one) there are three elements K^+, K^o, K^- in $\mathrm{Hom}_{\underline{k}}(V_1,R(U(g)))$, such that $K^{\alpha}(e_i)W_j \subset W_{j+\alpha 1}$, the defining formulas of $K^{\alpha}(e_i) = K_i$ being those of section 2.

This result has been used by the author to determine the unitary dual $\underline{sl}(3,\mathbb{R})$ [6] and $\underline{so}(3,2)$ [7]. It generalizes to any V_j - the defining formulas are of course different - and to any $\underline{k}$ (at least in what concerns the existence assertion).

Before going to generalizations, let us discuss another point. The three-vectors $\vec{K}$, $\vec{K} \wedge \vec{J}$ and $\vec{J}\, b = (\vec{K}.\vec{J})\vec{J}$ belong elementwise to $p.U(k)$, while the $\vec{K}^{\alpha}$ belong elementwise to $R(\underline{p}).R(U(k))$, that is they can be defined respectively as elements of $\mathrm{Hom}_{\underline{k}}(V_1,\underline{p}.U(k))$, $\mathrm{Hom}_{\underline{k}}(V_1,R(p).R(U(\underline{k})))$. On the other hand, consider the three mappings g, $\overline{M}$, JJ, sending the $\underline{g}$-module $\mathscr{L}(V_1) \approx V_1 \otimes V_1$ to $U(\underline{k})$, such that if E_{ij} is a basis of $\mathscr{L}(V_1)$ one has

(19a) $$g(E_{ih}) = g_{ih}$$

(19b) $$M(E_{ih}) = M_{ih} = \varepsilon_{ihk}\, J_k$$

(19c) $$JJ(E_{ih}) = J_i\, J_h$$

g being the 3 x 3 metric tensor which is $\mathbb{C}$-valued or $U(\underline{k})$-valued, since $\mathbb{C}$ is canonically imbeddable in $U(\underline{k})$. All of them belong to $\mathrm{Hom}_{\underline{k}}(\mathscr{L}(V_1),U(\underline{k}))$ as easily checked. Now, one has

(20) $$K_h = K_i\; g_{ih}\; ;\; (K \wedge J)_h = -K_i\; M_{ih}\; ;\; b\, J_h = K_i\, J_i\, J_h$$

and this strongly suggests matrix multiplication. Indeed one can show that $H = \mathrm{Hom}_{\underline{k}}(\mathscr{L}(V_1),U(\underline{k}))$ has a ring structure under the multiplication sending $H \times \overline{H}$ to H:

(21) $$(x,y) \to x.y\; ;\; (x.y)_{ih} = x_{ik}\, y_{kh}$$

The unity element of H is g and if M is a right $U(k)$-module then $\mathrm{Hom}_{\underline{k}}(V_1,M)$ becomes a right H-module; in particular one can take $M = \underline{p}.U(\overline{\underline{k}})$, $K_i \in M$ and (20) shows that the different 3-vectors used up to now can be interpretated as images of the same mapping, K, multiplied on the right by elements of H.

When one goes to g-modules (R,W) one can define similarly $R(H)$ by substituting $R(U(\underline{k}))$ to $U(\underline{k})$ in the definition of H. Full diagonalization of w sends $R(w)$ on j^2+j for every j , and one can write for every j :

(22) $$K^{\alpha}(e_i) = R(K_h)R(\lambda^{\alpha}(j)g_{hi} + \mu^{\alpha}(j)\, M_{ih} + \nu^{\alpha}(j)J_h\, J_i)$$

with adequate coefficients $\lambda^{\alpha}(j)$. This means that the reduction of $\underline{p} \otimes W_j$ into $\oplus_\alpha W_{j+\alpha 1}$ is effected by means of a partition of unity in $R_j(H)$, R_j being the restriction of R of $\underline{k}$ to W_j , or, in other words, for every j, $\mathrm{Hom}_{\underline{k}}(\mathcal{L}(V_1), R_j(U_k))$ is isomorpic to $\mathrm{End}_{\underline{k}}(V_1 \times W_k)$. This feature is also generalizable.

Last but not least, one must observe that K_i^{α} and the corresponding partition of unity can be defined for any $(R,W) = \oplus_j (R_j, W_j)$ but not in $\underline{p}.U_{\underline{k}}$ or in the ring H themselves. This is due to two things: first j is not in $U(\underline{k})$, secondly division is not admitted (for instance, $J_i b(w)^{-1} = K_i^{o}$ has no meaning in U). There is a simple method to discard this annoyance: one can consider the rational extension Rat $Z(\underline{k})$ of $Z(\underline{k})$ to its field of fractions, its (twofold) galois extension by j, $\overline{Z}(\underline{k}) = Z(\underline{k})[j]/(j^2+j-w)$ and the combined one Rat $\overline{Z}(\underline{k})$; then extend $U(\underline{k})$ to, say, Rat $\overline{U}(\underline{k}) = U(k) \otimes_{Z(\underline{k})}$ Rat $\overline{Z}(\underline{k})$. The partition of unity can then be defined in the extended ring $\mathrm{Hom}_{\underline{k}}(\mathcal{L}(V_i), \mathrm{Rat}\ \overline{U}(\underline{k}))$. The analog of (22) can then be written in $\underline{p}$.Rat $\overline{U}(\underline{k})$ by dropping symbols R in (22), so that K_i^{α} is defined inside $\underline{p}$.Rat $\overline{U}(\underline{k})$; finally, $U(\underline{g})$ itself can be extended by introducing. $[j, K_i^{\alpha}] = \alpha K_i^{\alpha}$. Working with such extensions enables to avoid back-and-forth reasonments from enveloping algebras to representations and vice-versa; however, if the final step concerns representations, formulas must be carefully handled to avoid dividing by some element of $\overline{Z}(\underline{k})$ which is sent to 0 . This feature is also subject to generalization.

4. Let us come now to the generalization promised. One first has:
Theorem 1: Let $\underline{g}$ be a reductive Lie algebra, U its enveloping algebra, (π, V) a finite-dimensional $\underline{g}$-module, $(\check{\pi}, \tilde{V})$ its contragredient, $\mathcal{L}(V) \approx V \otimes \tilde{V}$ the space of linear self-mappings of V , $\{e_A\}, \{e^A\}$ dual bases of $\tilde{V}$ and V , $\{E^{A}{}_{B}\}$ the basis of $\mathcal{L}(V)$ canonically related to them, the underlying field being $\mathbb{C}$. Then:
1) The space of intertwinning operators $\mathrm{Hom}_{\underline{g}}(\mathcal{L}(V), U)$ can be granted with an associate algebra structure, denoted $(\pi)_U$, the multiplication, called the contracted tensor product (ctp) , being defined by:

$$(23) \qquad (T.T')(E^{A}{}_{B}) = T(E^{A}{}_{C}).T'(E^{C}{}_{B})$$

If U is replaced by any associated algebra $\mathcal{A}$ such that $(D, \mathcal{A})$ is a $\underline{g}$-module, $D(\underline{g})$ being an algebra of derivations of $\mathcal{A}$, the statement remains valid.

2) If (R,W) is a $\underline{g}$-module and $u \to [R(X),u]$, $u \in \mathcal{L}(W)$ is the canonical $\underline{g}$-module structure on $\mathcal{L}(W)$, there is an associative algebra isomorphism σ from $(\pi)\mathcal{L}(W)$ to $\mathrm{End}_{\underline{g}}(\tilde{V} \times W)$ defined by:

(24) $$\sigma(T)(e_A \otimes f) = e_B \otimes T(E^B{}_A)f$$

3) If (R,W) is simple and finite dimensional the associative algebra homomorphism $\sigma o R$ from $(\pi)_U$ to $\mathrm{End}_{\underline{g}}(\tilde{V} \otimes W)$ through $(\pi)_{R(U)}$ is surjective.

The proof of this theorem is rather elementary: it lies upon the definitions of intertwinning operators and of tensor products of $\underline{g}$-modules for 1) and 2) and on Burnside's theorem (see, e.g. [8], ch. XVII, 3) implying that $R(U) = \mathcal{L}(W)$.

This theorem concerns tensor products of g-modules; it is irrelevant to specify that V and $\underline{g}$ lie in some bigger enveloping algebra, as it was the case with Cartan decompositions, though it has interesting applications in this topic. Notice that a couple of mutually contragredient representations needs to be introduced: this point was hidden in the Lorentz example because the representation considered was orthogonal, hence self-contragredient. The crucial hypothesis is that V is finite-dimensional: it is essential for both contraction of indices and for Burnside's theorem. But W may be any $\underline{g}$-module in part 2), finite-dimensional and simple (V need not) in part 3): both situations are dissymetric.

5. The main point of theorem 1 is that, once $\tilde{V}$ is fixed, the study of the reduction of its tensor products with other $\underline{g}$-modules can be greatly carried out by the study of the ctp ring $(\pi)_U$. Its ideals and its idempotents are related to those of $\mathrm{End}_{\underline{g}}(\tilde{V} \otimes W)$, hence to $\underline{g}$-invariant submodules of $\tilde{V} \otimes W$. So, the study of algebras $(\pi)_U$ has the goal to construct a partition of unity $1 = \Sigma_i P_i$, such that for every finite-dimensional simple (ρ, W) the idempotent element $\sigma o \rho(P_i) \in \mathrm{End}_{\underline{g}}(\tilde{V} \otimes W)$ projects $\tilde{V} \otimes W$ onto an isotypic component. As already seen in the example, one must extend the center Z to do so.

The technique which leads to the extension consists in establishing that $(\pi)_U$ is a Z-module of finite dimension, so that every $X \in (\pi)_U$ satisfies the equation of its characteristic polynomial, which is a polynomial with coefficients in Z : indeed, there is only a finite number of Z-linearly independent ctp powers of X. Extending Z to $\overline{Z}(\pi)$ so that the characteristic polynomial of every central X belongs to $\overline{Z}(\pi)$ solves the problem. $\overline{Z}(\pi)$ is a priori dependent on (π,V): one can show that there is a minimal π-independent extension $\overline{Z}$ which is a Galois one, its Galois group being the Weyl group of $\underline{g}$ (in fact, this may as well serve as the definition of the Weyl group, since all this can be done with no reference to Cartan subalgebras). The idempotents are then constructed by using standard techniques like Taylor expansion of polynomials.

Another way to see things is to say that $\overline{Z}$ is a parametrization of Z by some set of independent variables $\{x_i\}$, so that Z is the set of Weyl-symmetrie entire functions on $\{x_i\}$ and $\overline{Z}$ the full polynomial algebra on them. If the set $\{x_i\}$ is specified to design some basis of affine functions on a Cartan subalgebra of $\underline{g}$, this statement is a well-known theorem of Harish-Chandra (see, e.g. [9], ch. 7.4). It can be obtained, however, independently of any considerations on Cartan subalgebras, by using the extension techniques just described: the set $\{1\}\cup\{x_i\}$ can be defined as a basis of the complex vector space spanned by the roots of the characteristic polynomial C_X for X spanning $\mathrm{Hom}_{\underline{g}}(\mathcal{L}(V),\underline{g}\oplus\mathbb{C}.1)$; one can easily show that such elements satisfy a relation of the form:

$$(25)\qquad \check{\pi}\otimes R(w) = \check{\pi}(w)\otimes 1 - 2\ \sigma o R(X) + 1\otimes R(W),$$

for every simple R , w being a central element of degree 2 (in the example, X is -M , defined in (19b)).

The calculation of C_X and the corresponding partition of unity has been carried out by the author (in a paper to appear soon), in case (π,V) is the fundamental representation of a classical Lie algebra. This can be summarized as follows:

Theorem 2: Let $\underline{g} = \underline{gl}(N)^{\mathbb{C}}$, $\underline{so}(N)^{\mathbb{C}}$, $\underline{sp}(N)^{\mathbb{C}}$ and (π,V) a canonical representation of $\underline{g}$ on $\mathbb{C}^N$, w the Casimir element of $\underline{g}$, and $X\in(\pi)_U$ related to w by (25). Let m=1 if $\underline{g} = \underline{so}(2N'+1)$, m=0 otherwise. There is a polynomial C(t) with coefficients in Z , of degree N-m , even in t for $\underline{g}\neq\underline{gl}(N)$, irreducible in Z[t] for $\underline{g}\neq\underline{so}(2)$, satisfying (26) $((t+k)\delta-X)^A{}_B.T(t).{}^B{}_C = \delta^A{}_C(t+1/2)^m.C(t)$ where δ is the Kronecker symbol, $k\in R$ a $\underline{g}$-depending constant and $T(t)\in(\pi)_U$ the polynomial obtained by the Taylor expansion. The degree of the Galois extension $\overline{Z}$ of Z by the roots of C(t)=0 is r = rank $\underline{g}$. For every simple finite-dimensional (ρ,W), the set $\{u_i\}$ of roots of $\rho(C(t)) = 0$ determines completely ρ, unless $\underline{g} = \underline{so}$ (2N') and $(C(0)) \neq 0$, in which case it may correspond to either ρ or $\check{\rho}$. One always has $u_i - u_j \in \mathbb{Z}$, and there are no double roots except for $\underline{g} = \underline{so}(2N')$ which may have $t^2 = 0$ as a double root. Every isotypic component (ρ', W') of $V \otimes W$ has multiplicity one, and there may exist one component W' isomorphic to W only if $\underline{g} = \underline{so}(2N'+1)$; for non isomorphic W, W' the sets of roots $\{u_i\}, \{u'_i\}$ differ only by one root (if $\underline{g} = \underline{gl}(N)$), or by a unique couple of opposite roots (if $\underline{g} \neq \underline{gl}(N)$), the absolute value of the difference between the initial u_i and the shifted one being 1.

Remark: The labelling of finite-dimensional $\underline{g}$-modules by the set of roots (plus a $\pm$ sign for $\underline{g} = \underline{so}(2N')$) indicated in Theorem 2 is equivalent to the labelling by dominant weights: the u_i's are almost equal (module a fixed shift for every u_i) to suitably chosen coordinates of the dominant weight. The advantage lies elsewhere: characteristic polynomials and Taylor plynomials exist also in infinite-dimensional representations (R,W), and so do the u_i's, while dominant weights cannot, in general, be used. In fact, for $\underline{g} = \underline{so}(3,1)^{\mathbb{C}} = \underline{so}(4)^{\mathbb{C}}$ we have already met with the characteristic polynomial in (17) and (18), slightly modified: the set of roots $\{u_i\}$ corresponding to the one of Theorem 2 are $\{\pm j_o, \pm i\lambda\}$. One couple is related to the $\underline{k}$-lattice.

6. One can outline as follows the tools used to prove Theorem 2: once (π, V) of $\underline{g}$ is fixed, consider exterior powers of it, denoted π^k and $\wedge^k = \wedge^k(V)$, and introduce dual basis $\{e^K\}$, $\{e_K\}$ in $\wedge^k$, $\tilde{\wedge}^k = \wedge^k(\tilde{V})$. In order to manipulate together different powers, let $k = |K|$ be called the length of the symbol K if $e^K \in \wedge^k$ taking $K \in \mathcal{J}_{|K|}$, $\mathcal{J}_{|K|}$ being some indexing set of cardinality equal to dim $\wedge^{|K|}$, with $\mathcal{J}_k \cap \mathcal{J}_h = \emptyset$ if $k \neq h$. Using the fact that $\wedge = \oplus_k \wedge^k$ is an exterior algebra introduce the following two-row symbols:

$$(27) \qquad \left[{HK \atop LM}\right] = \langle e^H \wedge e^K, e_L \wedge e_M \rangle \in \mathbb{C}$$

which is nonzero only if $|H| + |K| = |L| + |M| \leq \dim V$. Using the properties of exterior algebra, one can develop a formalism with these two-row symbols, covering partial contractions or substitutions of indices; one easily proves, for instance, formulas like:

$$(28)\qquad \begin{bmatrix} H & K \\ L & M \end{bmatrix} = \sum_{0 \le |X| \le |H|} \begin{bmatrix} H \\ X\ Y \end{bmatrix} \begin{bmatrix} K \\ U\ V \end{bmatrix} \begin{bmatrix} X\ U \\ L \end{bmatrix} \begin{bmatrix} Y\ W \\ M \end{bmatrix} \cdot (-1)^{|U|\cdot|Y|}$$

$$(29)\qquad \begin{bmatrix} H & K \\ L \end{bmatrix} \cdot \begin{bmatrix} M \\ H\ K \end{bmatrix} = C_{|L|}^{|H|} \begin{bmatrix} M \\ L \end{bmatrix}$$

Once this formalism is established, one can define relations among the different spaces $H^k = \mathrm{Hom}_{\underline{g}}(\mathscr{L}(\wedge^k), U)$; in particular, for $k=N, \mathscr{L}(\wedge^N)$ is the trivial $\underline{g}$-module, so that the mapping

$$(30)\qquad z \in Z \to F(z) \in H^N;\ F(z)_P^Q = z[{}_P^Q]$$

is bijective. Similarly H^1 and H^{N-1} are isomorhic, and if $X \in H^1$, $Y \in H^{N-1}$, then $z_{B\hat{B}}^{A\hat{A}} = [{}_{C\hat{C}}^{A\hat{A}}]\, X_D^C\, Y_{\hat{D}}^{\hat{C}}\, [{}_{B\hat{B}}^{D\hat{D}}]$, with $|A|=|B|=|C|=|D|=1$, belongs to H^N. The study of the relations between the H^k's provides with techniques which enable to "extract" X_C^A out of some suitably constructed z , yielding formulas of type $z_{B\hat{B}}^{A\hat{B}} = X_C^A \cdot [{}_{B\hat{B}}^{C\hat{C}}]\, Y_{\hat{C}}^{\hat{B}}$ which can then be transcribed to a formula of type (26), that is a contracted tensor product.

The techniques involved consist in constructing operators sending H^k in $H^{k'}$, and on considerations on the algebraic structures generated by such operators. These structures depend on the family to which $\underline{g}$ belongs, so that the three families have to be treated separately.

7. The approach of the unitarizability problem relative to the family $\underline{g}_o = \underline{so}(p,2)$ by tensor calculus tools has been successful [4], and it is possible but not certain that it also gives complete results for algebras of arbitrarily high real rank. For the moment only partial results are available. The main point of this approach consists in using the extensions of both $Z(\underline{g})$ and $Z(\underline{k})$, for $\underline{g} = \underline{k} \oplus \underline{p}$, to obtain criteria of positive-definiteness.

We shall end with a comment on an apparently paradoxical situation: if calculating in U is difficult, then multiplying matrices with entries in U arises the difficulty to the square. How can one claim that calculations are more smooth?

A possible answer to this objection could be that one may always reformulate theoretically a known situation in a new formalism, which may be better or worse (whatever this means) than the old one, without effective gain in computations.

However, this is not the case, because the paradox is only apparent. Indeed, for any simple finite dimensional (ρ, W) the isotypic component of W in U (resp. the Z-module $\mathrm{Hom}_{\underline{g}}(W,U)$) has finite multiplicity modulo Z (resp. finite dimension over Z) but there are infinitely many W with nonzero multiplicity. If x and x' belong to the isotypic components of W and W', their product $xx' \in U$ decomposes into elements of isotypic components by following reduction of $W \otimes W'$. The use of $(\pi)_U$ restricts to $\underline{g}$-modules W which have also a nonzero isotypic component in $\mathcal{L}(V)$, and there is a very limited number of them (especially when V is "small"). But then x is a linear combination of elements of the form $T^A{}_B$, and the contracted tensor product essentially consists in discarding those components of $W \otimes W'$ which do not intertwin with $\mathcal{L}(V)$. Hence the ctp is not mere matrix multiplication, but also a trick to restrict ordinary multiplication in a small part of U. Therefore the claim of gain of computational power is at least nonparadoxical.

REFERENCES

[1] V. BARGMANN: Ann. Math., 48 (1947), 568.

[2] W. MILLER: Amer. Math. Soc. Memoirs, 50 (1964).

[3] M.A. NAIMARK: Amer. Math. Soc. Transl., 6 (1957), 379.

[4] E. ANGELOPOULOS: C.R. Acad. Sc. Paris I, 292 (1981), 469; Lett. Math. Phys., 7 (1983), 121 and unpublished work.

[5] D. VOGAN: Ann. Math., 109 (1960), 1.

[6] E. ANGELOPOULOS, J. Math. Phys., 19 (1978), 2108.

[7] E. ANGELOPOULOS: in A.O. Barut (ed.), Quantum Theory, Groups Fields and Particles, 101, Reidel, Dordrecht, (1983).

[8] S. LANG: Algebra, Addison-Wesley, Reading, Mass. (1978).

[9] J. DIXMIER: Algèbres enveloppantes, Gauthier-Villars, Paris (1974).

[10] E. ANGELOPOULOS: Comm. Math. Phys., 89 (1983), 41.

Representations of the Lorentz Algebra on the Space of its Universal Enveloping Algebra

by

Romuald Lenczewski * and Bruno Gruber

Physics Department

Southern Illinois University

Carbondale, Illinois 62901

1. Introduction

Irreducible representations of semisimple Lie algebras as well as of some non-semisimple ones have been known for a long time . Recently, however, considerable effort of many scientists has concentrated on other linear representations of physically significant Lie algebras . Those are so called indecomposable representations, in other words , reducible but not completely reducible . The carrier space for an indecomposable representation of a Lie algebra g has a g-invariant linear subspace , but its complement is not g-invariant. It has been pointed out that these representations might be of much greater importance in physical applications than it has been expected (see, for example, [6]).

Our motivation has been to try to develop an effective algebraic procedure that would lead to indecomposable representations of those Lie algebras that are of interest in physics . The natural chain of symmetries $so(3) \subset so(3,1) \subset iso(3,1) \subset so(4,2)$ has led to several papers on this subject (see [7] , [8] , [9]) . It is worth mentioning that all indecomposable representations of semisimple Lie algebras have to be infinite dimensional , whereas it is not a necessary condition for indecomposability of representations of other Lie algebras (see, for example,[9])

The approach employed in our previous work originates at the concept of the universal enveloping algebra (UEA) Ω of the Lie algebra g . It is an associative algebra whose basis can be chosen as an infinite set of standard monomials , the latter being the ordered tensor products of the basis elements of the Lie algebra g (this is so called Poincare-Birkhoff-Witt theorem) . The Lie algebra g is naturally embedded in its UEA .

The advantage of using the UEA of g instead of the Lie algebra g itself stems from the fact that the former is associative whereas the latter has multiplication defined in terms of Lie products . Moreover, it turns out that the left multiplication in Ω by elements of g defines a very general representation of g (using a slightly different language, it makes Ω a left g-module) . In the sequel this representation will be denoted by ρ .

Since it is, in general, indecomposable , one can identify its subrepresentations and form quotient representations . By the abuse of the language we will also denote them by ρ .

Usually, we will restrict our attention to the representations induced by ρ on Ω/I , where I is a certain left ideal of Ω . Those ideals will be generated by a finite number of elements from Ω , say, $\omega_1,\ldots,\omega_k$. We will denote it as follows:

$$I = \langle\omega_1,\ldots,\omega_k\rangle .$$

Since we are going to concentrate on representations of the Lorentz algebra so(3,1) , let us use the notational conventions introduced for semisimple Lie algebras . In particular , the Weyl canonical basis for g will comprise h's belonging to the Cartan subalgebra h of g , e_α's corresponding to positive roots and $e_{-\alpha}$'s corresponding to negative roots . Then Ω/I , where

$$I = \langle h_\alpha - \Lambda_\alpha ,\ldots,e_{-\alpha},\ldots \rangle$$

is th so called raising algebra and is denoted by Ω_+ . Similarly , Ω/J where

$$J = \langle h_\alpha - \Lambda_\alpha ,\ldots,e_\alpha,\ldots \rangle$$

is the so called lowering algebra and is denoted by Ω_- .

In both cases $\Lambda_\alpha,\ldots$ are complex numbers and are inherent in the definitions of Ω_+ and Ω_- . Sometimes , in literature, Ω_+ and Ω_- are referred to as Verma modules . Infinite dimensional representations induced on Verma modules are very significant as it has been demonstrated in our previous work since they establish a general framework for identifying the representations known in physics. At the same time they give new representations that have not been obtained before.

Depending on the values of $\Lambda_{\alpha},\ldots$, the above representations may turn out to be irreducible or indecomposable . In the case when they are indecomposable, the usual procedure of going to the quotient would lead to the finite dimensional representations .

In the process of going from the general to the particular , let us choose the Lorentz algebra so(3,1) as an example of a semisimple Lie algebra .illustrating the simplicity and generality of the method outlined above .

In the case of the Lorentz algebra , as in many others , its so(3) subalgebra corresponding to the rotation group becomes of special importance because of the role that the angular momentum beses play in physical applications of the representation theory . For that reason, a change of basis of $\Omega_{\pm}$ is required . Instead of the natural basis consisting of standard monomials one needs to construct another basis, in which the representations of so(3) take their standard form . We call this new basis the angular momentum basis .

After the change of basis mathematical induction is used in order to derive the formulae for the representation ρ in the angular momentum basis . This step is crucial in the analysis , but the calculations are lengthy . Nevertheless, the results are rewarding . One obtains a very general representation of the Lorentz algebra and the familiar representations are recovered easily as a special case.

2. Commutation relations .

The angular momentum basis for the Lie algebra D_2 is given by the basis elements:

D_2 : $\{ h_3 , h_+ , h_- , p_3 , p_+ , p_- \}$

(h's correspond to the rotations in 3-dimensional space and p's correspond to the Lorentz boosts) with the following Lie products :

$$[h_3, h_\pm] = \pm h_\pm \qquad [p_3, p\ ] = h$$

$$[h_3, p_\pm] = \pm p_\pm \qquad [p_3, h_\pm] = \pm p_\pm$$

$$[h_+, p_-] = [p_+, h_-] = 2p_3 \qquad [p_+, h_+] = [p_-, h_-] = [p_3, h_3] = 0$$

$$[h_+, h_-] = [p_-, p_+] = 2h_3$$

Using the above relations and the definition of the UEA one can obtain the following commutation relations within the UEA , that are necessary to bring the elements of UEA multiplied from the left by elements of g to their standard ordered form :

$$[h_3, h_\pm^m] = \pm m h_\pm^m$$

$$[p_3, h_\pm^m] = \pm m p_\pm h_\pm^{m-1}$$

$$[h_\mp, h_\pm^m] = \mp 2m h_\pm^{m-1} h_3 - m(m-1) h_\pm^{m-1}$$

$$[h_\mp, p_\pm^m] = \mp 2m p_\pm^{m-1} p_3 + m(m-1) h_\pm p_\pm^{m-2}$$

$$[p_\mp, p_\pm^m] = \pm 2m p_\pm^{m-1} h_3 + m(m-1) p_\pm^{m-1}$$

$$[p_\pm, h_\mp^m] = \pm 2m h_\mp^{m-1} p_3 - m(m-1) p_\mp h_\mp^{m-2}$$

$$[h_3, p_\pm^m] = \pm m p_\pm^m$$

$$[p_3, p_\pm^m] = \mp m h_\pm p_\pm^{m-1}$$

The above relations are valid if all upper signs or all lower signs are taken simultaneously .

3. The Verma modules Ω_+ in the standard basis .

In what follows we are going to restrict our attention to the representations induced on different Verma modules Ω_+ (depending on certain parameters) by the left multiplication in Ω . It turns out that Verma modules Ω_- have a very similar nature and give the same representations as Ω_+ up to an automorphism of the Lie algebra so(3,1) (see [8]) .

As we mentioned earlier , Ω_+ are quotient modules of Ω modulo certain left ideals of Ω . This , quite abstract definition , could be phrased in a different way . Let us treat the identity $\mathbb{1}$ of the UEA of so(3,1) as the vacuum. Then let us consider a carrier space for a representation of so(3,1) to be spanned by the states created by acting with raising operators p_+^s , h_+^n on the vacuum . Assume that p_- and h_- annihilate the vacuum and h_3 and p_3 , in turn , give complex numbers Λ_1 and Λ_2 , respectively . This will give precisely the Verma module Ω_+ corresponding to Λ_1 , Λ_2 (we should actually write $\Omega_+(\Lambda_1,\Lambda_2)$) .

The standard basis of Ω_+ in the explicit form is given by :

$$\{ \; X(s,n) = p_+^s \, h_+^n \, \mathbb{1} \quad , \; s,n \in \mathbb{N} \; \}$$

where $\mathbb{N}$ denotes nonnegative integers .

We assume that Λ_1, Λ_2 are fixed . Then the following relations are obtained :

$$\rho(h_3)\, X(s,n) = (\Lambda_1 + n + s)\, X(s,n)$$

$$\rho(p_3)\, X(s,n) = \Lambda_2 X(s,n) + nX(s+1,n-1) - sX(s-1,n+1)$$

$$\rho(h_-)\, X(s,n) = n(-2\Lambda_1 -2s - n + 1)X(s,n-1) - 2s\Lambda_2 X(s-1,n) + s(s-1)X(s-2,n+1)$$

$$\rho(h_+)\, X(s,n) = X(s,n+1) \qquad , \qquad \rho(p_+)X(s,n) = X(s+1,n)$$

$$\rho(p_-)\, X(s,n) = s(2\Lambda_1 + 2n + s - 1)X(s-1,n) - 2n\Lambda_2 X(s,n-1) - n(n-1)X(s+1,n-2)$$

One can notice that the standard basis is not the angular momentum basis . Therefore , a change of basis is required and will be performed in the next section.

4. The Verma modules Ω_+ in the angular momentum basis .

In order to carry out a change of basis from the basis of standard monomials $X(s,n)$ to the angular momentum basis one has to find the $\rho(h_-)$-extremal vectors and then act on them with the raising operator h_+ with positive integral powers .

The $\rho(h_-)$-extremal vectors are the vectors annihilated by $\rho(h_-)$. They should be the linear combinations of standard monomials having the same weight as far as the so(3)-representations are concerned . The most general form of such a vector is seen to be :

$$y_N = \sum_{k=o}^{N} c_k X(N-k,k)$$

where $N \in \mathbb{N}$. Notice that $\rho(h_3)y_N = (\Lambda_1 + N)y_N$ and $\rho(h_-)y_N = o$.

The new basis for the Verma module Ω_+ is obtained by recovering the bases for the so(3) - representations that give the so(3) content of Ω_+ . In other words , it is given by :

$$\{ \; h_+^n \, y_N = y_N^n \; , \; n,N \in \mathbb{N} \; \}$$

By finding several of the $\rho(h_-)$-extremal vectors (for small values of N) and using mathematical induction , one can derive the following relations on Ω_+ in the new basis :

$$\rho(h_3)y_N^n = (N + n + \Lambda_1) \, y_N^n$$

$$\rho(h_+)y_N^n = y_N^{n+1}$$

$$\rho(h_-)y_N^n = n(-2\Lambda_1 - 2N - n + 1) \, y_N^{n-1}$$

$$\rho(p_3)y_N^n = -\alpha_N(-2\Lambda_1 - 2N - n + 1)y_{N-1}^{n+1} + \beta_N(\Lambda_1 + N + n) \, y_N^n + n \, y_{N+1}^{n-1}$$

$$\rho(p_+)y_N^n = \alpha_N \, y_{N-1}^{n+2} + \beta_N \, y_N^{n+1} + y_{N+1}^n$$

$$\rho(p_-)y_N = -\alpha_N(-2\Lambda_1 - 2N - n + 1)(-2\Lambda_1 - 2N + 2 - n) \, y_{N-1}^n + \beta_N \, n(-2\Lambda_1 - 2N - n + 1) \, y_N^{n-1} - n(n-1) \, y_{N+1}^{n-2}$$

where

$$\alpha_N = \frac{(\Lambda_2^2 + (1-\Lambda_1-N)^2)N(-2\Lambda_1+2-N)}{(-\Lambda_1+1-N)(-2\Lambda_1-2N+3)(-2\Lambda_1-2N+1)}$$

$$\beta_N = \frac{-\Lambda_2(-\Lambda_1+1)}{(-\Lambda_1-N)(-\Lambda_1+1-N)}$$

Since the new basis is determined by two parameters (N,n) , we can represent it graphically as a lattice $\mathbb{N} \times \mathbb{N}$ in the 2-dimensional plane . If the vertical axis of the rectangular coordinate system is chosen to be representing N and the horizontal to be representing n , then the action of the operators $\rho(\;)$ can be pictured as follows:

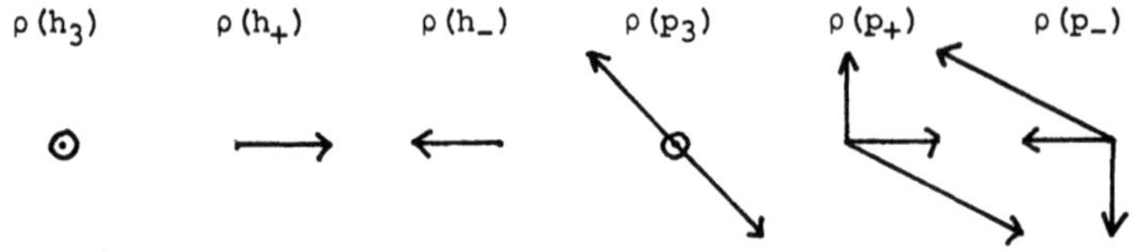

The key to identify the invariant subspaces of Ω_+ (if they exist) is to analyze the action of the operators $\rho(\ )$ for various values of Λ_1, Λ_2, N and n . In other words , one has to investigate what happens to the matrix elements of ρ when Λ_1, Λ_2, N and n are changed . In particular , our attention should concentrate on those values of Λ_1 and Λ_2 for which $\alpha_N = 0$ for some N . Namely , it happens when $N = -\Lambda_1 + 1 \pm i\Lambda_2$, $N = -2\Lambda_1 + 2$ and when $N = 0$. At the same time careful attention should be paid to singularities of α_N and β_N that occur when the denominators of α_N and β_N vanish , i.e., when $N = -\Lambda_1 + 1$, $N = -\Lambda_1 + 3/2$, $N = -\Lambda_1 + 1/2$, $N = -\Lambda_1$. As it was demonstrated in our previous work (see [8]) , if Λ_1 and $i\Lambda_2$ are both integers or half integers , one obtains invariant subspaces and if singular matrix elements exist , they can always be contained within invariant subspaces , so the quotient spaces still carry representations with finite matrix elements . One has to point out that in certain cases the values of α_N and β_N depend on the order in which the limits of parameters Λ_1, Λ_2 and N are taken . For a more detailed discussion of this feature , see [8] .

The matrix elements of ρ depend not only on the values of α_N and β_N . They also incorporate certain factors that are functions of n , In this way , new invariant subspaces are obtained by solving for n the following equation:

$$-2\Lambda_1 - 2N - n + 1 = 0$$

In the graphical presentation of ρ , the above equation leads to a <u>broken</u> line that separates the invariant subspace situated above and to the right of this line. The invariant subspaces that arise due to the vanishing of α_N are situated above <u>horizontal</u> lines .

As an example , take $\Lambda_1 = -3/2$, $\Lambda_2 = i/2$. Then , we obtain $\alpha_N = 0$ for $N = 0$, 2,4 , and α_N becomes singular for $N = 3$. This can be graphed as follows:

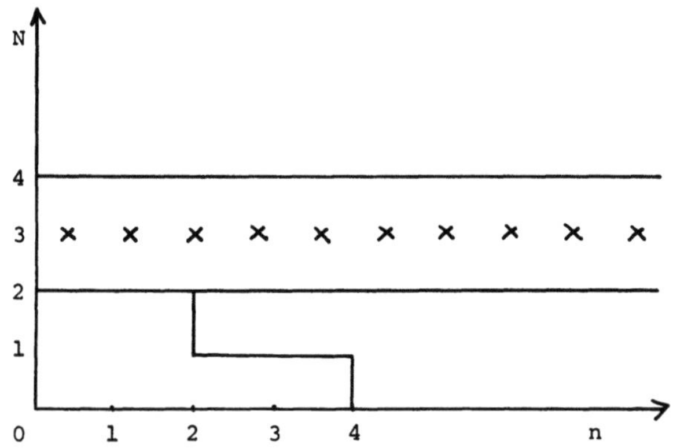

Denote: $W = \mathrm{sp}\ \{\ y_N^n\ ,\ n \in \mathbb{N}\ ,\ N \quad 4\ \}$

$V = \mathrm{sp}\ \{\ y_N^n\ ,\ N = 1\ ,\ 2\ ,\ n \in \mathbb{N}\ \}$

$S = \mathrm{sp}\ \{\ y_0^n\ ,\ n \quad 4\ \} \quad \{\ y_1^n\ ,\ n \quad 2\ \}$

$F = V/S = \mathrm{sp}\ \{\ y_0^0\ ,\ y_0^1\ ,\ y_0^2\ ,\ y_0^3\ ,\ y_1^0\ ,\ y_1^1\ \}$

where sp { ... } means the $\mathbb{C}$ - linear span of { ... } .

They are the carrier spaces for so(3,1) - representations . In particular , ρ induces an infinite dimensional irreducible representation on W , an infinite indecomposable representation on V , an infinite dimensional irreducible representation on S and finally , a finite dimensional irreducible representation on F . The so(3) content of those representations can be easily recovered once the bases are given .

Let us explicitly evaluate the matrix elements for the finite dimensional quotient representation on F .

$F\ :\ \{\ y_0^0\ ,\ y_0^1\ ,\ y_0^2\ ,\ y_0^3\ ,\ y_1^0\ ,\ y_1^1\ \}$

$\beta_0 = i/3\ ,\ \beta_1 = 5i/3\ ,\ \alpha_o = 0\ ,\ \alpha_1 = 4/9$

$$\rho(h_3)=\begin{bmatrix} -3/2 & & & & & \\ & -1/2 & & & & \\ & & 1/2 & & & \\ & & & 3/2 & & \\ & & & & -1/2 & \\ & & & & & 1/2 \end{bmatrix} \qquad \rho(p_3)=\begin{bmatrix} -i/2 & 0 & 0 & 0 & 0 & 0 \\ 0 & -i/6 & 0 & 0 & -8/9 & 0 \\ 0 & 0 & i/6 & 0 & 0 & -4/9 \\ 0 & 0 & 0 & i/2 & 0 & 0 \\ 0 & 1 & 0 & 0 & -5i/6 & 0 \\ 0 & 0 & 2 & 0 & 0 & 5i/6 \end{bmatrix}$$

$$\rho(h_+)=\begin{bmatrix} 0 & 0 & 0 & 0 & 0 & 0 \\ 1 & 0 & 0 & 0 & 0 & 0 \\ 0 & 1 & 0 & 0 & 0 & 0 \\ 0 & 0 & 1 & 0 & 0 & 0 \\ 0 & 0 & 0 & 0 & 0 & 0 \\ 0 & 0 & 0 & 0 & 1 & 0 \end{bmatrix} \qquad \rho(p_+)=\begin{bmatrix} 0 & 0 & 0 & 0 & 0 & 0 \\ i/3 & 0 & 0 & 0 & 0 & 0 \\ 0 & i/3 & 0 & 0 & 4/9 & 0 \\ 0 & 0 & i/3 & 0 & 0 & 4/9 \\ 1 & 0 & 0 & 0 & 0 & 0 \\ 0 & 1 & 0 & 0 & 5i/3 & 0 \end{bmatrix}$$

$$\rho(h_-)=\begin{bmatrix} 0 & 3 & 0 & 0 & 0 & 0 \\ 0 & 0 & 4 & 0 & 0 & 0 \\ 0 & 0 & 0 & 3 & 0 & 0 \\ 0 & 0 & 0 & 0 & 0 & 0 \\ 0 & 0 & 0 & 0 & 0 & 1 \\ 0 & 0 & 0 & 0 & 0 & 0 \end{bmatrix} \qquad \rho(p_-)=\begin{bmatrix} 0 & i & 0 & 0 & -8/3 & 0 \\ 0 & 0 & 4i/3 & 0 & 0 & -8/9 \\ 0 & 0 & 0 & i & 0 & 0 \\ 0 & 0 & 0 & 0 & 0 & 0 \\ 0 & 0 & -2 & 0 & 0 & 5i/3 \\ 0 & 0 & 0 & -6 & 0 & 0 \end{bmatrix}$$

This representation is obtained from the 6-dimensional representation listed in our previous work (see [8]) by an automorphism of the Lie algebra :

$$h_3 \to -h_3 \ , \ p_3 \to -p_3 \ , \ h_+ \to h_- \ , \ h_- \to h_+ \ , \ p_+ \to p_- \ , \ p_- \to p_+$$

The above example is very representative and it can be seen that other integral and half-integral values of Λ_1 and Λ_2 follow the same pattern .

In order to make a connection with the Gel'fand-Naimark basis (see [4],[5]) one more step into abstraction is necessary . It will be performed in the section that follows .

5. Connection with the Gel'fand-Naimark basis.

Inspection of the representation ρ in the angular momentum basis shows that the formulae are valid for all integral values of n and N. This clearly implies that we are dealing with an extension of the UEA to all integral exponents. Then the graphical presentation of $\{ y_N^n , N,n \in \mathbb{Z} \}$ is a two-dimensional lattice $\mathbb{Z} \times \mathbb{Z}$. Notice, that $\{ y_N^n , N \in \mathbb{N}, n \in \mathbb{Z} \}$ spans an invariant subspace, that has another invariant subspace spanned by $\{ y_N^n , N,n \in \mathbb{N} \}$. In a shorthand notation one could put it as follows:

$$\mathrm{sp} \{ \mathbb{Z} \times \mathbb{Z} \} \triangleright \mathrm{sp} \{ \mathbb{Z} \times \mathbb{N} \} \triangleright \mathrm{sp} \{ \mathbb{N} \times \mathbb{N} \}$$

The standard so(3,1) representations are obtained on the quotient space $\mathrm{sp} \{ \mathbb{Z} \times \mathbb{N} \} / \mathrm{sp} \{ \mathbb{N} \times \mathbb{N} \}$. The following picture emerges on $\mathbb{Z} \times \mathbb{N}$.

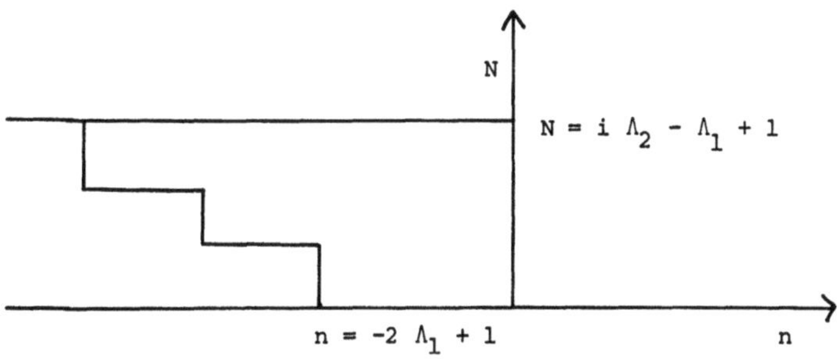

Let us now make the following redefinition of parameters:

$$n = m - \Lambda_1 - N , \quad N = \ell - \ell_0 , \quad \Lambda_1 = \ell_0 + 1 , \quad \Lambda_2 = i\ell_1$$

where m, ℓ_0, ℓ, and ℓ_1 take on the familiar values of ref.[4]. This definition yields the connection between the Gel'fand-Naimark basis expressed in terms of the parameters m, ℓ, ℓ_0, ℓ_1 and our basis expressed in terms of the parameters $n, N, \Lambda_1, \Lambda_2$. In fact, let us introduce the following renormalized basis elements:

$$| \ell , m \rangle = \left(\prod_{k=-\ell}^{m-1} (\ell - k)(\ell + k + 1) \right)^{-1/2} \left(\prod_{S=\ell_0+1}^{\ell} \alpha_S^{-1} \, 2S(2S-1) \right)^{1/2} y_N^n$$

where all the old parameters N , n Λ_1 , Λ_2 are expressed in terms of ℓ , m , ℓ_0 , ℓ_1 as indicated above ,. In the basis $|\ell , m\rangle$, we obtain exactly the same formulae as Gel'fand (see [4]) and Naimark (see [5]) for so(3,1) representations .

References

[1]. Humphreys J.E.,1972 " Introduction to Lie Algebras and Representation Theory" (New York : Springer)

[2]. Dixmier J.,1974 " Enveloping Algebras " (Amsterdam: North-Holland)

[3]. Varadarajan V.S., 1984 "Lie Groups , Lie Algebras and their Representations" (New York: Springer)

[4]. Gel'fand I.M., Minlos R.A., Shapiro Z.Ya, 1963 "Representations of the Rotation and Lorentz Groups and their Applications " (New York: Pergamon)

[5]. Naimark M.A., 1964 "Linear Representations of the Lorentz Group " (Oxford: Pergamon)

[6]. Dirac P.A.M.,1983 " The Future of Atomic Physics" , Int.J.Th.Phys.23 (8), 677-681 .

[7]. Gruber B., Klimyk A., 1984 "Matrix Elements for Indecomposable Representations of Complex su(2) " , J.Math,Phys.25 (4) ,755-764.

[8]. Gruber B., Lenczewski R., 1983 " Indecomposable Representations of the Lorentz Algebra in an Angular Momentum Basis " ,J.Phys.A 16 , 3703-3722.

[9], Lenczewski R.,Gruber B., 1986 " Indecomposable Representations of the Poincare Algebra " , J.Phys.A 19 ,1-20 .

* Permanent address : Institute of Physics , Technical University of Wroclaw , Wroclaw , Poland .

REDUCIBLE REPRESENTATIONS OF THE EXTENDED CONFORMAL SUPERALGEBRA AND INVARIANT DIFFERENTIAL OPERATORS

V.K. Dobrev and V.B. Petkova

Institute of Nuclear Research and Nuclear Energy
Bulgarian Academy of Sciences
72 Blvd. Lenin, 1784 Sofia, Bulgaria

Introduction.

The aim of this paper is to introduce the reader to some recent developments [1,2] in the representation theory of the conformal superalgebra and supergroup. The emphasis is on the reducible (and indecomposable) representations which are physically relevant and on the related invariant differential operators.

The paper is essentially selfcontained. In Section 1 we recall the definitions of a superalgebra in general, of the conformal superalgebra $\mathcal{G} = su(2,2/N)$, of its complexification $sl(4/N;\mathbb{C})$ and of the corresponding supergroups $G = SU(2,2/N)$ and $SL(4/N;\mathbb{C})$. In Section 2 we introduce the elementary representations (ER) of $\mathcal{G}$ and G using an indexless realization of the inducing irreducible finite-dimensional representations of $SL(2,\mathbb{C})$ and $SU(N)$. In Section 3 we demonstrate the lowest weight module associated with an ER and (adapting results of Kac [3]) the reducibility conditions. Then we present a canonical procedure (introduced earlier for the ordinary real semisimple Lie algebras and groups [4])for the construction of the invariant differential operators.

1. Preliminaries.

1.1. A superalgebra is a $\mathbb{Z}_2$ - graded algebra $\mathcal{A} = \mathcal{A}_0 \oplus \mathcal{A}_1$ (i.e., if $X \in \mathcal{A}_\alpha$, $Y \in \mathcal{A}_\beta$, $\alpha,\beta \in \mathbb{Z}_2=\{0,1\}$, then $XY \in \mathcal{A}_{\alpha+\beta}$). A Lie superalgebra is a superalgebra $\mathcal{G} = \mathcal{G}_0 \oplus \mathcal{G}_1$ with a bracket $[\ ,\ ]$ satisfying :

$$[X,Y] = -(-1)^{\alpha\beta}[Y,X]\ ,\ X\in\mathfrak{g}_\alpha\ ,Y\in\mathfrak{g}_\beta\ ;\ [X,[Y,Z]] = [[X,Y],Z] + (-1)^{\alpha\beta}[Y,[X,Z]].$$

The Lie superalgebra $\mathfrak{g}^{\mathbb{C}} = sl(4/N;\mathbb{C})$ will be realized as a matrix superalgebra

$$\mathfrak{g}^{\mathbb{C}} = \left\{ Y = \begin{pmatrix} a & b \\ c & d \end{pmatrix};\ \begin{pmatrix} a & 0 \\ 0 & d \end{pmatrix} \in \mathfrak{g}_0^{\mathbb{C}}\ ;\ \begin{pmatrix} 0 & b \\ c & 0 \end{pmatrix} \in \mathfrak{g}_1^{\mathbb{C}}\ ;\ \mathrm{str}Y \equiv \mathrm{tr}a - \mathrm{tr}d = 0 \right\},\quad (1)$$

where a,b,c,d are 4x4,4xN,Nx4,NxN matrices respectively.

The Lie superalgebra su(2,2/N) is the following $(N^2+8N+15)$-dimensional real noncompact form of $\mathfrak{g}^{\mathbb{C}}$:

$$\mathfrak{g} = su(2,2/N) = \left\{ Y \in \mathfrak{g}^{\mathbb{C}};\ Y_a^+\omega + (-i)^a\omega Y_a = 0 \right\}\ ;\ \omega = \begin{pmatrix} 0 & \mathbb{1}_2 & 0 \\ \mathbb{1}_2 & 0 & 0 \\ 0 & 0 & \mathbb{1}_N \end{pmatrix},\quad (2)$$

where Y^+ is the Hermitean conjugate of the matrix Y . (The matrix ω differs from the usual choice $\mathrm{diag}(\mathbb{1}_2,-\mathbb{1}_2,\mathbb{1}_N)$ by a real orthogonal transformation.) The even part of $\mathfrak{g}$ is the subalgebra

$$\mathfrak{g}_0 = su(2,2)\oplus u(1)\oplus su(N)\ .\quad (3)$$

1.2. We shall consider representations induced from the so-called maximal parabolic subalgebra

$$\mathfrak{p} = \mathfrak{m}\oplus\mathfrak{a}\oplus\mathfrak{n}\ ,\quad (4)$$

where $\mathfrak{a}$ is the 1-dimensional (dilatation) subalgebra

$$\mathfrak{a} = \mathfrak{a}_0 = \mathrm{l.s.} \begin{pmatrix} \mathbb{1}_2 & 0 & 0 \\ 0 & -\mathbb{1}_2 & 0 \\ 0 & 0 & 0 \end{pmatrix}\ ;\quad (5)$$

$\mathfrak{m}\oplus\mathfrak{a}$ is the centralizer of $\mathfrak{a}$ in $\mathfrak{g}$, $\mathfrak{m}=\mathfrak{m}_0=sl(2,\mathbb{C})\oplus u(1)\oplus su(N)$; $\mathfrak{n}$ is the subalgebra comprised of the negative restricted root spaces with respect to $\mathfrak{a}$, explicitly

$$\mathfrak{n} = \begin{pmatrix} 0_4 & & 0 \\ & & p \\ -ip^+ & 0 & 0_N \end{pmatrix} \oplus \mathrm{l.s.} \left\{ \begin{pmatrix} 0 & 0 & 0 \\ i\sigma_\mu & 0 & 0 \\ 0 & 0 & 0 \end{pmatrix};\ \mu=1,\dots,4 \right\},\quad (6)$$

$\dim\mathfrak{n}=4N+4$; $\mathfrak{n}$ generates the special superconformal transformations.

Finally we write a decomposition of $\mathfrak{g}$:

$$\mathfrak{g} = \tilde{\mathfrak{n}}\oplus\mathfrak{n}\oplus\mathfrak{a}\oplus\mathfrak{m}\ ,\quad (7)$$

where $\tilde{\mathfrak{n}}$ is the subalgebra (generating supertranslations) comprised of the positive restricted root spaces with respect to $\mathfrak{a}$.

1.3. The Lie supergroup $SL(4/N;\mathbb{C})$ can be realized as a matrix group with elements

$$g = (g_{mn}) = \begin{pmatrix} \text{even} & \text{odd} \\ \text{odd} & \text{even} \end{pmatrix} \begin{matrix} \}\,4 \\ \}\,N \end{matrix} = \begin{pmatrix} A & B \\ C & D \end{pmatrix},\ g_{mn}\in\mathcal{A}_a\ ,\quad (8)$$

$$\det(A-BD^{-1}C) = \det D\ ,$$

where $\mathcal{A} = \mathcal{A}_0\oplus\mathcal{A}_1$ is a complex Grassmann algebra with countably many odd generators, so that any element of $\mathcal{A}$ is a finite sum of monoms of these generators. (For the notions of supermanifold and Lie supergroup see [5].

The conformal Lie supergroup SU(2,2/N) is a real form of the Lie supergroup SL(4/N;$\mathbb{C}$)

$$G = SU(2,2/N) = \left\{ g \in SL(4/N;\mathbb{C}) \ ; \ g^{+}\omega g = \omega \right\} \quad ; \qquad (9)$$

where $g^{+}_{mn} = g^{*}_{nm}$ with $*$ - involution in $\mathcal{A}$; $*\circ* = id$.

In a neighbourhood of the identity $G = \exp \mathfrak{g}(\mathcal{A})$, where $\mathfrak{g}(\mathcal{A}) = $ $= \mathcal{A}_0(\mathbb{R})\otimes \mathfrak{g}_0 \oplus \mathcal{A}_1(\mathbb{R})\otimes \mathfrak{g}_1$ is the Grassmann envelope of $\mathfrak{g}$ with respect to $\mathcal{A}$, $\mathcal{A}(\mathbb{R}) = \left\{ \zeta = \zeta_0 + \zeta_1 \in \mathcal{A} \ ; \ \zeta_a^* - (-i)^a \zeta_a = 0 \right\}$.

Superspace can be identified with the subgroup of supertranslations $\mathcal{X}$ which is parametrized by 4 even and 4N odd elements of $\mathcal{A}$:

$$\mathcal{X} = \left\{ G \ni x = \begin{pmatrix} \mathbb{I}_2 & i\underset{\sim}{x} - 2\theta\bar{\theta} & 2\theta \\ 0 & \mathbb{I}_2 & 0 \\ 0 & -2\bar{\theta} & \mathbb{I}_N \end{pmatrix}; \begin{array}{l} x_\mu \in \mathcal{A}_0 \ , \ x_\mu^* = x_\mu , \ \theta_a^k \in \mathcal{A}_1 , \ \bar{\theta}_a^k = (\theta_a^k)^* , \\ \underset{\sim}{x} = \sigma_\mu x^\mu , \ (\theta\bar{\theta})_{ab} = \theta_a^k \bar{\theta}_b^k , \ k=1,\dots,N; \ a,b=1,2; \end{array} \right\} . \qquad (10)$$

The left action of G on $\mathcal{X}$ gives

$$g^{-1}\tilde{n} = \tilde{n}_g p(\tilde{n},g), \text{ or } g^{-1}\tilde{n} = \omega \tilde{n}^c_g p(\tilde{n},g) ,$$

where $\tilde{n},\tilde{n}_g \in \mathcal{X}$; $\tilde{n}^c \in \mathcal{X}^c \equiv \left\{ \tilde{n} \in \mathcal{X}; \ \mathbb{R} \ni x^2 - n(x^2) = 0; \ n(\zeta) \text{ - nilpotent part of } \zeta \in \mathcal{A}_0(\mathbb{R}) \right\}$ and $p(\tilde{n},g)$ is an element of the stationary subgroup P of the origin in superspace, P = $MA\mathcal{N}$ with $A = \exp \mathfrak{a}(\mathcal{A})$, $\mathcal{N} = \exp \mathfrak{n}(\mathcal{A})$, and MA is the centralizer of A in G. In the matrix realization (8)

$$M = \left\{ \begin{pmatrix} \ell\sigma_3^\nu e^{it/4} & 0 & 0 \\ 0 & \ell^{+-1}\sigma_3^\nu e^{it/4} & 0 \\ 0 & 0 & e^{it/N}u \end{pmatrix} \equiv \hat{\ell}\,\hat{\sigma}_3^\nu \mathfrak{z}(t)\hat{u} \ ; \right. \qquad (11)$$

$$\hat{\ell} = \begin{pmatrix} \ell & 0 & 0 \\ 0 & \ell^{+-1} & 0 \\ 0 & 0 & \mathbb{I}_N \end{pmatrix}, \ \ell \in SL(2/0;\mathbb{C}) \ (\text{i.e. } \ell_{ab} \in \mathcal{A}_0 \ , \ \det\ell = 1);$$

$$\hat{\sigma}_3^\nu = \begin{pmatrix} \sigma_3^\nu & 0 & 0 \\ 0 & \sigma_3^\nu & 0 \\ 0 & 0 & \mathbb{I}_N \end{pmatrix}, \ \nu = 0,1 \ ; \ \mathfrak{z}(t) = \begin{pmatrix} e^{it/4} & 0 & 0 \\ 0 & e^{it/4} & 0 \\ 0 & 0 & e^{it/N} \end{pmatrix}, \ t \in \mathcal{A}_0(\mathbb{R}) \ (\mathrm{mod}\ 2\pi) \ ;$$

$$\left. \hat{u} = \begin{pmatrix} \mathbb{I}_2 & 0 & 0 \\ 0 & \mathbb{I}_2 & 0 \\ 0 & 0 & u \end{pmatrix}, \ u \in SU(N/0) \right\};$$

$$A = \left\{ \begin{pmatrix} \sqrt{\delta} & 0 & 0 \\ 0 & \sqrt{\delta}^{-1} & 0 \\ 0 & 0 & \mathbb{I}_N \end{pmatrix} = a(\delta) \ , \ \delta = e^{\tau} \ , \ \tau \in \mathcal{A}_0(\mathbb{R}) \right\}. \qquad (12)$$

2. Induced representations.

2.1. We consider a class of P-induced representations of the conformal supergroup G and its Lie superalgebra called elementary representations (ER). They are induced by the finite-dimensional irreducible representations D_χ of MA ($\mathcal{N}$ acts trivially) :

$$\chi = [j_1 , j_2 ; d ; z ; r_1, \dots, r_{N-1}] , \qquad (13)$$

where $2j_1, 2j_2; r_1, \dots, r_{N-1}$ are nonnegative integers indexing the representations of $SL(2,\mathbb{C})$ and $SU(N)$ (the integers $m_1 \geq m_2 \geq \dots \geq m_{N-1} \geq 0$ such that $r_k = m_k - m_{k+1}$ are used also); d (the scale dimension) and z are complex numbers indexing the representations of A and $U(1)$ resp. For the ER of G the relation

$$z + j_1 - j_2 + (2/N) \sum_{i=1}^{N-1} m_i \in \mathbb{Z} \qquad (14)$$

should hold. For $N=4$ we may consider representations of the factorgroup $G/U(1)$; then $z=0$.

It is convenient to use an indexless realization of the finite-dimensional irreducible representations of $SL(2,\mathbb{C}) \times SU(N)$ (cf. [6]) instead of the more standard tensor fields realization. The representation space W_χ is given by

$$W_\chi = \Big\{ \text{polynomials } \varphi : \mathbb{C}^2 \times \mathbb{C}^2 \times U(N) \longrightarrow \mathbb{C} ;$$

$$\varphi(\lambda z, \bar\lambda \bar z; \dots e^{i(\alpha_k - \alpha_{k-1})} u^k \dots ; \dots e^{-i(\alpha_k - \alpha_{k-1})} \bar u^k \dots) = \qquad (15a)$$

$$= \lambda^{2j_1} \bar\lambda^{2j_2} \exp i \sum_{k=1}^{N-1} (\alpha_k - \alpha_{k-1}) m_k \cdot \varphi(z, \bar z; u^1, \dots, \bar u^N) ;$$

$$z = \binom{z_1}{z_2}, \ \bar z = (\bar z_1, \bar z_2); \ \lambda \in \mathbb{C}; \ (u^k)_i = \mathcal{U}_{iN-k+1}, \ \mathcal{U} \in SU(N);$$

$$D_{ik} \varphi \equiv (u^k \frac{\partial}{\partial u^i} - \bar u^i \frac{\partial}{\partial \bar u^k}) \varphi = 0 , \ 1 \leq k < i \leq N \Big\} . \qquad (15b)$$

The polynomiality requirement above can be equivalently replaced by the set of equations

$$D_{ii+1}^{r_k+1} \varphi = 0 = (w \frac{\partial}{\partial z})^{2j_1+1} \varphi = (\bar w \frac{\partial}{\partial \bar z})^{2j_2+1} \varphi ; i=1,2,\dots,N-1, w \in \mathbb{C}^2, \ w_1 z_2 - w_2 z_1 = 1. \qquad (16)$$

The set of equations (15b) (or (16)) imply that the elements of W_χ can be made functions of at most $N-1$ arguments, i.e. up to equivalence $\varphi = \varphi(u^1, \dots, u^{N-1})$; they are still subject to the remaining equations.

The representation D_{jzr} acts in W_χ according to

$$(D_{jzr}(\ell \sigma_3^\nu \chi(t) u) \varphi)(z, \bar z; \dots u^k \dots ; \dots \bar u^k \dots) = \qquad (17)$$

$$= \exp(itz/2)(-1)^{\varepsilon\nu} \varphi(\sigma_3^\nu \ell^{-1} z, \bar z \ell^{+-1} \sigma_3^\nu; \dots \mathcal{U}^{-1} u \dots ; \dots \bar u \, \mathcal{U} \dots)$$

$$\varepsilon \equiv z + j_1 - j_2 + (2/N) \sum m_i (\text{mod} 2).$$

The representations D_χ on W_χ can be extended to representations of the even supergroup $SL(2/0;\mathbb{C})\times SU(N/0)$.

2.2. The ER are defined by the following covariance property in a space $\mathcal{C}_\chi$ of W_χ - valued smooth in a certain sense functions on the supergroup

$$F(gman) = D_\chi^{-1}(ma)F(g) \qquad (18)$$

and the representation T_χ is given by the left regular action of G . Obviously the ER can be realized equivalently by functions on the homogeneous space $G/P = \mathcal{X}\cup\omega\mathcal{X}^c$, or in a space of functions on the superspace $\mathcal{X}$ (i.e. superfields) with certain asymptotic behaviour at infinity.

Another useful realization is provided by the $\mathcal{A}_0$- valued functions

$$\mathcal{F}(g\hat{\ell}\hat{u}) = F(g,\pi(\ell),\overline{\pi(\ell)},u) \quad , \quad (\pi(\ell))_a = \ell_{a2} \quad , \qquad (19)$$

satisfying the set of Eqs. (15b),(16). Note that

$$\mathcal{F}(\tilde{n}\hat{\ell}\hat{u}g_H\ell_- n) = \chi^{-1}(g_H)\mathcal{F}(\tilde{n}\hat{\ell}\hat{u}) \quad , \ g_H \in \exp\mathfrak{h}(\mathcal{A}), \ \ell_- = \begin{pmatrix} 1 & z \\ 0 & 1 \end{pmatrix} \ , \ n\in\mathcal{N} \ , \qquad (20)$$

where $\mathfrak{h}$ is a Cartan subalgebra of $\mathfrak{g}$, $\mathfrak{h} = \mathfrak{a}\oplus\mathfrak{b}$, $\mathfrak{b}$ is the Cartan subalgebra of $\mathfrak{m}$.

From now on we restrict ourselves to the elementary representations of the conformal superalgebra $\mathfrak{g}$ defined in a suitable space of functions by the infinitesimal version of (18). We only note that the elements of the resulting ER spaces (i.e. the superfields) have entirely different nature from their supergroup counterparts [5,2].

3. Invariant differential operators.

3.1. The ER are generically irreducible. We solve the problem of finding the reducible ones among them by adapting results of Kac on the theory of lowest weight representations of the basic classical Lie superalgebras. To every ER χ there is associated a lowest weight module (LWM) over $\mathfrak{g}^{\mathbb{C}}$ with lowest weight $\Lambda = \Lambda(\chi) \in (\mathfrak{h}^{\mathbb{C}})^*$ ($\mathfrak{h}^{\mathbb{C}}$ is the Cartan subalgebra of $\mathfrak{g}^{\mathbb{C}}$) and lowest weight vector v :

$$hv = (\Lambda+\rho)(h)v \ , \ h\in\mathfrak{h}^{\mathbb{C}} \ , \ 2\rho = \sum_{\alpha>0,\mathrm{even}} \alpha \ - \sum_{\alpha>0,\mathrm{odd}} \alpha \ , \qquad (21a)$$

$$Xv = 0 \ , \ X\in\mathfrak{g}_-^{\mathbb{C}} \ , \qquad (21b)$$

where $\mathfrak{g}_-^{\mathbb{C}}$ is comprised from the negative root spaces (recall the decomposition $\mathfrak{g}^{\mathbb{C}} = \mathfrak{g}_+^{\mathbb{C}}\oplus\mathfrak{h}^{\mathbb{C}}\oplus\mathfrak{g}_-^{\mathbb{C}}$). For $N=4$ $\Lambda\rightarrow\Lambda+\tilde{\Lambda}$, $\tilde{\Lambda}$ - weight of the $u(1)$ centre of $\mathfrak{g}$ (cf. [2]).Naturally it is enough to require (21b) for $X=e_{-\alpha}$ with α any simple root, $e_{-\alpha}$ is the root space vector of $-\alpha$. The LWM is displayed through the standard right action of $\mathfrak{g}^{\mathbb{C}}$ on the space $\mathcal{C}_\chi$

$$(\hat{X}\mathcal{F})(g) = \frac{d}{ds}\mathcal{F}(g\,\exp(s\circ X))\Big|_{s=0} \ , \ X\in\mathfrak{g}_a^{\mathbb{C}} \ , \ s\in\mathcal{A}_a \ , \qquad (22)$$

It is easy to see that every element of $\mathcal{C}_\chi$ can play the role of lowest weight vector. (In the dual (to the ER $\mathcal{E}_\chi$) LWM where $\mathcal{G}^{\mathbb{C}}$ acts from the left the lowest weight vector is the δ- fuction concentrated at the identity of G .) Indeed conditions (21) are an extension to $\mathcal{G}^{\mathbb{C}}$ of the infinitesimal right action in (20) (see also (15b)). In addition since we work with the real form $\mathcal{G}$ and with finite-dimensional representations of MA the functions from $\mathcal{E}_\chi$ satisfy (cf. (15c)) :

$$(e_{\alpha_i})^{\kappa_i}\mathcal{F} = 0 \ , \ \kappa_i = -2(\Lambda,\alpha_i)/(\alpha_i,\alpha_i) \in \mathbb{N} \ , \tag{23}$$

where α_i are the simple compact (i.e. $\alpha_i|_{\mathfrak{a}} = 0$) roots.

3.2. The results of Kac [3] adapted to our case give that the LWM is reducible only if at least one of the following 4+4N condititions is true :

$$2(\Lambda,\alpha) = -k(\alpha,\alpha) \ , \ k \in \mathbb{N} \ , \tag{24}$$

where α is some noncompact (i.e. $\alpha|_{\mathfrak{a}} \neq 0$) positive root. (For α - compact (24) is automatically satisfied for our LWM with $k = 2j_1+1, 2j_2+1, r_1+1, \ldots, r_{N-1}+1$ - cf. (23).) Explicitly the conditions (24) corresponding to the 4 even $((\alpha,\alpha)\neq 0)$ roots are :

$$-c \pm (1+j_1+j_2) \in \mathbb{N} \ , \quad -c \pm (j_1-j_2) \in \mathbb{N} \ , \quad c = d+N-2 \ , \tag{25}$$

while those corresponding to the 4N odd $((\alpha,\alpha)=0)$ roots are :

$$d = d^1_{Ns} - z\delta_{N4}, \ d^1_{Ns} = 4-2s+2j_2+z+2m_s-2m/N \ , \quad s=1,\ldots N \ , \tag{26.1}$$

$$d = d^2_{Ns} - z\delta_{N4}, \ d^2_{Ns} = 2-2s-2j_2+z+2m_s-2m/N \ , \tag{26.2}$$

$$d = d^3_{Ns} + z\delta_{N4}, \ d^3_{Ns} = 2+2s-2N+2j_1 -z-2m_s+2m/N \ , \tag{26.3}$$

$$d = d^4_{Ns} + z\delta_{N4}, \ d^4_{Ns} = 2s-2N-2j_1-z-2m_s+2m/N \ , \tag{26.4}$$

where $m = \sum_i m_i$. (For N=1 these conditions were found in [7].)

3.3. Whenever (24) is satisfied for some α and some k there arise an intertwining differential operator (in general nontrivial) $\mathcal{E}_\chi \to \mathcal{E}_{\chi'}$ where χ' is determined for α even from

$$\Lambda' = \Lambda - 2(\Lambda,\alpha)/(\alpha,\alpha) = \Lambda + k\alpha = w_\alpha \Lambda \ , \ \alpha \text{ even} \ , \tag{27a}$$

where w_α is the Weyl reflection of Λ with respect to α . For example

$$\chi' = [j_1+k/2, j_2+k/2, d+k; z; r_1, \ldots r_{N-1}] \quad \text{for} \quad -k = c+j_1+j_2+1 \ .$$

For α odd $((\alpha,\alpha)=0)$ χ' is determined from

$$\Lambda' = \Lambda + \alpha \ , \ (\Lambda,\alpha) = 0 \tag{27b}$$

which may be interpreted as odd Weyl reflections acting in the weight system (however one should bear in mind that the root system is not invariant under such reflections). Explicitly for the 4N cases in (26) we have (respectively) :

$$\chi' = [j_1, j_2-1/2, d+1/2; z+(N-4)/2N; r_1, \dots, r_{s-1}-1, r_s+1, \dots, r_{N-1}] ,$$
$$\chi' = [j_1, j_2+1/2, d+1/2; \quad \text{as above}] , \qquad (28)$$
$$\chi' = [j_1-1/2, j_2, d+1/2; z-(N-4)/2N; r_1, \dots, r_{s-1}+1, r_s-1, \dots, r_{N-1}] ,$$
$$\chi' = [j_1+1/2, j_2, d+1/2; \quad \text{as above}] ;$$

where we assume that the j_a and r_i entries are nonnegative; if not there is no nontrivial operator.

These maps are not onto for the ER (in general) hence the ER corresponding to the images should be added to the list of reducible ER; effectively $\chi' = \tilde{\chi}$, where χ satisfies some of the conditions (24) and $\tilde{\chi}$ is obtained from χ by the changes : $j_1 \leftrightarrow j_2$, $c \rightarrow -c$ ($\tilde{d}=4-2N-d$).

Every ER for which some condition from (24) holds appears actually in a sequence of ER connected by the various intertwining operators. This sequence is infinite if α is odd since then $\Lambda+\alpha$ satisfies the same condition as Λ (the restrictions for a nontrivial operator should be respected of course). Thus each resulting multiplet which groups together partially equivalent representations (which are reducible under some odd root) unlike the case N=0 contains an infinite number of members.

3.4. The explicit construction of the invariant differential operators also uses the information from the LWM picture. Whenever the LWM Λ is reducible under the root α , i.e. (24) holds for some $k \in \mathbb{N}$ then the LWM $\Lambda+k\alpha$ can be identified with a submodule of Λ . Moreover this implies the existence in the LWM Λ of a vector v_s , called singular vector, different from the lowest weight vector v of Λ and which has the characteristics of the lowest weight vector of $\Lambda+k\alpha$. In the Verma module realization V_χ of the LWM the singular vector can be represented by the formula (cf. [8]) :

$$v_s = P(e_{\alpha_1}, \dots, e_{\alpha_\ell})v \quad , \quad (\ell = \text{rank } \mathcal{G}) \quad , \qquad (29a)$$

where P is a homogeneous polynomial in the root space vectors e_{α_i} corresponding to the simple roots of degrees

$$kk_1, \dots, kk_\ell \ , \qquad \alpha = \sum k_i \alpha_i \quad , \ k_i = 0,1 \ ; \qquad (29b)$$

where the decomposition of α into simple roots is implemented.

The next step is to identify the differential operator D_α corresponding to the singular vector by replacing any root vector e_α with the corresponding right action of $\hat{e}_\alpha$ on $\mathcal{F} \in \mathcal{C}_\chi$

$$v_s \longrightarrow D_\alpha \mathcal{F} = P(\hat{e}_{\alpha_1}, \dots, \hat{e}_{\alpha_\ell}) \mathcal{F} \quad . \qquad (30)$$

In particular the differential operators corresponding to simple roots and to k=1 and only these are given exactly by the right infinitesimal action of $\mathfrak{g}$.

3.5. Let us give some examples of operators corresponding to the odd roots. The operators corresponding to the two odd (and noncompact) simple roots leading to conditions (26.4), s=N, and (26.2), s=1, respectively, are:

$$\mathcal{D}^N = \bar{u}^N_k D_{ka} z_a \quad , \quad D_{ka} = \frac{\partial}{\partial\theta^k_a} + i\bar{\theta}^k_b \tilde{\nabla}_{ba}, k=1,\dots,N; \ a,b=1,2; \tag{31a}$$

$$\tilde{\mathcal{D}}^1 = u^1_k \bar{D}_{ka} \bar{z}_a \quad . \tag{31b}$$

The operators corresponding to (26.4) for s=1,...,N-1 are

$$\mathcal{D}^s = \sum_{p=0}^{N-s-1} \sum_{N-1\geq j_1 > j_2 \dots > j_{p+1} = s} c_s(r) \left(\frac{\partial}{\partial u} {}_k D_{ka} z_a\right) D_{j_2 j_1} \dots D_{j_{p+1} j_p} , \tag{32}$$

(D_{ik} is defined in (15b)). The operators corresponding to (26.3) are obtained from (32) with the replacement $z \to \frac{\partial}{\partial z} \varepsilon$, $\varepsilon = \begin{pmatrix} 0 & 1 \\ -1 & 0 \end{pmatrix}$.

Because of the indexless realization these operators provide a compact form of the usual differential relations. Since they act irreducibly in SL(2,$\mathbb{C}$) - spin and SU(N) - isospin they can be useful also in the super-Poincare context where, of course, they are defined without the restrictive conditions (26) and correspond to the usual covariant derivatives. For instance

$$\tilde{\mathcal{D}}^1 f = 0 \ , \ \text{if} \quad d=z, \quad j_2=0=r_1=\dots=r_{N-1} \ ,$$

$$\mathcal{D}^N f = 0 \ , \ \text{if} \quad d=-z, \quad j_1=0=r_1=\dots=r_{N-1} \ ,$$

are recognized as the chirality and antichirality conditions on f .

For semisimple Lie algebras (and groups) all intertwining differential operators are exhausted by the compositions of those determined by the positive noncompact roots [4] . This is not the case here. In particular higher order odd operators corresponding to some elements of the positive root lattice arise. Apart from those needed for the description of the massless UIR (cf. [2]) we have only partial results. These operators are needed also for the more detailed analysis of the structure of the representations; in particular one needs them in the proof that the conditions of Kac are also sufficient for the reducibility of the infinite-dimensional representations as well.

References.

1. V.K. Dobrev and V.B. Petkova, Lett.Math.Phys. 9(1985)287.
2. V.K. Dobrev abd V.B. Petkova, ICTP Trieste preprint IC/85/29(1985)
3. V.G. Kac, in Lecture Notes in Math., Vol. 676 (Springer-Verlag, Berlin, 1978) p.597.
4. V.K. Dobrev, Talk at the IV Clausthal Workshop (1984), in: Proc.,

Ed. H.D. Doebner, to be published.

5. F. Berezin and D. Leites, Soviet.Math.Dokl. 16(1975)1218;
B. Kostant, in Lecture Notes in Math., Vol. 570 (Springer-Verlag, Berlin, 1977) p.177;
M. Batchelor, Trans.Amer.Math.Soc. 258(1980)257;
A.S. Schwarz, Teor.Mat.Fiz. 60(1984)37;
V.V. Molotkov, ICTP Trieste preprint IC/84/183 (1984);
6. D.P. Zhelobenko and A.I. Stern, Representations of Lie Groups (Nauka, Moscow, 1983, in Russian).
7. B.L. Aneva, S.G. Mikhov and D.Ts. Stoyanov, Teor.Mat.Fiz. 31(1977) 17.
8. J. Dixmier, Enveloping Algebras (North-Holland, N.Y., 1977).

ALL POSITIVE ENERGY UNITARY IRREDUCIBLE REPRESENTATIONS OF THE EXTENDED CONFORMAL SUPERALGEBRA

V.K. Dobrev and V.B. Petkova

Institute of Nuclear Research and Nuclear Energy
Bulgarian Academy of Sciences
72 Blvd. Lenin, 1784 Sofia, Bulgaria

Introduction.

In [1] we displayed the list of all positive energy unitary irreducible representations of the conformal superalgebra $\mathfrak{g} = su(2,2/N)$. In the present paper we give the proof of this result.

The paper is organized as follows. In Section 1 we introduce the representations of $\mathfrak{g}$ and discuss the superhermitian form used in the unitarity construction. In Section 2 we state the main result and in Section 3 we give the proof. An Appendix contains the relevant information on $\mathfrak{g}$, its complexification $\mathfrak{g}^{\mathbb{C}}$ and the roots and weights. Essentially the paper is self-contained although we use material from the adjacent paper [2] to which we refer as I ; also for formulae, sections and subsections, e.g., (I.29), I.3., I.3.2., respectively.

1. Representations and forms.

We consider (as in [1-4]) representations, called elementary representations (ER), of the conformal superalgebra $\mathfrak{g} = su(2,2/N)$ characterized by the signature

$$\chi = [j_1, j_2; d ; z ; r_1, \ldots, r_{N-1}] \quad , \tag{1}$$

where $2j_1, 2j_2 ; r_1, \ldots, r_{N-1}$ are nonnegative integers indexing the inducing representations of $sl(2,\mathbb{C})$ and $su(N)$; d and z are complex numbers indexing the representations of the dilatation subalgebra and of the $u(1)$ centre of the even subalgebra $\mathfrak{g}_0$ of $\mathfrak{g}$. For $N=4$

this u(1) is a centre of $\mathcal{G}$ itself and we may consider representations of the factoralgebra $\mathcal{G}/u(1)$; then $z=0$.

Let $\Phi = \Phi_0 + i\Phi_1$ be a superhermitian form on the Verma module $V_\chi = V_0 \oplus V_1$ associated (cf. I) with the ER χ . Superhermitian means that Φ_a is a hermitian (possibly degenerate) form on V_a ; V_0 and V_1 are orthogonal with respect to Φ :

$$\Phi(x,x') = \Phi_0(x_0,x_0') + i\Phi_1(x_1,x_1') \;, \quad x_a' \in V_a \;.$$

Let X^+ be the hermitian conjugate of a matrix $X \in \mathcal{G}^{\mathbb{C}}$. We specify the form Φ by the hermiticity condition

$$\Phi(\hat{X}_a x,y) = i^a(-1)^{a\deg x}\Phi(x,(\widehat{\beta_0 X_a^+ \beta_0})y) \tag{2a}$$

where $X_a \in \mathcal{G}_a^{\mathbb{C}}$, $a=0,1(\mathrm{mod}\ 2)$; $\hat{X}x$ denotes the module action of X on x ; $\deg x = a$ for $x \in V_a$; $\beta_0 = \mathrm{diag}(1,1,-1,-1,1,\dots,1)$; equivalently

$$\Phi_{\deg y}(\hat{X}_a x,y) = \Phi_{\deg x}(x,\widehat{\beta_0 X_a^+ \beta_0}y) \tag{2b}$$

We recall that the elements of V_χ belong to $U(\mathcal{G}_+^{\mathbb{C}})v$, where $U(\mathcal{G}_+^{\mathbb{C}})$ is the universal enveloping algebra of $\mathcal{G}_+^{\mathbb{C}}$ comprised from the positive root spaces , v is the lowest weight vector of the lowest weight module. Normalizing $\Phi(v,v) = \Phi_0(v,v) = 1$ and using (2) the norm $\Phi(x,x)$ of any state $x \in V_\chi$ is computed in a standard fashion moving the negative root space vectors $e_{-\alpha} = e_\alpha^+$ to the right with the help of the commutation relations (A.1) and the defining properties of the lowest weight vector v (I.21). For instance the "1-particle" norms are given by (cf. the Appendix):

$$\Phi_a(\hat{e}_\alpha v,\hat{e}_\alpha v) = (-i)^a\,\Phi(\hat{e}_\alpha v,\hat{e}_\alpha v) = \Phi_0(v,\widehat{\beta_0 e_{-\alpha}\beta_0}\hat{e}_\alpha v) =$$
$$= \begin{cases} -(\Lambda+\rho,\alpha) & \text{for } \alpha=\alpha_{12},\alpha_{43} \\ +(\Lambda+\rho,\alpha) & \text{otherwise} \end{cases} \quad ; \quad a = \deg e_\alpha \quad ; \quad \alpha > 0 \,. \tag{3}$$

For $\alpha > 0$ compact these are nonnegative integers (cf. I.3.2.). (For $N=4$ $\Lambda \longrightarrow \Lambda+\tilde{\Lambda}$ in (3), $\tilde{\Lambda}$ - weight of the u(1) centre of $\mathcal{G}$ (cf. (A.10) and [4]); $\tilde{\Lambda} = 0$ for representations of the factoralgebra $\mathcal{G}/u(1)$.) We give one more example which will be used in the unitarity proof below. For this we introduce ordering between the odd positive roots as follows. We first set that

$$\alpha_{1,4+N} > \alpha_{1,3+N} > \dots > \alpha_{15} > \alpha_{2,4+N} > \dots > \alpha_{25} \quad , \tag{4a}$$

$$\alpha_{53} > \alpha_{63} > \dots > \alpha_{4+N,3} > \alpha_{54} > \dots > \alpha_{4+N,4} \;. \tag{4b}$$

It follows from (4) that if $\alpha-\beta \in \Delta$ then $\alpha-\beta \in \Delta^+$ iff $\alpha > \beta$. We further choose $\alpha_{25} > \alpha_{53}$. Now we denote

$$v_k = e_{\gamma_1} \cdots e_{\gamma_k} v \; ; \; \gamma_i > 0, \text{ odd}; \; \gamma_1 > \cdots > \gamma_k \; ; \tag{5a}$$

then we have :

$$\phi_{\bar{k}}(v_k, v_k) = (-i)^{\bar{k}} \phi(v_k, v_k) = \prod_{i=1}^{k} (\Lambda + \rho + \sum_{s=1+i}^{k} \gamma_s, \gamma_i) \; , \; \bar{k} = k (\mathrm{mod}\, 2) = 0, 1; \tag{5b}$$

We note that the ordering (4) is essential for the simplicity of (5b) while the ordering between the two sets in (4a) and (4b) could be chosen also as $\alpha_{4+N,4} > \alpha_{1,N+4}$.

A (nondegenerate) superhermitian form ϕ is called positive definite if ϕ_0 is positive definite and ϕ_1 is either positive or negative definite. One can work equivalently with the hermitian forms $\phi_0 + \phi_1$ or $\phi_0 - \phi_1$, respectively, which is the usual physicists' convention.

2. Statement of the result.

Theorem. [1] (i) All unitary irreducible representations of the conformal superalgebra su(2,2/N) characterized by the signature in (1) are obtained for d and z real and are given in the following list:

$$\text{(a)} \quad d \geq d_{max} = \max(d1, d3) \quad , \quad j_1, j_2 \geq 0 \quad ; \tag{6a}$$

$$\text{(b)} \quad d = d4 \geq d1 \quad , \quad j_1 = 0 \; , \; j_2 \geq 0 \quad ; \tag{6b}$$

$$\text{(c)} \quad d = d2 \geq d3 \quad , \quad j_1 \geq 0 \; , \; j_2 = 0 \quad ; \tag{6c}$$

$$\text{(d)} \quad d = d2 = d4 \quad , \quad j_1 = j_2 = 0 \quad ; \tag{6d}$$

where

$$d1 = d^1_{N1} = 2 + 2j_2 + z + 2m_1 - 2m/N \; ,$$

$$d2 = d^2_{N1} = -2j_2 + z + 2m_1 - 2m/N \; ,$$

$$d3 = d^3_{NN} = 2 + 2j_1 - z + 2m/N \; ,$$

$$d4 = d^4_{NN} = -2j_1 - z + 2m/N \; ,$$

$$m_1 = r_1 + \ldots + r_{N-1} \; , \; m = r_1 + 2r_2 + \ldots + (N-1) r_{N-1} \; .$$

(The two signs $\geq$ in (6a) (also in (6b),(6c)) are not correlated.)

(ii) Case (d) $d = z = m_1 = m = 0$ is the trivial one-dimensional representation. In all other cases a UIR is realized as a subrepresentation of the corresponding ER.

Remarks. 1. The quantities d^a_{Nk} determine some of the conditions for the reducibility of the ER (cf. [3,4] and (I.26)). 2. In (6d) $d = m_1$, $z = 2m/N - m_1$ and thus case (d) is nontrivial only for $N \geq 2$ (since for $N=1$ $m_1 = m = 0$). 3. For $N=1$ statement (i) of the Theorem was announced in ref. [5]. 4. Excluding the one-dimensional case from conditions (6) follows :

$$d \geq d_M \equiv \begin{cases} 2 + j_1 + j_2 , & j_1 j_2 \neq 0 , \\ 1 + j_1 + j_2 , & j_1 j_2 = 0 , \end{cases} \tag{7}$$

which are the conditions for the positive energy UIR's of the conformal group $SU(2,2)$ [6]. The equality in (7) is achieved for $j_1 j_2 \neq 0$ from (6a) when $d = d_{max} = d1 = d3$ and $m_1 = 0$ and for $j_1 j_2 = 0$ from (6b) (respectively (6c)) when $d4 = d1$ (respectively $d2 = d3$) and $m_1 = 0$ and from (6d) when $m_1 = 1$ $(1 \leq m \leq N-1)$. The latter $(j_1 j_2 = 0)$ cases comprise the massless UIR's discussed in [I].

3. Proof of the Theorem.

The proof proceeds in two main steps. First we shall show that whenever conditions (6) are fulfilled the superhermitian form Φ defined above gives rise to a positive definite form on (a quotient of) V_χ with both Φ_0 and Φ_1 positive definite.

We recall that the Verma module V_χ can be reducible and the reducibility conditions were spelled out in [3,4] and (I.25,26). A reducible Verma module contains a submodule

$$I_\chi \cong U'(\mathfrak{g}_+^{\mathbb{C}})v \ , \ U'(\mathfrak{g}_+^{\mathbb{C}}) = \bigcup_i U(\mathfrak{g}_+^{\mathbb{C}})P_i \tag{8}$$

where the union is over the positive noncompact roots under which V_χ is reducible, P_i is the homogeneous polynomial in the simple root space vectors e_α such that $P_i(e_\alpha)v$ is the corresponding singular vector (cf. [4] and (I.29))). Finally the irreducible Verma module will be

$$V'_\chi = V_\chi / I_\chi \quad . \tag{9}$$

The module V'_χ inherits from V_χ the form Φ . The requirement of positive norm gives the unitarity conditions (6) as follows.

V_χ can be decomposed in $\mathfrak{m} \oplus \mathfrak{a}$ - finite-dimensional irreducible representations. Here $\mathfrak{a}$ is the 1-dimensional dilatation subalgebra and $\mathfrak{m} = sl(2,\mathbb{C}) \oplus u(1) \oplus su(N)$, (cf. I). The corresponding representation vectors provide an orthogonal basis in V_χ . The $U(\mathfrak{g}_+^{\mathbb{C}})_0$ - generated $\mathfrak{m} \oplus \mathfrak{a}$ states of the even part of V_χ reproduce conditions (7) which were seen to be implied by (6). To obtain (6) it is enough to look at the $\mathfrak{m} \oplus \mathfrak{a}$ - content generated by the action of the finite-dimensional Grassmann algebra $A = U(\mathfrak{g}_+^{\mathbb{C}})/U(\mathfrak{g}_+^{\mathbb{C}})_0$ on v [5,7]. As usually it is sufficient to check the norms of the "vacuum" vectors v' of the $\mathfrak{m} \oplus \mathfrak{a}$ - representions. These states are subject to the constraints (cf. Appendix for notation):

$$hv' = (\Lambda' + \rho)(h)v' \ , \ h \in \mathfrak{h}^{\mathbb{C}} \ , \ \Lambda' = \Lambda(\chi') \in (\mathfrak{h}^{\mathbb{C}})^* \ , \tag{10a}$$

$$e_{-\alpha} v' = 0 = (e_\alpha)^{n_\alpha} v' \quad , \; n_\alpha = -2(\Lambda',\alpha)/(\alpha,\alpha) \in \mathbb{N} \; , \qquad (10b)$$

where α is any simple compact root. It follows from the results of [4] that these states are given by $P_{i_1} \ldots P_{i_k} v$, where P_{i_j} are the polynomials producing the singular vectors (cf.(8)). However such a state, which we call quasi-singular, is not necessarily accompanied by some reducibility condition(s) from (I.26) on the signature χ' which condition(s) would arise if (10) is completed to (I.21,23) (with $\Lambda, v \rightarrow \Lambda', v'$). We recall that (I.21) means that v' is the lowest weight vector of $V_{\chi'}$ and a reducibility condition means that v' is a singular vector of a reducible V_χ in which $V_{\chi'}$ is embedded (via v'). The singular vectors were constructed in [4] and the computation of the norms of their quasisingular counterparts is straightforward (the true singular vectors having zero norms of course). The direct check shows that for $d > d_{max}$ all states have positive norm. For $d = d_{max}$ the states with zero norm (the would-be ghosts) are not contained in V'_χ (being contained in I_χ). Thus we obtain case (a).

For $d < d_{max}$ there are many states with negative norm. First we note that there is no positivity for $d < d'_{max} = \max(d2,d4)$ because $(d-d4)/2$, $(d-d2)/2$ are the norms of the "1-particle" states $e_\alpha v$ for $\alpha = \alpha_{25}$, $\alpha_{N+4,4}$, (i.e. the two odd simple roots),resp., which are absent from V'_χ only for $d=d4$, $d=d2$, resp. Further there are states with norm proportional (with positive coefficient) to $(d-d3)(d-d4)$, resp. to $(d-d1)(d-d2)$. (For example $e_{\alpha_{15}} e_{\alpha_{25}} v$, resp. $e_{\alpha_{N+4,3}} e_{\alpha_{N+4,4}} v$.) Because of this there can be no positivity in the open interval (d'_{max}, d_{max}). Thus we should try $d = d'_{max}$. However there are states with norm positively proportional to $d-d3+2 = d-d4-4j_1$ (e.g. $e_{\alpha_{15}} v$), resp. to $d-d1+2 = d-d2-4j_2$ (e.g. $e_{\alpha_{N+4,3}} v$),at least one of which would be negative for $d=d'_{max}$ if $j_1 j_2 \neq 0$. Obviously all of the above mentioned norms would be nonnegative <u>either</u> if $j_1 = 0$ and $d = d4 \geq d1$, <u>or</u> if $j_2 = 0$, and $d = d2 \geq d3$, <u>or</u> if $j_1 = j_2 = 0$ and $d = d2 = d4$. There are no more negative norm states under these conditions and again the zero norm states belong to I_χ and are factored out. Thus we obtain cases (b),(c) and (d).

The second step will be to relate the form Φ on V_χ to a $\mathfrak{g}$ - invariant form F on a subspace C'_χ of the ER space C_χ . A $\mathfrak{g}$ - invariant superhermitian form on a $\mathfrak{g}$ - representation space C'_χ is defined by the relation

$$F(T(X_a)u,u') = -(-1)^{a \deg u} F(u,T(X_a)u') \; , \; X_a \in \mathfrak{g}_a \; , \; u,u' \in C'_\chi \; . \qquad (11)$$

When extended to $X \in \mathfrak{g}^{\mathbb{C}}$ the relation (11) transforms into a definition of the hermitian conjugation of $T(X)$ analogous to (2) but with β_0 replaced by ω.

If F is positive definite with F_1 positive (resp. negative) definite a straightforward calculation shows that the superconformal Hamiltonian $\mathcal{P} = (P^0+K^0)/2 = (e_{14}+e_{23}+e_{31}+e_{42})/2$ is represented by a positive (resp. negative) operator on V_χ. (The metric $(-,+,+,+)$ in Minkowski space-time is assumed.) The same result applies to the energy operator P^0 itself.

The module V_χ was built by the right action of $U(\mathcal{G}_+^{\mathbb{C}})$ on the elements of the ER. One can define a lowest weight module of $\mathcal{G}^{\mathbb{C}}$ adapting the same abstract definition (I.21) but now identifying X with the left action $T(\mathcal{U}^{-1}X\mathcal{U})$ of $\mathcal{G}^{\mathbb{C}}$, i.e. with the ER itself. Here $\mathcal{U} = \frac{1}{\sqrt{2}}\begin{pmatrix} \mathbb{I}_2 & \mathbb{I}_2 & 0 \\ -\mathbb{I}_2 & \mathbb{I}_2 & 0 \\ 0 & 0 & \mathbb{I}_N \end{pmatrix}$ is the real orthogonal matrix relating the two realizations of $su(2,2/N)$, i.e. the matrices β_0 and ω.

In particular the new lowest weight Ω satisfies

$$T(\mathcal{P})\Omega = (\Lambda+\rho)(h_0)\Omega = d\cdot\Omega \ , \quad \mathcal{P} = \mathcal{U}^{-1}h_0\mathcal{U} \ , \ h_0 \text{ in (A.8)} \ , \tag{12a}$$

$$T(\mathcal{U}^{-1}h_{\alpha_{12}}\mathcal{U})\Omega = -2j_1\Omega \ , \ T(\mathcal{U}^{-1}h_{\alpha_{43}}\mathcal{U})\Omega = -2j_2\Omega \ .$$

These three generators belong to the maximal compact subalgebra $\mathcal{k} \subset \mathcal{G}$; they are diagonal in the standard realization of $su(2,2/N)$ using β_0. (Note that the identification above does not mean that we change our realization (I.2) of $\mathcal{G}$.) Similarly

$$T(\mathcal{U}^{-1}e_\alpha\mathcal{U})\Omega = 0 \qquad \text{for } \alpha<0 \ . \tag{12b}$$

The forms F and Φ are related according to

$$\begin{aligned} &F(\Omega,\Omega) = \Phi(v,v) = 1 \ , \\ &F(T(\mathcal{U}^{-1}X\mathcal{U})\Omega, T(\mathcal{U}^{-1}X'\mathcal{U})\Omega) = \Phi(\hat{X}v,\hat{X}'v) \ , \quad X,X' \in U(\mathcal{G}_+^{\mathbb{C}}) \ . \end{aligned} \tag{12c}$$

One easily checks that in this way (2) goes over to

$$F(T(X_a)u,u') - i^a(-1)^{adeg\,u}F(u,T(\omega X_a^+\omega)u') = 0 \ , \quad X_a \in \mathcal{G}_a^{\mathbb{C}} \ . \tag{13}$$

and thus the $\mathcal{G}$ - invariance (11) of F is recovered. Using the functional realization of the ER [4] one can easily find an explicit expression for Ω. The left-action lowest weight module realized in this way is nothing else but the $\mathcal{k}$ - induced subrepresentation in C_χ with $T(\mathcal{P})$ bounded from below. For the unitary weights listed above this is hence a positive energy ("holomorphic") subrepresentation of the ER.

It should be clear that all considerations above could be made directly in terms of F and the relevant subrepresentations. In particular the analysis of the $\mathcal{m}\oplus\mathcal{a}$ - content goes over to the $\mathcal{k}_0$ -content of the subrepresentation.

Finally we recall that whenever some reducibility condition from (I.25,26) is satisfied there arises an invariant differential operator in the ER space. To the factormodule V_χ/I_χ there corresponds a subrepresentation space of C_χ comprised from solutions of the resulting differential equation. This concludes the proof of the Theorem.

We should mention also that a negative energy ("antiholomorphic") subrepresentation of ER can be built starting from a highest weight vector Ω^- satisfying instead of (12)

$$T(\mathcal{U}^{-1}h\mathcal{U})\Omega^- = -(\Lambda+\rho)(h)\Omega^- \ , \tag{14a}$$

$$T(\mathcal{U}^{-1}e_\alpha\mathcal{U})\Omega^- = 0 \qquad \text{for } \alpha>0 \ . \tag{14b}$$

A positive definite $\mathcal{G}$- invariant form F^- with a negative definite odd part F_1^- can be defined for the same values of χ described in the Theorem.

Appendix.

We shall exploit a Cartan-Weyl basis in $\mathcal{G}^{\mathbb{C}} = sl(4/N;\mathbb{C})$ (cf. (I.1))

$$\begin{aligned} &[h,\ e_\alpha] = \alpha(h)e_\alpha \ , \quad [e_\alpha,\ e_{-\alpha}] = (e_\alpha, e_{-\alpha})h_\alpha \ , \\ &[e_\alpha,\ e_\beta] = N_{\alpha\beta}\, e_{\alpha+\beta} \ , \quad N_{\alpha\beta} = 0 \quad \text{if} \quad \alpha+\beta \notin \Delta \ , \end{aligned} \tag{A.1a}$$

where $[\ ,\ ]$ is the super-Lie bracket (cf. I), $h \in \mathcal{h}^{\mathbb{C}}$ (= the Cartan subalgebra of $\mathcal{G}^{\mathbb{C}}$), $\alpha,\beta \in \Delta$ (= the root system of $(\mathcal{G}^{\mathbb{C}},\mathcal{h}^{\mathbb{C}})$), e_α is the root space vector corresponding to the root α , $(\ ,\ )$ is the Killing form on $\mathcal{G}^{\mathbb{C}}$. Relations (A.1a) imply

$$(\alpha,\beta) \equiv \alpha(h_\beta) \equiv \beta(h_\alpha) = (h_\alpha,h_\beta) \equiv \mathrm{str}\, h_\alpha h_\beta \ . \tag{A.1b}$$

A standard choice satisfying (A.1) is provided by

$$\begin{aligned} &\alpha_{ij} = \sum_{s=i}^{j-1} \alpha_s > 0 \quad \text{for} \quad i<j \ ; \ \alpha_s \ , \ s=1,2,\dots,3+N, \text{ simple roots} \ ; \\ &e_{\alpha_{ik}} = e_{ik} \ ; \ (e_{ik})_{st} = \delta_{is}\delta_{kt} \ ; \\ &h_{\alpha_i} = \varepsilon(4-i)(e_{ii}-e_{i+1,i+1}) + \delta_{i4}(e_{44}+e_{55}) \ ; \ \varepsilon(x) = \begin{cases} \pm 1 & x \gtrless 0 \\ 0 & x=0 \end{cases} \ ; \\ &a_{ij} = (\alpha_i,\alpha_j) = (2\delta_{ij}-\delta_{i,j-1}-\delta_{i,j+1})(\varepsilon(4-j)+\delta_{j4}) - 2\delta_{ij}\delta_{j4} \ , \end{aligned} \tag{A.2}$$

which choice is described by the Dynkin diagram

```
1   2   3   4   5      3+N
o---o---o---⊗---o-...--o
```

where the white nodes o depict the even simple roots with $|(\alpha,\alpha)|=2$ and the black node ⊗ depicts the odd simple root with $(\alpha,\alpha)=0$; in addition for $N=4$ we have the relations

$$\sum_{s=1}^{3} s(h_{\alpha_s} + h_{\alpha_{8-s}}) + 4h_{\alpha_4} = \mathbb{1}_8 \ ; \quad \sum_{s=1}^{3} s(\alpha_s + \alpha_{8-s}) + 4\alpha_4 = 0$$

Our choice of a system of positive roots will differ from (A.2) (cf. I). Namely we define

$$\Delta^+ = \left\{ \beta_{ij} = \sum_{s=i}^{j-1} \beta_s ,\ i < j \right\} \tag{A.3a}$$

where the set of simple roots is provided by

$$\beta_1 = \alpha_1 \ ;\ \beta_2 = \alpha_{25} \ ;\ \beta_{2+s} = \alpha_{4+s} \ ,\ s=1,\ldots,N-1,$$

$$\beta_{2+N} = -\alpha_{4,N+4} = \alpha_{N+4,4} \ ;\ \beta_{3+N} = -\alpha_3 = \alpha_{43} \ . \tag{A.3b}$$

This implies a Dynkin diagram o—⊠—o—...—o—⊠—o and

$$\alpha_{13} = \beta_{1,N+4} \ ,\ \alpha_{14} = \beta_{1,N+3} \ ,\ \alpha_{24} = \beta_{2,N+3} \ ,\ \alpha_2 = \beta_{2,N+4} \ , \tag{A.3c}$$

$$\alpha_{a,k+4} = \beta_{a,k+2} \ ,\ -\alpha_{2+a,k+4} = \beta_{k+2,N+2+a} \ ,\ a=1,2,\ k=1,\ldots,N \ ,$$

$$\alpha_{4+k,4+s} = \beta_{2+k,2+s} \ ,\ k,s = 1,\ldots,N$$

We write down some useful formulae:

$$(\alpha_{a,4+s}, \alpha_{4+p,2+b}) = \delta_{sp} \ ;\ a,b=1,2;\ s,p=1,\ldots,N; \tag{A.4}$$

$$(\alpha_{a,4+s}, \alpha_{b,4+p}) = \delta_{ab} - \delta_{sp} \ ;\ (\alpha_{4+s,2+a}, \alpha_{4+p,2+b}) = \delta_{ab} - \delta_{sp} \ .$$

Let $2\rho = \sum_{\beta>0,\text{even}} \beta - \sum_{\beta>0,\text{odd}} \beta \in (\mathfrak{h}^{\mathbb{C}})^*$. Then we have

$$2\rho = (3-N)(\beta_1 + \beta_{3+N}) + \sum_{k=2}^{N+2} \beta_k (N-k)(k-4) =$$

$$= (3-N)\alpha_1 + 2(2-N)\alpha_2 + (1-N)\alpha_3 + \sum_{s=1}^{N-1} s(N-s)\alpha_{4+s} \ ; \tag{A.5}$$

$$(\rho, \alpha_{a,5+N-t}) = \delta_{a1} + t - N \ ;\ (\rho, \alpha_{5+N-t,2+a}) = \delta_{a1} + 1 - t. \tag{A.6}$$

The values of the weight $\Lambda \in (\mathfrak{h}^{\mathbb{C}})^*$ on the elements of the Cartan subalgebra in (A.2) are given by :

$$(\Lambda+\rho)(h_0; h_{\alpha_{12}}, h_{\alpha_{43}}; h' ; h_{\alpha_{3+N}}, \ldots, h_{\alpha_5}) =$$

$$= (d; -2j_1, -2j_2; z(1-\delta_{N4}); r_1, \ldots, r_{N-1}) \ ; \tag{A.7}$$

where h_0 is the dilatation generator

$$2h_0 = h_{\alpha_1} + 2h_{\alpha_2} + h_{\alpha_3} = \begin{pmatrix} \mathbb{1}_2 & 0 & 0 \\ 0 & -\mathbb{1}_2 & 0 \\ 0 & 0 & 0 \end{pmatrix} , \tag{A.8}$$

$$h' = \sum_{s=1}^{3} (s/2) h_{\alpha_s} + 2 \sum_{s=0}^{N-1} (1-s/N) h_{\alpha_{4+s}} = \frac{1}{2} \begin{pmatrix} \mathbb{1}_4 & 0 \\ 0 & \frac{4}{N}\mathbb{1}_N \end{pmatrix} . \tag{A.9}$$

For $N=4$ the weight $\tilde{\Lambda}$ mentioned after (3) takes the following values over the same elements of $\mathfrak{h}^{\mathbb{C}}$ $(h' = (1/2)\mathbb{1}_8)$:

$$\mathcal{K}(\ldots) = (0;0,0;z;0,\ldots,0) \quad . \tag{A.10}$$

References.

1. V.K. Dobrev and V.B. Petkova, Phys.Lett. 162B (1985) 127.
2. V.K. Dobrev and V.B. Petkova, preceding paper in these Proceed.
3. V.K. Dobrev and V.B. Petkova, Lett.Math.Phys. 9 (1985) 287.
4. V.K. Dobrev and V.B. Petkova, ICTP Trieste preprint IC/85/29(1985)
5. M. Flato and C. Fronsdal, Lett.Math.Phys. 8 (1984) 159.
6. G. Mack, Comm.Math.Phys. 55 (1977) 1.
7. W. Heidenreich, Phys.Lett. 110B (1982) 461.
   D.Z. Freedman and H. Nicolai, Nucl.Phys. B237 (1984) 342.

VI. INFINITE-DIMENSIONAL LIE ALGEBRAS

THE TWO-DIMENSIONAL QUANTUM CONFORMAL GROUP, STRINGS AND LATTICES

Yuval Ne'eman*

Sackler Faculty of Exact Sciences
Tel Aviv University, Tel Aviv, Israel+

and

University of Texas, Austin, Texas++

* Wolfson Chair Extraordinary in Theoretical Physics
+ Supported in part by the U.S.-Israel BNSF
++ Supported in part by the U.S. Departmentof Energy Grant No. DE-FG05-85ER40200 and by the Center for Theoretical Physics, University of Texas, Austin

1. Conformal Groups and the Virasoro Algebra

This conference centers on the Conformal Groups. Over any but 2-dimensional pseudo-Euclidean spaces, there are finite-dimensional pseudo-orthogonal Lie groups SO(s+1,t+1), for metrics with s space-like and t timelike dimensions. In 2-dimensions (x,y), defining a conformal transformation as an isogonal mapping which conserves the sense of an angle as well as its magnitude, it can be shown that it is equivalent to an analytical diffeomorphism $w = f(z)$, for $z = x+iy$. This is because the conditions for isogonality turn out to be just the Cauchy-Riemann differential equations[1)].

The 2-dimensional conformal group plays an important role in 2-dimensional quantum field theory[2)]. Because of the above correspondence between conformal and diffeomorphism groups, one may in fact define two analytical diffeomorphism groups Δ_2 and $\Delta_2^{\star}$ corresponding to the variables z and $\overline{z}$ in the above notation. These groups can be described algebraically, using a method due to Ogievetsky[3)]. One expands the infinitesimal variation $\delta z(z)$

$$z' = f(z) = z + \delta z \tag{1.1}$$

in powers of z, so that defining ($m \varepsilon \mathbb{Z}$)

$$L_m = -i \; z^{m+1} \frac{\partial}{\partial z} \tag{1.2}$$

we have

$$\delta z = (\sum_m c^m z^m) z = i \sum_m c^m (L_m z)$$

and the L_m form the classical infinite Ogievetsky algebra

$$[L_m, L_m] = (m-n) \; L_{m+n} \;, \quad m,n \varepsilon \mathbb{Z} \tag{1.3}$$

an algebra playing an important role in classical physics, (e.g. the Analytical Einstein Covariance group and its double-covering[4)]).

In Quantum Mechanics, the commutation relations (1.3) undergo a deformation through the insertion of a central element in the algebra. The grading L_o, a scale operator, is essential. The result is the Virasoro algebra[5)], deriving from both the equations of motion and the boundary conditions and representing the algebraic system of constraints for the Veneziano string. The spectrum is given by putting the vacuum as the highest weight

$$\begin{aligned} L_n \,| o \rangle &= o \;, \quad n \,\varepsilon\, \mathbb{Z} \;, \; n > o. \\ L_o \,| o \rangle &= \nu \,| o \rangle \end{aligned} \tag{1.4}$$

and using the lowering L_{-n} to construct the entire set of states. The

Virasoro algebra commutation relations are (d is a real number . $d \geq 1$)

$$[L_m, L_n] = (m-n)\, L_{m+n} + d\{\tfrac{1}{12}\,(m^3-m)\delta_{m,-n}\}$$
$$[L_m, d] = o \qquad (1.5)$$

and hermiticity determines that

$$L_n^+ = L_{-n} \qquad (1.6)$$

The representations are then characterized by (d,ν). In the Veneziano dual model[6,7] ν was the Regge intercept and d the dimensionality of the space. For the representation to be unitary, d=26 for a Minkowski-like metric (or 24 transverse dimensions). In the superstring of Ramond and Neveu-Schwarz and in the more restrictive version of Green-Schwarz[8], d=10. The correspondence between 2-dimensional Conformal symmetry and the Analytical Diffeomorphisms was exploited by Belavin et al[2] for the study of 2-dimensional systems in Statistical Mechanics and has produced a unifying algebraic treatment for a variety of problems in the Physics of Condensed Matter. The work was further developed by Friedan et al[9] and Goddard et al[10]. With d in (1.5) taking up a set of values within another allowed range, $o \leq d < 1$ (see e.g. (1.7)), one reproduces the critical behaviour parameters for the following problems,

d	model
1/2	Ising
7/10	tricritical Ising
4/5	three-state Potts
6/7	tri-critical three-state Potts

In these problems, the two algebras (for z and $\bar{z}$) are used, with common d. The values of $\nu(z) + \nu(\bar{z}) = \nu^{(+)}$ gives the "scaling" dimension of the field and $\nu^{(-)} = \nu(z) - \nu(\bar{z})$ gives its spin. Critical exponents are simple linear combinations of $\nu^{(+)}$ and $\nu^{(-)}$. The value of d is related to the conformal anomalies[11]. FQS[9] showed that unitarity (and a positive-definite Hilbert space scalar product) requires either $d \geq 1$ (a continuum) and $\nu \geq o$ or discrete values like the above, given by

$$d = 1 - 6/(m+1)(m+2) \quad (m \geq 1, \quad m \in \mathbb{Z})$$
$$\nu = \{[(m+2)p - (m+1)q]^2 - 1\}/4(m+1)(m+2) \qquad (1.7)$$
$$(1 \leq p \leq m \quad , \quad 1 \leq q \leq p \quad , \quad p,q \in \mathbb{Z})$$

Goddard[12] has emphasized the nature of the "group multiplication" for Δ , due to the group action (1.1)

$$z'' \otimes z'(z) = z''(z'(z)) \qquad (1.8)$$

This is thus <u>composition</u>, and Δ is non-abelian. The subalgebra with $m=\pm 1,o$ is an su(1,1) corresponding to the projective transformations

$$z' = \frac{az+b}{b^*z+a^*} \quad , \quad |a|^2 - |b|^2 = 1 \tag{1.9}$$

Similarly, there is an infinite sequence of su(1.1) algebras $\{\frac{1}{n} L_{-n} , \frac{1}{n} L_o , \frac{1}{n} L_n\}$, with

$$z' = \left(\frac{az^n+b}{b^*z^n+a^*}\right)^{1/n} \tag{1.10}$$

2. Local Currents and Affine Kac-Moody Algebras

Current algebras[13] consisted in a system of dynamical variables, the local charge-current densities ofthe SU(3) generators, or of their chiral[14] SU(3) x SU(3) extension, constrained by an angular condition [15,16] due to Lorentz-invariance. The local currents obeyed equal-time commutation relations,

$$[j_a^o(\vec{x},t) \ , \ j_b^o(\vec{x}',t)] = i \ f_{ab}{}^c \ \delta^3(\vec{x}-\vec{x}') \ j^o(\vec{x},t) \tag{2.1}$$

$$\begin{aligned}[j_a^o(\vec{x},t) \ , \ j_b^i(\vec{x}',t)] &= i \ f_{ab}{}^c \ \delta^3(\vec{x}-\vec{x}) \ \ j^i(\vec{x},t) \\ &+ i \ C \ \partial_i\{\delta^3(\vec{x}-\vec{x}')\} \qquad \text{etc.}\end{aligned} \tag{2.2}$$

The C "Schwinger" term in (2.2) is of the same nature as the central term in (1.5). The representations were classified[17]. H. Sugawara and C. M. Sommerfield constructed a candidate dynamical theory[18], including the commutation relations between the components of the local current densities and the components of the $\Theta^{\mu\nu}$ energy-momentum tensor. They constructed the Hamiltonian density as a bilinear in the currents

$$\begin{aligned}\Theta_{\mu\nu}(x) &= A\{j_\mu^a(x) \ j_\nu^a(x) + j_\nu^a(x) \ j_\mu^a(x)\} \\ &+ B \ g_{\mu\nu} \ \{j_\rho^a(x) \ j_\rho^a(x)\} \ + \ \dots\end{aligned} \tag{2.3}$$

$$[j_a^o(\vec{k},t), \ j_b^o(\vec{k}',t)] = i \ f_{ab}{}^c \ \ j_c^o(\vec{k}+\vec{k}',t) \tag{2.4}$$

with

$$j_a^o(\vec{k},t) = \int d^3x \ e^{i\vec{k}\vec{x}} \ \ j_a^o(\vec{x},t) \tag{2.5}$$

and a simple representation is given by

$$j_a^o(\vec{k},t) \sim \lambda_a(t)e^{i\vec{k}} \tag{2.6}$$

(λ_a an $su(3)$ matrix),

The mathematical structure of these algebras was described in Ref.19. Clearly, this was an infinite Lie algebra[20] with a close resemblence to finite Lie algebras. The mathematicians V. G. Kac and R. Moody[21] independently took up the simplest class of infinite Lie algebras, that of Affine infinite-dimensional algebras, admitting a $\mathbb{Z}$ grading, such as the one provided by "$-L_0$" in (1.5)

$$[-L_0,L_m] = m \ L_m \quad , \quad m \in \mathbb{Z} \tag{2.7}$$

i.e. through a scale-operator (from (1.2))

$$D = -L_0 = i \ z \ \frac{\partial}{\partial z} \tag{2.8}$$

and with "manageable" growth in the dimensionality of the graded subspaces

$$\dim\{V_{m+1}\}/\dim\{V_m\} \sim m^{g-1} , \quad g \le 2 \tag{2.9}$$

These are the Affine Kac-Moody algebras, and they resemble (2.4) very much . The Kac-Moody group K is generated by taking smooth maps from the circle S' into a simple finite Lie group G:

$$K : S' \to G \tag{2.10}$$

This is precisely (2.6), as

$$S' : z \in \mathbb{C} , |z| = 1 \tag{2.11}$$

i.e. $z = e^{i\phi} , \quad 0 \le \phi \le 2\pi$

$$z \to g(z) \in G \tag{2.12}$$

the multiplication rule here is pointwise multiplication,

$$g_1 \otimes g_2(z) = g_1(z) \cdot g_2(z) \tag{2.13}$$

K is also known as the "loop group" of G. Given the algebra of G,

$$g = \exp\{-i \, \theta^a(z) \, T_a\} \tag{2.14}$$

$$[T_a, T_b] = i \, f_{ab}{}^{c} \, T_c \tag{2.15}$$

write

$$\begin{aligned} g &= 1 - i \, \theta^a(z) \, T_a \\ &= 1 - i \sum_n (\theta^a)^n \, z^{-n} \, T_a \quad n \in \mathbb{Z} \\ &= 1 - i \sum_n (\theta^a)^n \, T_{a,-n} \end{aligned} \tag{2.16}$$

$$[T^a_m, T^b_n] = i \, f^{ab}{}_c \, T^c_{m+n} \tag{2.17}$$

and for hermitian T^a we find the unitarity condition

$$T^{a+}_n = T^a_{-n} \tag{2.18}$$

Quantizing from Poisson to Lie brackets

$$\begin{aligned} [T^a_m, T^b_n] &= i(\hbar) \, f^{ab}{}_c \, T^c + O(\hbar^2) \\ [T^a_m, T^b_n] &= i \, f^{ab}{}_c \, T^c_{m+n} + k \, m \, \delta_{m,-n} \, \delta^{ab} \end{aligned} \tag{2.19}$$

The Sugawara model[18] yields[12]

$$[L_m, T^a_n] = -n \, T^a_{M=n} \tag{2.20}$$

Ramond and Schwarz[22] had tried to classify all possible dual model gauge algebras. Their classification assigned the odd (i.e. fermionic) generators to non-trivial representations of G. This allowed only U(1) and SU(2), but for definitely unphysical values of d. However, once the odd generators were assumed to behave trivially under G, as was done in the new superstring of Green and Schwarz, it seemed possible to take an arbitrary group G. Still, the existence of anomalies interferes with renormalizability both in the (Yang-Mills generated) currents of (2.4) type, and in the energy-momentum "currents" (2.3). There are chiral, conformal and mixed anomalies. However, it was recently noticed by Green and Schwarz[23] that all anomalies cancel for G = $SO(32)/_{\mathbb{Z}(2)}$. J. Thierry-Mieg[24] showed that this is also true for G = $E_8 \times E_8$. The cancellations also occur in the "low-energy" field theory approximation. We shall see that the Cartan subalgebra of the above two candidate G make up the only two even unimodular lattices in 16 (Euclidean) dimensions (or 18 Minkowskian). Green and Schwarz further showed[25] that for these G, the perturbation expansion of the relevant superstring ("type I") is finite. The original method of introducing the group G in a string theory was based on attaching the internal quantum numbers to the extremities of "open" strings. Considerations at the level of "tree" diagrams[26,27] constrain G to the sets of $SO(n,\mathbb{R})$, USp(2n) and U(n). The latter are, however, unsuitable due to the appearance of anomalies in loop calculations. This classification thus does not allow the exceptional algebras, including E_8.

It was only through the development by Frenkel and Kac[28] of a construction for the unitary infinite representations of the K of (2.9), using string operator techniques[5,6,29], that it became possible to use the even-orthogonal Lie algebras D_r (generating $SO(2r,\mathbb{R})$, the traceless A_r (generating the SU(r+1) etc.) and the exceptionals of the E_r family (r=6,7,8) for G. A rather promising physical model has now been developed[30,31], using either $SO(32)/_{\mathbb{Z}(2)}$ or $E_8 \times E_8$. The latter group has great advantages in its fit with the phenomenology and has therefore attracted greater attention[32].

Several texts on Kac-Moody algebras have appeared in recent years[33,34].

3. Integral Lattices

In the above construction, the theory of Integral Lattices plays an essential role.

A lattice is a periodic array spanning the entire d-dimensional space. "Cubic" lattices, in any number of dimensions d, are like abelian Lie groups in Lie group theory. Lattices can be decomposed and the cubic arrays can be segregated. They are denoted $\mathbb{Z}^d$.

The simplest non-trivial integral lattice is the lattice of roots of the algebra A_2 (generating SU(3), SL(3,$\mathbb{R}$), SL(3,$\mathbb{C}$), etc.) The convention is that the length of the root vectors (six in this example) is normalized to $(p^i)^2=2$. We give an enumeration of root lengths: $u_o=1$ (this is the root at the origin; as a lattice, we count the origin only once, not as we do in counting the algebraic roots), $u_1=o$ (no odd-length vectors) $u_2=6$. For the A_n in general, $u_2=1$, $u_2=n(n+1)$.

The Dynkin diagram for the A_n is a straight line segment connecting n dots. Each dot represents one fundamental root, spanning the dimensionality of the Cartan subalgebra. In A_2, the fundamental roots are the I-spin raising and U-spin raising operators, with an angle of $2\pi/3$ between them. In the Dynkin diagram, the n dots of A_n are all marked "$p^2=2$", i.e. they all represent vectors with equal norms. The lattice has a determinant, that of the Cartan matrix of scalar products between fundamental roots. For A_2 as $\cos 2\pi/3 = \sin \pi/6$, the matrix has "2" in the diagonal and "-1" for the off-diagonal element. Thus the determinant will be equal to 3. For the A_n in general, all the angles between consecutive fundamental vectors are again $2\pi/3$. This is coded into the Dynkin diagram by the use of a single line to connect the dots. In other algebras, with other angles, one has double lines for $3\pi/4$ in the B_n, C_n and F_4, or a triple line for $5\pi/6$ in G_2. Between fundamental root vectors corresponding to unconnected dots, the angle is $\pi/2$ and the matrix gets no contribution.

Integral lattices Λ use only "single-laced" Dynkin diagrams, i.e. angles of $2\pi/3$. Thus only the A_n, D_n and E_6, E_7, E_8 contribute and represent the set of "component lattices" until we reach 24 dimensions. For the D_n (generating SO(2,n), etc.) det $D_n=4$, for all n. For E_6, E_7, E_8 the determinants are respectively 3,2,1. We thus learn that E_8 is a unimodular lattice. The determinant, incidentally, gives the density of lattice points per unit volume. The A_n, D_n and $E_{6,7,8}$ are the only component lattices (in any number of dimensions) generated by vectors of norm 2.

A lattice's dual Λ^* is another lattice, generated by all vectors whose scalar products with the component lattices' root vectors have an absolute value of one or zero. For A_2, this means adjoining the vectors for quarks and antiquarks. The new lattice A_2^* is in fact the lattice of all representations of A_2, i.e. all states in all representations lie on A_2^*. The same will be true of all A_n^*, D_n^* and E_n^*, except that we know from Lie algebra theory that all representations of E_8 are generated by the adjoint, so that E_8 is self-dual. One is interested[35] in unimodular lattices, and in particular in even ones (i.e. such that have $u_{2r+1}=o$, $r=o,1,\ldots$). Even unimodular lattices exist only (for Euclidean spaces) in d = 8k, k = 1,2,... For Minkowski type spaces[36] with m time-like dimensions out of a total of d, the signature is $s = d-2n$, and has to be a multiple of 8 to allow even unimodular lattices,

$$d = 8k + 2n \tag{3.1}$$

I have noted that this is precisely the condition <u>for the spinors in that space to allow both Weyl and Majorana conditions</u>[37]. These result from the dimensionalities of Clifford algebras for the various metrics.

Returning to the even unimodular lattices, we observe that there is one such lattice in k=1 (this is E_8), two in k=2, namely $E_8 \times E_8$ and D_{16} (generating $SO(32)/\mathbb{Z}_2$). In k=3, there are 24 such lattices. Of these, 23 correspond to various semi-simple or other direct product Lie algebras, and one, the Leech lattice[38] Λ_L, the only lattice up to 24 dimensions whose shortest vector has norm $p^2=4$. Note that in 32 dimensions there are more than 10^8 unimodular lattices.

The number u_{p^2} of vectors for each norm in the Leech lattice is given by the formula[39] (here $p^2=2\nu$)

$$u_{2\nu}(\Lambda_L) = \frac{65{,}520}{691}\,(\sigma_{11}(\nu) - \tau(\nu)) \tag{3.2}$$

$\sigma_{11}(\nu)$ is the sum of the eleventh powers of the divisors of ν, and $\tau(\nu)$ is Ramanujan's function. This yields

$$u_4 = 196{,}560 \;;\; u_6 = 16{,}773{,}120 \;;\; u_8 = 398{,}034{,}000 \tag{3.3}$$

Conway[40] defined the group "$\cdot 0$" (of order 8,315,553,613,086,720,000) of all Euclidean congruences (fixing the origin) of Λ_L. This order is derived as a product of

$$|\cdot 0| = u_4 \times 93{,}150 \times 2^{10} \times |M_{22}| \tag{3.4}$$

where $|M_{22}|$ is the order of that Mathieu group, one of the (now known to be) 26 "sporadic" simple finite groups. By taking the quotient of ·0 by its center {1,-1} (the reflection-parity), Conway got yet another simple "sporadic" group,

$$"\cdot 1" = "\cdot 0"/\mathbb{Z}(2) \tag{3.5}$$

Using "·1" and the properties of the Leech lattice, Griess and Fischer discovered[41] the "Monster" or "Friendly Giant" F_1, the last and largest of the "sporadic" simple finite groups, of order $\sim 10^{55}$. It contains 20 of the sporadics as quotients of F_1, by some subgroups ("the Happy Family"); 5 are clearly not contained (the "Pariahs") and for one (J_1, of order 175,560, the "wicked dwarf") it is still not known whether or not it is contained in F_1. The construction of F_1 has already been used to suggest a new superstring model in Physics[42] following a general "encouragement" in Ref.24.

By taking the quotient of the Leech lattice by its double $M = \Lambda L/_{2\Lambda_L}$ one defines a 24-dimensional vector space over a field of characteristic 2, F_2. For any prime p, there are "extraspecial" groups[43] p of order p^{2n+1}, with p^{2n} linear characters and (p-1) faithful irreducible characters of degree p^n, one for each primitive p^{th} root of unity. Here we use the extraspecial group Q, of order 2^{24+1}. It thus has an irreducible representation V_Q of order 2^{12} (note that this is the dimensionality of a Weyl-Majorana real spinor in 26-dimensional Minkowski space). A finite group C is constructed out of Q and (·1). The Friendly Giant F_1 is defined as a group containing an involution σ, with C as centralizer (i.e. the elements commuting with σ),

$$F_1 = \langle C, \sigma \rangle \tag{3.6}$$

F_1 and C have a representation of dimension 196,884, denoted B. This module B is endowed with the structure of a commutative non-associative algebra,

$$b_1 \oplus_B b_2 \to b_3 \quad , \quad \forall \quad b_i \in B \tag{3.7}$$

with a symmetric nondegenerate associative bilinear form. The product $\oplus_B$ is preserved by σ. In fact F_1 can be defined as $F_1 = \mathrm{Aut}(B, \oplus_B)$.

Note the split (in fact, B can be constructed as the union) of

$$\dim B = 196{,}884 = 24 \oplus 98{,}280 \oplus 98{,}280 \oplus 300 \tag{3.8}$$

The sum of the first two subspaces (the conventional notation we put in between quotation marks) "B_1"= $V_1 \oplus V_2$ is 24 x 2^{12}. The sum of the second and third dim $(V_2 \oplus V_3) = u_4(\Lambda_L)$. They are linked by a parity-like morphism exchanging "B_1"$=V_1$ with "B_2"$=V_3$. The last subspace V_4, dim $V_4 = 24(24+1)/_2$, corresponds to a symmetric SO(24,$\mathbb{R}$) tensor, or to the number of transverse (i.e. physical) components of a symmetric (massless) tensor (like $g_{\mu\nu}$) in 26 Minkowskian dimensions; it can in fact be split into $299\oplus 1$, by extraction of the trace, leaving dim "B_*"= 299. Note also that ifwe remove that singlet from (3.8) we find dim $\tilde{B}$ = 196.883 = 47 x 59 x 71, the product of three primes.

There are various ways of constructing Λ_2, either from the other 24-dimensional lattices (the Niemeier lattices)[44], or by combining various products[45] of E_8.

Returning to Kac-Moody algebras (2.19), we can now characterize the role of the integral lattice generating it, i.e. the lattice spanned by the root diagram of the generating Lie algebra (2.15). Frankel and Kac[28] used "vertex operators" (p^μ is the momentum in d-dimensional space, z is a Mandelstam-type invariant energy variable)

$$\begin{aligned}\tilde{V}(p,z) = z^{p/2} \exp(p^\mu \sum_{n>o} \frac{1}{n} \alpha^\mu_{-n} z^n)\, z^{p^\mu \alpha^\mu(o)} \times \\ \times \exp(-p^\mu \sum_{n>o} \frac{1}{n} \alpha^\mu_n z^{-n})\, e^{ip^\mu q^\mu}\end{aligned} \tag{3.9}$$

In the adaptation to Kac-Moody algebras, <u>the p^μ are identified with the lattice root vectors</u>. This means that <u>the dimensionality of the momenta is that of the lattice space</u>! Only the Cartan subalgebra $H(G)\subset G$ plays a role in the affinization of G into K. Moreover, the representation of the affinization of H,

$$K_H \; : \; S' \to H \tag{3.10}$$

already spans the entire K. For strings, the dimensionality of the embedding space is thus the dimensionality of the Cartan subalgebra, or of the lattice.

4. Physical Applications

The picture in which space-time is embedded in the Cartan subalgebra of a simple group was already present in Extended Supergravity. Cremmer Julia and the late Joel Scherk[46] constructed N=8 Supergravity by working out N=1 Supergravity in D=11 dimensions. They were surprized to discover that the theory possessed a symmetry under a non-compact form of E_7. When in the D=11 manifold we assume that space-time is 3- or 5- dimensional, the internal symmetry becomes E_8 or E_6 correspondingly. Julia and Thierry-Mieg[47] have checked that this is true for any space-time dimension $d \leq D$.

D=11 Supergravity has $SL(11,\mathbb{R})$ as covariance group, i.e. the Lie algebra is A_{10}. When we reduce space-time to d dimensions, its covariance involves $A_{d-1} \subset A_{10}$, leaving the possibility of an internal symmetry with rank 10-d+1.

Turning to superstrings, we have recently seen the exploitation of the anomaly-cancelling qualities of Γ_{16} (the $SO(32)/\mathbb{Z}(2)$ Cartan subalgebra) and $\Gamma_8 \times \Gamma_8$ (i.e. $E_8 \times E_8$) in the Princeton "heterotic" model[30,31]. It is constructed by working on the light-cone, and segregating the right-movers from the left-movers, treating the two sets of fields asymmetrically. The right-movers make a ten-dimensional superstring, with eight transverse bosonic $X^i(\tau-\sigma)$ and eight Majorana-Weyl $S^a(\tau-\sigma)$, with i=1.. a=1..8. The left-moving sector is bosonic and 26-dimensional, with eight transverse $\tilde{X}^i(\tau+\sigma)$ and sixteen "internal" $\tilde{X}^I(\tau+\sigma)$, I = 1,..,16,

$$\tilde{X}^I(\tau+\sigma) = x^I + p^I(\tau+\sigma) + \frac{i}{2}\sum_{n\neq 0} \frac{1}{n}\tilde{\alpha}^I_n e^{-2in(\tau+\sigma)}$$
$$[\tilde{\alpha}^I_n,\tilde{\alpha}^J_m] = n\,\delta_{m+n,o}\,\delta^{IJ} , \quad [x^I,p^J] = \frac{1}{2}\, i\, \delta^{IJ} \tag{4.1}$$

(the factor $\frac{1}{2}$ is due to the dependence on $(\tau+\sigma)$ only)
$\tilde{X}^I$ is made to parametrize a compact space, a torus. There are stable topological configurations where (R is a radius)

$$X(\sigma) = x + 2\,\alpha' p\,\tau + 2\,NR\sigma + .. \tag{4.2}$$

winds N times around the manifold as σ runs from o to π. The momenta p^I are quantized in R^{-1} units. The torus T is "maximal", a product of circles of equal radii $R = (\alpha')^{\frac{1}{2}} = \sqrt{\tfrac{1}{2}}$, identifying points according to

$$X^I \equiv x^I + \sqrt{2}\,\pi R \sum_{i=1} e^I_i\, n_i \tag{4.3}$$

where the e^I_i are the sixteen fundamental root vectors of an $E_8 \times E_8$ lattice, with norm

$$(e_i^I)^2 = 2 \tag{4.4}$$

This implies that

$$g_{ij} = \sum_{I=1}^{16} e_i^I e_j^I \quad , \quad \det g = 1 \tag{4.5}$$

(although a more "natural" identification would be to regard the e_i^I as inverse "terads"). The momenta are

$$p^I = \sum_{I=1}^{16} n_i e_i^I \quad (n_i \text{ integer}) \tag{4.6}$$

i.e. the momenta span the lattice vectors. They must equal the winding numbers N^I. The string mass operator is

$$\frac{1}{2}\alpha' M^2 = N + (\tilde{N}-1) + \frac{1}{2}\sum_I (p^I)^2 \tag{4.7}$$

N counts the right movers, $\tilde{N}$ the left movers, i.e.

$$N = p^+ \alpha_o^- \quad , \quad \tilde{N} = p^+ : \alpha_o^- :$$

The subtracted unit comes from normal ordering considerations and Lorentz invariance. In addition, one has to constrain

$$N = \tilde{N} - 1 + \frac{1}{2}\sum_I (p^I)^2 \tag{4.8}$$

implying that the unitary operator shifting σ to $\sigma+\Delta$ in X^i, $\tilde{X}^i$, $\tilde{X}^I$ does not affect physical states in the spectrum. This operator is $\exp 2 i \Delta[N-\tilde{N}+1-\frac{1}{2}\sum(p^I)^2]$. The constraint removes the left-mover tachyon from the physical Hilbert space.

The states $|i \text{ or } a>_R \times \tilde{\alpha}^j_{-1}|o>_L$ span N=1,D=10 supergravity. The states $|i \text{ or } a>_R \times \alpha^I_{-1}|\, o>_L$ and $|i \text{ or } a>_R \times |p^I>_L$ reproduce the N=1, D=10 super-Yang-Mills multiplet, with gauge group $E_8 \times E_8$ (or SO(32)). As the lattice has u_o=16, there are 16 neutral vector mesons (plus their supersymmetric partners), corresponding to a $U(1)^{16}$ isometry of the torus. The u_2= 480 other root vectors complete $E_8 \times E_8$. The model including interacting strings, is Lorentz and $E_8 \times E_8$ invariant. The one-loop diagrams are unitary (due to the self-dual lattice) and finite. The hexagonal anomalies are cancelled.

Chapline[42] has suggested a model in which the gravitational and Yang-Mills pieces of the superstring are constrained by the action of a finite group. The string is based on the Leech lattice Λ_L. Here the mass operator is (I=1...24), the transverse dimensions)

$$M^2 = \sum_{n>o} \alpha^I_{-n}\alpha^I_n - 1 + \frac{1}{2} \sum_I (p^I)^2 \tag{4.9}$$

with $u_{2\nu}$ as in (3.2) and (3.3). We see that for M^2=o we have the 24 states generated by α_{-1}^{I}. For M^2=1, we get contributions from

α_{-2}^{I} , 24 states

$\alpha_{-1}^{I}\,\alpha_{-1}^{J}$, 30 states (or 299+1)

$\frac{1}{2}(p^I)^2$, i.e. the 196.560 states for u_4 momentum vectors,

altogether 196.884 states, thus spanning the algebra B of (3.8). Supersymmetry is generated by the morphism exchanging V_2 and V_3. The B_1 structure fits a 26-dimensional gravitino. F_1 and σ of (3.6) constrain the entire system.

One cannot use compactification to a torus, as this will leave no possibility of having two "ordinary" space-time curved transverse dimensions.

The details of the model have not been worked out to date.

References

1. See, for example, E. T. Copson, Theory of Functions of a Complex Variable, Oxford Un. Press, pp. 180-201, London (1935), and references cited therein.
2. A. A. Belavin, A. M. Polyakov and A. B. Zamolodchikov, Nucl. Phys. B241, 333 (1984).
3. V. I. Ogievetsky, Lett. Nuovo Cim. 8, 988 (1973).
4. Y. Ne'eman, Ann. Inst. Henri Poincaré 28, 369 (1978); Ann. N.Y. Academy of Sciences, to be pub.
   Y. Ne'eman and Dj. Šijački, Phys. Lett. 157B, 275 (1985).
5. M. Virasoro, Phys. Rev. D1, 2933 (1969).
6. G. Veneziano, Nuovo Cim. 57A, 190 (1968).
7. R. Dolan, D. Horn and C. Schmid, Phys. Rev. 166, 1768 (1968).
8. P. Ramond, Phys. Rev. D3, 2415 (1971).
   A. Neveu and J. N. Schwarz, Nucl. Phys. B31, 86 (1971).
   M. B. Green and J. H. Schwarz, Nucl. Phys. B181, 502 (1981).
9. D. Friedan, Z. Qiu and S. Shenker, Phys. Rev. Lett. 52, 1575 (1984); same authors in Ref. 34.
10. P. Goddard, A. Kent and D. Olive, Phys. Lett. 152B, 88 (1985).
    P. Goddard and D. Olive, in Vertex Operators in Math. and Phys., J. Lepowsky, S. Mandelstam and I. M. Singer editors, MSRI Publications #3, Springer Verlag, New York pp. 51-96 (1985).
11. E. R. Huggins, Caltech thesis;
    C. G. Callan, S. Coleman and R. Jackiw, Ann. of Phys. (N.Y) 59, 42 (1970).
12. P. Goddard and D. Olive, in Ref. 34.
    P. Goddard, DAMTP 85/7, to be pub. in Proc. Srni 1985 School, Un. of Prague.
13. M. Gell-Mann, Phys. Rev. 125, 1067 (1962).
14. M. Gell-Mann and Y. Ne'eman, Ann. Phys. (N.Y) 30. 360 (1964).
15. R. Dashen and M. Gell-Mann, Phys. Rev. Lett. 17, 340 (1966).
16. M. Gell-Mann, D. Horn and J. Weyers, Proc. 1967 Heidelberg Intern. Conf. on Elem. Particles, North Holland Pub., Amsterdam, p.479 (1968).
17. S. Chang, R. Dashen and L. O'Raifertaigh, Phys. Rev. 182, 1805, 1819 (1969).
    A. Joseph, Comm. Math. Phys. 19, 106 (1970).
18. H. Sugawara, Phys. Rev. 170, 1659 (1968).
    C. M. Sommerfield, Phys. Rev. 176, 2019 (1968).
19. L. Ehrenpreis, B. Kostant, Y. Ne'eman and S. Sternberg, CNRS, RCP25, Strasbourg, 13 (1971).

20. I. I. Singer and S. Sternberg, J.Analyse Math. 15, 1-114 (1965).
21. V. G. Kac, Matt. USSR-Izv. 2, 1271 (1968).
R. V. Moody, Jour. of Algebra 10, 211 (1968).
22. P. Ramond and J. H. Schwarz, Phys. Lett. 64B, 75 (1976).
23. M. B. Green and J. H. Schwarz, Phys. Lett. 149B, 117 (1984).
24. J. Thierry-Mieg, Phys. Lett. 156B, 199 (1985).
25. M. Green and J. H. Schwarz, Phys. Lett. 151B, 21 (1985).
26. J. H. Schwarz, Phys. Reports 89, 223 (1982).
27. N. Marcus and A. Sagnotti, Phys. Lett. 119B, 97 (1982).
28. I. B. Frenkel and V. G. Kac, Inventiones Math. 62, 23 (1980).
I. B. Frenkel, Jour. Func. Analysis 44, 259 (1981).
G. Segal, Comm. Math. Phys. 80, 301 (1982).
29. See any review on strings.
30. D. J. Gross, J. A. Harvey, E. Martinec and R. Rohm, Phys. Rev. Lett. 54, 502 (1985).
31. P. Goddard, D. I. Olive and A. Schwimmer, Phys. Lett. 157B, 393 (1985).
32. P. Candelas, G. T. Horowitz, A. Strominger and E. Witten, Nucl. Phys. B. in press.
33. V. G. Kac, Infinite Dimensional Lie Algebras, Birkhauser Boston (1983).
34. J. Lepowsky, S. Mandelstam and I. I. Singer, editors, Vertex Operators in Mathematics and Physics, (MSRI publication No. 3), Springer Verlag 1984.
35. J. H. Conway and N. J. A. Sloane, Jour. of Number Theory 15, 83 (1982).
36. See, for example, J. H. Conway, Jour. of Algebra 80, 159 (1983) and references cited therein.
37. See, for example, M. F. Sohnius in Supersymmetry and Supergravity 84, B. deWitt et al eds., World Scientific Pub., Sinagpore, pp. 3-12 (1984).
38. J. Leech, Can. Jour. of Math. 19, 251 (1967).
39. J. H. Conway, Inventiones Math. 7, 137 (1969).
40. J. H. Conway, Bull. London Math. Soc. 1, 79 (1969).
41. See the reviews: (a) R. L. Griess Jr. Inventiones Math. 69, 1 (1982); (b) J. Tits, Seminaire Bourbaki 36e année, No. 620, 105 (1983/84).
42. G. Chapline, Phys. Letters 158B, 393 (1985).
43. D. Gorenstein, An Introduction to the Classification of Finite Simple Groups, Plenum Press, N.Y. (1982).

44. J. H. Conway and N. J. A. Sloane, Proc. R. Soc. Lonodn A381, 275 (1982).

45. J. Lepowsky and A. Meurman, Jour. of Algebra 77, 484 (1982).

46. E. Cremmer, B. Julia and J. Scherk, Phys. Lett. 76B, 409 (1978). E. Cremmer and B. Julia, Nucl. Phys. B159, 141 (1979).

47. B. Julia in Superspace and Supergravity, S. W. Hawking and M. Roček eds., Cambridge Un. Press, pp. 331-350, Cambridge (1981). J. Thierry-Mieg and B. Morel in Superspace and Supergravity, S. W. Hawking and M. Roček eds., Cambridge Un. Press, pp. 351-362, Cambridge (1981).

FINITE-SIZE SCALING AND IRREDUCIBLE REPRESENTATIONS OF VIRASORO ALGEBRAS

V. Rittenberg
Physikalisches Institut der Universität Bonn
D-5300 Bonn 1
West Germany

1. INTRODUCTION

It is the aim of this talk to show the connection between the spectra of one-dimensional quantum chains at the critical point with different boundary conditions and certain irreducible representations of Virasoro algebras. We confine ourselves to the finite-size scaling limit (to be defined later) of the spectra although in the last Section we will also consider correction terms. One-dimensional quantum chains are related to the transfer matrix of two-dimensional spin systems[1,2] and are relevant for the understanding of critical phenomena in statistical mechanics.

In order to illustrate the problem, I will take an example which is the three-states Potts model which is simple enough in order to be handled with the available numerical methods and has a very rich structure.

The model is defined by the Hamiltonian

$$H=-\frac{2}{3\sqrt{3}}\sum_{i=1}^{N}\left[\sigma_i+\sigma_i^{+}+\lambda\left(\Gamma_i\Gamma_{i+1}^{+}+\Gamma_i^{+}\Gamma_{i+1}\right)\right] \tag{1.1}$$

where λ is the inverse of the temperature and N represents the number of sites, σ and Γ are the matrices

$$\sigma=\begin{pmatrix}1&0&0\\0&\omega&0\\0&0&\omega^2\end{pmatrix},\quad \Gamma=\begin{pmatrix}0&0&1\\1&0&0\\0&1&0\end{pmatrix} \tag{1.2}$$

and $\omega = e^{\frac{2\pi i}{3}}$. The Hamiltonian has a critical point at $\lambda=1$. We specify the boundary

$$\Gamma_{N+1}=\omega^{\tilde{Q}}\Gamma_1\ ,\ (\tilde{Q}=0,1,2) \tag{1.3}$$

and get the Hamiltonians $H^{(\tilde{Q})}$. If we take $\Gamma_{N+1}=0$ (free boundary conditions) we have $H^{(F)}$. The overall factor $2/(3\sqrt{3})$ in the Hamiltonian (1.1) (which fixes the euclidean time scale in a conformal theory) is taken from Ref. [3].

Since the Hamiltonian (1.1) is Z_3 invariant, each of the matrices $H^{(\tilde{Q})}$ and $H^{(F)}$ has a block-diagonal form $H_Q^{(\tilde{Q})}(H_Q^{(F)})$ corresponding to the charge sector Q=0,1 and 2 of $H^{(\tilde{Q})}(H^{(F)})$. At $\lambda=1$ self-duality[2] and the invariance under charge conjugation of the Hamiltonian (1.1) give the following relations among the spectra:

$$H_Q^{(\tilde{Q})} = H_{\tilde{Q}}^{(Q)}, \quad H_1^{(\tilde{Q})} = H_2^{(\tilde{Q})}, \quad H_1^{(F)} = H_2^{(F)} \tag{1.4}$$

and we are thus left with five independent spectra:

$$H_0^{(0)}, \quad H_1^{(0)}, \quad H_1^{(1)}, \quad H_0^{(F)}, \quad \text{and} \quad H_1^{(F)} \tag{1.5}$$

In the case of periodical ($H_0^{(0)}$, $H_1^{(0)}$) and twisted ($H_1^{(1)}$) boundary conditions we can further prediagonalize the spectra using translational invariance and we will denote by $E_Q^{(\tilde{Q})}(P)$ the energy levels corresponding to the momentum P(P=0,1,2...). For free boundary condition one can use the invariance under parity of the Hamiltonian $H^{(F)}$ to prediagonalize the spectra and we denote the levels by $E_Q^{(F)}(+)$ ($E_Q^{(F)}(-)$) corresponding to the positive (negative) parity states. We denote by $E^{(P)}$ the ground-state energy of $H_0^{(0)}$ and $E^{(F)}$ the ground-state energy of $H_0^{(F)}$.

We now consider the following quantities which are relevant for finite-size scaling:

$$\mathcal{E}_Q^{(\tilde{Q})}(P) = \lim_{N\to\infty} \frac{N}{2\pi}\left(E_Q^{(\tilde{Q})}(P) - E^{(P)}\right) \tag{1.6a}$$

$$\mathcal{E}_Q^{(F)}(\pm) = \lim_{N\to\infty} \frac{N}{\pi}\left(E_Q^{(F)}(\pm) - E^{(F)}\right) \tag{1.6b}$$

These quantities can be evaluated numerically diagonalizing the Hamiltonian for different number sites N and using Van den Broeck-Schwartz approximants[4] to evaluate the limits in Eqs.(1.6a,b). Notice the appearance of a factor 2 in Eq. (1.6a) and its absence in Eq. (1.6b). This is no misprint and the ground for this mismatch will be explained later.

Other quantities of interest are the large N behaviour of the ground-state energies pro site:

$$\frac{E^{(P)}}{N} = -A_0 - \frac{A_2^{(P)}}{N^2} - \dots \qquad (1.7a)$$

$$\frac{E^{(F)}}{N} = -A_0 - \frac{A_1^{(F)}}{N} - \frac{A_2^{(F)}}{N^2} - \dots \qquad (1.7b)$$

The numbers $A_2^{(P)}$, $A_1^{(F)}$, and $A_2^{(F)}$ can be estimated numerically. A_0 is known exactly[5]:

$$A_0 = \frac{8}{9\sqrt{3}} + \frac{4}{3\pi}$$

As will be shown in the next two Sections, the quantities $\mathcal{E}_Q^{(\tilde{Q})}$, $\mathcal{E}_Q^{(F)}$, $A_2^{(P)}$, and $A_2^{(F)}$ can be understood algebraically.

2. CORRELATION FUNCTIONS ON A STRIP (PERIODIC AND TWISTED BOUNDARY CONDITIONS)

In a conformal invariant theory in two dimensions, to each primary field $\varphi(X,Y)$ one associates two numbers Δ, $\bar{\Delta}$ and the two-point correlation function is completely determined[6]:

$$\langle \varphi(z_1,\bar{z}_1)\,\varphi(z_2,\bar{z}_2)\rangle = \frac{1}{(z_1-z_2)^{2\Delta}\,(\bar{z}_1-\bar{z}_2)^{2\bar{\Delta}}} \qquad (2.1)$$

where $z=X+iY$, $\bar{z}=X-iY$. The quantities $x=\Delta+\bar{\Delta}$, $s=\Delta-\bar{\Delta}$ are called the scaling dimension and the spin of the operator φ. Note that the right hand side of Eq. (2.1) fixes the normalization of the field φ. The three-point function is also fixed by conformal invariance[6]:

$$\begin{aligned}\langle \varphi_{\Delta_1,\bar{\Delta}_1}(z_1,\bar{z}_1)\,\varphi_{\Delta_2,\bar{\Delta}_2}(z_2,\bar{z}_2)\,\varphi_{\Delta_3,\bar{\Delta}_3}(z_3,\bar{z}_3)\rangle &= \\ = C_{\Delta_1,\Delta_2,\Delta_3}\, C_{\bar{\Delta}_1,\bar{\Delta}_2,\bar{\Delta}_3}\, (z_1-z_2)^{\Delta_3-\Delta_1-\Delta_2} (\bar{z}_1-\bar{z}_2)^{\bar{\Delta}_3-\bar{\Delta}_1-\bar{\Delta}_2} &\cdot \qquad (2.2) \\ (z_2-z_3)^{\Delta_1-\Delta_2-\Delta_3} (\bar{z}_2-\bar{z}_3)^{\bar{\Delta}_1-\bar{\Delta}_2-\bar{\Delta}_3} (z_3-z_1)^{\Delta_2-\Delta_3-\Delta_1} (\bar{z}_3-\bar{z}_1)^{\bar{\Delta}_2-\bar{\Delta}_3-\bar{\Delta}_1}\end{aligned}$$

The $c_{\Delta_1,\Delta_2,\Delta_3}$'s are called expansion coefficients are also in principle fixed by the conformal theory[6].

Under a conformal transformation w=w(z) the correlation function of primary fields transforms asfollows:

$$\langle \varphi_{\Delta_1,\bar{\Delta}_1}(w_1,\bar{w}_1)\ldots\rangle = \left(w'(z_1)\right)^{-\Delta_1}\left(\overline{w'(z_1)}\right)^{-\bar{\Delta}_1}\langle \varphi_{\Delta_1,\bar{\Delta}_1}(z_1,\bar{z}_1)\ldots\rangle \tag{2.3}$$

We consider now the conformal transformation[7]:

$$w = \frac{N}{\pi}\ln z = \tau + i v \tag{2.4}$$

which maps the plane into the strip ($-\frac{N}{2}\leq v\leq \frac{N}{2}$, $-\infty<\tau<\infty$).
As the result of the transformation (2.4), the two-point function (2.1) has the following form on the strip:

$$\langle \varphi(v_1,\tau_1)\,\varphi(v_2,\tau_2)\rangle = \left(\frac{2\pi}{N}\right)^{2x}\sum_{n,\bar{n}=0}^{\infty} a_n(2\Delta)\,a_{\bar{n}}(2\bar{\Delta})\,\cdot \\ \cdot\, e^{-\frac{2\pi}{N}(x+n+\bar{n})(\tau_2-\tau_1)-\frac{2\pi i}{N}(s+n-\bar{n})(v_2-v_1)} \tag{2.5}$$

where

$$a_n(\alpha) = \frac{\Gamma(\alpha+n)}{n!\,\Gamma(\alpha)} \tag{2.6}$$

Assume now that the euclidean time (τ) evolution of the system is given by a Hamiltonian H with eigenstates $|E_t(\mathcal{P})\rangle$:

$$H\,|E_t(\mathcal{P})\rangle = E_t(\mathcal{P})\,|E_t(\mathcal{P})\rangle \tag{2.7a}$$

$$\hat{\mathcal{P}}\,|E_t(\mathcal{P})\rangle = \mathcal{P}\,|E_t(\mathcal{P})\rangle \tag{2.7b}$$

where $\hat{\mathcal{P}}$ is the momentum operator. Using the standard spectral decomposition, the correlation function (2.5) can be reexpressed as follows:

$$\langle \varphi(v_1,\tau_1)\varphi(v_2,\tau_2)\rangle = \langle 0|\varphi(v_1,\tau_1)\varphi(v_2,\tau_2)|0\rangle =$$
$$= \sum_{t,P} e^{-(E_t(P)-E^{(P)})(\tau_2-\tau_1)-\frac{2\pi i}{N}P(v_2-v_1)} \cdot$$
$$\cdot\ |\langle 0|\varphi(0,0)|E_t(P)\rangle|^2 \qquad (2.8)$$

here $E^{(P)}$ is the energy of the ground-state $|0\rangle$. We now compare Eqs. (2.5) and (2.8) and get the following identities:

$$\frac{N}{2\pi}(E_t(P)-E^{(P)}) = (\Delta+r)+(\bar{\Delta}+\bar{r}) \qquad (2.9)$$

$$P = r-\bar{r} = P-s \qquad (2.10)$$

Eq. (2.9) is of the form (1.6a). Eq. (2.10) needs further interpretation. When dealing with a quantum chain, the boundary condition is fixed (see Eq. (1.3)), and the momenta P will have the values:

$$P = (r+\Delta)-(\bar{r}+\bar{\Delta}) - a \qquad (2.11)$$

where a is fixed by the boundary oondition. The values of a are $\frac{\tilde{Q}}{3}$ for the boundary condition of Eq. (1.3).

We are now in the position to make the connection with the Virasoro algebras. We assume[6] that the spectrum (2.9) is given by two Virasoro algebras with the same central charge c, Δ and $\bar{\Delta}$ corresponding to the lowest weights of the two irreducible representations one corresponding to one Virasoro algebra, the second to the other Virasoro algebra. r ($\bar{r}$) represent then the descendents of $\Delta(\bar{\Delta})$. If the descendent $r(\bar{r})$ has the degeneracy $d(\Delta,r)(d(\bar{\Delta},\bar{r}))$, the energy level $(\Delta+r+\bar{\Delta}+\bar{r})$ has obviously the degeneracy $d(\Delta,r)d(\bar{\Delta},\bar{r})$.

For the three-states Potts model, the central charge is $c=\frac{4}{5}$ and the possible values for Δ are[8,9]:

$$\Delta = 0,\ 1/40,\ 1/15,\ 1/8,\ 2/5,\ 21/40,\ 2/3,\ 7/5,\ 13/8,\ 3 \qquad (2.12)$$

the possible values for $\bar{\Delta}$ cover clearly the same range. The degeneracies $d(\Delta,r)$ can be computed using the character formula of Rocha-Caridi[10] and this calculation was done for us by Altschüler and Lacki[11] for the first 10 values of r. In Table 1 we show the results

	r										
Δ	0	1	2	3	4	5	6	7	8	9	10
0	1	0	1	1	2	2	4	4	7	8	12
2/5	1	1	1	2	3	4	6	8	11	15	20
1/40	1	1	2	3	4	6	9	12	17	23	31
7/5	1	1	2	2	4	5	8	10	15	19	26
21/40	1	1	2	3	5	7	10	14	19	26	35
1/15	1	1	2	3	5	7	10	14	20	26	36
3	1	1	2	3	4	5	8	10	14	18	24
13/8	1	1	2	3	4	6	9	12	16	22	29
2/3	1	1	2	2	4	5	8	10	15	19	27
1/8	1	1	1	2	3	4	6	8	11	15	20

Table 1 The function $d(\Delta,r)$ representing the degeneracy of the level (Δ,r) of the irreducible representation with lowest weight Δ.

We now formulate our problem. For each of the numerically determined spectra $\mathcal{E}_0^{(0)}(P)$, $\mathcal{E}_1^{(0)}(P)$ and $\mathcal{E}_1^{(1)}(P)$ (see Eq. (1.6a)) one has to find the irreducible representations $(\Delta,\bar{\Delta})$ (there are one hundred of them) which build the spectrum. This is done using Eqs.(2.9) and (2.11) and Table 1. The answer to this problem is given in Sec. 4.

As the reader might have noticed up to now we have only considered differences of energy levels. It was shown however (see Ref. [12]) that the finite-size correction to the ground-state energy $E^{(P)}$ (see Eq. (1.7a)) has an algebraic meaning: it is related to the central charge c of the Virasoro algebra:

$$A_2^{(P)} = \frac{\pi}{6} c \tag{2.13}$$

This relation was checked and the results are also given in Sec. 4.

3. CORRELATION FUNCTIONS ON A STRIP (FREE BOUNDARY CONDITIONS)

We consider the two point correlation function in a half plane ($-\infty<X<\infty$, $Y\geq 0$) with free boundary conditions. If we are at the critical point and the operator has scaling dimensions x, it was shown by Cardy[13] using conformal invariance that the correlation function has the form:

$$\langle \varphi(X_1,Y_1)\,\varphi(X_2,Y_2)\rangle = (Y_1 Y_2)^{-x} F(\rho) \tag{3.1}$$

where

$$\rho = \frac{Y_1 Y_2}{(X_1-X_2)^2+(Y_1-Y_2)^2} \tag{3.2}$$

The function $F(\rho)$ depends on the operator φ which appears in the correlation function and has the following asymptotic behaviours:

$$\underset{\rho\to 0}{F(\rho) =} \rho^{x_s} R(\rho) \quad , \quad \underset{\rho\to\infty}{F(\rho) \sim} \rho^{x} \tag{3.3}$$

where the function $R(\rho)$ is regular at $\rho = 0$ and x_s is the surface exponent[7,14]. We now perform the conformal transformation:

$$w = \frac{N}{\pi}\ln z = \frac{N}{\pi}\ln(X+iY) = \tau + iv \tag{3.4}$$

which maps the half-plane on the strip $(-\infty<\tau<\infty, -\frac{N}{2}<v<\frac{N}{2})$. Here τ can be interpreted as the Euclidean time. The corrělation function on the strip reads (see Eq. (2.3))

$$\begin{aligned}\langle \varphi(v_1,\tau_1)\,\varphi(v_2,\tau_2)\rangle &= \left(\frac{\pi}{N}\right)^{2x} |z_1 z_2|^{x} \rho^{x_s} R(\rho) \\ &= \left(\frac{\pi}{N}\right)^{2x} \left(\sin\frac{\pi}{N}v_1 \sin\frac{\pi}{N}v_2\right)^{x_s-x} \sum_{n=0}^{\infty} a_n(v_1,v_2)\, e^{-\frac{\pi}{N}(x_s+n)(\tau_2-\tau_1)}\end{aligned} \tag{3.5}$$

where a_0 is independent of v_1 and v_2. Assume now that the Euclidean time evolution of the system is described by a Hamiltonian H with eigenvalues $E^{(F)}(r)$ $(E^{(F)}<E^{(F)}(0)<$ $<E^{(F)}(1)<\dots)$:

$$H|r\rangle = E^{(F)}(r)|r\rangle \tag{3.6}$$

$E^{(F)}$ being the ground-state energy. Using the spectral decomposition, the correlation function of Eq. (3.5) can be reexpressed as follows:

$$\langle \varphi(v_1,\tau_1)\varphi(v_2,\tau_2)\rangle = \sum_{r=0}^{\infty} e^{-(E^{(F)}(r)-E^{(F)})(\tau_2-\tau_1)} \langle 0|\varphi(v_1,0)|r\rangle\langle r|\varphi(v_2,0)|0\rangle \tag{3.7}$$

In the sum over the r's one has to keep in mind that several states can correspond to the same energy level $E^{(F)}(r)$.

Comparing Eqs. (3.5) and (3.6) we notice the relation

$$\frac{N}{\pi}\left(E^{(F)}(r) - E^{(F)}\right) = x_s + r, \quad (r = 0,1,\ldots) \tag{3.8}$$

We now recall that the known surface exponents x_s coincide with lowest weights Δ of irreducible representations of the Virasoro algebra with the central charge c fixed by the universality class[13,15]. The relation (3.8) then suggests that the finite-size limit of the spectrum of the Hamiltonian with free boundary conditions is given by the lowest weight and the descendents of irreducible representations of the Virasoro algebra. Notice that in this case we have only one Virasoro algebra instead of the two which occur for periodic and twisted boundary conditions (see Sec. 2).

We now return to Eq. (1.6b) and formulate the problem in the case of free boundary conditions. We first compute numerically the spectra $\mathcal{E}_0^{(F)}(\pm)$ and $\mathcal{E}_1^{(F)}(\pm)$ we thus get the spectra

$$\begin{aligned} \mathcal{E}_0^{(F)} &= \mathcal{E}_0^{(F)}(+) + \mathcal{E}_0^{(F)}(-) \\ \mathcal{E}_1^{(F)} &= \mathcal{E}_1^{(F)}(+) + \mathcal{E}_1^{(F)}(-) \end{aligned} \tag{3.9}$$

We next look which irreducible representations Δ the list (2.12) give contributions to $\mathcal{E}_0^{(F)}$ and $\mathcal{E}_1^{(F)}$. The answer is found in Sec. 5.

Finally, as in the periodic case, part of the finite-size corrections to the ground-state energy $E^{(F)}$ (see Eq. (1.7b)) are again controlled by the central charge c. One has[15]:

$$A_2^{(F)} = \frac{\pi}{24}\, c \tag{3.10}$$

How this relation works will be seen in Sec. 5.

4. IDENTIFICATION OF THE IRREDUCIBLE REPRESENTATIONS (PERIODIC AND TWISTED BOUNDARY CONDITIONS)[16)]

We start with $\mathcal{E}_0^{(0)}(P)$ (see Eq. 1.6a)). The lowest excitations of the quantum chains of lengths up tp 14 sites have been used in order to determine the Van den Broeck-Schwartz approximants. The results are shown in the right-hand side of Table 2. One notices:

a) The approximants degeneracy of levels (see for example the cluster at $(\Delta+r+\bar{\Delta}+\bar{r}) = 2.8$ at P=2)

b) The exact degeneracy for any number of sites of other level (see the level 2.79 at P=0). This reflects a supplementary internal symmetry of the problem which is unknown to us.

The approximate degenerate levels and the exact degenerate levels coincide for large N and build the Hamiltonian in the finite-size scaling limit. Using Eqs. (2.9), (2.11), (2.12) and Table 1 we have found that the irreducible representations (0,0),(2/5,2/5), (7/5,2/5),(2/5,7/5), (7/5,7/5), (3,0) and (0,3) give a perfect description of the levels. A sure bet is that the representation (3,3) is also present but it would have shown up at $(\Delta+r+\bar{\Delta}+\bar{r}) = 6$ (P=0) and this goes beyond our numerical ability

P	$\Delta+r+\bar{\Delta}+\bar{r}$	(0,0)	$(\frac{2}{5},\frac{2}{5})$	$(\frac{7}{5},\frac{2}{5})$	$(\frac{2}{5},\frac{7}{5})$	$(\frac{7}{5},\frac{7}{5})$	(3,0)	$\mathcal{E}_0^{(0)}(P)$ ("Exp")
0	0.8	-	1	-	-	-	-	0.820(3)
	2.8	-	1	1	1	1	-	2.79(1)*;2.81(2);2.832(2)
	4.0	1	-	-	-	-	-	3.996(4)
	4.8	-	1	1	1	1	-	4.77(2)*;4.82(2);4.83(1)
1	1.8	-	1	1	-	-	-	1.798(3);1.824(4)
	3.8	-	1	1	1	1	-	3.78(2);3.78(1);3.82(1);3.83(2)
2	2	1	-	-	-	-	-	1.99998(4)
	2.8	-	1	1	-	-	-	2.77(8);2.8(1)
	4.8	-	2	1	2	1	-	4.75(6);4.77(5);4.82(4)
3	3	1	-	-	-	-	1	2.995(5);2.999(1)
	3.8	-	2	1	-	-	-	

Table 2 The spectrum $\mathcal{E}_0^{(0)}(P)$. The Van den Broeck-Schwartz approximants for the level are denoted by $\mathcal{E}_0^{(0)}(P)$ ("Exp."). The contributions of each irrep. $(\Delta,\bar{\Delta})$ to the spectrum is shown. The numbers under each $(\Delta,\bar{\Delta})$ indicates the degeneracy. The levels marked by an asterisk are doubly degenerate (parity doublets) even for finite chains. The spectrum for negative momenta is the same as for positive momenta. The figure in brackets in the last column indicates the estimated error.

The same exercise was repeated in Table 3 for $\mathcal{E}_1^{(0)}(P)$ and in Table 4 for $\mathcal{E}_1^{(1)}(P)$

P	$\Delta+r+\bar{\Delta}+\bar{r}$	$(\frac{1}{15},\frac{1}{15})$	$(\frac{2}{3},\frac{2}{3})$	$\mathcal{E}_1^{(0)}(P)$ ("Exp")
0	2/25 ≈ 0.133	1	-	0.1333(1)
	4/3 ≈ 1.333	-	1	1.3333(5)
	32/15 ≈ 2.333	1	-	2.139(1)
	10/3 ≈ 3.333	-	1	3.333(4)
	62/15 ≈ 4.133	4	-	4.13(2)*;4.138(5);4.18(1)
	16/3 ≈ 5.333	-	4	5.31(7);5.333(1)*;5.334(5)
1	17/15 ≈ 1.133	1	-	1.1344(5)
	7/3 ≈ 2.333	-	1	2.332(5)
	47/15 ≈ 3.133	2	-	3.10(5);3.13(5)
	13/3 ≈ 4.333	-	2	4.329(5);4.332(3)
2	32/15 ≈ 2.133	2	-	2.132(1);2.134(3)
	10/3 ≈ 3.333	-	2	3.32(2);3.332(6)
	62/15 ≈ 4.133	3	-	4.12(3);4.124(5)
3	47/15 ≈ 3.133	3	-	3.13(2);3.13(1);3.13(3)
	13/3 ≈ 4.333	-	2	4.3(1);4.33(2)
	77/15 ≈ 5.133	5	-	5.15(5)

Table 3 The spectrum $\mathcal{E}_1^{(0)}(P)$. Double degeneracy marked by * as in Table 2.

P	$\Delta+r+\bar{\Delta}+\bar{r}$	$(\frac{2}{3},\frac{1}{15})$	$(0,\frac{2}{3})$	$(\frac{7}{5},\frac{1}{15})$	$(3,\frac{2}{3})$	$\mathcal{E}_1^{(1)}(P)$ ("Exp")
-2	5/3 ≈ 1.666	-	1	-	-	1.656(5)
	37/15 ≈ 2.466	2	-	-	-	2.39(1);2.5(1)
	67/15 ≈ 4.466	3	-	3	-	4.28(2);4.4(2);4.4(1)
-1	2/3 ≈ 0.666	-	1	-	-	0.66666(3)
	27/15 ≈ 1.466	1	-	-	-	1.47(2)
	52/15 ≈ 3.466	2	-	2	-	3.45(2);3.47(2);3.48(1);3.51(3)
	14/3 ≈ 4.666	-	2	-	-	
0	7/15 ≈ 0.466	1	-	-	-	0.4667(3)
	37/15 ≈ 2.466	1	-	1	-	2.460(5);2.478(2)
	11/3 ≈ 3.666	-	1	-	-	3.665(8)
	67/15 ≈ 4.466	2	-	2	-	4.45(2);4.44(1)
1	22/15 ≈ 1.466	1	-	1	-	1.466(5);1.469(2)
	8/3 ≈ 2.666	-	1	-	-	2.667(2)
	52/15 ≈ 3.466	1	-	1	-	3.45(2);3.47(2)
	14/3 ≈ 4.666	-	1	-	1	4.67(5)
2	37/15 ≈ 2.466	1	-	1	-	2.43(3);2.44(6)
	11/3 ≈ 3.666	-	1	-	1	3.65(2);3.663(3)
	67/15 ≈ 4.466	2	-	2	-	4.4(1);4.46(4);4.47(5)

Table 4 The spectrum $\mathcal{E}_1^{(1)}(P)$. In this case the spectrum for negative momenta is different than for positive momenta.

In conclusion the following 18 irreducible representations describe the spectrum with periodic and twisted boundary conditions:

$$\mathcal{E}_0^{(0)}(P):\ (0,0),\ (2/5,2/5),(7/5,2/5),(2/5,7/5),(7/5,7/5),$$
$$(3,0),(0,3),(?)\ (3,3) \tag{4.1a}$$

$$\mathcal{E}_1^{(0)}(P):\ (1/15,1/15),(2/3,2/3) \tag{4.1b}$$

$$\mathcal{E}_1^{(1)}(P):\ (2/5,1/15),(0,2/3),(7/5,1/15),(3,2/3) \tag{4.1c}$$

$$\mathcal{E}_1^{(2)}(P):\ (1/15,2/5),(2/3,0),(1/15,7/5),(2/3,3) \tag{4.1d}$$

The spectrum $\mathcal{E}_1^{(2)}(P)$ was obtained directly from $\mathcal{E}_1^{(1)}(P)$ using Eq. (1.4).

Finally the prediction (2.13) on the finite-size correction to the ground state energy was checked. One obtains

$$\frac{6}{\pi} A_2^{(P)} = 0.80008(1) \tag{4.2}$$

in excellent agreement with the expected value c=4/5.

5. IDENTIFICATION OF THE IRREDUCIBLE REPRESENTATIONS (FREE BOUNDARY CONDITIONS)[17)]

The lowest excitations for the Hamiltonian $H^{(F)}$ have been determined using the Lanczos method considering chains up to 12 sites. Van den Broeck approximants for $\mathcal{E}_0^{(F)}(\pm)$ are shown in Table 5 and those for $\mathcal{E}_1^{(F)}(\pm)$ (see Eq. (1.6b) for the definition) are shown in Table 6.

Δ+r	(0)	(3)	$\mathcal{E}_0^{(F)}(+)$ ("Exp")	$\mathcal{E}_0^{(F)}(-)$ ("Exp")
2	1	-	2.000(5)	-
3	1	1	-	2.99(2),2.98(3)
4	2	1	4.004(6),3.995(8),4.01(3)	-
5	2	2	-	4.98(3),4.98(2),5.00(3),4.99(3)
6	4	3	5.97(5),5.98(4),5.99(6),5.8(2)	-
7	4	4	-	7.0(2),6.9(3),>6.6(?)

Table 5 The spectrum $\mathcal{E}_0^{(F)}$ for the 3-states Potts model in the charge zero sector. The Van den Broeck-Schwartz approximants for the levels with positive parity ($\mathcal{E}_0^{(F)}(+)$) and negative parity ($\mathcal{E}_0^{(F)}(-)$) are given. The figure in brackets in the last two columns indicates the estimated error. On the left side of the table we indicate the number of states having $\mathcal{E}_0^{(F)}=\Delta+r$ generated by the irreducible representations Δ=0 and 3,

$\Delta+r$	$(\frac{2}{3})$	$\mathcal{E}_1^{(F)}(+)$ ("Exp")	$\mathcal{E}_1^{(F)}(-)$ ("Exp")
$\frac{2}{3}\simeq 0.6666...$	1	0.6662(4)	-
1.6666...	1	-	1.668(2)
2.6666...	2	2.66(1),2.68(4)	-
3.6666...	2	-	3.64(4),3.66(2)
4.6666...	4	4.65(4),4.66(3),4.68(3),4.67(2)	-
5.6666...	5	-	5.58(8),5.65(7),5.65(5), 5.66(6),5.66(4)
6.6666...	8	6.6(2),6.55(10)	-
7.6666...	10	-	>7.5(?)

Table 6 The spectrum $\mathcal{E}_1^{(F)}$ for the 3-states Potts model in the charge one sector. The Van den Broeck-Schwartz approximants for the levels with positive parity ($\mathcal{E}_1^{(F)}(+)$) and negative parity ($\mathcal{E}_1^{(F)}(-)$) are given. On the left side of the table we indicate the number of states having $\mathcal{E}_1^{(F)}=\frac{2}{3}$ + r generated by the irreducible representation $\Delta=\frac{2}{3}$.

We have then used Eq. (3.8) and checked from the possible values of Δ(Eq. (2.12)) and the degeneracies given in Table 1 which irreducible representations build the spectra. The conclusion is:

$$\mathcal{E}_0^{(F)} : (0), (3) \tag{5.1a}$$

$$\mathcal{E}_1^{(F)} : (\tfrac{2}{3}) \tag{5.1b}$$

Finally, we have checked the prediction (3.10) on the finite-size corrections to the ground-state energy $E^{(F)}$ (see Eq. (3.10)).
One finds:

$$\frac{24}{\pi} A_2^{(F)} = 0.792(1) \tag{5.2}$$

again in excellent agreement with c=4/5.

6. CORRECTIONS TO FINITE-SIZE SCALING (PERIODIC AND TWISTED BOUNDARY CONDITIONS)[18,19]

The reader might wonder why in Sec. 2 we have introduced the three-point function (see Eq. (2.2)) and the expansion coefficients $C_{\Delta_1,\Delta_2,\Delta_3}$ without further use. In this section we will show that they are essential in the understanding of the corrections to finite-size scaling.

Let us consider an energy level corresponding to the lowest weight of an irreducible representation $(\Delta,\bar{\Delta})$, with scaling dimensions $x=\Delta+\bar{\Delta}$ and spin $s=\Delta-\bar{\Delta}$. This corresponds to taking $r=\bar{r}$ in Eqs. (2.9) and (2.11). Our task is to understand the nature of the correction terms:

$$\mathcal{E}(\Delta,\bar{\Delta};N) = \frac{2\pi}{N}(E_o(\Delta-\bar{\Delta}-a)-E^{(P)}) \tag{6.1}$$

$$= x+c_1N^{-\alpha}$$

where $E_o(\Delta-\bar{\Delta}-a)$ is the lowest energy level from the set $E_t(P)$ with momentum $P=\Delta-\bar{\Delta}-a$. In Table 7 we show the values of α and c_1 determined from the knowledge of the energy levels for various number of sites N. Notice that $\alpha \approx 0.8$ for $x=2/15, 4/5$ and $7/15$ and it is much larger for $x=4/3$ and $2/3$. We will be able to explain this difference. We will also be able to provide predictions for ratios of several c_1's. In order to do so we first return to the two point function (see Eqs. (2.5) and (2.8)).

$(\Delta,\bar{\Delta})$	x	c_1	α
(1/15,1/15)	$\frac{2}{15}$	0.00657(1)	+0.795(10)
(2/5,2/5)	$\frac{4}{5}$	0.2364(1)	+0.7998(3)
(2/5,1/15)	$\frac{7}{15}$	-0.03947(5)	+0.82(2)
(2/3,2/3)	$\frac{4}{3}$	-1.003(5)	+1.6961(2)
(0,2/3)	$\frac{2}{3}$	-0.2681(1)	+2.10(6)

Table 7 Values of c_1 and α defined by Eq. (6.1)

Let us assume that we consider only spinless operators and we are in the case of periodic boundary conditions and let $|1\rangle$ with energy E_1 be the first excited state next to the ground-state $|0\rangle$ (energy $E^{(P)}$). We also assume that the states $|0\rangle$ and $|1\rangle$ have momentum zero. Taking the large $(\tau_2-\tau_1)$ limit in Eqs. (2.5) and (2.8) we get:

$$\langle 0|\varphi(0,0)|1\rangle = \left(\frac{2\pi}{N}\right)^x \tag{6.2}$$

We now consider the effect of the conformal transformation (2.4) on the three-point function (2.2). Using Eq. (2.3) we obtain:

$$\langle \varphi_{\Delta_1,\Delta_1}(v_1,\tau_1)\,\varphi_{\Delta_2,\Delta_2}(v_2,\tau_2)\,\varphi_{\Delta_3,\Delta_3}(v_3,\tau_3)\rangle = \left(\frac{2\pi}{N}\right)^{x_1+x_2+x_3} \cdot |\xi_1|^{x_1}\,|\xi_2|^{x_3}$$

$$(C_{\Delta_1,\Delta_2,\Delta_3})^2\,|1-\xi_1|^{x_3-x_1-x_2} \cdot |1-\xi_2|^{x_1-x_2-x_3} \cdot$$

$$|1-\xi_1\xi_2|^{x_2-x_3-x_1} \tag{6.3a}$$

where

$$\xi_1 = e^{-\frac{2\pi}{N}(\tau_2-\tau_1)-\frac{2\pi i}{N}(v_2-v_1)}$$

$$\xi_2 = e^{-\frac{2\pi}{N}(\tau_3-\tau_2)-\frac{2\pi i}{N}(v_3-v_2)} \tag{6.3b}$$

In the limit $\tau_2-\tau_1\to\infty$, $\tau_3-\tau_2\to\infty$, the three-point function has the expression:

$$\langle \varphi_{\Delta_1,\Delta_1}(v_1,\tau_1)\,\varphi_{\Delta_2,\Delta_2}(v_2,\tau_2)\,\varphi(v_3,\tau_3)\rangle \approx \left(\frac{2\pi}{N}\right)^{x_1+x_2+x_3} \cdot$$

$$\cdot\,(C_{\Delta_1,\Delta_2,\Delta_3})^2\, e^{-\frac{2\pi}{N}x_1(\tau_2-\tau_1)-\frac{2\pi}{N}x_3(\tau_3-\tau_2)} \tag{6.4}$$

Using the spectral decomposition and taking $\Delta_3=\Delta_1=\Delta$, we find in the same limit (using Eq. (6.2)):

$$\langle 0|\,\varphi_{\Delta,\Delta}(v_1,\tau_1)\,\varphi_{\Delta_2,\Delta_2}(v_2,\tau_2)\,\varphi_{\Delta,\Delta}(v_3,\tau_3)\,|0\rangle \simeq$$

$$\left(\frac{2\pi}{N}\right)^{2x} \langle 1|\,\varphi_{\Delta_2,\Delta_2}(0,0)\,|1\rangle\, e^{-(E_1-E^{(g)})(\tau_3-\tau_1)} \tag{6.5}$$

Comparing Eqs. (6.4) and (6.5), we find

$$\frac{N}{2\pi}(E_1-E^{(g)}) = x$$

and

$$\langle 1|\,\varphi_{\Delta_2,\Delta_2}(0,0)\,|1\rangle = \left(\frac{2\pi}{N}\right)^{x_2}(C_{\Delta,\Delta_2,\Delta})^2 \tag{6.6}$$

In order to compute the correction term in Eq. (6.1) we assume that the Hamiltonian of the perturbed systems is:

$$\bar{H} = H + g \int_{-\frac{N}{2}}^{\frac{N}{2}} \varphi_{\Delta_2,\Delta_2}(v,0)\, dv \tag{6.7}$$

where H is the conformal invariant Hamiltonian, g is a small coupling constant and $\varphi_{\Delta_2,\Delta_2}$ is a spinless primary field with scaling dimensions $x_2=2\Delta_2$. Applying standard perburbation theory and Eq. (6.6) we find:

$$\mathcal{E}(\Delta,\Delta;N) = x + g(2\pi)^{x_2-1}\left(C_{\Delta,\Delta_2,\Delta}\right)^2 N^{2-x_2} \tag{6.8}$$

This expression generalizes obviously in the case where $\Delta\neq\bar{\Delta}$ (see Eq. (6.1)) to

$$\mathcal{E}(\Delta,\bar{\Delta};N) = x + g(2\pi)^{x_2-1} C_{\Delta,\Delta_2,\Delta}\, C_{\bar{\Delta},\Delta_2,\bar{\Delta}}\, N^{2-x_2} \tag{6.9}$$

We are now close to give an interpretation of the results shown in Table 7. The levels 2/5, 4/5 and 7/5 have

$$\alpha = 2 - x_2 \approx 0.8 \tag{6.10}$$

This implies:

$$\Delta_2 = 7/5 \tag{6.11}$$

In Table 8 we show which expansion coefficients $c_{\Delta_1,\Delta_2,\Delta}$ might be different of zero[6]. and we notice that $c_{2/3,2/3,7/5} = 0$ which explains why there are no $N^{-0.8}$ corrections for the x = 2/3 and 4/3 levels.

From the knowledge of some known four-point functions[9] we have[19] determined the following values for the square of the expansion coefficients:

$$\left(C_{\frac{2}{5},\frac{7}{5},\frac{2}{5}}\right)^2 = \frac{6}{7}\left[\frac{\left(\Gamma(\frac{3}{5})\right)^3}{\left(\Gamma(\frac{2}{5})\right)^3}\,\frac{\Gamma(\frac{1}{5})}{\Gamma(\frac{4}{5})}\right]^{1/2}$$

$$\left(C_{\frac{2}{5},\frac{7}{5},\frac{2}{5}}\right)^2 = 36\left(C_{\frac{1}{15},\frac{7}{5},\frac{1}{15}}\right)^2 \tag{6.12}$$

$\Delta_1 \backslash \Delta_2$	(0)	(3)	(7/5)	(2/5)	(2/3)	(1/15)
(0)	(0)	(3)	(7/5)	(2/5)	(2/3)	(1/15)
(3)	x	(0)	(2/5)	(7/5)	(2/3)	(2/3⊕(1/15)
(7/5)	x	x	(0)⊕(7/5)	(2/5)	(1/15)	(2/3⊕(1/15)
(2/5)	x	x	x	(0)⊕(7/5)	(1/15)	(2/3)⊕ (1/15)
(2/3)	x	x	x	x	(0)⊕(3)⊕(2/3)	(7/5)⊕(2/5)⊕ (1/15)
(1/15)	x	x	x	x	x	(3)⊕(7/5)⊕(2/5) ⊕(2/3)⊕(1/15)

Table 8 Possible nonzero expansion coefficients $c_{\Delta_1,\Delta_2,\Delta}$ for various values of Δ_1 and Δ_2 ($c_{\Delta_1,\Delta_2,\Delta} = c_{\Delta_2,\Delta_1,\Delta}$).

From Eq. (6.9) and (6.12) we derive

$$\frac{C_1(x=\frac{4}{5})}{C_1(x=\frac{2}{15})} = 36 \quad , \quad \left(C_1(x=\tfrac{7}{5})\right)^2 = C_1(x=\tfrac{4}{5})\, C_1(x=\tfrac{2}{5}) \tag{6.13}$$

From Table 7 we find

$$\frac{C_1(x=\frac{4}{5})}{C_1(x=\frac{2}{5})} = 35.99 \quad ; \quad \frac{C_1(x=\frac{7}{5})}{C_1(x=\frac{4}{5})C_1(x=\frac{2}{5})} = 1.003 \tag{6.14}$$

in excellent agreement with Eq. (6.13).

REFERENCES

1. E. Fradkin and L. Susskind Phys. Rev. D17 2637 (1978)
2. J. Kogut Rev. Mod. Phys. 51 659 (1979)
3. G. v. Gehlen, V. Rittenberg and H. Ruegg J.Math. Phys. A 19, 107 (1985)
4. J.M. Van den Broeck and L.W. Schwartz SIAM J.Math.Anal 10, 639 (1979)
5. C.J. Hamer J.Phys. A 14, 298 (1981)
6. A.A. Belavin, A.M. Polyakov and A.B. Zamolodchikov Nucl.Physics B241, 333 (1984)
7. J.L. Cardy in "Phase Transitions and Critical Phenomena" vol. 11 (C. Domb and J.L. Lebowitz Eds.) Academic Press, London, to be published
8. D. Friedan, Z. Qiu and S. Shenker Phys.Rev. Letters 52, 1575 (1984)
9. Vl.S. Dotsenko Nucl.Phys. B235, 54 (1984)

10. A. Rocha-Caridi, in "Vertex Operators in Mathematics and Physics" (J. Lepovsky Ed.) 1985, Springer Berlin, p. 451
11. D. Altschüler and J. Lacki (private communication)
12. H.W. Blöte, J.L. Cardy and M.P. Nightingale, to be published
13. J.L. Cardy Nuclear Physics B240, 514 (1984)
14. K. Binder in "Phase Transitions and Critical Phenomena" vol. 8 (C. Domb and J.L. Lebowitz, Eds.) 1983, Academic Press, London, p. 1
15. G. v. Gehlen and V. Rittenberg, to be published, preprint Bonn HE-86-02
16. G. v. Gehlen and V. Rittenberg, to be published, preprint Bonn HE-86-03
17. G. v. Gehlen and V. Rittenberg, to be published
18. J.L. Cardy, to be published, preprint UCSB-TH-75-1985
19. G. v. Gehlen, V. Rittenberg and T. Vescan, to be published

UNITARIZABLE HIGHEST WEIGHT REPRESENTATIONS
OF THE VIRASORO, NEVEU-SCHWARZ AND RAMOND ALGEBRAS

Victor G. Kac
IHES and MIT

Minoru Wakimoto
Hiroshima University

§0. The Virasoro algebra Vir is the universal central extension of the complexified Lie algebra of vector fields on the circle with finite Fourier series. Its (irreducible) highest weight representations $\sigma_{z,h}$ are parametrized by two numbers, the central charge z, and the minimal eigenvalue h of the energy operator ℓ_o. These representations play a fundamental rôle in statistical mechanics [1,5,6] and string theory [16].

The study of representations $\sigma_{z,h}$ was started by the first author [8], [9] with the computation of the determinant of the contravariant Hermitian form lifted to the corresponding "Verma module", on each eigenspace of ℓ_o. This led to a criterion of inclusions of Verma modules and the computation of the characters $\operatorname{tr} q^{\ell_o}$ in some cases, in particular, for the critical value $z = 1$ [9]. Feigin and Fuchs [3] succeeded in proving the fundamental fact (conjectured in [10]) that Verma modules over Vir are multiplicity-free, which led them, in particular, to the computation of the characters of all representations $\sigma_{z,h}$.

Using the determinantal formula, it is not difficult to show that $\sigma_{z,h}$ is unitarizable (i.e. the contravariant Hermitian form is positive definite) for $z \geq 1$ and $h \geq 0$ [10]. It is obvious that $V(z,h)$ is not unitarizable if $z < 0$ or $h < 0$. The case $0 \leq z < 1$ was analysed, using the determinantal formula, by Friedan-Qiu-Shenker [5]. They found the remarkable fact that the only possible places of unitarity in this region are $(z_m, h^{(m)}_{r,s})$, where

$$(0.1)\quad z_m = 1 - \frac{6}{(m+2)(m+3)}\ ;\quad h^{(m)}_{r,s} = \frac{[(m+3)r-(m+2)s]^2-1}{4(m+2)(m+3)}\ .$$

Here $m,r,s \in \mathbb{Z}_+ = \{0,1,2,\dots\}$ and $1 \leq s \leq r \leq m+1$. (Actually, the series (0,1) was discovered by Belavin-Polyakov-Zamolodchikov [1].)

On the other hand, according to the Goddard-Kent-Olive (GKO) construction [7], Vir acts on the tensor product of two unitarizable highest weight representations of an affine (Kac-Moody) Lie algebra $\hat{\mathfrak{g}}'$ commuting with $\hat{\mathfrak{g}}'$. This construction was applied in [7] to the tensor product of the basic representation with a highest weight representation of level m of $\hat{s\ell}'_2$ to show that all the z_m indeed occur as central charges of unitarizable representations of Vir.

In the present paper we show that the "discrete series" representations $\sigma_{z,h}$ of Vir described by (0.1) appear with multiplicity one in the space of highest weight vectors of the tensor product of the basic representation and the sum of all unitarizable highest weight representations of $\hat{s\ell}_2'$, and hence are unitarizable. This is derived by a simple calculation with the Weyl-Kac character formula for $\hat{s\ell}_2'$ (see e.g. [11, Chapter 12]) and the Feigin-Fuchs character formula for Vir [3].

A similar result for the Neveu-Schwarz and Ramond superalgebras is obtained by applying the same argument to the super-symmetric extensions of $\hat{s\ell}_2'$ and their minimal representations (in place of the basic representation) constructed in [13]. (The list analogous to (0.1) was found in [6], and it was shown in [13] that all corresponding central charges indeed occur).

All the discrete series unitarizable representations $\sigma_{z,h}$ are degenerate (i.e. correspond to the zeros of the determinant). The only other degenerate unitarizable representations (apart from the "non-interesting" case $z > 1$, $h = 0$) are $\sigma_{1,m^2/4}$, where $m \in \mathbb{Z}_+$, and all of them appear with multiplicity one on the space of highest weight vectors for $s\ell_2$ in the sum of (two) fundamental representations of $\hat{s\ell}_2'$ [9]. We show that a similar result holds in the super case as well.

Finally, the above construction of the discrete series representations, allowed us to give a very simple proof of all determinantal formulas (cf. [2], [6], [9], [17]).

Geometrically, the main result of the paper concerning Vir can be stated as follows. Let G be the "minimal" group associated to $\hat{s\ell}_2'$ and let U_+ and U_- be the "opposite maximal unipotent" subgroups of G [19]. Let V be the space of the basic representation of G. Then Vir acts on the space of regular U_+-equivariant maps $\mathrm{Map}_{U_+}(U_-\backslash G, V)$, and all its unitarizable representations $\sigma_{z,h}$ with $z < 1$ appear with multiplicity 1.

The first author acknowledges the support of IHES and the NSF grant DMS-8508953.

After this work was completed, we received two preprints, "Unitary representations of the Virasoro algebra" by A. Tsuchiya and Y. Kanie, and "Unitary representations of the Virasoro and super Virasoro algebras" by P. Goddard, A. Kent and D. Olive, which overlap considerably with the present paper.

We added several Appendices to the paper. Appendix 1 provides a simple self-contained proof of the determinantal formulas for the Neveu-Schwarz and Ramond superalgebras Vir_ε. Appendix 2 contains multiplicative formulas for characters of Vir and Vir_ε; we hope that these formulas will provide a clue to more explicit constructions of the discrete series representations of Vir and Vir_ε (cf. Remark 8.2). Finally, in Appendix 3 we uncover a mysterious connection between "exceptional"

Lie algebras E_8, E_7, A_2 and E_6, and the representations of Vir corresponding to the following two dimensional models : Ising, tricritical Ising, 3-state Potts and tricritical 3-state Potts respectively (see Remark 8.3).

The first author acknowledges the hospitality of TIFR.

§1. Here we recall some necessary facts about affine Kac-Moody algebras in the simplest case of $\widehat{s\ell}_2$.

Let $g = s\ell_2(\mathbb{C})$ be the Lie algebra of complex traceless 2×2-matrices, and let

$$e = \begin{pmatrix} 0 & 1 \\ 0 & 0 \end{pmatrix} , \quad \alpha = \begin{pmatrix} 1 & 0 \\ 0 & -1 \end{pmatrix} , \quad f = \begin{pmatrix} 0 & 0 \\ 1 & 0 \end{pmatrix}$$

be its standard basis.

Let $\mathbb{C}[t,t^{-1}]$ be the algebra of Laurent polynomials over $\mathbb{C}$ in an indeterminate t . We regard the loop algebra $\tilde{g} = s\ell_2(\mathbb{C}[t,t^{-1}])$ as an (infinite-dimensional) complex Lie algebra. It has a central extension $\hat{g}' = \tilde{g} \oplus \mathbb{C}c$ by a 1-dimensional center $\mathbb{C}c$ with the bracket

$$(1.1) \qquad [x,y] = xy-yx+(\mathrm{Res}_{t=o}\ \mathrm{tr}\ \frac{dx}{dt}\ y)c$$

for $x,y \in \tilde{g}$. One includes $\hat{g}'$ as a subalgebra of codimension 1 in a larger algebra $\hat{g} = \hat{g}' \oplus \mathbb{C}d$, where

$$(1.2) \qquad [d,x] = t\,\frac{dx}{dt} \ \text{ for } x \in \tilde{g} \ ; \ [d,c] = 0 .$$

The Lie algebra $\hat{g}$ (and often its subalgebra $\hat{g}'$) with bracket defined by (1.1) and (1.2) is called an affine (Kac-Moody) Lie algebra associated to g . This is the simplest example of an infinite-dimensional Kac-Moody algebra (cf. [11, Chapter 7]). Putting $x(k) = t^k x$ for $x \in g$ and $k \in \mathbb{Z}$, we have an equivalent form of (1.1) and (1.2) :

$$(1.3) \qquad [x(k),y(n)] = (xy-yx)(k+n)+k\delta_{k,-n}(\mathrm{tr}\ xy)c;[d,x(k)] = kx(k);[c,\hat{g}] = 0 .$$

The (commutative 3-dimensional) subalgebra $\hat{h} = \mathbb{C}\alpha+\mathbb{C}c+\mathbb{C}d$ of $\hat{g}$ is called the Cartan subalgebra. Introduce the "upper triangular" subalgebra $\hat{n} = \mathbb{C}e+ \sum_{k>o} t^k g$. Define a symmetric bilinear form $(.|.)$ on $\hat{h}$ by :

$$(1.4) \qquad (\alpha|\alpha) = 2 \ ; \ (c|d) = 1; \ (\alpha|c) = (\alpha|d) = (d|d) = (c|c) = 0 .$$

(It extends to a non-degenerate invariant symmetric bilinear form on $\hat{\mathfrak{g}}$ by $(x(k)|y(n)) = \delta_{k,-n}\operatorname{tr} xy$, $(x(k)|c) = (x(k)|d) = 0)$. Introduce the following subsets of $\hat{h}$: $P_+^o = \{md+\frac{1}{2} n\alpha | m,n \in \mathbb{Z}_+, n \leq m\}$; $P_+ = P_+^o+\mathbb{R}c$.

Given $\lambda \in P_+$, there exists a unique (up to equivalence) irreducible representation π_λ of $\hat{\mathfrak{g}}$ on a complex vector space $L(\lambda)$ which admits a non-zero vector $v_\lambda \in L(\lambda)$ such that

(1.5) $\pi_\lambda(\hat{n})v_\lambda = 0$; $\pi_\lambda(\mu)v_\lambda = (\lambda|\mu)v_\lambda$ for all $\mu \in \hat{h}$.

This is called the integrable representation with highest weight λ (cf. [11, chapter 10]), v_λ being called the highest weight vector. The number $m = (\lambda|c)$ is called the level of $L(\lambda)$; we have : $\pi_\lambda(c) = mI$. Recall that $m \in \mathbb{Z}_+$, furthermore, $m = 0$ if and only if $\dim L(\lambda) = 1$. Note that viewed as a representation of $\hat{\mathfrak{g}}'$, π_λ remains irreducible and is independent of the c-component of λ .

All representations π_λ are unitarizable in the sense that there exists a positive definite Hermitian form $<.|.>$ on $L(\lambda)$ such that (cf. [11, Theorem 11.7b]) :

(1.6) $<\pi_\lambda(x(k))u|v> = <u|\pi_\lambda({}^t\bar{x}(-k))v>$ for all $u,v \in L(\lambda)$.

(Actually, property (1.6) together with $<v_\lambda|v_\lambda> = 1$ determines the Hermitian form uniquely; a Hermitian form satisfying (1.6) exists for any $\lambda \in \hat{h}$, but is positive definite only for $\lambda \in P_+$) .

With respect to $\pi_\lambda(d)$ we have the eigenspace decomposition :

(1.7) $L(\lambda) = \bigoplus_{k \in \mathbb{Z}_+} L^{((\lambda|d)-k)}$, where $\dim L^{((\lambda|d)-k)} < \infty$.

Consider the domain $D = \{z\alpha+\tau d+uc \in \hat{h} \,|\, \tau,u,z \in \mathbb{C}$ and $\operatorname{Im}\tau > 0\}$. Define the character of the representation π_λ by :

$$\mathrm{ch}_\lambda(\tau,z,u) = \sum_{k \in \mathbb{Z}_+} \operatorname{tr} \exp 2\pi i(\pi_\lambda(\tfrac{1}{2} z\alpha-\tau d+uc))|_{L^{((\lambda|d)-k)}} .$$

This is an absolutely convergent series defining a holomorphic function on D . It can be written in terms of elliptic theta functions $\Theta_{n,m}$ as follows [11, Chapter 12]. For a positive integer m and an integer n put

$$\Theta_{n,m}(\tau,z,u) = e^{2\pi imu} \sum_{k \in \mathbb{Z}+\frac{n}{2m}} q^{mk^2} e^{2\pi imkz} .$$

Here and further on, $q = e^{2\pi i\tau}$. For $\lambda \in P_+$, $\lambda = md+\frac{1}{2} n\alpha+rc$, $r \in \mathbb{R}$, put

$$s_\lambda = \frac{(n+1)^2}{4(m+2)} - \frac{1}{8} + r.$$

Then we have the following special case of the Weyl-Kac character formula :

(1.8) $ch_\lambda = q^{-s_\lambda}(\Theta_{n+1,m+2}-\Theta_{-n-1,m+2}) / (\Theta_{1,2}-\Theta_{-1,2})$.

In the following three simplest cases there are simpler formulas (cf. [12, p.218]) :

(1.9a) $ch_d = \Theta_{0,1}/\varphi(q)$, where

(1.9b) $\varphi(q) = \prod_{k=1}^{\infty} (1-q^k)$;

(1.10a) $ch_{2d}+ q^{1/2} ch_{2d+\alpha} = (\Theta_{0,2}+\Theta_{2,2})/\varphi_{1/2}(q)$, where

(1.10b) $\varphi_{1/2}(q) = \varphi(q^{1/2})\varphi(q^2)/\varphi(q)$;

(1.11a) $ch_{2d+1/2\alpha} = q^{-1/8}(\Theta_{1,2}+\Theta_{-1,2})/\varphi_0(q)$, where

(1.11b) $\varphi_0(q) = \varphi(q)^2/\varphi(q^2)$.

§2. We now recall a special case of the Goddard-Kent-Olive construction [7]. Let $\{u_i\}$ and $\{u^i\}$ be dual bases of $\mathfrak{g}$, i.e. tr $u_i u^i = \delta_{ij}(i,j = 1,2,3)$. Pick $\lambda,\lambda' \in P_+$ of levels m and m' and define the following operators L_k on the space $L(\lambda)\otimes L(\lambda')$ $(k \in \mathbb{Z})$:

$$
\begin{aligned}
(2.1)\quad L_k = & \frac{-1}{m+m'+2} \sum_{j\in\mathbb{Z}} \sum_i \pi_\lambda(u_i(-j))\otimes\pi_{\lambda'}(u^i(j+k)) \\
& + \left(\frac{1}{2(m+2)} - \frac{1}{2(m+m'+2)}\right) \sum_{j\in\mathbb{Z}} \sum_i \pi_\lambda(:u_i(-j)u^i(j+k):)\otimes 1 \\
& + \left(\frac{1}{2(m'+2)} - \frac{1}{2(m+m'+2)}\right) \sum_{j\in\mathbb{Z}} \sum_i 1\otimes \pi_{\lambda'}(:u_i(-j)u^i(j+k):)
\end{aligned}
$$

Let Ω be the Casimir element of $\hat{\mathfrak{g}}$ (cf. [11, Chapter 2 and Exercise 7.16]). We will need only the following property of Ω . If (π,V) is a representation of $\hat{\mathfrak{g}}$ on which Ω acts and $v \in V^{\hat{n}}$, then

(2.2) $\pi(\Omega)v = \pi(2(c+2)d+ \frac{1}{2} \alpha^2+\alpha)v$.

Here and further on V^{a} stands for $\{v \in V|\pi(a)v = 0$ for all $a \in \mathfrak{a}\}$.

The proof of the following formulas is straightforward (cf. [12, §2.5] or [18]) :

(2.3a) $[L_k,L_n] = (k-n)L_{k+n}+3\delta_{k,-n}\frac{k^3-k}{12}p(m,m')$, where

(2.3b) $p(m,m') = \frac{m}{m+2} + \frac{m'}{m'+2} - \frac{m+m'}{m+m'+2}$

(2.4) $L_o = \frac{1}{2}\left(\frac{(\lambda|\lambda+\alpha)}{m+2} + \frac{(\lambda'|\lambda'+\alpha)}{m'+2} - \frac{\Omega}{m+m'+2}\right)$

(2.5) $[L_k,\hat{\mathfrak{g}}'] = 0$,

i.e. the L_k are intertwining operators for the representation $\pi_\lambda \otimes \pi_{\lambda'}$ of $\hat{\mathfrak{g}}'$.

<u>Remark</u>. Formulas (2.3-5) hold for all non-twisted affine algebras $\hat{\mathfrak{g}}$ with the following changes : m+2 , m'+2 and m+m'+2 are replaced by m+g , m'+g and m+m'+g , where g is the dual Coxeter number [11, Chapter 6] , the coefficient 3 is replaced by dim $\mathfrak{g}$, and α is replaced by 2ρ . In the twisted case, formulas are somewhat more complicated (see Appendix 3).

§3. Now we turn to the Virasoro algebra Vir. Recall that this is a complex Lie algebra with a basis $\{\tilde{c};\ell_j, j \in \mathbb{Z}\}$ with commutation relations

(3.1) $[\ell_i,\ell_j] = (i-j)\ell_{i+j}+\frac{1}{12}(i^3-i)\delta_{i,-j}\tilde{c};[\tilde{c},\ell_j] = 0$.

Given two numbers z and h , there exists a unique irreducible representation $\sigma_{z,h}$ of Vir on a complex vector space V(z,h) which admits a non-zero vector $v = v_{z,h}$ such that

(3.2) $\sigma_{z,h}(\ell_j)v = 0$ for $j > 0$; $\sigma_{z,h}(\ell_o)v = hv$; $\sigma_{z,h}(\tilde{c}) = zI$.

Note an analogy of this definition with that of highest weight representation of $\hat{\mathfrak{g}}$. Similarly, provided that z and h are real numbers, V(z,h) carries a unique Hermitian form $\langle .|.\rangle$ such that $\langle v_{z,h}|v_{z,h}\rangle = 1$ and

(3.3) $\langle\sigma_{z,h}(\ell_j)u|v\rangle = \langle u|\sigma_{z,h}(\ell_{-j})v\rangle$ for all $u,v \in V(z,h)$.

The representation $\sigma_{z,h}$ is called <u>unitarizable</u> if this Hermitian form is positive definite.

With respect to $\sigma_{z,h}(\ell_o)$ we have the eigenspace decomposition

(3.4) $V(z,h) = \bigoplus_{k\in h+\mathbb{Z}_+} V(z,h)_k$, where $\dim V(z,h)_k < \infty$.

We define the character of the representation $\sigma_{z,h}$ by

$$((3.5)\quad ch_{z,h} = \sum_{k \in h+\mathbb{Z}_+} (\dim V(z,h)_k) q^k \ (= \operatorname{tr} q^{\ell_o}) .$$

Note that putting (cf. §2) :

$$(3.6)\quad \pi(\ell_j) = L_j \ , \quad \pi(\tilde{c}) = 3p(m,m')I,$$

we obtain a unitarizable representation of the Virasoro algebra on the space $L(\lambda') \otimes L(\lambda)$. It decomposes into a direct sum of unitarizable highest weight representations of Vir with "central charge" $3p(m,m')$. Note that the central charge z_m (defined by (0.1)) occurs if one takes $\lambda' = d$ and λ of level m [7]. In the next section we show that all $h_{r,s}^{(m)}$ from (0.1) occur in this construction as well and, moreover, we "locate" the corresponding representations of Vir .

§4. Fix $\lambda = md + \frac{1}{2} n\alpha \in P_+^o$, and put $J_\lambda = \{k \in \mathbb{Z} \mid -\frac{1}{2}(m+1-n) \leq k \leq \frac{1}{2} n\}$. Define the following subspace for $k \in J_\lambda$:

$$U_{\lambda,k} = \{v \in (L(d) \otimes L(\lambda))^{\mathfrak{n}} \mid (\pi_d \otimes \pi_\lambda)(\alpha)v = (n-2k)v\} .$$

Note that this is the subspace spanned by highest weight vectors of $\mathfrak{g}'$ in $L(d) \otimes L(\lambda)$ with weight $d+\lambda-k\alpha$. In particular, $(L(d) \otimes L(\lambda))^{\mathfrak{n}}$ decomposes into a direct sum of the $U_{\lambda,k}$. Furthermore, $U_{\lambda,k}$ is invariant with respect to d and hence decomposes into a direct sum of its eigenspaces $U_{\lambda,k}^{(j)}$ (with eigenvalue $j \in \mathbb{Z}$) . Note that every non-zero vector of $U_{\lambda,k}^{(j)}$ is a highest weight vector for $\hat{\mathfrak{g}}$ with highest weight $d+\lambda-k\alpha+jc$. In other words, $\dim U_{\lambda,k}^{(j)}$ is the multiplicity of occurence of $L(d+\lambda-k\alpha+jc)$ in $L(d) \otimes L(\lambda)$. Here and further on we use the fact that all representations in question are completely reducible with respect to $\hat{\mathfrak{g}}$ and Vir (since they are unitarizable).

Putting $m_{\lambda,k}(q) = \sum_j (\dim U_{\lambda,k}^{(j)}) q^{-j}$, we have :

$$(4.1)\quad ch_d \, ch_\lambda = \sum_{k \in J_\lambda} m_{\lambda,k} \, ch_{d+\lambda-k\alpha} .$$

To compute the $m_{\lambda,k}$ we multiply formulas (1.9) and (1.8) and use the following multiplication formula of theta functions [12, p.188] :

$$(4.2)\quad \Theta_{n,m}\Theta_{n',m'} = \sum_{j \in \mathbb{Z} \bmod (m+m')\mathbb{Z}} d_j^{(m,m',n,n')} \Theta_{n+n'+2mj,m+m'} \text{ , where}$$

$$d_j^{(m,m',n,n')}(q) = \Theta_{m'n-mn'+2jmm',mm'(m+m')}(\tau,0,0) .$$

We obtain :

$$(4.3)\quad m_{\lambda,k} = \varphi(q)^{-1}(f_k^{(m,n)} - f_{n+1-k}^{(m,n)}) \quad ,$$

where

$$(4.3a)\quad f_k^{(m,n)} = \sum_{j\in\mathbb{Z}} q^{(m+2)(m+3)j^2+((n+1)+2k(m+2))j+k^2} \, .$$

(Formula (4.3) may be also derived from [4]).

On the other hand, it follows from (2.5) that the subspace $U_{\lambda,k}$ is invariant with respect to Vir and thus carries a unitary representation of Vir . Putting $m' = 1$ in (3.6) and (2.3) we find (as GKO did) that the central charge of this representation is z_m (see (0.1)). Furthermore, it is clear from (4.3) that the minimal eigenvalue of $-d$ on $U_{\lambda,k}$ is k^2 . But we have by (2.4) and (2.2) :

$$(4.4)\quad L_o = -d + \frac{n(n+2)}{4(m+2)} - \frac{(n-2k)(n-2k+2)}{4(m+3)} \quad \text{on} \quad U_{\lambda,k} \, .$$

Defining numbers r_λ and $s_{\lambda,k}$ by $r_\lambda = n+1$, $s_{\lambda,k} = n+1-2k$ if $k \geq 0$ and $r_\lambda = m-n+1$, $s_{\lambda,k} = m-n+2+2k$ if $k < 0$, we arrive at the following

Lemma 4.1. The minimal eigenvalue of L_o on $U_{\lambda,k}$ is $h^{(m)}_{r_\lambda,s_{\lambda,k}}$.

Thus, $U_{\lambda,k}$ contains the unitary representation of Vir, which we denote by σ for short, with highest weight $(z_m, h^{(m)}_{r_\lambda,s_{\lambda,k}})$. But actually it coincides with this representation. Indeed $\operatorname{tr} q^{L_o}$ on $U_{\lambda,k}$ is equal to $m_{\lambda,k}(q)$ (given by (4.3)) multiplied by a power of q equal to the constant in the right-hand side of (4.4). Comparing this with the Feigin-Fuchs character formula for σ [3] (see [15] for an exposition of their results) we find that the character of σ coincides with $\operatorname{tr} q^{L_o}$ on $U_{\lambda,k}$!

We summarize the results obtained in the following theorem.

Theorem 4.1. (a) All highest weight representations of the Virasoro algebra with highest weights $(z_m, h^{(m)}_{r,s})$ given by (0.1) are unitary. Moreover, all these representations appear with multiplicity 1 in $\bigoplus_{\lambda\in P_+^o} (L(d) \otimes L(\lambda))^{\mathfrak{n}}$.

(b) With respect to the direct sum of $\mathfrak{g}'$ and Vir , we have the following decomposition, for $\lambda \in P_+^o$ of level m :

$$L(d) \otimes L(\lambda) = \bigoplus_{k\in J_\lambda} (L(d+\lambda-k\alpha) \otimes V(z_m, h^{(m)}_{r_\lambda,s_{\lambda,k}})) \, .$$

Remark 4.1. The characters $ch_{z_m,h^{(m)}_{r,s}}$ become holomorphic modular forms in τ of weight 0 on the upper half-plane when multiplied by a suitable power of q . Since they coincide with $m_{\lambda,k}$ multiplied by a power of q, it follows from [12, p.243] that the linear span of these "corrected" characters for fixed m and all $h^{(m)}_{r,s}$ from(0.1) form an $(m+1)(m+2)/2$-dimensional space invariant with respect to the usual action of $SL_2(\mathbb{Z})$ $(f(\tau)\mapsto f((a\tau+b)/(c\tau+d)))$.

Remark 4.2. Theorem 4.1(a) gives us what is called a model (i.e. a space where each representation of a given family appears once) for all unitary representations of the Virasoro algebra with $z<1$. A model for all degenerate representations with $z=1$ was constructed in [9]. Namely, the space $(L(d)\oplus L(d+\frac{1}{2}\alpha))^e$ contains exactly once all representations $V(1,\frac{m^2}{4})$, $m\in\mathbb{Z}_+$, so that with respect to the direct sum of $\mathfrak{g}$ and Vir we have [9] :

$$L(d)\oplus L(d+\frac{1}{2}\alpha) = \bigoplus_{m\in\mathbb{Z}_+} (T_{m+1}\otimes V(1,\frac{m^2}{4})),$$

where T_m denotes the m-dimensional irreducible representation of $\mathfrak{g}=s\ell_2(\mathbb{C})$.

§5. We now turn to the supersymmetric extensions of the above results. The terminology and conventions of Lie superalgebra theory adopted here are that of [14, §1.1].

Fix $\varepsilon=\frac{1}{2}$ or 0. Take the superloop algebra $\tilde{\mathfrak{g}}_\varepsilon = s\ell_2(\mathbb{C}[t,t^{-1},\theta])$, where $\theta^2=0$, and put $x(k+\varepsilon)' = t^k\theta x$ for $x\in\mathfrak{g}$ and $k\in\mathbb{Z}$. Define the affine superalgebra [13] $\hat{\mathfrak{g}}_\varepsilon=\tilde{\mathfrak{g}}_\varepsilon\oplus\mathbb{C}c\oplus\mathbb{C}d$ with the (super)bracket defined by (1.3) and

(5.1a) $[x(k)',y(n)']_+ = \delta_{k,-n}(\mathrm{tr}\,xy)c$ for $k,n\in\varepsilon+\mathbb{Z}$;

(5.1b) $[x(k),y(n)'] = (xy-yx)(k+n)'$ for $k\in\mathbb{Z}$, $n\in\varepsilon+\mathbb{Z}$;

(5.1c) $[d,x(k)'] = kx(k)'$ for $k\in\varepsilon+\mathbb{Z}$; $[c,\hat{\mathfrak{g}}_\varepsilon]=0$.

The Lie superalgebra $\hat{\mathfrak{g}}_\varepsilon$ contains $\hat{\mathfrak{g}}$ as the even part and $\hat{h}$ is called the Cartan subalgebra of $\hat{\mathfrak{g}}_\varepsilon$. Also, $\hat{\mathfrak{g}}_\varepsilon=\tilde{\mathfrak{g}}_\varepsilon+\mathbb{C}c$ is a subalgebra of $\hat{\mathfrak{g}}_\varepsilon$. Put $\hat{n}_{1/2} = \hat{n}+\sum_{k>0}\theta t^k\mathfrak{g}$ and $\hat{n}_0=\hat{n}+\mathbb{C}\theta e+\sum_{k>0}\theta t^k\mathfrak{g}$. For $\lambda\in\hat{h}$ define the ($\mathbb{Z}_2$-graded) irreducible highest weight representation $(\pi_{\lambda;\varepsilon},L_\varepsilon(\lambda))$ of $\hat{\mathfrak{g}}_\varepsilon$ by the property (1.5) where $\hat{n}$ is replaced by $\hat{n}_\varepsilon$. Unitarizability of $\pi_{\lambda;\varepsilon}$ and its character $ch_{\lambda;\varepsilon}$ are defined in the same way as for π_λ [13].

The representation of $\hat{\mathfrak{g}}_\varepsilon$ with highest weight $\lambda_\varepsilon = 2d+(\frac{1}{2}-\varepsilon)\alpha$ is called minimal [13]. With respect to $\hat{\mathfrak{g}}$ it decomposes as follows :

(5.2) $L_{1/2}(\lambda_{1/2}) = L(2d)\oplus L(2d+\alpha-\frac{1}{2}c)$; $L_0(\lambda_0) = L(\lambda_0)\oplus L(\lambda_0)$.

Denote the right-hand sides of (5.2) by F_ε. Given a representation (π,V) of $\hat{\mathfrak{g}}$, one can construct its "supersymmetrization" $(\pi^\varepsilon,V^\varepsilon)$ [13], which with respect to $\hat{\mathfrak{g}}$

is just $F_\varepsilon \otimes V$. It is shown in [13] that all unitarizable highest weight representations of $\hat{\mathfrak{g}}_\varepsilon$ are of the form $\pi_{\lambda+\lambda_\varepsilon;\varepsilon}$, $\lambda \in P_+$, and that $\pi_{\lambda+\lambda_\varepsilon;\varepsilon} = \pi_\lambda^\varepsilon$. It follows that with respect to $\hat{\mathfrak{g}}$ we have :

(5.3) $$L_\varepsilon(\lambda_\varepsilon) \otimes L_\varepsilon(\lambda+\lambda_\varepsilon) \simeq (F_\varepsilon \otimes L(\lambda))^\varepsilon \quad , \lambda \in P_+ .$$

We denote by Vir_ε the complex Lie superalgebra with a basis $\{\tilde{c}; \ell_j, j \in \mathbb{Z}$, and g_j , $j \in \varepsilon+\mathbb{Z}\}$ with commutation relations (3.1) and

(5.4a) $$[g_m, \ell_n] = (m-\frac{n}{2})g_{m+n} \quad ; \quad [g_m, \tilde{c}] = 0 ;$$

(5.4b) $$[g_m, g_n]_+ = 2\ell_{m+n} + \frac{1}{3}(m^2-\frac{1}{4})\delta_{m,-n}\tilde{c} .$$

(For $\varepsilon = \frac{1}{2}$ or 0 , Vir_ε is called the Neveu-Schwarz and Ramond superalgebras, respectively). The highest weight representation $(\sigma_{z,h;\varepsilon}, V_\varepsilon(z,h))$ of Vir_ε is defined by (3.2) and $\sigma_{z,h;\varepsilon}(g_j)v_{z,h} = 0$ for $j > 0$. Its unitarizability and character $\mathrm{ch}_{z,h;\varepsilon}$ are defined in the same way as for $\sigma_{z,h}$ in §3.

The analysis of the unitarizability of the representations $\sigma_{z,h;\varepsilon}$ is similar to that of $\sigma_{z,h}$ [5], [6], [9], [10], [13]. It turned out that these representations are unitarizable for $z \geq \frac{3}{2}$ and $h \geq 0$ [6], [10]. (Note that $\mathrm{ch}_{z,h;\varepsilon} = (2-2\varepsilon)q^h/\varphi_\varepsilon(q)$, the character of the Verma module, if $z > \frac{3}{2}$ and $h \geq 0$). Furthermore, the only other possible places of unitarity are $(z_{m;\varepsilon}, h^{(m)}_{r,s;\varepsilon})$, where [5], [6] :

(5.5) $$z_{m;\varepsilon} = \frac{3}{2}\left(1-\frac{8}{(m+2)(m+4)}\right) ; \quad h^{(m)}_{r,s;\varepsilon} = \frac{((m+4)r-(m+2)s)^2-4}{8(m+2)(m+4)} + \frac{1}{8}\left(\frac{1}{2}-\varepsilon\right) .$$

Here $m,r,s \in \mathbb{Z}_+$, $1 \leq s \leq r+1-2\varepsilon \leq m+2-2\varepsilon$ and $r-s \in 2\varepsilon+1+2\mathbb{Z}$, $r \neq 0$.

Let $\lambda, \lambda' \in P_+$ be of level m and m' . In the same way as in §2, one can construct intertwining operators $L_j^{(\varepsilon)}$ and $G_j^{(\varepsilon)}$ on the space $L_\varepsilon(\lambda+\lambda_\varepsilon) \otimes L_\varepsilon(\lambda'+\lambda_\varepsilon)$ (see [13]) which satisfy commutation relations (5.4) with central charge

(5.6) $$3\left(\frac{m}{m+2} + \frac{m'}{m'+2} - \frac{m+m'+2}{m+m'+4}\right) + \frac{3}{2}$$

and with the following expression for $L_0^{(\varepsilon)}$ on the kernel of $\hat{n}_\varepsilon$:

(5.7) $$\frac{1}{2}\left(\frac{(\lambda|\lambda+\alpha)}{m+2} + \frac{(\lambda'|\lambda'+\alpha)}{m'+2} - \frac{\frac{1}{2}\alpha^2+\alpha}{m+m'+4}\right)-d+\frac{3}{8}\left(\frac{1}{2}-\varepsilon\right) .$$

Now take $\lambda' = 0$ (so that $m' = 0$) and $\lambda = md+\frac{1}{2}n\alpha \in P_+^o$. Then (as pointed out in [13]) , we get all the central charges $z_{m;\varepsilon}$. We proceed as for the Virasoro algebra, to show that all the h's from (5.5) occur as well. Put $J_{\lambda;\varepsilon} = \{k \in \mathbb{Z} \mid -\frac{m-n+1}{2} -\varepsilon \le k \le \frac{n+1}{2} -\varepsilon\}$ and, for $k \in J_{\lambda;\varepsilon}$, put

$$U_{\lambda,k;\varepsilon} = \{v \in (F_\varepsilon \otimes L(\lambda))^{\mathfrak{n}} \mid (\pi_{\lambda_\varepsilon} \otimes \pi_\lambda)(\alpha)v = (n-2k+1-2\varepsilon)v\} .$$

Then the subspace spanned by all highest weight vectors of $\hat{\mathfrak{g}}'_\varepsilon$ in $L_\varepsilon(\lambda_\varepsilon) \otimes L_\varepsilon(\lambda+\lambda_\varepsilon)$ of weight $2\lambda_\varepsilon+\lambda-k\alpha$ coincides with $v_{\lambda_\varepsilon} \otimes U_{\lambda,k;\varepsilon}$, where v_{λ_ε} is the highest weight vector os F_ε (see (5.3)), and $(L_\varepsilon(\lambda_\varepsilon) \otimes L_\varepsilon(\lambda+\lambda_\varepsilon))^{\hat{\mathfrak{n}}_\varepsilon}$ decomposes into a direct sum of these subspaces with $k \in J_{\lambda;\varepsilon}$. Each subspace $U_{\lambda,k;\varepsilon}$ decomposes with respect to d into a direct sum of eigenspaces $U^{(j)}_{\lambda,k;\varepsilon}$ with eigenvalue $j \in \varepsilon+\mathbb{Z}$. Putting $m_{\lambda,k;\varepsilon} = \sum_j (\dim U^{(j)}_{\lambda,k;\varepsilon})q^{-j}$, we have

$$\text{(5.8)} \quad ch_{\lambda_\varepsilon;\varepsilon} ch_{\lambda+\lambda_\varepsilon;\varepsilon} = \sum_{k \in J_{\lambda;\varepsilon}} m_{\lambda,k;\varepsilon} ch_{\lambda+2\lambda_\varepsilon-k\alpha;\varepsilon} .$$

To compute the $m_{\lambda,k;\varepsilon}$, we multiply formulas (1.10) (resp. (1.11)) and (1.8) and use (4.2).

We obtain :

$$\text{(5.9)} \quad m_{\lambda,k;\varepsilon} = (2-2\varepsilon)\varphi_\varepsilon(q)^{-1} (f^{(m,n)}_{k,\varepsilon} - f^{(m,n)}_{n+1-k,\varepsilon}) ,$$

where

$$\text{(5.9a)} \quad f^{(m,n)}_{k,\varepsilon} = \sum_{j \in \mathbb{Z}} q^{\frac{1}{2}(m+2)(m+4)j^2+((n+1)+(k+\varepsilon-\frac{1}{2})(m+2))j+\frac{1}{2}k^2-(\frac{1}{2}-\varepsilon)k} .$$

Using (5.7) and (5.9), we find that the lowest eigenvalue of $L_o^{(\varepsilon)}$ on $U_{\lambda,k;\varepsilon}$ is

$$\text{(5.10)} \quad \frac{[(n+1)+(k+\varepsilon-1/2)(m+2)]^2-1}{2(m+2)(m+4)} + \frac{1}{8}(\frac{1}{2}-\varepsilon) .$$

Define numbers r_λ and $s_{\lambda,k}$ by $r_\lambda = n+1$, $s_{\lambda,k} = n+2-2\varepsilon-2k$ if $k \ge 0$, and $r_\lambda = m-n+1$, $s_{\lambda,k} = m-n+2k+2+2\varepsilon$ if $k < 0$. We arrive at the following theorem.

Theorem 5.1. (a) All highest weight representations of the Neveu-Schwarz and Ramond superalgebras Vir_ε with highest weights (5.5) are unitary. All these representations appear in

$$\bigoplus_{\lambda \in P^o_+} (L_\varepsilon(\lambda_\varepsilon) \otimes L_\varepsilon(\lambda+\lambda_\varepsilon))^{\hat{\mathfrak{n}}_\varepsilon}$$

with multiplicity one, except for $(z_{m;\varepsilon}, h^{(m)}_{r+2\varepsilon,r+1;\varepsilon})$ with $m \neq 2r$, which appears twice.

(b) Given $\lambda \in P^o_+$ of level m, we have the following decomposition with respect to the direct sum of $\hat{\mathfrak{g}}'_\varepsilon$ and Vir_ε :

$$L_\varepsilon(\lambda_\varepsilon) \otimes L_\varepsilon(\lambda+\lambda_\varepsilon) = \bigoplus_{k\in J_{\lambda;\varepsilon}} L_\varepsilon(\lambda+2\lambda_\varepsilon-k\alpha) \otimes V(z_{m;\varepsilon}, h^{(m)}_{r_\lambda, s_{\lambda,k};\varepsilon}).$$

Remark 5.1. The proof of Theorem 5.1 (b) and the part of 5.1(a) concerning multiplicities require showing that, up to multiplication by a suitable power of q, we have the following equality :

$$(5.11)\quad ch_{z_{m;\varepsilon}, h^{(m)}_{r_\lambda, s_{\lambda,k};\varepsilon}} = m_{\lambda,k;\varepsilon} \quad \text{(given by (5.9))}.$$

This can be done by applying the Feigin-Fuchs analysis [3] to Vir_ε. Let us say that a number from the set $\{h^{(m)}_{r_\lambda, s_{\lambda,k};\varepsilon} | k \in J_{\lambda;\varepsilon}\}$ is good if adding to it a positive integer never gives a number from this set. It follows from (5.5) and (5.9) that for $(z_{m;\varepsilon}, h)$ with good h, (5.11) holds automatically. This observation proves (5.11) in most of the cases (but not in all of them). Similar remark holds, of course, for Vir.

Remark 5.2. Taking integral and half-integral powers of q in $m_{\lambda,k;1/2}$ gives the characters of the even and odd part for the Neveu-Schwarz superalgebra. For the Ramond superalgebra these two characters are both equal to the half of $m_{\lambda,k;o}$, since g_o is invertible and hence permutes the even and odd parts of all representations in question (since $g_o^2 = \ell_o - \frac{1}{24}\tilde{c}$ and the spectrum of ℓ_o on all unitarizable representations from (5.5) with $\varepsilon = 0$ is greater than $\frac{1}{24}$).

Remark 5.3. Vir_ε acts on $L_\varepsilon(\lambda_\varepsilon)$, commuting with $\mathfrak{g}(\subset\hat{\mathfrak{g}}_\varepsilon)$, hence on $L_\varepsilon(\lambda_\varepsilon)^e$, with central charge $z = \frac{3}{2}$ [13]. It is not difficult to show that $L_\varepsilon(\lambda_\varepsilon)^e$ is a model for degenerate highest weight representations of Vir_ε with $z = \frac{3}{2}$. More precizely, with respect to the direct sum of $\mathfrak{g}$ and Vir_ε we have the following decomposition :

$$L_\varepsilon(\lambda_\varepsilon) = \sum_{k\in \mathbb{Z}_+} T_{2k+2-2\varepsilon} \otimes V_\varepsilon\left(\frac{3}{2}, \frac{k^2+(1-2\varepsilon)k}{2} + \frac{3}{8}\left(\frac{1}{2}-\varepsilon\right)\right).$$

Remark 5.4. Using the above construction, we can give a very simple proof of the formulas for $\det_n(z,h)$ of the determinant of the contravariant form on the subspace of elements of degree n of the Verma module with highest weight (z,h) (cf. [9], [2], [3], [6], [17],...). Consider, for example, the case of Vir (the argument for Vir_ε is exactly the same). It follows from (4.3) and the fact that Vir acts on $U_{\lambda,k}$, that

$$ch_{z_m, h^{(m)}_{r,s}} \leq q^{h^{(m)}_{r,s}} \varphi(q)^{-1}(1-q^{rs}-q^{(m+2-r)(m+3-s)} + \dots).$$

Hence the kernel of the contravariant form on the Verma module with highest weight

$(z_m,h^{(m)}_{r,s})$ contains non-zero vectors of degree rs and $(m+2-r)(m+3-s)$. Hence, $h = h^{(m)}_{r,s}$ are roots of $\det_{rs}(z_m,h)$ for all $r,s > 0$. So, as a polynomial in two variables, $\det_{rs}(z,h)$ vanishes at infinitely many points of the curve $\phi_{r,s}(z,h)=0$, where $\phi_{r,s}$ is defined by $\phi_{r,s}(z_m,h) = (h-h^{(m)}_{r,s})(h-h^{(m)}_{s,r})$. Thus, $\det_{rs}(z,h)$ is divisible by $\phi_{r,s}(z,h)$ if $r \neq s$ or by its square root if $r = s$. An easy induction on n , as in [2, §4.2] , completes the proof of the formula [8], [9]:

$$(\det_n(z,h))^2 = \text{const} \prod_{a=1}^{n} \prod_{j|a} \phi_{j,a/j}(z,h)^{p(n-a)} ,$$

where const $\neq 0$ depends only on the choice of basis. The argument for Vir_ε is given in Appendix 1.

Appendix 1. A proof of the determinantal formulas.

We give here, for the convenience of the reader, a selfcontained proof of the determinantal formulas for Vir_ε.

Given numbers z and h, there exists a unique ($\mathbb{Z}_2$- graded) module $M_\varepsilon(z,h)$ over Vir_ε, called Verma module, which admits a non-zero vector $v_{z,h}$, such that $\ell_0 v_{z,h} = hv_{z,h}$, $\tilde{c}v_{z,h} = zv_{z,h}$ and the vectors

$$v(i_1,\dots,i_\alpha;j_1,\dots,j_\beta) = g_{-j_\beta}\cdots g_{-j_1}\ell_{-i_\alpha}\cdots\ell_{-i_1}v_{z,h}$$

with $0 < i_1 \leq \dots \leq i_\alpha$ and $0 \leq j_1 < \dots < j_\beta$ form a basis of $M_\varepsilon(z,h)$ (in particular, $\ell_j z_{v,h} = 0$ and $g_j v_{z,h} = 0$ for $j > 0$). The space $M_\varepsilon(z,h)$ carries a unique Hermitian form $\langle\cdot|\cdot\rangle$ such that the norm[1)] of $v_{z,h}$ is 1 and $\ell_j^* = \ell_{-j}$, $g_j^* = g_{-j}$, called the contravariant Hermitian form. With respect to ℓ_0, $M_\varepsilon(z,h)$ decomposes into an orthogonal direct sum of eigenspaces $M_\varepsilon(z,h)_n$ with eigenvalues $h + n$, where $n \in (1-\varepsilon)\mathbb{Z}_+$. We say that vectors from $M_\varepsilon(z,h)_n$ have degree n. Let $M_\varepsilon(z,h)^+_n$ and $M_\varepsilon(z,h)^-_n$ denote the even (resp. odd) part of $M(z,h)_n$. We have : $M_{\frac12}(z,h)_n = M_{\frac12}(z,h)^+_n$ (resp. $= M_{\frac12}(z,h)^-_n$) if $n \in \mathbb{Z}_+$ (resp. $n \in \frac{1}{2}+\mathbb{Z}_+$) and $M_0(z,h)_n$ is an orthogonal direct sum of subspaces $M_0(z,h)^+_n$ and $M_0(z,h)^-_n$. Let $p_\varepsilon(n)$ be the coefficient of q^n in the power series expansion of $\varphi_\varepsilon(q)^{-1}$. Note that

(6.1) $\dim M_{\frac12}(z,h)_n = p_{\frac12}(n)$; $\dim M_0(z,h)^{\pm}_n = p_0(n)$.

We put $p_0^{\pm}(n) = p_0(n) \mp \delta_{n,0}$, and

$$\phi_{r,s;\varepsilon}(z_{m;\varepsilon},h) = (h-h^{(m)}_{r,s;\varepsilon})(h-h^{(m)}_{s,r;\varepsilon}).$$

1) We assume that the even and odd subspaces are orthogonal (this is not satisfied automatically if $\varepsilon = 0$).

Note that $\phi_{r,s;\varepsilon}(z,h)$ is a polynomial (of degree 2) in h and z. Given $n \in (2-2\varepsilon)\mathbb{Z}_+$, let $\det^{\pm}_{\frac{1}{2}n}(z,h)_\varepsilon$ denotes the determinant of the contravariant Hermitian form on $M_\varepsilon(z,h)^{\pm}_n$. The aim of this appendix is to prove the following formula (cf. [9] and [6]) :

$$(6.2)\quad \det^{\pm}_{\frac{1}{2}n}(z,h)^2_\varepsilon = \text{const}\,(h-\tfrac{1}{24}z)^{(1-2\varepsilon)p^{\pm}_0(\frac{1}{2}n)} \prod_{\substack{a,b\in\mathbb{Z}_+ \\ 1\le ab\le n \\ a-b\in 2\varepsilon+1+2\mathbb{Z}}} \phi_{a,b;\varepsilon}(z,h)^{p_\varepsilon(\frac{1}{2}(n-ab))}$$

where const. is a non-zero constant, independent of z and h.

As in Remark 5.4, it follows from (5.9) and the fact that Vir_ε acts on $U_{\lambda,k;\varepsilon}$, that

$$\mathrm{ch}_{z_{m;\varepsilon},h^{(m)}_{r,s;\varepsilon}}(q) \le q^{h^{(m)}_{r,s;\varepsilon}}\varphi_\varepsilon(q)^{-1}(1-q^{\frac{1}{2}rs}-q^{\frac{1}{2}(m+2-r)(m+4-s)}+\ldots).$$

Since $L_\varepsilon(z,h)$ is the quotient of $M_\varepsilon(z,h)$ by the kernel of the contravariant form, it follows that for $M(z_{m;\varepsilon},h^{(m)}_{r,s;\varepsilon})$ this kernel contains non-zero vectors of degree $\frac{1}{2}rs$ and $\frac{1}{2}(m+2-r)(m+2-s)$. It follows that for all a and b as in (6.2), $\det_{\frac{1}{2}ab}(z,h)_\varepsilon$ is divisible by $\phi_{a,b;\varepsilon}(z,h)$ if $a \ne b$ or by its square root if $a=b$. Furthermore, it is clear that $g_0 v_{z+h}$ is in the kernel of $\langle\cdot|\cdot\rangle$ if $h=\frac{1}{24}z$ (and $\varepsilon=0$); also g_0 is invertible on $M_0(z,h)$ if $h>\frac{1}{24}z$. It follows that for all a and b as in (6.2), $\det^{\pm}_{\frac{1}{2}ab}(z,h)_0$ is divisible by $\phi_{a,b;0}(z,h)$ and that $\det^-_0(z,h)$ is divisible by $h-\frac{1}{24}z$. An induction on n, using (6.1) and well-known elementary properties of Verma modules, proves that the left-hand side of (6.2) is divisible by its right-hand side.

We will show that, for a fixed z, the degree of $Q^{\pm}_{n;\varepsilon}(h) = \det^{\pm}_{\frac{1}{2}n}(z,h)_\varepsilon$, viewed as a polynomial in h, is exactly the half of the degree of the polynomial on the right of (6.2). Recall that the vectors $v(i_1,\ldots,i_\alpha;j_1,\ldots,j_\beta)$ with $i_1+\ldots+i_\alpha+j_1+\ldots+j_\beta = n$ and β even (resp. odd) form a basis of $M_\varepsilon(z,h)^+_n$ (resp. $M_\varepsilon(z,h)^-_n$), so that $Q^{\pm}_{n;\varepsilon}(h)$ is the determinant of the matrix of the inner products of these vectors. It is clear that only the product of the diagonal entries of this matrix gives a non-zero contribution to the highest power of h, and that $\langle v(i_1,\ldots,i_\alpha;j_1,\ldots,j_\beta|v(i_1,\ldots,i_\alpha;j_1,\ldots,j_\beta)\rangle$ has degree $\alpha+\beta$ in h. It is easy to deduce now that :

$$\deg Q_{n;\frac{1}{2}}(h) = \sum_{\substack{s>0\\ s \text{ even}}}\sum_{m>0} p_{\frac{1}{2}}(n-\tfrac{1}{2}ms) + \sum_{\substack{s>0\\ s\text{ odd}}}\sum_{m>0}(-1)^{m+1}\,p_{\frac{1}{2}}(n-\tfrac{1}{2}ms),$$

$$\deg Q^{\pm}_{n;0}(h) = \tfrac{1}{2}(p^{\pm}_0(n) + \sum_{s>0}\sum_{m>0}(p_0(n-ms) + (-1)^{m+1}\,p_0(n-ms)),$$

where s and m are integers. This completes the proof of (6.2).

Appendix 2. Multiplicative formulas for characters.

We present here formulas connecting the characters of discrete series representations of Vir and Vir_ε with specialized characters of affine Kac-Moody algebras of type $A_1^{(1)}$ and $A_2^{(2)}$. In many cases this gives simple product decompositions of characters of Vir and Vir_ε. In what follows we use freely notation and results of the book [11].

Let A be the generalized Cartan matrix of type $A_1^{(1)}$ or $A_2^{(2)}$. Let $g(A)$ be the associated Kac-Moody algebra. Let Δ_+ be the set of positive roots and let α_0,α_1 be simple roots (in the case $A_1^{(1)}$, $\alpha_0 = c-\alpha$ and $\alpha_1 = \alpha$). Let Λ_0,Λ_1 be fundamental weights (in the case $A_1^{(1)}$, $\Lambda_0 = d$ and $\Lambda_1 = d+\frac{1}{2}\alpha)$ and let $P_+^0 = \{k_0\Lambda_0+k_1\Lambda_1 | k_i \in \mathbb{Z}_+\}$. Given $\Lambda = k_0\Lambda_0+k_1\Lambda_1 \in P_+^0$, which is usually written as $\Lambda = (k_0,k_1)$, we have the integrable representation $L(\Lambda;A)$ of $g(A)$ with highest weight Λ.

Let $W(A)$ be the Weyl group and let $\rho = \Lambda_0 + \Lambda_1$. Given $\lambda \in \rho + P_+^0$, put

$$N_\lambda^{(A)} = \sum_{w \in W(A)} \operatorname{sgn}(w)\, e^{w\cdot\lambda-\lambda} .$$

Then the Weyl-Kac character and denominator formulas read [11, Chapter 10] :

$$e^{-\Lambda} \operatorname{ch} L(\Lambda;A) = N_{\Lambda+\rho}^{(A)}/N_\rho^{(A)} ; \tag{7.1}$$

$$N_\rho^{(A)} = \prod_{\alpha \in \Delta_+} (1-e^{-\alpha}) . \tag{7.2}$$

(Note that in the case $A_1^{(1)}$, formula (7.1) is another form of formula (1.8); note that in our cases, mult $\alpha = 1$ for all $\alpha \in \Delta_+$).

Given a pair of positive integers $t = (t_0,t_1)$, the algebra homomorphism $F_t^{(A)} : \mathbb{C}[[e^{-\alpha_0},e^{-\alpha_1}]] \longrightarrow \mathbb{C}[[q]]$ defined by $F_t^{(A)}(e^{-\alpha_i}) = q^{t_i}$ $(i = 0,1)$ is called the specialization of type t. In what follows we shall often write 1 and 2 in place of $A_1^{(1)}$ and $A_2^{(2)}$ respectively.

Fix $\Lambda = (M-1,N-1)$, where M and N are positive integers. Using that $W(A) = \{(r_0r_1)^n, (r_0r_1)^n r_0;\ n \in \mathbb{Z}\}$, one easily deduces the following formulas :

$$F_t^{(1)}(N_{\Lambda+\rho}^{(1)}) = \sum_{j \in \mathbb{Z}} q^{|t|(M+N)j^2+(|t|N-t_1(M+N))j} - \sum_{j \in \mathbb{Z}} q^{|t|(M+N)j^2+(|t|N+t_1(M+N))j+t_1N} \tag{7.3}$$

$$F_t^{(2)}(N_{\Lambda+\rho}^{(2)}) = \sum_{j \in \mathbb{Z}} q^{\frac{1}{2}\|t\|(M+2N)j^2+\frac{1}{2}(2\|t\|N-t_1(M+2N))j} - \sum_{j \in \mathbb{Z}} q^{\frac{1}{2}\|t\|(M+2N)j^2+\frac{1}{2}(2\|t\|N+t_1(M+2N))j+t_1N} , \tag{7.4}$$

where $|t| = t_0+t_1$ and $\|t\| = 2t_0+t_1$.

One knows the following general product decomposition [11, Chapter 10] :

$$F^{(A)}_{(1,1)}(N^{(A)}_{\Lambda+\rho}) = F^{(A)}_{(M,N)}(N^{(A)}_{\rho}). \tag{7.5}$$

Furthermore, there are the following special product decompositions[20] :

$$F^{(1)}_{(1,2)}(N^{(1)}_{\Lambda+\rho}) = F^{(2)}_{(M,2N)}(N^{(2)}_{\rho}) \tag{7.6a}$$

$$F^{(1)}_{(2,1)}(N^{(1)}_{\Lambda+\rho}) = F^{(2)}_{(N,2M)}(N^{(2)}_{\rho}) \tag{7.6b}$$

$$F^{(1)}_{t}(N^{(1)}_{(n,2n)}) = F^{(2)}_{(nt_0,2nt_1)}(N^{(2)}_{\rho}) \tag{7.6c}$$

$$F^{(1)}_{t}(N^{(1)}_{(2n,n)}) = F^{(2)}_{(nt_1,2nt_0)}(N^{(2)}_{\rho}). \tag{7.6d}$$

We put

$$d^{(t;A)}_{\Lambda}(q) = F^{(A)}_{t}(e^{-\Lambda}\ \mathrm{ch}\ L(\Lambda;A)).$$

In the case $t = \mathbb{1} = (1,1)$, $d^{(\mathbb{1};A)}_{\Lambda}(q)$ is called the q-dimension of $L(\Lambda;A)$; due to (7.5), it has a product decomposition.

We turn now to the product decompositions of the characters of the Virasoro algebra. For the sake of simplicity, we put

$$\chi^{(m)}_{r,s} = q^{-h^{(m)}_{r,s}}\ \mathrm{ch}_{z_m,h^{(m)}_{r,s}}(q).$$

Comparing formula (4.3) (which gives the character of a discrete series representation of Vir) with (7.3) and using (7.1) and (7.2), we arrive at the following beautiful formula.

<u>Proposition 7.1</u>. Take $1 \le s \le r \le m+1$, and put $\Lambda = (m+2-s,s-1)$ and $t = (m+2-r,r)$ (or $\Lambda = (m+1-r,r-1)$ and $t = (m+3-s,s)$ respectively). Then

$$\chi^{(m)}_{r,s}(q) = d^{(t;1)}_{\Lambda}(q) \prod_{\substack{j\geq 1 \\ j\not\equiv 0,\pm r \bmod (m+2)\ \ (\text{or } 0,\ \pm s \bmod (m+3) \text{ resp.})}} (1-q^j)^{-1}. \tag{7.7}$$

(If 2r=m+2 (or 2s=m+3 resp.), the product on the right should be interpreted in a usual way).

<u>Remark 7.1</u>. Formula (7.7) shows that $V(z_m,h^{(m)}_{r,s})$ is a tensor product of the $(m+2-r,r)$-graded space $L(m+2-s,s-1;A^{(1)}_1)$ and $(1,1)$-graded space $L(m+1-r,r-1;A^{(1)}_1)^{s_+}$, where s_+ is the "positive part" of the principal Heisenberg subalgebra of $\hat{s\ell}_2$. This suggests that there may be some more explicit constructions of the discrete series representations of the Virasoro algebra.

Using formulas (7.6), we can obtain, in some cases, from (7.7) multiplicative formulas. They are collected in Table 1, where, for simplicity, we use the abbreviated product symbol

$$\prod_j (1-q^{uj\pm v}) = \prod_{j\geq 0} (1-q^{uj+v}) \prod_{j\geq 1} (1-q^{uj-v}),$$

and similarly for "-" replaced by "+". If r and s do not satisfy the condition $1 \leq s \leq r \leq m+1$, it is assumed further on that they are brought to this form by transformation $r' = k(m+2)\pm r$, $s' = k(m+3)\pm s$, with some $k \in \mathbb{Z}$ (which leave $h^{(m)}_{r,s}$ unchanged).

Table 1

$$\chi^{(2r-2)}_{r,s}(q) = \frac{\varphi(q^{r(2r+1)})}{\varphi(q)} \prod_j (1-q^{r(2r+1)j\pm rs})$$

$$\chi^{(3r-2)}_{r,s}(q) = \frac{\varphi(q^{2r(3r+1)})}{\varphi(q)} \prod_j (1-q^{r(3r+1)j\pm rs})$$

$$\times \prod_{j=\mathrm{odd}} (1+q^{r(3r+1)j\pm rs})$$

$$\chi^{(3r-2)}_{2r,s}(q) = \frac{\varphi(q^{2r(3r+1)})}{\varphi(q)} \prod_j (1-q^{r(3r+1)j\pm rs})$$

$$\times \prod_{j=\mathrm{even}} (1+q^{r(3r+1)j\pm rs})$$

$$\chi^{(2s-3)}_{r,s}(q) = \frac{\varphi(q^{s(2s-1)})}{\varphi(q)} \prod_j (1-q^{s(2s-1)j\pm rs})$$

$$\chi^{(3s-3)}_{r,s}(q) = \frac{\varphi(q^{2s(3s-1)})}{\varphi(q)} \prod_j (1-q^{s(3s-1)j\pm rs})$$

$$\times \prod_{j=\mathrm{odd}} (1+q^{s(3s-1)j\pm rs})$$

$$\chi^{(3s-3)}_{r,2s}(q) = \frac{\varphi(q^{2s(3s-1)})}{\varphi(q)} \prod_j (1-q^{s(3s-1)j\pm rs})$$

$$\times \prod_{j=\mathrm{even}} (1+q^{s(3s-1)j\pm rs})$$

Next, we put

$$\psi^{(m)\pm}_{r,s} = q^{-h^{(m)}_{r,s}}(\mathrm{ch}_{z_m,h^{(m)}_{r,s}}(q) \pm \mathrm{ch}_{z_m,h^{(m)}_{m+2-r,s}}(q)) .$$

Then, in a similar way, we obtain the following table :

Table 2

$$\psi_{r,s}^{(4r-2)\pm}(q) = \frac{1}{\varphi(q)} \prod_j (1-(\mp 1)^j q^{\frac{r(4r+1)}{2} j})$$

$$\times \prod_j (1-(\mp 1)^j q^{\frac{r(4r+1)}{2} j \pm rs})$$

$$\psi_{r,s}^{(4s-3)\pm}(q) = \frac{1}{\varphi(q)} \prod_j (1-(\mp 1)^j q^{\frac{s(4s-1)}{2} j})$$

$$\times \prod_j (1-(\mp 1)^j q^{\frac{s(4s-1)}{2} j \pm rs})$$

$$\psi_{r,s}^{(3r-2)-}(q) = \frac{\varphi(q^{\frac{r(3r+1)}{2}})}{\varphi(q)} \prod_j (1-q^{\frac{r(3r+1)}{4} j \pm \frac{rs}{2}})$$

$$\times \prod_j (1+q^{\frac{r(3r+1)}{2} j \pm \frac{rs}{2}})$$

$$\psi_{r,s}^{(6r-2)-}(q) = \frac{\varphi(q^{r(6r+1)})}{\varphi(q)} \prod_j (1-q^{r(6r+1)j \pm rs})$$

$$\times \prod_{j=\mathrm{odd}} (1-q^{r(6r+1)j \pm 2rs})$$

$$\psi_{r,s}^{(3s-3)-}(q) = \frac{\varphi(q^{\frac{s(3s-1)}{2}})}{\varphi(q)} \prod_j (1-q^{\frac{s(3s-1)}{4} j \pm \frac{rs}{2}})$$

$$\times \prod_j (1+q^{\frac{s(3s-1)}{2} j \pm \frac{rs}{2}})$$

$$\psi_{r,s}^{(6s-3)-}(q) = \frac{\varphi(q^{s(6s-1)})}{\varphi(q)} \prod_j (1-q^{s(6s-1)j \pm rs})$$

$$\times \prod_{j=\mathrm{odd}} (1-q^{s(6s-1)j \pm 2rs})$$

Note that formulas from Tables 1 and 2 cover all cases for small m. The case $m = 1$ is well-known; the case $m = 2$ was worked out in [15].

In a similar way, one finds product decompositions for the characters of Vir_ε. Put [2)]

$$\chi^{(m)}_{r,s;\varepsilon}(q) = \frac{1}{2-2\varepsilon}\, q^{-h^{(m)}_{r,s;\varepsilon}}\, \mathrm{ch}_{z_{m;\varepsilon},h^{(m)}_{r,s;\varepsilon}}(q)\ .$$

Then we have

$$(7.8)\quad \chi^{(m)}_{r,s;\varepsilon}(q) = \frac{1}{\varphi_\varepsilon(q)}\, d_\Lambda^{(t;1)}(q^{\frac{1}{2}}) \prod_{\substack{j\geq 1\\ j\equiv 0,\pm r \bmod (m+2)}} (1-q^{j/2})\ ,$$

where $\Lambda = (m+3-s, s-1)$, $t = (m+2-r, r)$.

There are other formulas, similar to (7,8), which involve only integral j, and also, in some cases, multiplicative formulas for Vir_ε, similar to that from Tables 1 and 2 for Vir. We present some of these formulas in Tables 3 and 4.

Table 3

$$\chi^{(2r-2)}_{r,s;\varepsilon}(q) = \frac{\varphi(q^{r(r+1)})}{\varphi_\varepsilon(q)} \prod_j (1-q^{r(r+1)j\pm rs/2})$$

$$\chi^{(3r-2)}_{r,s;\varepsilon}(q) = \frac{\varphi(q^{r(3r+2)})}{\varphi_\varepsilon(q)} \prod_j (1-q^{r(3r+2)j\pm rs/2}) \times \prod_{j=\mathrm{odd}} (1-q^{r(3r+2)j\pm rs})$$

$$\chi^{(3r-2)}_{2r,s;\varepsilon}(q) = \frac{\varphi(q^{r(3r+2)})}{\varphi_\varepsilon(q)} \prod_j (1-q^{2r(3r+2)j\pm rs}) \times \prod_{j=\mathrm{odd}} (1-q^{(r(3r+2)/2)j\pm rs/2})$$

$$\chi^{(2s-4)}_{r,s;\varepsilon}(q) = \frac{\varphi(q^{s(s-1)})}{\varphi_\varepsilon(q)} \prod_j (1-q^{s(s-1)j\pm rs/2})$$

$$\chi^{(3s-4)}_{r,s;\varepsilon}(q) = \frac{\varphi(q^{s(3s-2)})}{\varphi_\varepsilon(q)} \prod_j (1-q^{s(3s-2)j\pm rs/2}) \times \prod_{j=\mathrm{odd}} (1-q^{s(3s-2)j\pm rs})$$

$$\chi^{(3s-4)}_{r,2s;\varepsilon}(q) = \frac{\varphi(q^{s(3s-2)})}{\varphi_\varepsilon(q)} \prod_j (1-q^{2s(3s-2)j\pm rs}) \times \prod_{j=\mathrm{odd}} (1-q^{(s(3s-2)/2)j\pm rs/2})$$

Table 4

$$\psi^{(4r-2)-}_{r,s;\varepsilon}(q) = \frac{\varphi(q^{r(r+\frac{1}{2})})}{\varphi_\varepsilon(q)} \prod_j (1-q^{r(r+\frac{1}{2})j\pm rs/2})$$

$$\psi^{(6r-2)-}_{r,s;\varepsilon}(q) = \frac{\varphi(q^{r(3r+1)})}{\varphi_\varepsilon(q)} \prod_j (1-q^{r(3r+1)j\pm rs/2}) \times \prod_{j=\mathrm{odd}} (1-q^{r(3r+1)j\pm rs})$$

$$\psi^{(6r-2)-}_{2r,s;\varepsilon}(q) = \frac{\varphi(q^{r(3r+1)})}{\varphi_\varepsilon(q)} \prod_j (1-q^{(r(3r+1)/2)j\pm rs/2}) \times \prod_j (1+q^{r(3r+1)j\pm rs/2})$$

$$\psi^{(4s-4)-}_{r,s;\varepsilon}(q) = \frac{\varphi(q^{s(s-\frac{1}{2})})}{\varphi_\varepsilon(q)} \prod_j (1-q^{s(s-\frac{1}{2})j\pm rs/2})$$

---

[2] The definition of $\psi^{(m)\pm}_{r,s;\varepsilon}$ is completely similar to that of $\psi^{(m)\pm}_{r,s}$.

$$\psi^{(6s-4)-}_{r,s;\varepsilon}(q) = \frac{\varphi(q^{s(3s-1)})}{\varphi_\varepsilon(q)} \prod_j (1-q^{s(3s-1)j\pm rs/2}) \times \prod_{j=odd} (1-q^{s(3s-1)j\pm rs})$$

$$\psi^{(6s-4)-}_{r,2s;\varepsilon}(q) = \frac{\varphi(q^{s(3s-1)})}{\varphi_\varepsilon(q)} \prod_j (1-q^{(s(3s-1)/2)j\pm rs/2}) \times \prod_j (1+q^{s(3s-1)j\pm rs/2})$$

$$\psi^{(4r-2)+}_{r,s;\varepsilon}(q) = \frac{1}{\varphi_\varepsilon(q)} \prod_j (1-(-1)^j q^{r(r+\frac{1}{2})j}) \times \prod_j (1-(-1)^j q^{r(r+\frac{1}{2})j\pm rs/2})$$

$$\psi^{(4s-4)+}_{r,s;\varepsilon}(q) = \frac{1}{\varphi_\varepsilon(q)} \prod_j (1-(-1)^j q^{s(s-\frac{1}{2})j}) \times \prod_j (1-(-1)^j q^{s(s-\frac{1}{2})j\pm rs/2})$$

Remark 7.2. It is always possible to write $\chi^{(m)}_{r,s}$ and $\chi^{(m)}_{r,s;\varepsilon}$ as a sum of two infinite products (using the Jacobi triple product identity) :

(7.9a) $$\chi^{(m)}_{r,s}(q) = \frac{\varphi(q^{2(m+2)(m+3)})}{\varphi(q)}$$

$$\times \Big[\prod_{\substack{j\geq 1\\ j=odd}} (1+q^{(m+2)(m+3)j\pm((m+3)r-(m+2)s)}) - q^{rs} \prod_{\substack{j\geq 1\\ j=odd}} (1+q^{(m+2)(m+3)j\pm((m+3)r+(m+2)s)}) \Big]$$

(7.9b) $$\chi^{(m)}_{r,s;\varepsilon}(q) = \frac{\varphi(q^{(m+2)(m+4)})}{\varphi_\varepsilon(q)}$$

$$\times \Big[\prod_{\substack{j\geq 1\\ j=odd}} (1+q^{\frac{(m+2)(m+4)}{2}j\pm\frac{(m+4)r-(m+2)s}{2}}) - q^{\frac{rs}{2}} \prod_{\substack{j\geq 1\\ j=odd}} (1+q^{\frac{(m+2)(m+4)}{2}j\pm\frac{(m+4)r+(m+2)s}{2}}) \Big]$$

Appendix 3. An application to the decomposition of tensor products of two level 1 representations of exceptional affine algebras.

In this appendix we will show that the affine Lie algebras $E_8^{(1)}$, $E_7^{(1)}$, $A_2^{(1)}$ and $A_2^{(2)}$, $E_6^{(1)}$ and $E_6^{(2)}$ provide a model for discrete series representations of the Vivasoro algebra with central charge z_m, where $m = 1,2,3,4$ respectively. Namely we will prove the following remarkable fact : taking tensor products of the basic representation with all level 1 fundamental representations of the affine algebras listed above, one gets (in the space of highest weight vectors) all discrete series representations of Vir for $m = 1,2,3,4$ and exactly once. Turning the point of view, "generalized string functions" [12, § 4.9] of the tensor product of two level 1 fundamental representations of the above affine algebras turn out to be nothing else but the characters of the corresponding discrete series representations of Vir.

As in Appendix 2, we will use freely the notation, conventions and results of the book [11]. In particular, the enumeration of the vertices of the Dynkin diagrams of affine algebras adopted here is that of [11, Chapters 4 and 6].

First, we will prove a few facts about Kac-Moody algebras which are used later on.

Lemma 8.1. Let $\mathfrak{g}(A)$ be a Kac-Moody algebra with a symmetrizable Cartan matrix. Let $\Lambda, \Lambda' \in P_+$ and $\sigma \in W$ be such that $M = \sigma\cdot\Lambda + \Lambda' \in P_+$. Then

(a) $\mathrm{mult}_\Lambda\,(M+\rho-w(\Lambda'+\rho))$ is 1 if $w = 1$ and is 0 if $w \in W$, $w \neq 1$.

(b) The multiplicity of $L(M)$ in $L(\Lambda)\otimes L(\Lambda')$ is 1.

Proof. Claim (a) for $w = 1$ is clear. If $w \neq 1$, then $(M+\rho|\Lambda'+\rho-w(\Lambda'+\rho)) > 0$, and we have :

$$|M+\rho-w(\Lambda'+\rho)|^2-|\Lambda|^2=|M+\rho|^2+|w(\Lambda'+\rho)|^2-2(M+\rho|w(\Lambda'+\rho)-(\Lambda'+\rho))-2(M+\rho|\Lambda'+\rho)-|\Lambda|^2$$

$$>|M+\rho|^2+|\Lambda'+\rho|^2-2(M+\rho|\Lambda'+\rho)-|\Lambda|^2=|M+\rho-(\Lambda'+\rho)|^2-|\Lambda|^2=|\sigma\cdot\Lambda|^2-|\Lambda|^2=0.$$

Thus, $|M+\rho-w(\Lambda'+\rho)|^2-|\Lambda|^2 > 0$ and hence (by [11, Proposition 11.4]), $M+\rho-w(\Lambda'+\rho)$ is not a weight of $L(\Lambda)$, which completes the proof of (a). Claim (b) follows from (a) and the Racah "outer multiplicity" formula (cf. [4]) : the multiplicity of $L(M)$ in $L(\Lambda)\otimes L(\Lambda')$ is $\sum_{w\in W} \varepsilon(w)\,\mathrm{mult}_\Lambda(M+\rho-w(\Lambda'+\rho))$.

Further on, S^2V and Λ^2V stand for the symmetric and antisymmetric square of the space V, respectively.

Lemma 8.2. Let $\mathfrak{g}(A)$ be an affine algebra of A-D-E type all of whose exponents are odd, and let $\Lambda \in P_+$ be of level 1. Suppose that $L(M)$ occurs in $L(\Lambda)\otimes L(\Lambda)$. Then $L(M) \subset S^2L(\Lambda)$ (resp. $\subset \Lambda^2L(\Lambda)$) if and only if $\mathrm{ht}(2\Lambda-M)$ is even (resp. odd).

Proof. Using a diagram automorphism of $\mathfrak{g}(A)$, we may assume that $\Lambda = \Lambda_0$. The (basic) representation $L(\Lambda_0)$ of $\mathfrak{g}(A)$ is realized on the space of polynomials $\mathbb{C}[u_j; j \in E_+]$, where $E_+ = \mathbb{Z}_+ \cap E$ and E is the set of exponents of $\mathfrak{g}(A)$, so that the principal gradation is given by $\deg u_j = j$, and $u_j \in n_-$ and $\frac{\partial}{\partial u_j} \in n_+ (\subset \mathfrak{g}(A))$, $j \in E_+$ (cf. [11], Chapter 14). But then

$$L(\Lambda_0)\otimes L(\Lambda_0) = \mathbb{C}[u_j^{(1)},u_j^{(2)};\ j\in E_+] = \mathbb{C}[x_j,y_j; j\in E_+],$$

where we put $x_j = u_j^{(1)}+u_j^{(2)}$ and $y_j = u_j^{(1)}-u_j^{(2)}$, so that $x_j \in n_-$ and $\frac{\partial}{\partial x_j} \in n_+$. Thus a highest weight vector of $L(M)$ is a polynomial in y_j's whose principal degree is equal to $\mathrm{ht}(2\Lambda_0-M)$.

Since E_+ consists of odd numbers, we deduce that

$$S^2L(\Lambda_0) = \mathbb{C}[x] \otimes \mathbb{C}_{even}[y]; \ \Lambda^2L(\Lambda_0) = \mathbb{C}[x] \otimes \mathbb{C}_{odd}[y],$$

where $\mathbb{C}_{even}[y]$ (resp. $\mathbb{C}_{odd}[y]$) denotes the subspace spanned by all monomials in y_j's of even (resp. odd) principal degree. This completes the proof of the lemma.

Let now A be an affine generalized Cartan matrix of type $X_N^{(k)}$, let $\mathfrak{g}(A)$ be the corresponding affine (Kac-Moody) algebra and let $d = \dim \mathfrak{g}(X_N)$ be the dimension of the "underlying" simple finite dimensional Lie algebra. Let $L(\Lambda')$ and $L(\Lambda'')$ be two highest weight representations of levels $m' = \Lambda'(c)$ and $m'' = \Lambda''(c)$, such that m', m'' and $m'+m'' \neq -g$, where g is the dual Coxeter number. Then (as has been mentioned in § 2), Vir acts on $L(\Lambda') \otimes L(\Lambda'')$ commuting with $\mathfrak{g}'(A)$, and formulas, corresponding to (2.3 a,b) and (2.4) generalize as follows (cf. [12],[18]) :

(8.1a) the central charge = $dp(m',m'')$, where

(8.1b) $p(m',m'') = \frac{m'}{m'+g} + \frac{m''}{m''+g} - \frac{m'+m''}{m'+m''+g}$

(8.2) $L_0 = \frac{1}{2k}\left[\frac{(\Lambda'|\Lambda'+2\rho)}{m'+g} + \frac{(\Lambda''|\Lambda''+2\rho)}{m''+g} - \frac{\Omega}{m'+m''+g}\right] + \left[\frac{d}{24} - \frac{|\rho|^2}{2gk}\right] p(m',m'')$.

Note that the second term on the right in (8.2) vanishes if $k = 1$ due to the Frendenthal-de Vries strange formula, whereas in case $k > 1$ it is "alive" and will play an important role.

The main result of this Appendix is the following theorem.

Theorem 8.1. One has the following decompositions with respect to the direct sum of $\mathfrak{g}'(A)$ and Vir :

1) $A = E_8^{(1)}$:

$$S^2L(\Lambda_0) = L(2\Lambda_0) \otimes V(\tfrac{1}{2},0) + L(\Lambda_7) \otimes V(\tfrac{1}{2},\tfrac{1}{2}), \ \Lambda^2L(\Lambda_0) = L(\Lambda_1) \otimes V(\tfrac{1}{2},\tfrac{1}{16}).$$

2) $A = E_7^{(1)}$:

$$S^2L(\Lambda_0) = L(2\Lambda_0) \otimes V(\tfrac{7}{10},0) + L(\Lambda_5) \otimes V(\tfrac{7}{10},\tfrac{3}{5}),$$

$$\Lambda^2L(\Lambda_0) = L(2\Lambda_6) \otimes V(\tfrac{7}{10},\tfrac{3}{2}) + L(\Lambda_1) \otimes V(\tfrac{7}{10},\tfrac{1}{10}),$$

$$L(\Lambda_0) \otimes L(\Lambda_6) = L(\Lambda_0+\Lambda_6) \otimes V(\tfrac{7}{10},\tfrac{3}{80}) + L(\Lambda_7) \otimes V(\tfrac{7}{10},\tfrac{7}{16}).$$

3) $A = A_2^{(1)}$:

$$S^2L(\Lambda_0) = L(2\Lambda_0) \otimes V(\tfrac{4}{5},0) + L(\Lambda_1+\Lambda_2) \otimes V(\tfrac{4}{5},\tfrac{7}{5}),$$

$$\Lambda^2L(\Lambda_0) = L(2\Lambda_0) \otimes V(\tfrac{4}{5},3) + L(\Lambda_1+\Lambda_2) \otimes V(\tfrac{4}{5},\tfrac{2}{5}),$$

$$L(\Lambda_0) \otimes L(\Lambda_1) = L(2\Lambda_2) \otimes V(\tfrac{4}{5},\tfrac{2}{3}) + L(\Lambda_0+\Lambda_1) \otimes V(\tfrac{4}{5},\tfrac{1}{15}).$$

$A = A_2^{(2)}$:

$$S^2L(\Lambda_0) = L(2\Lambda_0) \otimes V(\tfrac{4}{5},\tfrac{1}{40}) + L(\Lambda_1) \otimes V(\tfrac{4}{5},\tfrac{13}{8}),$$

$\Lambda^2 L(\Lambda_0) = L(2\Lambda_0) \otimes V(\frac{4}{5},\frac{21}{40}) + L(\Lambda_1) \otimes V(\frac{4}{5},\frac{1}{8})$.

4) $A = E_6^{(1)}$:

$S^2 L(\Lambda_0) = L(2\Lambda_0) \otimes V(\frac{6}{7},0) + L(\Lambda_1+\Lambda_5) \otimes V(\frac{6}{7},\frac{5}{7}) + L(\Lambda_6) \otimes V(\frac{6}{7},\frac{22}{7})$,

$\Lambda^2 L(\Lambda_0) = L(2\Lambda_0) \otimes V(\frac{6}{7},5) + L(\Lambda_1+\Lambda_5) \otimes V(\frac{6}{7},\frac{12}{7}) + L(\Lambda_6) \otimes V(\frac{6}{7},\frac{1}{7})$,

$L(\Lambda_0) \otimes L(\Lambda_1) = L(2\Lambda_5) \otimes V(\frac{6}{7},\frac{4}{3}) + L(\Lambda_0+\Lambda_1) \otimes V(\frac{6}{7},\frac{1}{21}) + L(\Lambda_4) \otimes V(\frac{6}{7},\frac{10}{21})$.

$A = E_6^{(2)}$:

$S^2 L(\Lambda_0) = L(2\Lambda_0) \otimes V(\frac{6}{7},\frac{1}{56}) + L(\Lambda_1) \otimes V(\frac{6}{7},\frac{33}{56}) + L(\Lambda_4) \otimes V(\frac{6}{7},\frac{23}{8})$,

$\Lambda^2 L(\Lambda_0) = L(2\Lambda_0) \otimes V(\frac{6}{7},\frac{85}{56}) + L(\Lambda_1) \otimes V(\frac{6}{7},\frac{5}{56}) + L(\Lambda_4) \otimes V(\frac{6}{7},\frac{3}{8})$.

The proof of the theorem is based on the following observations. Let $\Lambda \in P_+$ be of level 1 and let $M \in P_+$ be such that $L(M)$ occurs in $L=L(\Lambda_0) \otimes L(\Lambda)$. Note that M has level 2 and $M \in \Lambda_0 + \Lambda + Q$, where Q is the root lattice of $g(A)$. Let U_M denote the sum of all subrepresentations in L of the form $L(M+s\delta)$, $s \in \mathbb{Z}$. Then L decomposes into a direct sum of subspaces of the form U_M. Vir acts on $U_M^{n_+}$ with central charge z_m, where $m = 1,2,3$ or 4 is the number of claim of Theorem 8.1, and with respect to the direct sum of $g'(A)$ and Vir we have : $L = \bigoplus_{M \bmod \mathbb{C}\delta} (L(M) \otimes U_M^{n_+})$. The eigenvalues of L_0 on $U_M^{n_+}$ are, due to (8.2), of the form $h_M^{(\Lambda)} + \frac{1}{k}\mathbb{Z}$, where

$$(8.3) \qquad h_M^{(\Lambda)} = \frac{1}{2k}\left[\frac{(\Lambda|\Lambda+2\rho)}{g+1} - \frac{(M|M+2\rho)}{g+2}\right] + \left[\frac{d}{24} - \frac{|\rho|^2}{2gk}\right]p(1,1).$$

On the other hand, since the representation of $g(A)$ on L is unitary, so is the representation of Vir on $U_M^{n_+}$, hence the eigenvalues of L_0 on $U_M^{n_+}$ are of the form $h_{r,s}^{(m)} + \mathbb{Z}$.

The values of $h_M^{(\Lambda)} \bmod \frac{1}{k}\mathbb{Z}$ for all $\Lambda \in P_+$ of level 1 and all $M \in P_+$ of level 2 such that $M \in \Lambda_0 + \Lambda + Q$ are listed in the Table M below.

The proof of Theorem 8.1 in all cases, except for the representation $L(\Lambda_0) \otimes L(\Lambda_0)$ of $E_6^{(1)}, A_2^{(2)}$ and $A_2^{(1)}$, is obtained now directly by making use of Lemmas 8.1 and 8.2.

The remaining cases require more calculations. We shall demonstrate them in the case of $A_2^{(1)}$. From Table M we see that $L(\Lambda_0) \otimes L(\Lambda_0)$ for $A_2^{(1)}$ decomposes as follows :

$$(8.4)\ L(\Lambda_0)\otimes L(\Lambda_0)=L(2\Lambda_0)\otimes V(\tfrac{4}{5},0)+b_1L(2\Lambda_0)\otimes V(\tfrac{4}{5},3)+b_2L(\Lambda_1+\Lambda_2)\otimes V(\tfrac{4}{5},\tfrac{7}{5})+b_3L(\Lambda_1+\Lambda_2)\otimes V(\tfrac{4}{5},\tfrac{2}{5}),$$

where $b_i \in \mathbb{Z}_+$.

In order to show that $b_i = 1$ and to distribute each term in the right hand side of (8.4) to the symmetric or the skew-symmetric part, we compute the q-dimension of each component. In doing this, it suffices to know only coefficients of q^i for

$0 \le j \le 9$, since the lowest among leading weights $2\Lambda_0$, $2\Lambda_0-3\delta$, $\Lambda_1+\Lambda_2$, $\Lambda_1+\Lambda_2-\delta$ is $2\Lambda_0-3\delta$ and $ht(3\delta)$ is equal to 9. The coefficients of q^j of q-dimensions are listed on the following Table Q, where $\psi(q) = \varphi(q)/\varphi(q^3)$. They are computed using [11, Proposition 10.10].

Table M

A	Λ	M	$h_M^{(\Lambda)} \bmod \frac{1}{k}\mathbb{Z}$	1'st level S^2	1'st level Λ^2
$E_8^{(1)}$	Λ_0	$2\Lambda_0$	0	0	
m=1	Λ_0	Λ_1	1/16		1
$z_1=\frac{1}{2}$	Λ_0	Λ_7	1/2	14	
$E_7^{(1)}$	Λ_0	$2\Lambda_0$	0	0	
m=2	Λ_0	$2\Lambda_6$	3/2		27
$z_2=\frac{7}{10}$	Λ_0	Λ_1	1/10		1
	Λ_0	Λ_5	3/5	10	
	Λ_6	$\Lambda_0+\Lambda_6$	3/80	0	
	Λ_6	Λ_7	7/16	7	
$A_2^{(1)}$	Λ_0	$2\Lambda_0$	$0 \equiv 3$	0	9
m=3	Λ_0	$\Lambda_1+\Lambda_2$	$2/5 \equiv 7/5$	4	1
$z_3=\frac{4}{5}$	Λ_1	$2\Lambda_2$	2/3	2	
	Λ_1	$\Lambda_0+\Lambda_1$	1/15	0	

A	Λ	M	$h_M^{(\Lambda)} \bmod \frac{1}{k}\mathbb{Z}$	1'st level S^2	1'st level Λ^2
$A_2^{(2)}$	Λ_0	$2\Lambda_0$	$1/40 \equiv 21/40$	0	3
m=3	Λ_0	Λ_1	$1/8 \equiv 13/8$	10	1
$E_6^{(1)}$	Λ_0	$2\Lambda_0$	$0 \equiv 5$	0	60
m=4	Λ_0	$\Lambda_1+\Lambda_5$	$5/7 \equiv 12/7$	8	20
$z_4=\frac{6}{7}$	Λ_0	Λ_6	$1/7 \equiv 22/7$	37	1
	Λ_1	$2\Lambda_5$	4/3	16	
	Λ_1	$\Lambda_0+\Lambda_1$	1/21	0	
	Λ_1	Λ_4	10/21	5	
$E_6^{(2)}$	Λ_0	$2\Lambda_0$	$1/56 \equiv 85/56$	0	27
m=4	Λ_0	Λ_1	$5/56 \equiv 33/56$	10	1
	Λ_0	Λ_4	$3/8 \equiv 23/8$	52	7

Table Q

	q^0	q^1	q^2	q^3	q^4	q^5	q^6	q^7	q^8	q^9
$\psi(q)\ \dim_q S^2L(\Lambda_0)$	1	0	1	1	2	2	4	4	7	8
$\psi(q)\ \dim_q \Lambda^2L(\Lambda_0)$	0	1	1	1	2	3	3	5	6	8
$\psi(q)\ \dim_q L(2\Lambda_0)\cdot\chi_{1,1}^{(3)}(q^3)$	1	0	1	1	1	1	3	2	4	5
$q^9\psi(q)\ \dim_q L(2\Lambda_0)\cdot\chi_{4,1}^{(3)}(q^3)$	0	0	0	0	0	0	0	0	0	1
$q\psi(q)\ \dim_q L(\Lambda_1+\Lambda_2)\cdot\chi_{2,1}^{(3)}(q^3)$	0	1	1	1	2	3	3	5	6	7
$q^4\psi(q)\ \dim_q L(\Lambda_1+\Lambda_2)\cdot\chi_{3,1}^{(3)}(q^3)$	0	0	0	0	1	1	1	2	3	3

In Table Q, $\chi_{r,s}^{(m)}(x)$ is as defined in Appendix 2, and we put $x = q^3$ since $\mathrm{ht}\,\delta = 3$. The statements for $A_2^{(1)}$ in Theorem 8.1 follow immediately from Table Q. A similar proof works also for $A_2^{(2)}$ and $E_6^{(1)}$; one has to compute the concerned q-dimensions up to the 10-th and 60-th power of q respectively.

Remark 8.1. Theorem 8.1 covers all cases when tensor products of level 1 representations of affine algebras produce representations of Vir with $z < 1$, except for $A_1^{(1)}$, covered by Theorem 4.1, $G_2^{(1)}$ and $F_4^{(1)}$. Specifically, for $A_1^{(1)}$ we have :

$$S^2L(\Lambda_0) = L(2\Lambda_0) \otimes V(\tfrac{1}{2},0),\ \Lambda^2L(\Lambda_0) = L(2\Lambda_1) \otimes V(\tfrac{1}{2},\tfrac{1}{2}),$$

$$L(\Lambda_0) \otimes L(\Lambda_1) = L(\Lambda_0+\Lambda_1) \otimes V(\tfrac{1}{2},\tfrac{1}{16}).$$

For $G_2^{(1)}$ the central charge is $z_7 = \frac{14}{15}$; putting $U_1 = S^2L(\Lambda_0)$, $U_3 = S^2L(\Lambda_2)$, $U_5 = L(\Lambda_0) \otimes L(\Lambda_2)$, $U_7 = \Lambda^2L(\Lambda_2)$, $U_9 = \Lambda^2L(\Lambda_0)$, and $L_1 = L(2\Lambda_0)$, $L_3 = L(2\Lambda_2)$, $L_5 = L(\Lambda_0+\Lambda_2)$, $L_7 = L(\Lambda_1)$, we have:

$$U_s = \sum_{r=1,3,5,7} L_r \otimes V(z_7,h_{r,s}^{(7)}) .$$

For $F_4^{(1)}$ the central charge is $z_8 = \frac{52}{55}$; putting $U_1 = S^2L(\Lambda_0)$, $U_3 = S^2L(\Lambda_4)$, $U_5 = L(\Lambda_0) \otimes L(\Lambda_4)$, $U_7 = \Lambda^2L(\Lambda_4)$, $U_9 = \Lambda^2L(\Lambda_0)$, and $L_1 = L(2\Lambda_0)$, $L_3 = L(2\Lambda_4)$, $L_5 = L(\Lambda_0+\Lambda_4)$, $L_7 = L(\Lambda_3)$, $L_9 = L(\Lambda_1)$, we have:

$$U_r = \sum_{s=1,3,5,7,9} L_s \otimes V(z_8,h_{r,s}^{(8)}) .$$

Theorem 8.1 can be written in a similar compact form.

Remark 8.2. It is fairly well-known that all unitarizable representations of Vir with $z = \frac{1}{2}$ can be constructed as follows. Fix $\varepsilon = 0$ or $\frac{1}{2}$. Consider the "superoscillator" algebra A_ε on generators ψ_m, $m \in \varepsilon + \mathbb{Z}$, and defining relations

$$[\psi_m,\psi_n]_+ = \delta_{n,-m} .$$

Let $V_\varepsilon = \Lambda[\xi_j \mid j \geq 0,\ j \in \varepsilon+\mathbb{Z}]$ be a Grassmann algebra. Define a representation of A_ε on V_ε by $(n > 0)$:

$$\psi_n \to \frac{\partial}{\partial\xi_n},\ \psi_{-n} \to \xi_n,\ \ \psi_0 \to \frac{1}{\sqrt{2}}(\xi_0 + \frac{\partial}{\partial\xi_0}) .$$

Define a Hermitian form on V_ε by taking monomials for an orthonormal basis. Let V_ε^+ (resp. V_ε^-) denote the subspace of V_ε spanned by monomials of even (resp. odd) degree, where $\deg \xi_j = 1$, all j. Put

$$L_0 = \frac{1}{8}(\frac{1}{2} - \varepsilon) + \sum_{j \in \varepsilon+\mathbb{Z}_+} j\psi_{-j}\psi_j ,$$

$$L_n = \frac{1}{4} \sum_{j \in \varepsilon+\mathbb{Z}} (2j-n)\ \psi_{-j+n}\psi_j \quad \text{for } n \neq 0.$$

This gives irreducible representations of Vir with $z = \frac{1}{2}$ on $V_\varepsilon^\pm$. Explicitly :

$$V_{\frac{1}{2}}^{+} = V(\tfrac{1}{2},0),\ V_{\frac{1}{2}}^{-} = V(\tfrac{1}{2},\tfrac{1}{2}),\ V_{0}^{+} = V_{0}^{-} = V(\tfrac{1}{2},\tfrac{1}{16}).$$

No such simple construction is known (so far) for other discrete series representations of Vir.

Remark 8.3. Note the following remarkable coincidence. Let g be a simple Lie algebra of type E_8, E_7, A_2 or E_6 and let $\hat{g}$ be the associated affine algebra. Then all highest weights of the representations of Vir that occur in all pairwise tensor products of all level 1 representations of $\hat{g}$ are of the form (z_m, h), where $m = 1,2,3$ or 4 respectively and h is precisely one of the critical exponents of the Ising, tricritical Ising, 3-state Potts and tricritical 3-state Potts models respectively (cf. [5]). In other words, the $h_{r,s}^{(m)}$ that occur in 2-dimensional statistical models are precisely those which correspond to non-twisted affine algebras.

Remark 8.4. The same argument as above can be applied to the study of the problem of restriction of a unitary highest weight representation of an affine algebra $\hat{g}$ to an affine subalgebra $\hat{p}$, where p is a reductive subalgebra of reductive algebra g. In our next publication we will classify the pairs (g,p) for which the central charge of the Virasoro algebra is less than 1 and calculate the corresponding generalized string functions.

Références

[1] Belavin A.A., Polyakov A.M., Zamolodchikov A.B., Infinite conformal symmetry of critical fluctuations in two dimensions, J. Stat. Phys. 34 (1984), 763-774. Infinite conformal symmetry in two dimensional quantum field theory, Nucl. Physics B241 (1984), 333-380.

[2] Feigin B.L., Fuchs D.B., Skew-symmetric invariant differential operators on a line and Verma modules over the Virasoro algebra, Funct. Anal. Appl. 16 (1982), no.2, 47-63 (in Russian).

[3] Feigin B.L., Fuchs D.B., Verma modules over the Virasoro algebra, Funct. Anal. Appl., 17 (1983), 91-92 (in Russian). Representations of the Virasoro algebra (1983), preprint.

[4] Feingold A.J., Tensor products of certain modules for the generalized Cartan matrix Lie algebra $A_1^{(1)}$, Comm. in Alg., 9 (1981), 1323-1361.

[5] Friedan D., Qiu Z., Shenker S., Conformal invariance, unitarity and two dimensional critical exponents, MSRI Publ., 3 (1985), 419-449.

[6] Friedan D., Qiu Z., Shenker S., Superconformal invariance in two dimensions and the tricritical Ising model, Phys. Lett. 151B (1985) 37-43.

[7] Goddard P., Kent A., Olive D., Virasoro algebras and coset space models, Phys. Lett. 152B (1985), 88-93.

[8] Kac V.G., Highest weight representations of infinite-dimensional Lie algebras, Proceedings of ICM, 299-304, Helsinki, 1978.

[9] Kac V.G., Contravariant form for infinite dimensional Lie algebras and superalgebras, Lect. Notes in Phys., 94 (1979), 441-445.

[10] Kac V.G., Some problems in infinite dimensional Lie algebras and their representations, Lect. Notes in Math., 933 (1982), 117-126.

[11] Kac V.G., Infinite Dimensional Lie Algebras, Progress in Math. 44, Birkhäuser, Boston, 1983. Second edition : Cambridge University Press, 1985.

[12] Kac V.G. and Peterson D.H., Infinite-dimensional Lie algebras, theta functions and modular forms, Advances in Math., 53 (1984), 125-264.

[13] Kac V.G., Todorov I.T., Superconformal current algebras and their unitary representations, Comm. Math. Phys 102 (1985), 337 - 347.

[14] Kac V.G., Lie superalgebras, Adv. Math. 26 (1977), 8-96.

[15] Rocha-Caridi A., Vacuum vector representations of the Virasoro algebra, MSRI Publ., 3 (1985), 451-473.

[16] Schwarz J.H., Superstring theory, Physics Rep. 83 (1982), 223-322.

[17] Thorn C.B., Computing the Kac determinant using dual model techniques and more about the no-ghost theorem, Nucl. Phys. B248 (1984), 551-569.

[18] Wakimoto M., Basic representations of extended affine Lie algebras, RIMS-Kokyuroku 503 (1983), 36-46. Affine Lie algebras and the Virasoro algebra I (1984), preprint.

[19] Kac V.G., Peterson D.H., Regular functions on certain infinite dimensional groups, in Arithmetics and Geometry, 141-166, Progress in Math. 36, Birkhaüser, Boston, 1983.

[20] Wakimoto M., Two formulae for specialized characters of Kac-Moody Lie algebras (1983), preprint.

STRUCTURE OF KAC-MOODY GROUPS

J. Mickelsson

Department of Mathematics
University of Jyväskylä
SF-40100 Jyväskylä
Finland

For a physicist, a Kac-Moody algebra is the current algebra of a quantum field theory model in $1+1$ space-time dimensions with an internal symmetry group $\underline{G}$ [1]. More precisely, let $\underline{g}$ be the Lie algebra of $\underline{G}$. The Kac-Moody algebra $\hat{\underline{g}}$ is a one-dimensional central extension of the loop algebra $\mathrm{Map}(S^1, \underline{g})$. If $f_1, f_2 \in \mathrm{Map}(S^1, \underline{g})$, then the commutator is defined point-wise,

$$[f_1, f_2](\varphi) := [f_1(\varphi), f_2(\varphi)] . \tag{1}$$

The central extension is defined by a 2-cocycle $\theta: \mathrm{Map}(S^1, \underline{g}) \times \mathrm{Map}(S^1, \underline{g}) \to \mathbb{R}$ (for an introduction to the use of group cohomology in physics, see refs. [2], [3]; for a more general mathematical treatise, see [4])

$$\theta(f_1, f_2) = \frac{1}{4\pi} \int_0^{2\pi} \langle f_1(\varphi), \frac{d}{d\varphi} f_2(\varphi) \rangle \, d\varphi , \tag{2}$$

where $\langle \cdot, \cdot \rangle$ is the Killing form on $\underline{g}$; in the adjoint representation $\langle X, Y \rangle = \mathrm{tr}\, XY$. The modified commutator is

$$[f_1, f_2](\varphi) = [f_1(\varphi), f_2(\varphi)] + i p \,\theta(f_1, f_2) . \tag{1)'$$

Here p is a constant; if there is a group $\hat{\underline{G}}$ corresponding to the algebra $\hat{\underline{g}}$, then p is an integer. An alternative way to write the commutation relations (1)' is obtained using the Fourier components $X_a^{(n)}$ ($n \in \mathbb{Z}$) of a map $f: S^1 \to \underline{g}$ in an ortonormal basis $X_1, \ldots, X_N$ of $\underline{g}$,

$$[X_a^{(n)}, X_b^{(m)}] = \lambda_{ab}^c X_c^{(n+m)} + \frac{ip}{4\pi} n \, \delta_{n,-m} \, \delta_{ab} , \tag{3}$$

where the λ_{ab}^c's are the structure constants of $\underline{g}$.

In this lecture I shall explain the structure of the group $\hat{\underline{G}}$ having as its Lie algebra the algebra $\hat{\underline{g}}$. The discussion will be rather formal in the sense that any definitions regarding differentiable (Banach manifold) structures on $\hat{\underline{G}}$ will be avoided (these have been studied in detail in [5], [6]). It would be natural to start from the Ansatz $\underline{\hat{G}} = \mathrm{Map}(S^1, \underline{G}) \times U(1)$ with the multiplication

$$(F_1, \lambda_1)(F_2, \lambda_2) = (F_1 F_2, \lambda_1 \lambda_2 \exp 2\pi i \omega(F_1, F_2)), \tag{4}$$

where $F_1, F_2 \in \mathrm{Map}(S^1, \underline{G})$, $\lambda_1, \lambda_2 \in U(1)$ and $\omega(F_1, F_2)$ is some real valued function of F_1 and F_2; here $(F_1 F_2)(\varphi) := F_1(\varphi) F_2(\varphi)$. However, it turns out that in general (for an arbitrary $\underline{G}$) there does not exist any 2-cocycle ω which would give infinitesimally the Lie algebra cocycle θ. As shown by Pressley and Segal [5], there is topological reason for this fact; one has to think of the group $\underline{\hat{G}}$ as a non-trivial fiber bundle sitting over the base space $\mathrm{Map}(S^1, \underline{G})$ with fiber $U(1)$. The cocycle ω can be defined only locally in $\mathrm{Map}(S^1, \underline{G})$. In this lecture I will give an alternative way to [5], [6] for the construction of the bundle $\underline{\hat{G}}$. We shall study the structure of $\underline{\hat{G}}$ directly in terms of local coordinate charts and local cocycles. Our treatment is motivated by the recent research on cohomology of gauge anomalies [7]. In fact, the bundle $\underline{\hat{G}}$ is homotopically equivalent to the determinant line bundle of a massless Dirac operator in $1+1$ dimensions (the base space there is the set A/G of gauge group orbits in the space of all $\underline{g}$-valued vector potentials) [8].

We shall start from the topological density

$$C(F) = \frac{1}{24\pi^2} \operatorname{tr}(dFF^{-1})^3 . \tag{5}$$

It is known that the integral of the 3-form $C(F)$ over S^3 is an integer (that is, when F is a smooth map from S^3 to $\underline{G}$). By a simple computation, $dC = 0$, and therefore $C = dH$, locally; H is some 2-form. In fact, there is an explicit formula for $H(F)$ when $F = \exp X$ (X is a $\underline{g}$-valued function) [9],

$$H(X) = \operatorname{tr} dX\, h(\operatorname{ad} X)\, dX , \tag{6}$$

$$h(z) = -\frac{1}{8\pi^2} \frac{\sinh z - z}{z^2} . \tag{7}$$

Let us consider the case $\underline{G} = SU(2)$ as an (typical) example. For each $a \in \underline{G}$ let $V_a \subset \mathrm{Map}(S^1, \underline{G})$ consist of loops F such that $F(x) \neq -a$ for all $x \in S^1$. For smooth loops $\mathrm{Map}(S^1, \underline{G}) = \bigcup_{a \in \underline{G}} V_a$. Let $B = \{X \in \underline{g} \mid \frac{1}{2} \operatorname{tr} X^2 < \pi^2\}$. Then the exponential map $\exp: B \to SU(2) \setminus \{-1\}$ is one-to-one; the sphere $\operatorname{tr} X^2 = 2\pi^2$ is mapped onto the point -1. Any loop F in V_a can be extended to a map $\tilde{F}: D \to SU(2) \setminus \{-a\}$, where D is the unit disc with boundary $S^1 = \partial D$. Now $a^{-1} \tilde{F}$ is a map into $SU(2) \setminus \{-1\}$, and therefore $a^{-1} \tilde{F} = \exp X$, where X is a uniquely defined B-valued function in D. Next we can define the transition functions

$$h_{ab}: V_a \cap V_b \to U(1), \quad h_{ab} = \exp 2\pi i \beta_{ab}$$

for the $U(1)$ bundle $\underline{\hat{G}}$,

$$\beta_{ab}(F) = \int_D (H(\ln a^{-1} \tilde{F}) - H(\ln b^{-1} \tilde{F})) . \tag{8}$$

We shall show that $h_{ab}(F)$ does not depend on the extension $\tilde{F}$ for a given F. If $\tilde{F}_\pm$ are two extensions taking values in $SU(2) \setminus \{-a, -b\}$, then we can define a smooth map $\tilde{F}: S^2 = D_+ \cup D_- \to SU(2) \setminus \{-a, -b\}$ by joining $\tilde{F}_\pm$ along the common boundary $\partial D_+ = S^1 = \partial D_-$. The difference for the two extensions gives

$$\beta_{ab}^{(+)}(F) - \beta_{ab}^{(-)}(F) = \int_{S^2} (H(\ln a^{-1} \tilde{F}) - H(\ln b^{-1} \tilde{F})) = \int_{D_3} (C(F_a) - C(F_b)) , \quad (9)$$

where $D^3 \subset \mathbb{R}^3$ is the closed unit ball, $\partial D^3 = S^2$, and $F_a = a \exp X_a$, $F_b = b \exp X_b$ for some extensions X_a, X_b of $\ln a^{-1} \tilde{F}$, $\ln b^{-1} \tilde{F}$ to the unit ball D^3. We can define a map $F_{ab}: S^3 \to G$ from F_a and F_b by joining two balls D^3 along the common boundary $S^2 = \partial D^3$ (since $F_a|_{S^2} = F_b|_{S^2}$). We obtain

$$\beta_{ab}^{(+)}(F) - \beta_{ab}^{(-)}(F) = \int_{S^3} C(F_{ab}) . \quad (10)$$

The right-hand side is always an integer and it follows that $\exp 2\pi i\beta$ is well-defined.

In order to construct the 2-cocycle ω and to understand its relation to anomalies in two dimensional Yang-Mills theories we shall consider the second Chern class of the field $F = dA + [A, A]$, where A is a $\underline{g}$-valued vector potential in four dimensions,

$$c_2 = \operatorname{tr} F \wedge F = \frac{1}{4} \operatorname{tr} \varepsilon^{\mu\nu\alpha\beta} F_{\mu\nu} F_{\alpha\beta} . \quad (11)$$

Locally, in the domain of definition of A,

$$c_2 = d\omega_3^1 , \; \omega_3^1 = \operatorname{tr}(A\,dA + \frac{2}{3} A^3) . \quad (12)$$

Let g be a gauge transformation and denote $A^g = g^{-1} Ag + g^{-1} dg$. We can define

$$(\delta\omega_3^1)(A; g) = \omega_3^1(A^g) - \omega_3^1(A) = d \operatorname{tr} A\,dg\,g^{-1} - \frac{1}{3} \operatorname{tr}(g^{-1} dg)^3$$

$$= d \operatorname{tr} A\,dg\,g^{-1} - 8\pi^2 C(g) = d(\operatorname{tr} A\,dg\,g^{-1} - 8\pi^2 H(X)) ; \quad (13)$$

the last equation holds for $g = \exp X$. Thus $\delta\omega_3^1 = d\omega_2^1$ for a 2-form ω_2^1. Next we consider the variation $\delta\omega_2^1$,

$$(\delta\omega_2^1)(A; g_1, g_2) = \omega_2^1(A; g_1) + \omega_2^1(A^{g_1}; g_2) - \omega_2^1(A; g_1 g_2)$$

$$= \operatorname{tr} g_1^{-1} dg_1\, dg_2\, g_2^{-1} + 8\pi^2(H(X_1) + H(X_2) - H(X_{12})) , \quad (14)$$

where $g_1 = \exp X_1$, $g_2 = \exp X_2$ and $g_1 g_2 = \exp X_{12}$. Note that ω_2^1 is the anomalous variation of the Fermion determinant in a two dimensional field theory (left-handed massless fermions minimally coupled to A). By a direct computation $d(\delta\omega_2^1) = 0$, so $\delta\omega_2^1 = d\omega_1^2$, at least locally for some 1-form ω_1^2. In particular, if the $\underline{G}$-valued functions g_1 and g_2 are defined in the unit disc, then

$$\frac{1}{8\pi^2} \int_D \operatorname{tr} g_1^{-1} dg_1\, dg_2\, g_2^{-1} - \int_D (H(X_1) + H(X_2) - H(X_{12})) \quad (15)$$

depends only on the boundary values on S^1, modulo an integer: If $F_\pm$ and $G_\pm$ are defined in D such that $F_+ = F_-$ and $G_+ = G_-$ on the boundary, then we can again define $F, G: S^2 \to \underline{G}$ by joining along the equator S^1. Let us assume for the moment that all maps have an exponential representation. Denoting the expression (15) by $\omega(g_1, g_2)$ we get

$$\begin{aligned}\omega(F_+, G_+) - \omega(F_-, G_-) &= \frac{1}{8\pi^2}\int_{S^2} \operatorname{tr} F^{-1}\,dF\,dG\,G^{-1} + \int_{S^2}(H(\ln FG) - H(\ln F) - H(\ln G)) \\ &= \frac{1}{8\pi^2}\int_{D^3} (\operatorname{tr} F^{-1}\,dF\,(dG\,G^{-1})^2 + \operatorname{tr}(F^{-1}\,dF)^2\,dG\,G^{-1}) + \int_{D^3}(C(E) - C(F) - C(G)) \\ &= \int_{D^3} (C(E) - C(FG)) \,,\end{aligned} \tag{16}$$

where $E: D^3 \to \underline{G}$ is an extension of $FG: S^2 \to \underline{G}$ obtained using any extension of $\ln FG$ to D^3. Now $E = FG$ on the boundary $S^2 = \partial D^3$ and we can define $\tilde{E}: S^3 \to \underline{G}$ by joining along S^2. Thus the difference (16) is equal to

$$\int_{S^3} C(\tilde{E}) = n \in \mathbb{Z} \,,$$

and we conclude that $\exp 2\pi\omega$ is well-defined.

There is still a problem: The formula (15) for ω relies on the exponential representation of the loops. In general, a loop in $\underline{G}$ does not have a smooth lift to a loop in the Lie algebra $\underline{g}$. Let us return to our example $\underline{G} = SU(2)$. If F and G are two loops in $SU(2)$, then $F \in V_a$, $G \in V_b$ and $FG \in V_c$ for some $a, b, c \in SU(2)$. Choosing extensions $\tilde{F}, \tilde{G}$ to the unit disc D such that $\tilde{F}, \tilde{G}$ and $\widetilde{FG}$ do not meet a, b and c, respectively, we can define

$$\omega_{abc}(F, G) = \frac{1}{8\pi^2}\int_D \operatorname{tr} \tilde{F}^{-1}\,d\tilde{F}\,d\tilde{G}\,\tilde{G}^{-1} + \int_D (H(\ln c^{-1}\tilde{F}\tilde{G}) - H(\ln a^{-1}\tilde{F}) - H(\ln b^{-1}\tilde{G})) \,, \tag{17}$$

where all the logarithms are chosen in $B \subset \underline{g}$. Exactly as in (16) we can show that the right-hand side does not depend on the extensions $\tilde{F}$ and $\tilde{G}$, modulo an integer (note that $C(a^{-1} f) = C(f)$ for a constant $a \in \underline{G}$ and for any map f into $\underline{G}$).

Next one can show by a direct computation that for $E \in V_a$, $F \in V_b$, $G \in V_c$, $EF \in V_k$, $FG \in V_l$ and $EFG \in V_m$,

$$\omega_{abk}(E, F) + \omega_{kcm}(EF, G) = \omega_{bcl}(F, G) + \omega_{alm}(E, FG) \,. \tag{18}$$

Equation (18) is in fact the definition of a real valued 2-cocycle on $\mathrm{Map}(S^1, \underline{G})$. With the help of the local cocycles ω_{abc} and the transition functions h_{ab} we can now define the group $\hat{\underline{G}}$. As a set, $\hat{\underline{G}}$ consists of equivalence classes of triples (F, λ, a), where $a \in SU(2)$, $\lambda \in U(1)$ and $F \in V_a$, with respect to the relations

$$(F, \lambda, a) \sim (F, \lambda h_{ba}(F), b) \tag{19}$$

when $F \in V_a \cap V_b$. The product of the equivalence classes $[(F, \lambda, a)]$ is given by

$$[(F, \lambda, a)][(G, \mu, b)] = [(FG, \lambda\mu \exp 2\pi i\, \omega_{abc}(F, G), c] , \tag{20}$$

where $c \in SU(2)$ is any element such that $FG \in V_c$. From the equations (8) and (17) it follows that

$$\omega_{a'b'c'}(F, G) = \omega_{abc}(F, G) + \beta_{aa'}(F) + \beta_{bb'}(G) - \beta_{cc'}(FG) \tag{21}$$

when $F \in V_a \cap V_{a'}$, $G \in V_b \cap V_{b'}$ and $FG \in V_c \cap V_{c'}$. As a consequence, the product of the equivalence classes in (20) is well-defined. The associativity of the group multiplication law in $\hat{\underline{G}}$ follows at once from the cocycle condition (18).

There is an alternative more geometric method to construct a local 2-cocycle, near the unit element in $\underline{G}$. First, one can define a 3-cocycle $\omega^{(3)}(g_1, g_2, g_3)$ on $\underline{G}$ by setting

$$\omega^{(3)}(g_1, g_2, g_3) = \frac{1}{24\pi^2} \int_{\Delta_{123}} \operatorname{tr}(g^{-1} dg)^3 , \tag{22}$$

where $\Delta_{12\ldots n} = \{(x_1, \ldots, x_n) \in (\mathbb{R}_+)^n \mid x_1 + x_2 + \ldots + x_n \leq 1\}$ is an n-simplex and $g = g(r, s, t) = g_1(r)\, g_{12}(s)\, g_{123}(t)$ is defined by the one-parameter subgroups $g_1(r)$, $g_{12}(s)$ and $g_{123}(t)$ such that $g_1(1) = g_1$, $g_{12}(1) = g_1 g_2$ and $g_{123}(1) = g_1 g_2 g_3$. Note that near the unit element in $\underline{G}$ for each element g there is a unique one-parameter subgroup $g(t)$ such that $g(1) = g$. The coboundary $\delta\omega^{(3)}$ is by definition [3]

$$\delta\omega^{(3)}(g_1, g_2, g_3, g_4) = \omega^{(3)}(g_1, g_2, g_3) + \omega^{(3)}(g_2, g_3, g_4)$$
$$- \omega^{(3)}(g_{12}, g_3, g_4) + \omega^{(3)}(g_1, g_{23}, g_4) - \omega^{(3)}(g_1, g_2, g_{34}) .$$

We denote $g_{i_1 \ldots i_\nu} = g_{i_1} \cdots g_{i_\nu}$. By eq. (22) the right-hand side is equal to the integral

$$\frac{1}{24\pi^2} \int_{\partial\Delta_{1234}} \operatorname{tr}(g^{-1} dg)^3 , \tag{23}$$

where g is defined in the four dimensional simplex Δ_{1234}, $g(t_1, t_{12}, t_{123}, t_{1234}) = g_1(t_1) \ldots g_{1234}(t_{1234})$. According to Stokes' theorem the integral (23) is an integral over Δ_{1234} of the 4-form $d \operatorname{tr}(g^{-1} dg)^3$; by a simple computation, this form is identically zero and thus $\delta\omega^{(3)} = 0$, which means that $\omega^{(3)}$ is indeed a 3-cocycle.

Suppose next that the elements g_1, g_2 and g_3 depend on a variable $x \in S^1$. Consider the 1-form

$$\hat{d}\omega^{(3)} = \frac{3}{24\pi^2}\, dx \int_{\Delta_{123}} \operatorname{tr}(g^{-1} dg)^2\, \partial_x (g^{-1} dg)$$
$$= \frac{1}{8\pi^2}\, dx \int_{\partial\Delta_{123}} \operatorname{tr}(g^{-1} dg)\, \partial_x(g^{-1} dg) , \tag{24}$$

where $\hat{d}$ is the exterior derivative with respect to x . On the other hand, $\hat{d}\omega^{(3)}(g_1, g_2, g_3) = (\delta\omega^{(2)})(g_1, g_2, g_3) = -\omega^{(2)}(g_1, g_2) + \omega^{(2)}(g_2, g_3) - \omega^{(2)}(g_1 g_2, g_3) + \omega^{(2)}(g_1, g_2 g_3)$, where

$$\omega^{(2)}(g_1, g_2) = -\frac{1}{8\omega^2} \int_{\Delta_{12}} \mathrm{tr}(g^{-1}\,dg)\,dx\,\partial_x(g^{-1}\,dg) \quad . \tag{25}$$

The integral

$$\omega(g_1, g_2) = \int_{S^1} \omega^{(2)}(g_1, g_2) \tag{26}$$

defines a local 2-cocycle on $\mathrm{Map}(S^1, \underline{G})$. Writing $g_1 = \exp Z_1$, $g_{12} = \exp Z_{12}$ one gets by integration from (25) the formula

$$\omega(g_1, g_2) = \frac{1}{4\pi^2} \int_{S^1} \mathrm{tr} \sum_{n=0}^{\infty} \frac{(-1)^n}{(n+2)!} (\partial_x Z_{12})(\mathrm{ad}\, Z_{12})^n\, Z_1 \tag{27}$$

This form of the 2-cocycle was derived in [10]; it is not equal to the form (15) but it represents the same cohomology class. There is also a closed explicit formula for ω , which is valid for $\underline{G} = SU(2)$, given in [11].

The Lie algebra cocycle θ defines a symplectic form on the subgroup $\mathrm{Map}_0(S^1, \underline{G})$ consisting of maps F with $F(1) = 1 \in \underline{G}$. Namely, the tangent vectors to $\mathrm{Map}_0(S^1, \underline{G})$ can be thought of as elements of $\mathrm{Map}(S^1, \underline{g})$ which vanish at $1 \in S^1$. Thus the cocycle θ is a 2-form on $\mathrm{Map}_0(S^1, \underline{G})$, with constant coefficients, and therefore $d\theta = 0$. It is clearly non-degenerate; consequently θ is symplectic. What is important from the standpoint of geometric quantization, is that θ is <u>integral</u>: Take any set $F(\hat{r}, \cdot)$ of loops in $\mathrm{Map}_0(S^1, \underline{G})$ parametrized by $\hat{r} \in S^2$. Then

$$\int_{S^2} \theta = \frac{1}{4\pi} \int_{S^2 \times S^1} \mathrm{tr}(F^{-1}\partial_t F)\,\partial_x(F^{-1}\partial_s F)\,dt\,ds\,dx = \frac{1}{12\pi} \int_{S^2 \times S^1} \mathrm{tr}(F^{-1}dF)^3 = 2\pi n \ ,$$

where $n \in \mathbb{Z}$. Because of $d\theta = 0$ there are local 1-forms α on $\mathrm{Map}(S^1, \underline{G})$ such that $d\alpha = \theta$; in fact, one can take $\alpha(F; X) = \frac{1}{2}\frac{d}{dt}\,\omega(e^{tX}, F)\big|_{t=0}$, where $X \in \mathrm{Map}(S^1, \underline{g})$ is a tangent vector at $F \in \mathrm{Map}(S^1, \underline{G})$. Finally, there is a natural inner product in the space of sections of the associated complex line bundle to the principal bundle $\hat{\underline{G}}$, [6], therefore we have the essential parts of a machinery for quantization in the phase space $(\mathrm{Map}_0(S^1, \underline{G}), \theta)$, in the spirit of Kostant and Souriau, [12].

In physics literature the Lie algebra cocycle θ is often called a "Schwinger term". The geometry of Schwinger terms can be studied also in higher space-time dimensions in the spirit of the present lecture. However, there is one essentially new phenomenon (as compared to the present case). Namely, the Lie algebra (and group) extensions are infinite-dimensional. For example, in the case of $\mathrm{Map}(S^3, \underline{G})$ the group $U(1)$ has to be replaced by the group $\mathrm{Map}(A_3, U(1))$, where A_3 is the space of $\underline{g}$-valued vector potentials in S^3 (point-wise multiplications); this reflects the fact that in $3+1$ dimensions the Schwinger terms depend on the vector potential, [13].

REFERENCES

[1] I.B. Frenkel, J. Funct. Anal. <u>44</u>, 259 (1981)

[2] R. Jackiw, Lecture at the "Symposium on Anomalies, Geometry and Topology" in Argonne, March 1985 (to be publ. in the proceedings)

[3] B. Zumino, Lecture at the above meeting

[4] S. MacLane, *Homology*. Springer-Verlag (1963)

[5] A.N. Pressley and G.B. Segal, *Loop Groups*. Oxford University Press (in press)

D. Freed, *The Geometry of Loop Groups*. Ph. D. thesis, UCLA (Berkeley, 1985)

[6] G.B. Segal and G. Wilson, IHES Publ. Math., N° 61, 5 (1985)

[7] M. Atiyah and I. Singer, Proc. Natl. Acad. Sci. (USA), 81, 2597 (1984)

L. Alvarez-Gaumé and P. Ginsparg, Nucl. Phys. B 243, 449 (1984)

[8] I. Singer, Commun. Math. Phys. 60, 7 (1978); Physica Scripta 24, 817 (1981)

[9] C. Cronström and J. Mickelsson, J. Math. Phys. 24, 2528 (1983)

[10] Han-ying Guo, Shi-kun Wang and Ke Wu, ITP-SB-850-44 (Stony Brook, 1985)

[11] M.L. Laursen, G. Schierholz and U.-J. Wiese, DESY 85-062

[12] D.J. Simms and N.M.J. Woodhouse, *Lectures on Geometric Quantization*, Springer Lecture Notes in Physics 53. Springer-Verlag (1976)

[13] J. Mickelsson, Phys. Rev. Lett. 55, 2099 (1985); HU-TFT-85-50 (Helsinki, 1985)

INFINITE DIMENSIONAL LIE ALGEBRAS CONNECTED WITH THE FOUR-DIMENSIONAL LAPLACE OPERATOR

D. T. STOYANOV
INRNE, BULGARIAN ACADEMY OF SCIENCES
SOFIA 1784, BULGARIA

INTRODUCTION

The infinite-dimensional Lie algebras are among the most interesting objects which appear in a series of models in two-dimensional space-time (see e.g. [1], [2]). Such algebras connected with some differential operators can also arise in other spaces, with higher dimensions [3].

In particular, in the present work we find two infinite dimensional Lie algebras which act on and preserve the set of solutions of the Laplace equation, in four-dimensional Euclidean space. The first one is connected with all the substitutions preserving those solutions and it is constructed on the base of the Fueter analyticity condition for the functions of quaternion variables [4], [5], [6] and [7]. The second algebra is composed from a set of differential operators of arbitrary orders.

§1. As well known, the Witt type infinite-dimensional algebra may be considered as a Lie algebra corresponding to the conformal transformations of the complex plane. Analogously representing the four-dimensional Euclidean space as a hypercomplex space $\mathbb{H}$ (a space of the quaternions) we can construct some infinite Lie algebra using the Fueter analyticity. The method we are going to describe in this section may be generalized for the cases of spaces with dimension higher than four. A very suitable object for these generalizations is the so called Clifford analysis [8], [9], [10]. Note that this analysis is based on the Clifford algebras and the quaternion division algebra, in particular, is a Clifford algebra. Let e_i $(i = 1,2,3)$ be the three pure imaginary quaternion units:

(1) $$e_i e_j = -\delta_{ij} + \varepsilon_{ijk}\, e_k$$

Then let us denote the operator

(2) $$D = \frac{\partial}{\partial x_o} + e_i\frac{\partial}{\partial x_i} \equiv \partial_o + e_i\partial_i$$

By definition the quaternion function $f(x_o, x_1, x_2, x_3)$ is Fueter analytic if and only if the following condition is satisfied:

(3) $$Df = 0$$

It is easy to verify the identity

(4) $$\Box = D\tilde{D}$$

where $\Box = \partial_o^2 + \partial_1^2 + \partial_2^2 + \partial_3^2$ is the Laplace operator and $\tilde{D} = \partial_o - e_i\partial_i$ is a quaternion conjugate to D (D and $\tilde{D}$ commute). From (3) and (4) we obtain that every Feuter analytic function satisfies the Laplace equation

(5) $$D\tilde{D}f \equiv \tilde{D}Df \equiv \Box f = 0$$

<u>Remark:</u> Let us note that the Feuter condition (3) coincides with the free Maxwell equations in quaternionic form. In this case the function f is pure imaginary.

We can obtain also the so called Cauchy-Riemann-Feuter conditions [5] from the expression (3). To do this, we must write the function f in the form:

(6) $$f = R + e_i J$$

where e_i is one fixed quaternion unit. In particular, we may select $i = 3$. In this case the functions R and J depend on

(7) $$\xi_{\pm} = x_o \pm e_1 x_1 \qquad \text{and} \qquad \eta_{\pm} = x_3 \pm e_1 x_2$$

only. These new variables are commutative.

Then the operator D takes the form:

(8) $$D = 2\frac{\partial}{\partial \xi_-} + 2e_3\frac{\partial}{\partial \eta_-}$$

From (3), (6) (with $i = 3$) and (8) we obtain the Cauchy-Riemann-Feuter conditions:

(9) $$\frac{\partial R}{\partial \xi_-} = \frac{\partial J}{\partial \eta_+} ; \qquad \frac{\partial R}{\partial \eta_-} = \frac{\partial J}{\partial \xi_+}$$

Analogously, if we select $i = 2$ or $i = 1$, then we may obtain also two other forms of these conditions. But we shall consider the upper case only, because all three cases are equivalent to one another.

Now we are going to propose a method to construct those functions which will satisfy the conditions (9). For this, let us consider any arbitrary analytic function of two complex variables $w(z_1 z_2)$, where

(10) $$z_1 = u_1 + iv_1; \quad z_2 = u_2 + iv_2 \quad (i = \sqrt{-1})$$

(u_1, u_2, v_1 and v_2 are real).

Let $U(u_1,v_1;\ u_2\ v_2)$ and $V(u_1,v_1;\ u_2\ v_2)$ be the real and imaginary parts of $w(z_1,z_2)$ respectively. Then it is easy to verify that the following functions

(11) $$R = U(\xi_-,\eta_+;\xi_+,\eta_-) + cU(\xi_+,\eta_-;\xi_-,\eta_+)$$

(12) $$J = V(\xi_-,\eta_+;\xi_+,\eta_-) + cV(\xi_+,\eta_-;\xi_-,\eta_+)$$

where c is an arbitrary constant, satisfy conditions (9).
Remark: The function $U(\xi_-,\eta_+;\xi_+,\eta_-)$ is obtained from $U(u_1,v_1;u_2,v_2)$ with the change

(13) $$u_1 \to \xi_-;\ v_1 \to \eta_+;\ u_2 \to \xi_+;\ v_2 \to \eta_-$$

and so on.

One must use the usual Cauchy-Riemann conditions to prove this statement.

The correspondence between the two kinds (Feuter and usual) of analytic functions allows to define a set of nonlinear transformations preserving the Feuter analyticity. This means that such transformations will preserve also the solutions of the equation (5). To construct this transformations, let us consider the general conformal transformations in the set W of all usually analytic functions $w(z_1,z_2)$ of two complex variables. If

(14) $$w_1(z_1,z_2) \in W; \qquad w_2(z_1,z_2) \in W$$

then we also have

(15) $w'(z_1,z_2) = w[w_1(z_1,z_2),\ w_2(z_1,z_2)] \in W$

The change

(16) $z_1 \to w_1(z_1,z_2); \quad z_2 = w_2(z_1,z_2)$

is some conformal transformation $t_{w_1w_2}$ in W:

(17) $w(z_1,z_2) \to w'(z_1,z_2)$

Because the functions $w(z_1,z_2)$ and $w[w_1(z_1,z_2),w_2(z_1,z_2)]$ are both analytic, we may construct the corresponding Feuter analytic functions f_w and f'_w. Then the transformation $t_{w_1w_2}$ between the two functions from (17) will appear between the two Feuter analytic functions f_w and f'_w:

(18) $f_w \to f'_w \equiv T_{w_1w_2} f$

Therefore, the transformation $T_{w_1w_2}$ is a nonlinear realization of $t_{w_1w_2}$ in the set of Feuter analytic functions. $t_{w_1w_2}$ and, of course, $T_{w_1w_2}$ are infinite parameter transformations and their Lie algebra $\mathcal{a}_T$ has infinite dimension.

The generators of the real Lie algebra may be denoted as

$$X_\alpha^{n_\alpha;n_{-\alpha}} \quad \text{and} \quad iX_\alpha^{n_\alpha;n_{-\alpha}} \qquad (i = \sqrt{-1}$$

where $\alpha = \pm 1$ and n_1 and n_{-1} are integer numbers. Then the commutation relations have the form:

$$\left[X_\alpha^{n_\alpha;n_{-\alpha}}, X_\beta^{m_\beta;m_{-\beta}}\right] = \delta_{\alpha\beta}(m_\alpha - n_\alpha)\, X_\alpha^{n_\alpha+m_\alpha;n_{-\alpha}+m_{-\alpha}+1} +$$

$$+\delta_{\alpha,-\beta}(m_\alpha + 1)\, X_\beta^{n_\beta+m_\beta+1;n_\alpha+m_\alpha} -$$

$$-\delta_{\alpha,-\beta}(n_\beta + 1)\, X^{n_\alpha+m_\alpha+1;n_\beta+m_\beta}$$

The obtained algebra is a D type algebra according to Kac classification [11]. It contains a finite subalgebra coinciding with the Lie algebra sl(3,C).

§2. In this section we are going to consider another possibility to construct an infinite Lie algebra connected with the Laplace operator. We may imagine that equation (5) has arisen from D'Alembert equattion (free massless scalar field theory in Minkowski space-time) after the compactification by Cayley transformation. Then we may consider the stress-energy tensor $\Theta_{\mu\nu}(z)$ in the compactified form. In the corresponding quantum theory this tensor also determines some infinite Lie algebra. To find some representation of this algebra, we introduce the free scalar field $\phi(z)$.

Then the infinitesimal operator

(19) $$\Theta = \int \delta_\mu(x)\Theta_{\mu o}(x)d^3x$$

commutes with $\phi(z)$ in the following way:

(20) $$[\Theta,\phi(z)] = L(\phi(z))$$

where L is a differential operator.

Because $\phi(z)$ satisfies equation (5), $L(\phi(z))$ also satisfies it.

The differential operator L depends on an infinite number of arbitrary parameters, because the infinitesimal functions $\delta_\mu(x)$ are completely arbitrary. The general form of the operator L is the following:

(21) $$L = \sum_{n=o}^{\infty} A_{\mu_1\mu_2\ldots\mu_n}(z)\ \partial_{\mu_1}\partial_{\mu_2}\ldots\partial_{\mu_n} \qquad (\mu_i = 0,1,2,3)$$

The functions $A_{\mu_1\mu_2\ldots\mu_n}(z)$ are assumed to depend on the infinite number of parameters. In general, we may suppose that $A_{\mu_1\mu_2\ldots\mu_n}(z)$ are completely traceless, because the operator L acts on the functions satisfying (5). Then we will obtain the following equations:

(22) $$\frac{2}{n}\,\partial_{\{\mu_1} A_{\mu_2\mu_3\ldots\mu_n\}}(z) = \delta_{\{\mu_1\mu_2}\, g_{\mu_3\ldots\mu_n\}}(z) - \Box A_{\mu_1\mu_2\ldots\mu_n}(z)$$

where the curly brackets mean a symmetrization of the indices inside them, and $g_{\mu_3\ldots\mu_n}(z)$ are some functions expressed by the divergences from $A_{\mu_2\mu_3\ldots\mu_{n-1}}(z)$.

The solution of the latter equations can be obtained and the part of the operator L that acts only on the positive frequency part of $\phi(z)$ has the form:

$$L^{++} = \sum_{n,k} a_{\mu_1\mu_2\ldots\mu_n;\,\nu_1\nu_2\ldots\nu_k}\, z_{\mu_1} z_{\mu_2}\ldots z_{\mu_n} :e^{-z_\rho\partial_\rho}: \partial_{\nu_1}\partial_{\nu_2}\ldots\partial_{\nu_k}$$

$:\ldots:$ denotes normal ordering; $a_{\mu_1\mu_2\ldots\mu_n;\nu_1\nu_2\ldots\nu_k}$ - arbitrary constants having two (μ's and ν's) groups of indices, each of them is symmetric and traceless. We may take the basis elements of the algebra in the form:

$$t_{\mu_1\mu_2\ldots\mu_n\,\nu_1\nu_2\ldots\nu_k} = \mathrm{sytl}(z_{\mu_1} z_{\mu_2}\ldots z_{\mu_n}) :e^{-z_\rho\partial_\rho}: \partial_{\nu_1}\partial_{\nu_2}\ldots\partial_{\nu_k}$$

(sytl = symmetric + traceless).

To write down the commutation relations, we introduce the following notations:

$$(23) \qquad \underset{n\geq 0}{E_{n,\ell,m}(\xi_\mu)} = \left(\frac{1}{2}\xi^2\right)^{n/2} \underset{u^2=1}{D_{n,\ell,m}(u_\mu)}; \quad \xi_\mu = \sqrt{\xi^2}\, u_\mu;$$

where

$$D_{n,\ell,m}(u_\mu) = \sqrt{\pi}\; 2^{\ell+1} \ell! \sqrt{\frac{(n-\ell)!}{(n+\ell+1)!}}\; \sin^\ell\lambda\; C^{1+\ell}_{n-\ell}(\cos\lambda)\; Y_{\ell,m}(\theta,\varphi)$$

and

$$Y_{\ell,m}(\theta,\varphi) = \sqrt{\frac{(2\ell+1)(\ell-m)!}{4\pi(\ell+m)!}}\; P^{|m|}_\ell(\cos\theta)\; e^{im\varphi}$$

In these formulae λ, θ, φ are the polar angles in four-dimensional space; $C^k_n(x)$ are the Gegenbauer polynomials and $Y_{\ell m}(\theta,\varphi)$ are the usual spherical functions. The functions (23) are linear independent and they form a O(4) irreducible basis in the set of the solutions of the Laplace equation.

Then we may rewrite the generators in the form

$$(24) \qquad X_{k\ell m;n\ell'm'} = E_{k,\ell,m}(z) :e^{-z_\rho\partial_\rho}: E_{n,\ell',m'}(\partial)$$

with the following commutation relations:

(25) $$\left[X_{n,\ell,m;n'''\ell'''m'''},\ X_{n''\ell''m'';n'\ell'm'}\right]$$

$$= n'''!\ \delta_{n''n'''}\ \delta_{\ell''\ell'''}\ \delta_{m''m'''}\ X_{n\ell m;n'\ell'm'} -$$

$$- n'!\ \delta_{nn'}\ \delta_{\ell\ell'}\ \delta_{mm'}\ X_{n''\ell''m'';n'''\ell'''m'''}$$

For the sake of simplicity, we shall consider some part of our algebra only. Let us define the new operators

(26) $$Q_{n,\ell,m} = \sum_{k,j,s} \frac{1}{k!} X_{n+k+1,\ell+j,m+s;k+1,j,s}$$

Obviously, $Q_{n\ell m}$ belong to our algebra. Using the relations (25), we may obtain the following expression:

(27) $$\left[Q_{n\ell m},\ Q_{n'\ell'm'}\right] = (n'-n)\ Q_{n+n',\ell+\ell',m+m'}$$

This means that $Q_{n\ell m}$ form the infinite dimensional subalgebra of the Witt type.

It has a central extension, which we obtain using the usual methods in the following form:

(28) $$\left[Q_{n\ell m},\ Q_{n'\ell'm'}\right] =$$

$$= (n'-n)\ Q_{n+n',\ell+\ell',m+m'} + ¢_{\ell+\ell',m+m'}\ n(n^2-1)\,\delta_{n+n',0}$$

This algebra is a Virasoro type algebra, but the coefficient $¢_{\ell+\ell',m+m'}$ depends on the ℓ's and m's . The usual Virasoro algebra is a subalgebra of our algebra and may be obtained if we put $\ell=\ell'=m=m'=0$.

REFERENCES

[1] I.T.Todorov. Infinite Lie algebras in 2-dimensional conformal field theory, Lectures given in ISAS, Trieste, November 1984.

[2] D.T.Stoyanov, L.L.Hadjiivanov. Communication JINR, Dubna, P2-84-466, 1984.

[3] D.T.Stoyanov, Preprint ISAS, Trieste, 1985.

[4] F.Gursey, H.G.Tze, Ann. of Phys., 128, 29 (1980).

[5] A.Sudbery, Math. Proc. Camb. Phil. Soc., 85, 199 (1979).

[6] R.Fueter, Comment. Math. Helv., 7, 307 (1935).

[7] R.Fueter, Comment. Math. Helv., 8, 371 (1936).

[8] J.Ryan, Funct. Anal., 61, 117 (1985).

[9] J.Ryan, Complex Variables, 1, 119 (1983).

[10] J.Ryan, Complex Variables, 2, 177 (1983).

[11] V.G.Kac, Izvestija A.N., SSSR, Math. Ser., 32, 1323 (1968).

INFINITE DIMENSIONAL LIE ALGEBRAS IN CONFORMAL QFT MODELS *)

I. T. Todorov

International School for Advanced Studies (ISAS), Trieste, Italy

Institute for Nuclear Research and Nuclear Energy, Sofia, Bulgaria **)

Abstract

The "minimal theories" of critical behaviour in two dimensions of Belavin, Polyakov and Zamolodchikov are reviewed. Conformally invariant operator product expansions (OPEs) are written down in terms of composite quasiprimary fields.

A new version of conformal quantum field theory (QFT) on compactified Minkowski space $\overline{M} = U(2)$ is developed. Light cone OPEs in four dimensions are shown to follow the same pattern as (2-or) 1-dimensional OPEs.

*) Expanded version of a talk, presented at the International Symposium on Conformal Groups and Structures, Technische Universität Clausthal, August 1985. Lectures presented at the International School for Advanced Studies in Trieste in January and at Collège de France, Paris, in February 1986.

**) Permanent address.

Contents

I. INTRODUCTION. PRELIMINARIES

I.1. Infinite dimensional Lie algebras are inherent to any quantum theory of an infinite system. The classification of inequivalent representations of the algebra of canonical commutation relations (the Heisenberg algebra) by Gårding and Wightman and by Segal is an early memorable result about such (generalized) Lie algebras. It was not, however, until physicists' attempt to use current algebras in the mid 60's and their success in studying the Virasoro algebra (first in the framework of dual resonance models - see [G2, V1, M1], then also in 2-dimensional quantum field theory (QFT) - see e.g. [F2,9]), and the parallel mathematical development of Kac-Moody algebras (for recent reviews and further references see [K2] and [VOMP]) and the study of central extensions of the algebra of diffeomorphisms of the circle (see [G1]), that infinite dimensional Lie algebras became an essential part of modern mathematical physics. With the revival of (super)string fashion the field became so croweded that one has to make a choice (for a small sample of current papers in which the applications of (super)Virasoro and Kac Moody algebras to strings is a dominating theme see [VOMP] as well as [A1-4; C1,2; F5,6; G9; N1,4,5; T1]).

We shall restrict our attention to QFT models including the field theoretical description of critical phenomena of two-dimensional statistical systems, greatly advanced recently in the work of Belavin, Polyakov and Zamolodchikov [B1] which initiated a flow of papers ([D2,3, 4, F7,8, G7, K7, I1, O1, T4,5, Z1]). Since the theory of Kac-Moody and Virasoro algebras has been worked out (and reviewed) in a number of mathematical publicaitons ([F1,3,4, G9, K1-4]) and in lecture notes by (and for) physicists [G4,O2, T5] I shall only briefly sketch - in Sec. 2 of this Introduction - some facts about these infinite dimensional (graded) Lie algebras and their relation to 2-dimensional models.

A new view on the minimal theories of [B1] is explored in Part II. Composite conformal ('quasiprimary') fields are constructed and used to write down explicit global OPEs. The notion of a pure primary field is introduced in Sec. 4B and used to associate a finite cyclic group of conformal families for the Ising model and its 3-critical extension ,the group multiplication being related to the 'fusion rule' of [B1] for OPEs.

Some steps in extending the 2-dimensional conformal techniques to four space-time dimensions are presented in Part III. A complex 0-torsion and 0-curvature frame bundle on compactified Minkowski space $\bar{M} \simeq U(2)$ is introduced in Sec. III.1. A new treatment of free 0-mass fields on $\bar{M}$ is given in Sec. III.2 using expansion in homogeneous harmonic polynomials (of a complex 4-vector z_α and $\frac{z_\alpha}{z^2}$). Composite conformal fields and light-cone OPEs are studied in Sec. III.3 using the techniques of Part II.

Formulas of the introductory Part I are labelled by (I.1), (I.30). Equations in Parts II and III are numbered by sections (like (1.1),(4.19)). References in Part III to formulas of Part II are given as (II.4.6).

I.2 PRELIMINARIES ON THE ALGEBRA OF CONSERVED TENSOR CURRENTS IN 2-DIMENSIONAL QFT

2.A Conformal stress-energy tensor and conserved chiral currents as fields on a circle. The traceless stress tensor $\Theta_{\mu\nu}$ in 1+1 dimensions has two independent components, which, as a consequence of the conservation law $\partial_\mu \Theta^{\mu\nu} = 0$, can each be taken as a function of a single light-cone variable

$$\Theta(\xi) = \frac{1}{2}\left[\Theta_{10}(x) - \Theta_{00}(x)\right], \quad \xi = x^1 - x^0 \tag{I.1a}$$

$$\bar{\Theta}(\bar{\xi}) = \frac{1}{2}\left[\Theta_{10}(x) + \Theta_{00}(x)\right], \quad \bar{\xi} = -x^1 - x^0 \tag{I.1b}$$

(the signs in the definition of ξ and $\bar{\xi}$ are so chosen that they both decrease with increasing time). Θ and $\bar{\Theta}$ transform under 1-dimensional representations of the Lorentz group SO(1,1); for $x \to \Lambda x$,

$$\Lambda = \begin{pmatrix} ch\alpha & sh\alpha \\ sh\alpha & ch\alpha \end{pmatrix},$$

$$\Theta(\xi) \to e^{-2\alpha}\,\Theta(e^{-\alpha}\xi), \quad \bar{\Theta}(\bar{\xi}) \to e^{2\alpha}\,\bar{\Theta}(e^{\alpha}\bar{\xi}).$$

They are called right and left moving fields, respectively.(Note that the bar on $\bar{\xi}$ and on $\bar{\Theta}$ is <u>not</u> related to complex conjugation.) According to a Lüscher-Mack theorem, reviewed in [T5], dilation invariance (with $\Theta_{\mu\nu}$ having dimension 2) allows to compute the local commutators of Θ and $\bar{\Theta}$ and to find all their vacuum expectation values which turn out to be invariant under the <u>projective conformal group</u> $SL(2,\mathbb{R})\times SL(2,\mathbb{R})$. Their asymptotic behaviour allows - and makes it advantageous - to formulate the theory on compactified space-time - i.e., on the torus $\bar{M}=S^1\times S^1$. To this end one uses a Cayley transformation for each of the light-cone variables:

$$\xi = \frac{z-i}{1-iz}\,, \qquad z = \frac{\xi + i}{1+i\xi} \qquad \text{etc.} \tag{I.2}$$

It maps the real line ξ onto the unit circle $|z| = 1$ with $\pm\infty \leftrightarrow z=-i$. Then the compact picture stress energy tensor is defined by

$$T(z) = -2\pi\left[\xi'(z)\right]^2 \Theta(\xi(z)) = -2\pi\frac{4}{(1-iz)^4}\Theta\left(\frac{z-i}{1-iz}\right) \tag{I.3}$$

and a similar expression for the left moving components. Local commutativity implies

$$\left[T(z),\ \bar{T}(\bar{z}')\right] = 0, \tag{I.4}$$

so that, as far as the stress energy tensor is concerned, the theories on the two circles ($|z| = 1$ and $|\bar{z}| = 1$) completely decouple.

Similarly, a current $j^{\mu}(x)$ of dimension 1 (in a dilation invariant theory), that is conserved togehter with its dual,

$$\partial_{\mu} j^{\mu}(x) = 0 = \partial_{\mu}\varepsilon^{\mu\nu} j_{\nu}(x), \tag{I.4}$$

has a right and a left moving components given by

$$j(\xi) = \frac{1}{2}\left(j^0(x) + j^1(x)\right), \quad \bar{j}(\bar{\xi}) = \frac{1}{2}\left(j^0(x) - j^1(x)\right). \tag{I.5}$$

The corresponding compact picture current is

$$J(z) = \frac{4\pi}{(1-iz)^2}\, j\left(\frac{z-i}{1-iz}\right), \quad \bar{J}(\bar{z}) = \frac{4\pi}{(1-i\bar{z})^2}\, \bar{j}\left(\frac{\bar{z}-i}{1-i\bar{z}}\right). \tag{I.6}$$

We assume that the vacuum state in the two pictures is the same so that we can 'translate' the vacuum expectation values from the non-compact to the compact picture.

2.B The Virasoro algebra for the stress tensor; current algebras. One immediate benefit of using the compact space picture arises from the fact that the counterpart of the Fourier integral - at least for fields of (half)integer dimension - is the Fourier-Laurent series:

$$T(z) = \sum_{n\in\mathbb{Z}} L_n z^{-n-2}, \qquad J(z) = \sum_{n\in\mathbb{Z}} Q_n z^{-n-1}. \tag{I.7}$$

To exhibit the hermiticity properties of the fields (I.7) one should rememeber that the image of the real line (ξ) is the unit circle, the upper and lower half-planes are mapped onto the exterior and the interior of the circle. The involution which gives the relevant conjugation in the complex z plane is the inversion

$$z \to \frac{1}{z^*} \tag{I.8}$$

(where z^* stands for ordinary complex conjugation).The hermiticity condition for a field $\varphi(z)$ of dimension Δ reads

$$\varphi(z^*)^* \equiv \frac{1}{z^{2\Delta}} \varphi^*\left(\frac{1}{z}\right) = \varphi(z). \tag{I.9}$$

It is straightforward to verify that T(z) and J(z) are hermitian iff

$$L_n^* = L_{-n}, \qquad Q_n^* = Q_{-n}. \tag{I.10}$$

This is the property which justifies the choice of labelling of the Laurent coefficients in (I.7).

The above mentioned Lüscher-Mack theorem implies that the L_n satisfy the Virasoro commutation relations

$$[L_n, L_m] = (n-m) L_{n+m} + \frac{c}{12} n(n^2-1) \delta_{n+m} \tag{I.11}$$

where c is the 'central charge' ($[c, L_n] = 0$) and δ_{n+m} stands for the Kronecker symbol $\delta_{n,-m}$. For a (compact) internal symmetry group with hermitian Lie algebra generators I_a satisfying

$$[I_a, I_b] = if_{abc} I_c \tag{I.12}$$

the 'charges' Q_n^a (defined by (I.7) for $J = J^a$) obey the commutation rules

$$[Q_n^a, Q_m^b] = i f_{abc} Q_{n+m}^c + \frac{\lambda}{2} n \delta_{ab} \delta_{n+m} . \tag{I.13}$$

For a simple (non-abelian) group G these are the defining rleations for a Kac-Moody Lie algebra; in the mathematical literature one often uses the notation

$$Q_n^a = I_a \otimes t^n,$$

which allows to write (I.13) in a basis independent form:

$$[I \otimes f(t), J \otimes g(t)] = [I, J] \otimes f(t)\, g(t) + \lambda \langle I, J \rangle \operatorname{Res}(f'(t) g(t)).$$

Here I,J belong to dG, the Lie algebra of G; f,g are elements of $\mathbb{C}[t, t^{-1}]$, the algebra of polynomials in t and t^{-1}; $\langle I,J \rangle$ is the Killing form, normalized in such a way that $\langle I,J \rangle$ = tr IJ for the fundamental (lowest dimensional, faithful) representation of dG.

The infinitesimal space-time transformations of the current are generated by the Virasoro operators:

$$[L_n, J(z)] = \frac{d}{dz}\left(z^{n+1} J(z)\right) \tag{I.14a}$$

or

$$[Q_m, L_n] = m\, Q_{m+n} . \tag{I.14b}$$

Thus, the operator algebra of a 2-dimensional conformal theory with a continuous internal symmetry includes the semi-direct product of (two commuting copies of) the Virasoro algebra with the (corresponding copies of the chiral) current algebra, which is (just as well as each factor) a graded Lie algebra.

The commutation relations (I.11) (I.13) (I.14) allow to write down all vacuum expectation values of products of T's and J's; we have

$$\langle T(z_1)\,T(z_2)\rangle = \frac{c}{2\,z_{12}^4}\,,\quad \langle T(z_1)T(z_2)\,T(z_3)\rangle = \frac{c}{z_{12}^2\,z_{23}^2\,z_{13}^2}\,,$$

$$\langle T(z_1)\,T(z_2)\,T(z_3)\,T(z_4)\rangle = \frac{c^2}{4}\left(z_{12}^{-4}\,z_{34}^{-4} + z_{13}^{-4}\,z_{24}^{-4} + z_{14}^{-4}\,z_{23}^{-4}\right) +$$

$$+c\left(z_{12}^{-2}\,z_{23}^{-2}\,z_{34}^{-2}\,z_{14}^{-2} + z_{12}^{-2}\,z_{24}^{-2}\,z_{34}^{-2}\,z_{13}^{-2} + z_{13}^{-2}\,z_{23}^{-2}\,z_{24}^{-2}\,z_{14}^{-2}\right)\ etc., \tag{I.15}$$

$$\langle J^a(z_1)\,J^b(z_2)\rangle = \frac{\lambda\,\delta_{ab}}{2\,z_{12}^2}\,,\quad \langle J^a(z_1)\,J^b(z_2)\,J^c(z_3)\rangle =$$

$$(z_{ij} = z_i - z_j) \qquad\qquad = \frac{\lambda}{2i}\,\frac{f_{abc}}{z_{12}\,z_{13}\,z_{23}}\,,\ etc.$$

2.C Verma modules and lowest weight (LW) unitary irreducible representations (UIRs) of the Virasoro algebra. Energy positivity implies that the spectrum of L_o on physical states should be bounded below. That is why we are interested in LW representation of the Virasoro algebra*) $\hat{W}$. If $|\Delta\rangle$ is a LW vector of $\hat{W}$,

$$(L_o - \Delta)\,|\Delta\rangle = 0\,, \qquad L_o \geq \Delta\,, \tag{I.16}$$

then the relation $(L_o + n - \Delta)L_n|\Delta\rangle = 0$, which follows from (I.11), implies

$$L_n\,|\Delta\rangle = 0 \quad for \quad n = 1, 2, \ldots \tag{I.17}$$

(Δ is the intercept of the corresponding Regge trajectory in the terminology of dual resonance models). A Verma module is the representation space $V_{c,\Delta}$ for $\hat{W}$ for fixed (real) c spanned by all the vectors of the form

$$L_{-n_1}\,L_{-n_2}\cdots L_{-n_k}|\Delta\rangle \quad \text{with} \quad n_1 \geq n_2 \geq \ldots \geq n_k \geq 1. \tag{I.18}$$

$V_{c,\Delta}$ is an inner product space, the inner product of any two vectors of the basis being evaluated from (I.11) (I.16) (I.17) under the assumption that

$$\langle\Delta|\Delta\rangle = 1.$$

---

*) The notation $\hat{W}$ for the Virasoro algebra reminds that it is a central extension of the Witt algebra W, i.e. the algebra of diffeomorphisms of the circle generated by the first order differential operators $-z^{n+1}\frac{d}{dz}$.

The vacuum $|0\rangle$ is a special case of a LW vector characterized by being $sl(2)$-invariant (and having zero charge in the presence of a continuous internal symmetry):

$$L_n\,|0\rangle = 0 \text{ for } n \geq -1 \qquad (Q_n|0\rangle = 0 \text{ for } n \geq 0). \qquad \text{(I.20)}$$

We note, that for $c \neq 0$, the Virasoro commutation relations (I.11) imply that $L_{-n}|0\rangle \neq 0$. Since, on the other hand, the condition $c > 0$ is a consequence of the non-vanishing of the stress-energy tensor in a positive metric Hilbert space, we deduce that $\hat{W}$ is never a symmetry of the quantum theory. Only the projective conformal group, generated by L_0 and $L_{\pm 1}$, can be a symmetry of QFT Green functions. Instead, the algebra of operator products for LW UIRs of $\hat{W}$ with a fixed central charge c can be regarded as a mathematical expression of the corresponding conformal QFT. This observation leaves room for the hope to extend the 2-dimensional techniques of this and the following chapters to higher number of space-time dimensions.

Two questions arise: (i) when is the inner product in $V_{c,\Delta}$ positive definite? (ii) when is the representation of $\hat{W}$ acting in $V_{c,\Delta}$ irreducible?

The answers to both questions come from the analysis of the Kac determinant M_N of inner products of all vectors of the form (I.18) for fixed $N = n_1 + \ldots + n_k$. (The number of such vectors is clearly equal to the partition function P(N), i.e. to the number of different ways in which the positive integer N can be split into a sum of positive integers.) Kac [K1] has demonstrated that for a fixed c the possible zeroes of all $M_N(c,\Delta)$ are labelled by two positive integers p and q and take the values

$$\Delta_{p,q} = \frac{1}{24}(c-1) + \frac{1}{4}(\alpha_+ p + \alpha_- q)^2 \qquad \text{(I.21a)}$$

where

$$\alpha_\pm = \sqrt{\frac{1-c}{24}} \pm \sqrt{\frac{25-c}{24}}\ . \qquad \text{(I.21b)}$$

It follows that for $c > 1$ and $\Delta \geq 0$ all M_N are positive and the corresponding representation of W are unitary and irreducible. (For $1 < c < 25$ there are no real zeroes of M_N; for $c \geq 25$ such zeroes exist, but correspond to non-positive dimensions.) On the contrary, for $0 \leq c \leq 1$ there are non-negative $\Delta_{p,q}$ for which M_N vanishes. The representations of $\hat{W}$ at such points are reducible and unitarity, in general, fails for $c<1$. It is remarkable that for a series of such exceptional values,

$$c = 1 - \frac{6}{(m+2)(m+3)}, \qquad m = 1, 2, \ldots \tag{I.22}$$

$$\Delta^{(m)}_{p,q} = \frac{[(m+3)p - (m+2)q]^2 - 1}{4(m+2)(m+3)}, \qquad \begin{matrix} 1 \leq p \leq m+1 \\ 1 \leq q \leq m+2, \end{matrix} \tag{I.23}$$

the space $V_{c,\Delta}$ admits an invariant subspace $V^{(o)}_{c,\Delta}$ of zero length vectors and the representation of $\hat{W}$ in the factor space $V_{c,\Delta}/V^{(o)}_{c,\Delta}$ is unitary and irreducible. These exceptional points correspond to physically interesting 2-dimensional models of phase transition. Moreover, the correlation functions of the primary fields of dimensions $\Delta_{p,q}$ satisfy linear differential equations which make the models exactly soluble. To see how this comes about we consider the condition for having a "null-vector" at level 2. In other words, we would like to find a relation between c and Δ for which a vector of the form $|\Delta+2, c\rangle = (L_{-2} - \alpha L_{-1}^2)|\Delta\rangle$ is LW vector. To this end it is sufficient to find conditions for which a vector of this type is annihilated by L_1 and L_2. We have $L_1|\Delta+2, c\rangle = (3 - 2(2\Delta+1)\alpha) L_{-1}|\Delta\rangle = 0$ or $\alpha = \frac{3}{2}(1+2\Delta)^{-1}$; $L_2|\Delta+2, c\rangle = (4\Delta + \frac{c}{2} - 6\alpha\Delta)|\Delta\rangle = 0$; this gives $c = 18\Delta(1+2\Delta)^{-1} - 8\Delta$ or $16\Delta = 5 - c \pm \sqrt{(25-c)(1-c)}$, i.e. $4\Delta_{2,1} = 1 + \frac{3}{m+2}$, $4\Delta_{1,2} = 1 - \frac{3}{m+3}$ (for $1 - c = \frac{6}{(m+2)(m+3)}$) in accord with (I.23).

<u>Remark</u>. Each dimension in the rectangular range of p and q in (I.23) is encountered excatly twice, since $\Delta^{(m)}_{p,q}$ does not change under the substitution $p \to m+2-p$, $q \to m+3-q$. For this reason it suffices to consider the range $1 \leq q \leq p \leq m+1$ in order to obtain every dimension Δ exactly

once (cf. [F7]). In studying the so-called "fusion rules" [B1], however, it is advantageous to have the Δ's defined for all points in the rectangle (taking the symmetry into account - see Sec. II.4).

The unitary postulate for $c<1$ was explored in [F7] where it was argued on the basis of some numerical computations that the points (I.22) (I.23) are the only ones that may correspond to Verma modules with positive semi-definite inner product. The fact that the factor representations at these points (modulo the invariant subspaces of zero norm vectors) are indeed unitary was demonstrated - in an elegant application of the Sugawara formula - by Goddard, Kent and Olive [G5] (see also [K4,5]).

It is clear that such a unitarity postulate is necessary for a QFT (Minskowski space) interpretation of the corresponding critical models. its relevance for statistical mechanics, however, (in which the postulate appears as Osterwalder - Schrader reflection positivity in the Euclidean formulation) is open to discussion.

We shall say more about the simplest example of the discrete series (I.22), the case m = 1 ($c = \frac{1}{2}$) — the critical Ising model, in Part II.

2D. Few words about the super Virasoro algebra. It was noted that the Virasoro algebra $\hat{W}$ appears as a central extension of the algebra of first order differential operators in a complex variable z with polynomial coefficients in z and z^{-1}. Similarly, one can define the N=1 super Virasoro algebra $\hat{W}_k$ as a central extension of the Witt algebra of super differential operators, in the even variable z and the odd variable θ, that preserve the conformal class of the 1-form [K3]

$$\omega_\kappa = z^{2\kappa-1}dz - \theta d\theta \qquad \kappa = 0,\ \tfrac{1}{2}\,. \tag{I.24}$$

A basis for this "super Witt" algebra is given by the differential operators

$$\hat{\ell}_n(\kappa,\Delta) = -z^n\left\{z\frac{\partial}{\partial z} + (n+2\kappa)\left(\Delta + \frac{1}{2}\theta\frac{\partial}{\partial\theta}\right)\right\} \tag{I.25a}$$

$$\hat{g}_{n+\kappa}(\kappa,\Delta) = z^n\left\{z^{2\kappa}\frac{\partial}{\partial\theta} - \theta\left(z\frac{\partial}{\partial z} + 2(n+2\kappa)\Delta\right)\right\}. \tag{I.25b}$$

The parameter Δ labels the conformal weight of the representation of this infinite dimensional Lie superalgebra. The central extension again adds just one new generator, the central charge c. The supercommutation relations of the operators L_n and G_ρ, corresponding to $\hat{l}_n$ and $\hat{g}_\rho$ after the central extension, are

$$[L_n, L_m] = (n-m)L_{n+m} + \frac{c}{12} n(n^2-1)\delta_{n+m}, \tag{I.26a}$$

$$[G_{m+\kappa}, L_n] = (m+\kappa-\frac{n}{2}) G_{m+n+\kappa}, \tag{I.26b}$$

$$[G_{m+\kappa}, G_{n-\kappa}]_+ = 2L_{m+n} + \frac{c}{3}\{(\kappa+m)^2 - \frac{1}{4}\}\delta_{m+n}. \tag{I.26c}$$

The case $\kappa = 0$ coincides with the Ramond algebra [R1] , the case $\kappa = ½$ is the Neveu-Schwarz algebra [N2] .

We note that for $\kappa = ½$ the operator $\hat{g}_{-½} = \frac{\partial}{\partial\theta} - \theta\frac{\partial}{\partial z}$ appears as square root of $\hat{l}_{-1} = -\frac{\partial}{\partial z}$:

$$\hat{g}^2_{-\frac{1}{2}} = -[\theta, \frac{\partial}{\partial\theta}]_+ \frac{\partial}{\partial z} = -\frac{\partial}{\partial z} = \hat{l}_{-1} \tag{I.27}$$

($\hat{g}_{-½}$ has the properties of iD where D is the superderivative $D = \frac{\partial}{\partial\theta} + \theta\frac{\partial}{\partial z}$, satisfying $D^2 = \frac{\partial}{\partial z}$).

In the Neveu-Schwarz case the odd operators $G_{n+½}$ are generated by a (conserved) local Fermi current

$$G(z) = \sum_{n\in\mathbb{Z}} G_{n+\frac{1}{2}} z^{-n-2}, \tag{I.28}$$

with a translation invariant 2-point function:

$$\langle G(z_1) G(z_2)\rangle = \frac{2c}{3}(z_1-z_2)^{-3}. \tag{I.29}$$

The unitary representations of the super Virasoro algebra are obtained for $c \geq 3/2$, $\Delta \geq 0$ or

$$c = 3/2\,(1 - \frac{8}{(m+2)\,(m+4)}),\quad m = 0,1,2,... \tag{I.30}$$

and a suitable spectrum of allowed conformal weights Δ for each m (see [F7, 8, G5, K3,4,5]). The LWUIRs of the semidirect product of the superVirasoro algebra with a supersymmetric current algebra were constructed in [K3] . Concerning recent field theoretical and string theoretic applications of the super Virasoro and the supercurrent algebra see e.g., [B2, D2,E1,F6,8] (the second reference [F6] also contains bibliography). An N=2 supersymmetric extension of the Virasoro algebra is considered in [A1] .

II. PRIMARY AND QUASIPRIMARY FIELDS. OPE

II.1 Basic notions

We started with a field theoretical notion - the notion of a conserved tensor current (in particular, with the stress energy tensor T) and ended up with an infinite dimensional Lie algebra (the Virasoro algebra $\hat{W}$, in the case of T). Then we apparently forgot about field theory and started constructing Verma modules and LWUIR's of $\hat{W}$. The notion of a primary field, introduced in [B1] , provides a link between LW representations of $\hat{W}$ and local QFT.

Roughly speaking, a primary field $\psi(z,\bar{z})$ is a field that transforms homogeneously under reparametrization and gauge transformations.

Since neither the current J nor the stress tensor T (I.7) mix z and $\bar{z}$ we can consider the transformation properties in each variable separately. The homogeneous law for an infinitesimal reparametrization of the first argument (z) for a field of conformal weight Δ in z reads

$$[L_n, \psi(z)] = z^n \left(z\frac{\partial}{\partial z} + (n+1)\Delta \right) \psi(z). \tag{1.1}$$

If furthermore ψ transforms under a finite dimensional representation of an internal symmetry group G with infinitesimal (hermitian) generators I_a associated with a conserved current $J_a(z)$, then the homogeneous gauge law for a primary ψ is

$$[\psi(z), Q_n^a] = z^n I_a \psi(z); \tag{1.2}$$

here Q_n^a are the Laurent coefficients in the expansion (I.7) of J_a. In particular, if ψ is a complex field that carries an electric charge e (corresponding to an U(1)-gauge group) then (1.2) is true with

$Q_n^a = Q_n$ and $I_a = e$.

The current $J_a(z)$ provides an example of a primary field - of weight (1,0) - under reparametrization, which is not primary under gauge transformations, since eq.(I.7) and (I.13) imply

$$[Q_n^a , J_b(z)] = i f_{abc} z^n J_c(z) + \frac{\lambda}{2} n \delta_b^a z^{n-1} \tag{1.3}$$

($(I_b)_{ac} = if_{abc}$ in this case).

Remark For non-integer Δ the notation $\psi(z)$ might be misleading, since ψ is not a single valued function on the circle (respectively, on the torus). Instead, it can be regarded as a function on the real line θ (for $z = e^{i\theta}$) or, alternatively, as a local section of a fibre bundle on S^1. The following proposition shows, however, that the vector function $\psi(z)|0\rangle$ is single valued analytic in z for $|z| < 1$.

Proposition 1.1 If ψ is a local field in a theory in which the energy of all states is positive, then the vector function $\psi(z,\bar{z})|0\rangle$ admits analytic continuation in the 2-disk $|z|<1$, $|\bar{z}|<1$. If ψ is primary of weight $(\Delta,\bar{\Delta})$ under reparametrization, then

$$\psi(0,0)|0\rangle = |\Delta, \bar{\Delta}\rangle \tag{1.4}$$

is a LW vector in the corresponding Verma module for $\hat{W}\otimes\hat{W}$.

Proof. The analiticity property of $\psi(z,\bar{z})|0\rangle$ is a consequence of the known analyticity of $\psi_M(\xi,\bar{\xi})|0\rangle$ in the product of upper half planes in any theory satisfying spectral conditions; here ψ_M is the Minkowski-space field corresponding to the compact picture field under the Cayley transformation, described in Sec. I.2A. (In order to have analyticity in the upper half plane for both ξ and $\bar{\xi}$ it was necessary to use the awkward looking sign in (I.1b).) The second property is a consequence of the definition (1.1) of a primary field and of the characterization (I.17-18) of a LW vector of $\hat{W}$.

Warning: An operator field with a nontrivial Laurent expansion like J(z) (I.7) is not defined for z=0; however, $J(0)|0\rangle$ exists nevertheless since the operator coefficients to the negative powers of z annihilate the vacuum.

As the symmetry group of the class of theories under consideration is the 6-parameter projective conformal group SU(1,1)xSU(1,1) (or its covering group), not the infinite Virasoro algebra, it is desirable to have a name also for the fields which are covariant under UIRs of this group. We say that the field $\psi(z,\bar{z})$ is quasiprimary*) if it satisfies (1.1) for n=0, ±1. The stress energy tensor T(z) provides an example of a quasiprimary field which is not primary, since

$$[L_n, T(z)] = z^{n+1} T'(z) + 2(n+1) z^n T(z) + + \frac{c}{12} z^{n-2} n(n^2-1). \tag{1.5}$$

Correlation functions of quasiprimary fields are conformally invariant; 2- and 3-point functions are determined from this property up to constant factors. In particular, the 2-point function of a pair of quasiprimary fields transforming under disjoint representations of the projective conformal group vanish. This allows, as we shall see, to expand products of (quasi)primary fields in a series of integrals of mutually orthogonal quasiprimary composite fields of different dimensions.

II.2 Two concepts for "frequency parts" of conserved currents

We can also write (as in [T4,5]) an integrated form of the definition (1.1) (1.2) of a primary field, which makes use of a notion of frequency part for J and T. For a field with a Laurent expansion, like (I.7) we can define a negative frequency part by the sum of the negative powers of z in its Laurent series; thus,

$$J^{(-)}(z) = \sum_{n=0}^{\infty} Q_n z^{-n-1}, \quad T^{(-)}(z) = \sum_{n=-1}^{\infty} L_n z^{-n-2}. \tag{2.1a}$$

---

*) This is the term adopted in [B1] (see Appendix A). The concept is older and applies to any number of space-time dimensions - these are the 'basic' (as opposed to derivative) conformal fields of [C3, T3] (see also earlier work cited there). Note also that the term 'primary' is only used in [B1] for fields transforming homogeneously under reparametrization. It was extended to gauge covariant fields in [T4] where also the relevance of this extension for the algebraic treatment of the Wess-Zumino model [K7] and for the Thirring model [D1] was exhibited.

Similarly, we define the positive frequency parts of J and T by

$$J_{(+)}(z) = J(z) - J^{(-)}(z) \left(= \sum_{n=1}^{\infty} Q_{-n} z^{n-1} \right), \quad T_{(+)}(z) = T(z) - T^{(-)}(z). \tag{2.1b}$$

Eqs.(1.1) and (1.2) then give (for $\psi(z_1, \bar{z}_1)$ abbreviated by $\psi(1)$)

$$\left[T^{(-)}(z), \psi(1) \right] = \frac{1}{z - z_1} \frac{\partial \psi(1)}{\partial z_1} + \frac{\Delta}{(z-z_1)^2} \psi(1), \tag{2.2}$$

$$\left[J_a^{(-)}(z), \psi(1) \right] = -\frac{I_a}{z - z_1} \psi(1) \quad \left(\text{for } |z| > |z_1| \right). \tag{2.3}$$

Simple and useful as these formulas are, they also have a shortcoming: they are not conformally invariant. Indeed, if a current is a quasiprimary field its frequency parts are not. We have, for instance,

$$\left[L_1, J^{(-)}(z) \right] = \frac{d}{dz} \left(z^2 J^{(-)}(z) \right) - Q_0 \tag{2.4}$$

instead of (I.14). This defect has been turned into a virtue in refs. [T4,5] where it was demonstrated that the non-invariance of the normal product

$$:J(z)\, \psi(z,\bar{z}): \; = J_{(+)}(z)\, \psi(z,\bar{z}) + \psi(z,\bar{z})\, J^{(-)}(z)$$

is exactly compensated by the non-invariance of the derivative $\frac{\partial \psi}{\partial z}$ for a field ψ of positive Δ, so that an equation of the type $\partial_z \psi + :J\psi: = 0$ (for a suitably normalized current) is conformally invariant.

<u>Remark</u> There is one exception to the non-invariance of frequency parts: a free (say, charged) field ψ of conformal weight (½,0) can be split into two SU(1,1)-covariant components

$$\psi_-(z) \left(= \psi^{(-)}(z) \right) = \sum_{n=1}^{\infty} c_{n-\frac{1}{2}} z^{-n}, \quad \psi_+(z) = \psi(z) - \psi_-(z) = \sum_{n=0}^{\infty} c_{-n-\frac{1}{2}} z^{n}. \tag{2.5}$$

The point is, that Eq.(1.1) for Δ = ½ implies

$$\left[L_n, c_{k-\frac{1}{2}} \right] = -\left(\frac{n-1}{2} + k \right) c_{n+k-\frac{1}{2}}, \tag{2.6}$$

so that $\left[L_1, c_{-1/2} \right] = 0 = \left[L_{-1}, c_{1/2} \right]$. Moreover, we can construct a current out of ψ and ψ^* with frequency parts

$$J_+^{\psi}(z) = \frac{e}{2}\left(\psi_+^*(z)\,\psi(z) - \psi_+(z)\,\psi^*(z)\right) \qquad J_-^{\psi}(z) = J^{\psi}(z) - J_+^{\psi}(z) =$$
$$= \frac{e}{2}\left(\psi^*(z)\,\psi_-(z) - \psi(z)\,\psi_-^*(z)\right), \tag{2.7}$$

different from (2.1) and SU(1,1)-covariant. Their commutation relations with the field ψ also change and involve its frequency parts:

$$[J_-^{\psi}(z), \psi(z_1)] = \frac{-e}{z-z_1}\left(\frac{1}{2}\psi(z) + \psi_-(z)\right) \tag{2.8}$$
$$\text{for} \quad [c_{m-\frac{1}{2}}, c^*_{n-\frac{1}{2}}] = \delta_{mn}.$$

If we set

$$\sum_{n\in\mathbb{Z}} z_2^n z_1^{-n-1} = \delta(z_1 - z_2) \qquad \left(\oint_{S^1} f(z_2)\delta(z_1-z_2)\frac{dz_2}{2\pi i} = f(z_1)\right), \tag{2.9}$$

then we can also write the following $\hat{W}$-covariant counterpart of (1.1) and (1.2):

$$[T(z), \psi(1)] = \delta(z-z_1)\frac{\partial\psi(1)}{\partial z_1} + \Delta\,\delta'(z-z_1)\,\psi(1) \tag{2.10}$$

$$[J_a(z), \psi(1)] = -\delta(z-z_1)\,I_a\,\psi(1). \tag{2.11}$$

II.3 Renormalized composite quasiprimary fields and OPE on S^1.

A QFT on the circle is defined by a reducible positive energy representation of the algebra of observables which includes the Virasoro algebra $\hat{W}$. The representation space is spanned by vectors of the form $T(z_1)\ldots T(z_n)\,\varphi(z)|0\rangle$ where φ runs over the set of all primary fields of the theory, n=0,1,... In particular, the vector valued distribution $\varphi_1(z_1)\,\varphi_2(z_2)|0\rangle$ can be expanded in a sum of such conformal families for any pair of primary fields φ_1 and φ_2. Accorinding to [I1] the set of LWs for $\hat{W}$ should be such that the corresponding characters transform under a linear representation of the modular group SL(2,**Z**) (cf. [K4]).

We shall single out the simplest cases, in which the vacuum OPE of $\varphi_1\varphi_2$ contains a single conformal family of dimensions Δ_3+n, n = 0,1,.., $\Delta_3 \geq 0$. If we define the superselection operator $U = e^{2\pi i L_0}$ (which commutes with algebra of all conserved conformal currents), then $\varphi_1\varphi_2|0\rangle$

would span a coherent subspace of U, $(U - e^{2\pi i \Delta_3})\varphi_1(z_1)\varphi_2(z_2)|0\rangle = 0$ in such models.

Remark The existence of conformal OPEs should be regarded as a basic postulate in the present approach. It can be justified in a Lagrangian framework for a renormalization group fixed point - in any number of space-time dimensions - provided that the product of fields is applied to a finite energy state (say, to the vacuum) - see [D3, C4] and references therein.

We first consider the case in which the minimal dimension Δ_3 appearing in the OPE of $\varphi_1 \varphi_2$ is positive. The following simple fact was recognized by 1970.

Let φ_i, i = 1,2,3 be three (not necessarily different) quasiprimary fields of weights Δ_i. Their conformally invariant 2- and 3- point functions are determined up to constant factors and have the form

$$\langle \varphi_i(z_1)\varphi_i(z_2)\rangle = \frac{N_i}{z_{12}^{2\Delta_i}}, \quad \langle \varphi_1(z_1)\varphi_2(z_2)\varphi_3(z_3)\rangle = \frac{N_{123}}{z_{23}^{\delta_1} z_{13}^{\delta_2} z_{12}^{\delta_3}} \tag{3.1}$$

where δ_i are determined from the 'conservation of dimension' law:

$$\delta_2 + \delta_3 = 2\Delta_1, \quad \delta_1 + \delta_3 = 2\Delta_2, \quad \delta_1 + \delta_2 = 2\Delta_3. \tag{3.2}$$

(For a proof, see e.g. [T5,6] where earlier work is also cited.)

Proposition 3.1 [S4] Under the above assumptions if $\Delta_3 > 0$ and the bilocal field

$$B_{12}(z_1, z_2) = \frac{N_3}{N_{123}} z_{12}^{\delta_3} \varphi_1(z_1)\varphi_2(z_2) \tag{3.3a}$$

satisfies

$$(U - e^{2\pi i \Delta_3}) B_{12}(z_1, z_2)|0\rangle = 0 \tag{3.3b}$$

then its OPE has the form

$$B_{12}(z+\varepsilon, z) = \sum_{n=0}^{\infty} \frac{\Gamma(\delta_1+\delta_2+2n)\,\varepsilon^n}{\Gamma(\delta_1+n)\Gamma(\delta_2+n)} \int_0^1 du\,(1-u)^{\delta_1+n-1} u^{\delta_2+n-1} \times O^{(\delta_1,\delta_2)}_{\Delta_3+n}(z+u\varepsilon), \tag{3.4}$$

where $O(z)$ are composite quasiprimary fields given by

$$n!\, O^{(\delta_1,\delta_2)}_{\Delta_3+n}(z) = \lim_{z_1,z_2\to z} D^{(\delta_1,\delta_2)}_n(\partial_1,\partial_2)\, B_{12}(z_1,z_2); \tag{3.5}$$

$D_n(\alpha,\beta)$ is a homogeneous polynomial degree n expressed in terms of a Jacobi polynomial:

$$\binom{\delta_1+\delta_2+2n-2}{n} D^{(\delta_1,\delta_2)}_n(\alpha,\beta) = (\alpha+\beta)^n\, P^{(\delta_1-1,\delta_2-1)}_n\left(\frac{\alpha-\beta}{\alpha+\beta}\right) \tag{3.6a}$$

$$= \sum_{k=0}^{n} \binom{n+\delta_1-1}{n-k}\binom{n+\delta_2-1}{k} \alpha^{n-k}(-\beta)^k . \tag{3.6b}$$

The normalization in (3.4) is chosen in such a way that $O_{\Delta_3} = \varphi_3$ (for φ_3 satisfying (3.1)).

Proof. The form of the differential operator D_n is determined from the requirement that the composite field (3.5) is quasiprimary, so that, in particular,

$$\left[L_1, O_{\Delta_3+n}(z)\right] = \left\{z^2\frac{d}{dz} + 2(\Delta_3+n)z\right\} O_{\Delta_3+n}(z) . \tag{3.7}$$

Using the fact that the bilocal field (3.3) has infinitesimal conformal transformation law

$$\begin{aligned}[L_1, B_{12}(z_1,z_2)] = \big(&z_1^2\partial_1 + z_2^2\partial_2 + \\ &+ \delta_2 z_1 + \delta_1 z_2\big) B_{12}(z_1,z_2),\end{aligned} \tag{3.8}$$

condition (3.7) is then satisfied iff D_n satisfies the partial differential equation

$$\left(\alpha\frac{\partial^2}{\partial\alpha^2} + \beta\frac{\partial^2}{\partial\beta^2} + \delta_2\frac{\partial}{\partial\alpha} + \delta_1\frac{\partial}{\partial\beta}\right) D^{(\delta_1,\delta_2)}_n(\alpha,\beta) = 0. \tag{3.9}$$

Eq.(3.6) gives the unique homogeneous polynomial solution of (3.9) normalized by

$$D^{(\delta_1,\delta_2)}_n(1,-1) = 1. \tag{3.10}$$

The u-dependent weight in the integral in each term of the expansion (3.4) is fixed by the relation between 2- and 3- functions which is reduced to the integral formula

$$z^{-\mu}(z+\varepsilon)^{-\nu} = \frac{\Gamma(\mu+\nu)}{\Gamma(\mu)\Gamma(\nu)} \int_0^1 du \frac{(1-u)^{\mu-1} u^{\nu-1}}{(z+\varepsilon u)^{\mu+\nu}} . \tag{3.11}$$

(The expansion of both sides in ε gives all the moments of the weight and hence determines it in a unique way.)

If φ_1 and φ_2 are the same hermitian field so that $\Delta_3 = 0$ we must subtract the vacuum expectation value in the definition of the bilocal field B_{12}. We shall illustrate the necessary changes in Proposition 3.1 for the important special case in which $\varphi_1 = \varphi_2 = T(z)$.

Proposition 3.2 The bilocal operator

$$T_2(z_1,z_2) = \tfrac{1}{2} z_{12}^2 \left\{ T(z_1)T(z_2) - \frac{c}{2 z_{12}^4} \right\} \tag{3.12}$$

(where c is the central charge of the Virasoro algebra) admits the following OPE:

$$T_2\left(z+\tfrac{\varepsilon}{2}, z-\tfrac{\varepsilon}{2}\right) = \frac{3}{4}\int_{-1}^{1} d\lambda\,(1-\lambda^2)\,T\left(z+\lambda\tfrac{\varepsilon}{2}\right) + \sum_{n=1}^{\infty} \varepsilon^{2n} \int_{-1}^{1} d\lambda\, p_{2n+2}(\lambda)\, T_{2n+2}\left(z+\lambda\tfrac{\varepsilon}{2}\right) \tag{3.13}$$

where $p_k(\lambda)$ are the normalized weights

$$p_k(\lambda) = \frac{(2k-1)!!}{2^k (k-1)!!} (1-\lambda^2)^{k-1} , \qquad \int_{-1}^{1} d\lambda\, p_k(\lambda) = 1, \tag{3.14}$$

and the composite fields T_{2n} are given by

$$(2n-2)!\, T_{2n}(z) = \lim_{z_1, z_2 \to z} D_{2n-2}^{(2,2)}(\partial_1, \partial_2)\, T_2(z_1,z_2). \tag{3.15}$$

(In particular, $T_2(z) = \lim_{z_1, z_2 \to z} T_2(z_1,z_2) = T(z)$.)

The proof follows the argument of Proposition 3.1. We have here no free normalization (of the type appearing in Eq. (3.3)), since the n-point functions of T are determined by the commutation relations

$$\left[T^{(-)}(z_1),\ T(z_2) \right] = \frac{c}{2 z_{12}^4} + \frac{2}{z_{12}^2} T(z_2) + \frac{1}{z_{12}} T'(z_2). \qquad (3.16)$$

It is remarkable that Eqs.(3.5) and (3.15) define renormalized composite fields whose matrix elements are free of divergences. The latter property follows from the above construction for matrix elements of products of O_{Δ_3+n} and T_{2n} with the fields $\varphi_1 \varphi_2$ and T's, respectively. A closer look shows that it is actually true in general. We shall content ourselves by verifying that the limit (3.15) exists for the 4-point function $\langle T(z_1)T(z_2)\varphi(z_3)\varphi(z_4)\rangle$ where φ is an arbitrary primary field of dimension Δ. This 4-point function is determined from the Ward identity (2.2) to be

$$\langle T(z_1) T(z_2) \varphi(z_3) \varphi(z_4) \rangle = \Big\{ \frac{c}{2 z_{12}^4} + \\ + \frac{\Delta z_{34}^2}{z_{12}^2 z_{13} z_{24} z_{23} z_{14}} \Big(2 + \frac{\Delta z_{12}^2 z_{34}^2}{z_{13} z_{24} z_{23} z_{14}} \Big) \Big\} \langle \varphi(z_3) \varphi(z_4) \rangle ; \qquad (3.17)$$

hence,

$$\langle T_2(z_1, z_2) \varphi(z_3) \varphi(z_4) \rangle = \\ = \frac{\Delta z_{34}^2 \langle \varphi(z_3) \varphi(z_4) \rangle}{z_{13} z_{24} z_{23} z_{14}} \Big(1 + \frac{\Delta z_{12}^2 z_{34}^2}{2 z_{13} z_{24} z_{23} z_{14}} \Big) \qquad (3.18)$$

does have a limit (together with its partial derivatives) for $z_1 \to z_2$. (The general result follows from the fact that the singularities of n-point functions of $T_2(z_1, z_2)$ in z_{12} are also determined from the Ward identity (2.2) and do not exceed those of the 4-point function.)

We shall see in Chapter III that the above construction of renormalized composite conformal operators also extends to 4-dimensional models. Its importance is enhanced by the fact that the small distance behaviour of massive theories is expected to coincide with that of a conformally invariant massless limit (like in the case of the 2-dimensional Thirring model).

II.4 Fusion rules and OPEs in minimal theories. The Ising model.

4A. OPEs involving conserved tensor currents.

For generic values of c and Δ there are just two cases for which the minimal value Δ_3 in the OPE (3.4) is known. One, when the fields φ_1 and φ_2 are hermitian conjugate to each other so that $\Delta_3 = 0$, and another, when one of the factors, say φ_1, is a conserved current (for instance T(z)) of the family of the unit operator, so that $\Delta_3 = \Delta_2$. It is instructive to write down the OPE in these special cases.

The 3-point function of the "electromagnetic" current J(z) with a pair of conjugate charged fields ψ^* and ψ is determined by the Ward identity (2.3) (with I_a replaced by the charge e) to be

$$\langle \psi^*(z_1)\, \psi(z_2)\, J(z_3)\rangle = \frac{e\, z_{12}}{z_{13}\, z_{23}} \langle \psi^*(z_1)\psi(z_2)\rangle . \tag{4.1}$$

If ψ is a pure primary field of dimension Δ (i.e. if only the conformal family of the unit operator appears in the OPE of $\psi^*\psi$) then the bilocal field

$$B_{\psi^*\psi}(z_1, z_2) = \frac{1}{z_{12}}\left\{\frac{\psi^*(z_1)\, \psi(z_2)}{\langle \psi^*(z_1)\psi(z_2)\rangle} - 1\right\} \tag{4.2}$$

has an OPE of the form

$$\begin{aligned} B_{\psi^*\psi}\left(z+\frac{\varepsilon}{2}, z-\frac{\varepsilon}{2}\right) &= \frac{1}{2N_J e}\int_{-1}^{1} d\lambda\, J\left(z+\lambda\frac{\varepsilon}{2}\right) + \\ &+ \frac{3\Delta}{2c}\,\varepsilon \int_{-1}^{1} d\lambda\,(1-\lambda^2)\, T\left(z+\lambda\frac{\varepsilon}{2}\right) + \\ &+ \sum_{n=3}^{\infty} \varepsilon^{n-1}\frac{(2n-1)!!}{2^n (n-1)!!}\int_{-1}^{1} d\lambda\,(1-\lambda^2)^{n-1}\, O_n^{\psi^*\psi}\left(z+\lambda\frac{\varepsilon}{2}\right). \end{aligned} \tag{4.3}$$

Here N_J is a normalization constant appearing in the current 2-point function,

$$\langle J(z_1)\, J(z_2)\rangle = \frac{N_J^2 e^2}{z_{12}^2} . \tag{4.4}$$

The current, the stress energy tensor and the higher rank conserved tensor currents O_n are expressed as (renormalized) composite fields of $\psi^*(1)\,\psi(2)$ in terms of limits of derivatives of the bilocal operator $B_{\psi^*\psi}$:

$$J(z) = N_J\, e \lim_{\varepsilon\to 0} B_{\psi^*\psi}\left(z+\frac{\varepsilon}{2}, z-\frac{\varepsilon}{2}\right), \tag{4.5a}$$

$$T(z) = \frac{c}{2\Delta} \lim_{\varepsilon\to 0} \frac{\partial}{\partial\varepsilon} B_{\psi^*\psi}\left(z+\frac{\varepsilon}{2}, z-\frac{\varepsilon}{2}\right), \tag{4.5b}$$

$$(n-1)!\, O_n^{\psi^*\psi}(z) = \lim_{z_1,z_2\to z} D_{n-1}^{(1,1)}(\partial_1,\partial_2) B_{\psi^*\psi}(z_1,z_2). \tag{4.5c}$$

Similarly, for the product of a current with a charged field we have

$$z_{12}\, J(z_1)\,\psi(z_2) = -e\left\{(2\Delta-1)\int_0^1 du\,(1-u)^{2\Delta-2}\,\psi(z_2+uz_{12}) + \right.$$
$$\left. + \sum_{n=1}^{\infty} \frac{\Gamma(2\Delta+2n)\, z_{12}^n}{n!\,\Gamma(2\Delta+n-1)} \int_0^1 du\; u^n (1-u)^{2\Delta+n-2}\, O_{\Delta+n}(z_2+uz_{12})\right\}. \tag{4.6}$$

An expansion of type (4.6) is also valid for the bilocal field $z_{12}^2 T(z_1)\,\psi(z_2)$. The leading term then is

$$\Delta(2\Delta-1)(2\Delta-2)\int_0^1 du\; u\,(1-u)^{2\Delta-3}\,\psi(z_2+uz_{12}). \tag{4.7}$$

Eqs.(4.6) and (4.7) imply the identities

$$\lim_{\varepsilon\to 0}\left(\varepsilon\, J(z+\varepsilon)\,\psi(z)\right) = e\,\psi(z), \quad \lim_{\varepsilon\to 0}\left(\varepsilon^2 T(z+\varepsilon)\,\psi(z)\right) = \Delta\,\psi(z). \tag{4.8}$$

4B. Fusion rules for minimal theories. For c and Δ given by (I.22) (I.23) the following "fusion rule" has been established in [B1] .

$$[p_1,q_1]\times[p_2,q_2] =$$

$$= \sum_{k=|p_1-p_2|+1}^{p_1+p_2-1} \sum_{l=|q_1-q_2|+1}^{q_1+q_2-1} [k,l] . \qquad (4.9)$$

Here, p, q are the two integers in the rectangle

$$1 \leq p \leq m+1 , \quad 1 \leq q \leq m+2 \text{ for } c = 1 - \frac{6}{(m+2)(m+3)}, \quad m=1,2,\ldots \qquad (4.10)$$

that label the allowed lowest weights $\Delta^{(m)}_{p,q}$ (I.23); k and ℓ run over those integers of the same parity

$$|p_1-p_2|+1,\ |p_1-p_2|+3,\ \ldots,\ p_1+p_2-3,\ p_1+p_2-1, \qquad (4.11)$$
$$|q_1-q_2|+1,\ |q_1-q_2|+3,\ \ldots,\ q_1+q_2-3,\ q_1+q_2-1,$$

which belong to the rectangle (4.10). Moreover, because of the equivalence between the representations [p,q] and [m+2 - p, m+3 - q] , only the intersection of the sets of [k,l] appearing for various products of equivalent representations should be taken into account.

These rules indicate that the space of states spanned by $\varphi(1)\varphi(2)|0\rangle$, $\varphi(i)\equiv\varphi^{(z_i)}_{k_i,q_i}$ is not a coherent space (for the superselection operator $U = e^{2\pi i L_0}$). We shall demonstrate that in the two simplest Ising type models, for m=1,2, a coherent basis of fields can be chosen (on the expense of allowing a greater than one multiplicity for some of the LWs $\Delta_{p,q}$. It turns out that the OPE algebra of "pure conformal families" for m=1,2 is isomorphic to the cyclic group $\mathbb{Z}_{2m(m+1)}$.

Indeed, identifying for both m=1,2, the field φ with a (complex) field of dimension $\Delta_{2,2}(m)$ we can write symbolically the fusion rules for pure primary fields as follows:

$$m=1: \quad \varphi \Rightarrow (\tfrac{1}{16}), \quad \varphi^2 \Rightarrow (\tfrac{1}{2}), \quad \varphi^3 \Rightarrow (\tfrac{1}{16}), \quad \varphi^4 \Rightarrow (0) ; \qquad (4.12)$$

$$m = 2: \quad \varphi \Rightarrow \left(\tfrac{3}{80}\right), \quad \varphi^2 \Rightarrow \left(\tfrac{1}{10}\right), \quad \varphi^3 \Rightarrow \left(\tfrac{7}{16}\right), \quad \varphi^4 \Rightarrow \left(\tfrac{3}{5}\right),$$
$$\varphi^5 \Rightarrow \left(\tfrac{3}{80}\right), \quad \varphi^6 \Rightarrow \left(\tfrac{3}{2}\right), \quad \varphi^7 \Rightarrow \left(\tfrac{3}{80}\right), \quad \varphi^8 \Rightarrow \left(\tfrac{3}{5}\right),$$
$$\varphi^9 \Rightarrow \left(\tfrac{7}{16}\right), \quad \varphi^{10} \Rightarrow \left(\tfrac{1}{10}\right), \quad \varphi^{11} \Rightarrow \left(\tfrac{3}{80}\right), \quad \varphi^{12} \Rightarrow (0), \tag{4.13}$$

where the corresponding weights are written in parantheses. It is easily verified that if we set $[2,2] = \varphi + \varphi^5 + \varphi^7 + \varphi^{11}$, $[3,1] = \varphi^6$, $[2,1] = \varphi^3 + \varphi^9$, $[3,3] = \varphi^2 + \varphi^{10}$, $[3,2] = \varphi^4 + \varphi^8$, then the multiplication rules in (4.13) agree with those in (4.9). (The easiest way to derive (4.13) is to start with a group of two generators, say f and g, setting $[1,2] = f + f^7$, $[2,1] = g + g^5$ and to factor with respect to the identities $g^6 = 1 = f^8$, $g^3 = f^4$.)

4C. <u>The critical Ising model</u>. We shall display the field theoretical construction behind the "fusion group" in the simplest case m=1, in which it is isomorphic to the cyclic group $\mathbb{Z}_4$. The OPE algebra in this case is generated by a complex primary field $\varphi(z)$ of weight 1/16. Its powers correspond to the composite primary fields

$$\psi(z) = \lim_{\varepsilon \to 0} \left\{ \frac{\langle \psi(z)\, \psi(z') \rangle}{\langle \varphi(z+\frac{\varepsilon}{2})\, \varphi(z-\frac{\varepsilon}{2})\, \psi(z') \rangle}\, \varphi\left(z+\frac{\varepsilon}{2}\right) \varphi\left(z-\frac{\varepsilon}{2}\right) \right\} \tag{4.14}$$

$$\varphi^*(z) = \lim_{\varepsilon \to 0} \left\{ \frac{\langle \varphi(z')\, \varphi^*(z) \rangle}{\langle \varphi(z')\, \varphi(z+\varepsilon)\, \psi(z) \rangle}\, \varphi(z+\varepsilon)\, \psi(z) \right\}, \tag{4.15}$$

which do not depend on the choice of the point z'. Normalizing the 2-point functions of φ and ψ,

$$\langle \varphi(z_1)\, \varphi^*(z_2) \rangle = z_{12}^{-1/8}, \qquad \langle \psi(z_1)\, \psi(z_2) \rangle = z_{12}^{-1}, \tag{4.16}$$

we can characterize the model by a single constant N appearing in the 3-point function:

$$N \langle \varphi(z_1)\, \varphi(z_2)\, \psi(z_3) \rangle = z_{12}^{3/8} \cdot (z_{13}\, z_{23})^{-1/2}. \tag{4.17}$$

The 2-dimensional fields, the energy density $\varepsilon(z,\bar{z})$ and the local spin $\sigma(z,\bar{z})$, of the critical Ising model are expressed in terms of the pure primary fields $\varphi(z)$ and $\bar{\varphi}(\bar{z})$ and their composites ψ and $\bar{\psi}$ by

$$\varepsilon(z,\bar{z}) = \psi(z)\, \bar{\psi}(\bar{z}), \quad \sigma(z,\bar{z}) = \frac{1}{\sqrt{2}} \left(\varphi(z)\, \bar{\varphi}(\bar{z}) + \varphi^*(z)\, \bar{\varphi}^*(\bar{z}) \right). \tag{4.18}$$

It is remarkable that the splitting of σ into two factorized terms corresponds to a similar splitting of the 4-point function of σ. For small z_{12} and z_{34} the first two terms in the expansion of this 4-point function are (in accord with the exact expression evaluated in Appendix E to ref. [B1]):

$$\langle\sigma(1)\sigma(2)\sigma(3)\sigma(4)\rangle \simeq (z_{12}\bar{z}_{12}z_{34}\bar{z}_{34})^{-\frac{1}{8}} + + \frac{(z_{12}\bar{z}_{12}z_{34}\bar{z}_{34})^{3/8}}{4z_{13}\bar{z}_{13}} + \dots ; \tag{4.19}$$

thus $N = \sqrt{2}$.

We recall that the constant N is fixed in [B1] from either crossing symmetry or singlevaluedness of the Euclidean 4-point function.

III. THE ALGEBRA OF FREE MASSLESS FIELDS ON COMPACTIFIED SPACE-TIME

III.0 Introduction

The motivation for studying compact space models of the Universe is multifold. Reasons vary from alleged explanation of Olbers' paradox (of why the night sky is dark) to attempts to get rid of infrared divergences .

There are special reasons for considering conformal field algebras on compactified Minkowski space $\bar{M}$. Among other things, (i) it gives room for global conformal transformations; (ii) it provides a natural discrete basis for conformal fields and thereby a refined machinery for defining renormalized normal products of quantum fields.

The theory of free massless fields on $\bar{M}$ (and on its universal cover, the globally causal cosmos $\tilde{M} \simeq \mathbb{R}^1 x S^3$) has been studied systematically by Segal and Paneitz [P1,2] within their analysis of conformal space-time bundles. We offer here a different approach to this - basically elementary - problem which can be characterized as follows.

If one only gives physical meaning to the conformal structure of space-time - i.e., to the causal order of events (in a given neighbourhood), or to the ratio of Lorentz intervals in the tangent space at each point, then one should allow for complex conformal factors in a conformally flat space-time (for definition and basic properties of conformal mappings and conformally flat spaces see, e.g. [P3] Sec. 1.8 and chapter 6 or sec. 1 of [T2]). It turns out that there are two distinguished (complex conjugate) conformal factors on $\bar{M} \simeq U(2)$, which lead to a zero Riemann curvature tensor (Proposition 1.4 of sect. 1C). Our choice corresponds to a metric form

$$dz^2 = d\underline{z}^2 + dz_4^2 \qquad (d\underline{z}^2 = dz_1^2 + dz_2^2 + dz_3^2) \tag{0.1}$$

on $\bar{M}$ where z_α are expressed in terms of Minkowski space Cartesian coordinates x^μ by

$$\underline{z} = \left(1 + \frac{x^2}{4} - ix^0\right)^{-1} \underline{x}\,, \quad z_4 = \left(1 + \frac{x^2}{4} - ix^0\right)^{-1}\left(1 - \frac{x^2}{4}\right) \qquad (x^2 \equiv \underline{x}^2 - x_0^2). \tag{0.2}$$

(These expressions are related to the Cayley transform of complex quaternions in Sec. 1B.; see also [U1,R2].)

It may be instructive to add a couple of remarks about the reasons behind the conventions concerning numerical factors and signs in Eq.(0.2) (which differ from those used in our previous work [T2,3,6]).

Clearly, Eq.(0.2) presupposes that the coordinates x^μ (and z_α) are dimensionless. There is an inherent conformal invariant scale, associated with compactified Minkowski space $\bar{M}$: this is the radius R of the 3-space S^3 (for fixed z^2). Indeed, if we replace in (0.2) z by $R^{-1}z$ and x by $R^{-1}x$, we shall have (for real x) $z^2 = R^2 e^{2i\tau}$ so that the 4-vector $ze^{-i\tau}$ belongs to a real 3-sphere of radius R. Identifying - in accord with Segal [S1-3] R with the radius of the Universe, it is natural to assume that for x^μ small compared to R the compact picture coordinates $\underline{\rho} = e^{-i\tau}\underline{z}$ and $R\tau$ coincide with the Minkowski space coordinates $\underline{x}$ and x^0 up to terms of order $\left(\frac{x^\mu}{R}\right)^2$. This is indeed the case for z given by (0.2); we have

$$\underline{\rho} = e^{-i\tau}\underline{z} = \left\{\left(1 + \frac{x^2}{4R^2}\right)^2 + \left(\frac{x^0}{R}\right)^2\right\}^{-\frac{1}{2}} \underline{x} \simeq \underline{x}\,,$$

$$R\tau = \frac{R}{2i} \ln \frac{z^2}{R^2} = \frac{R}{2i}\left\{\ln\left(1 + \frac{x^2}{4R^2} + i\frac{x^0}{R}\right) - \ln\left(1 + \frac{x^2}{R^2} - i\frac{x^0}{R}\right)\right\} \simeq x^0. \tag{0.3}$$

It is tempting to identify - following Segal - the parameter τ with the physical time variable and to equate the corresponding conjugate variable, the compact conformal generator $i\frac{\partial}{\partial\tau}$, with the true Hamiltonian.

Our second remark concerns the choice of sign of ix^0 in (0.2). If $\varphi(z)$ is a compact picture local quantum field and $|0\rangle$ is the corresponding conformal invariant vacuum vector*), then we demand that the primitive analyticity domain of the vector function $\varphi(z)|0\rangle$ (the image under the Cayley transform (0.2) of the forward tube T_+ - see Eq.(1.19) below) contains the origin so that we can again write down "lowest weight vectors" in the form $\varphi(0)|0\rangle$ used in chapter I for 2-dimensional models. This is achieved just with the choice (0.2) - see Proposition 1.2 below.

A free massless scalar field $\varphi_M(x)$ in Minkowski space - a solution of the wave equation $\Box\varphi_M(x)=0$ - goes in the (campact-z-picture into a solution of the 4-dimensional Laplace equation $\Delta_4\varphi(z) = 0$ on $\bar{M}$ (as well as in a 4-dimensional complex neighbourhood of $\bar{M}$ - if we restrict the quantum field operator $\varphi(z)$ to, say, finite energy vectors). The free field $\varphi(z)$ splits into a creation and an annihilation part $\varphi^{(\pm)}(z)$ satisfying

$$\varphi(z) = \varphi^{(+)}(z) + \varphi^{(-)}(z)\,, \qquad \varphi^{(-)}(z)|0\rangle = 0 = \langle 0|\varphi^{(+)}(z) \tag{0.4}$$

(for a hermitian φ we have $\varphi^{(-)}(z) = 1/z^2\,\varphi^{(+)*}(z/z^2)$). The counterpart of the Fourier integral is a discrete expansion in homogeneous harmonic polynomials for the creation part $\varphi^{(+)}(z)$ and in homogeneous polynomials of z/z^2 (times $1/z^2$) for $\varphi^{(-)}$ (Sec. 2A). A similar expansion in homogeneous polynomial solutions of an elliptic (Dirac-Weyl) system of first order partial differential equations is written for a 2-component massless spinor field in Sec. 2B.

We briefly discuss in Sec. 3 the extension of 2 dimensional techniques of the previous chapter to the construction of composite conformal fields and light-cone OPEs in four dimensions.

---

*) The reader with background in mathematics will find all basic notions and facts about quantum fields, used without explanation, in any text on axiomatic QFT (see, e.g. [B3]).

1. COMPLEX, ZERO-CRUVATURE REALIZATION OF COMPACTIFIED MINKOWSKI SPACE

1A. Mappings of M onto the Lie algebra of U(2). Complex quaternions.

We shall use two mappings of 4-dimensional Minkowski space M onto the Lie algebra u(2) realized in terms of pure imaginary quaternionrs:

$$\begin{aligned} &x \to \tfrac{i}{2}\tilde{x}, \quad i\tilde{x} = ix^0 + \underset{\sim}{x}\,\underset{\sim}{q} \\ &x \to -\tfrac{i}{2}\underset{\sim}{x}, \quad i\underset{\sim}{x} = ix^0 - \underset{\sim}{x}\,\underset{\sim}{q} \end{aligned} \qquad (\underset{\sim}{x}\,\underset{\sim}{q} = x^1q_1 + x^2q_2 + x^3q_3) \quad . \tag{1.1}$$

Here q_i are the imaginary quaternion units satisfying

$$q_i\, q_j = \varepsilon_{ijk}\, q_k - \delta_{ij}\,,$$

so that

$$[\tfrac{1}{2}q_i\,,\, \tfrac{1}{2}q_j] = \varepsilon_{ijk}\,\tfrac{1}{2}q_k\,, \quad (\underset{\sim}{x}\,\underset{\sim}{q})^2 = -\underset{\sim}{x}^2\,, \tag{1.2}$$

while i is the complex imaginary unit which commutes with q_j. A Lorentz transformation $\Lambda \in SL(2,\mathbb{C})$ is defined as a complex unit quaternion:

$$\Lambda = \underset{\sim}{\lambda}\,\underset{\sim}{q} + \lambda_4\,, \quad \Lambda\Lambda^+ = \underset{\sim}{\lambda}^2 + \lambda_4^2 = 1 \quad \text{for } \Lambda^+ = \lambda_4 - \underset{\sim}{\lambda}\,\underset{\sim}{q} \quad (\lambda_\alpha \in \mathbb{C}). \tag{1.3}$$

Its action on M in the u(2) picture is given by

$$i\underset{\sim}{x} \to \Lambda\, i\underset{\sim}{x}\,\Lambda^*, \quad i\tilde{x} \to \Lambda^{*-1}\, i\tilde{x}\,\Lambda^{-1} \qquad (\Lambda^* = \bar{\Lambda}^+ = \bar{\lambda}_4 - \bar{\underset{\sim}{\lambda}}\,\underset{\sim}{q}). \tag{1.4}$$

These transformations leave invariant - as they should - the Lorentz square of x,

$$x^2\,(= \underset{\sim}{x}^2 - x_0^2) = i\tilde{x}\,i\underset{\sim}{x} = \Lambda^{*-1}\, i\tilde{x}\,\Lambda^{-1}\Lambda\, i\underset{\sim}{x}\,\Lambda^*. \tag{1.5}$$

The standard matrix realizations of u(2) are recovered if we express the quaternion units in terms of the Pauli matrices:

$$q_j = -i\sigma_j\,, \quad \sigma_1 = \begin{pmatrix} 0 & 1 \\ 1 & 0 \end{pmatrix}, \quad \sigma_2 = \begin{pmatrix} 0 & -i \\ i & 0 \end{pmatrix}, \quad \sigma_3 = \begin{pmatrix} 1 & 0 \\ 0 & -1 \end{pmatrix}. \tag{1.6}$$

1B. Compactification of Minkowski space as a Cayley transformation.

The Lorentz transformations (1.4) are a special case of the following (local) conformal action on M = u(2).

Let

$$G = SU(2,2) = \left\{ g = \begin{pmatrix} A & B \\ C & D \end{pmatrix} \in SL(4,\mathbb{C});\ \ g^* \beta g = \beta,\ det(\beta - \lambda) = (\lambda^2 - 1)^2 \right\} \quad (1.7)$$

where A, B, C, D are complex 2x2 matrices, and

$$\beta = -\begin{pmatrix} 0 & 1 \\ 1 & 0 \end{pmatrix} \ \left(\equiv - \begin{pmatrix} 0 & 0 & 1 & 0 \\ 0 & 0 & 0 & 1 \\ 1 & 0 & 0 & 0 \\ 0 & 1 & 0 & 0 \end{pmatrix} \right) \quad (1.8)$$

be the 4-fold cover of the conformal group of space-time. For a fixed (bounded) neighbourhood $\mathcal{O}$ of the origin in M there exists a neighbourhood $G(\mathcal{O})$ of the unit element of G such that $g \in G(\mathcal{O})$ iff $\det(Ci\tilde{x}+D) > 0$ for all $x \in \mathcal{O}$. For $x \in \mathcal{O}$ the conformal transformations from $G(\mathcal{O})$ are defined by

$$g: \ \frac{i}{2}\tilde{x} \rightarrow \frac{i}{2}\tilde{x}_g = \left(A \frac{i}{2}\tilde{x} + B\right)\left(C \frac{i}{2}\tilde{x} + D\right)^{-1}. \quad (1.9)$$

Global conformal transformations can be defined without restrictions on compactified Minkowski space

$$\bar{M} = U(2) \ \left(= \frac{S^3 \times S^1}{Z_2} \right)$$

$$= \left\{ Z = \underset{\sim}{z}\underset{\sim}{q} + z_4;\ \ z_\alpha = e^{i\tau}\hat{z}_\alpha,\ \hat{z} \in S^3 \ (i.e.\ \hat{z} \in \mathbb{R}^4,\ \hat{z}^2 = 1) \right\}. \quad (1.10)$$

The imbedding of the Lie algebra $u(2)(=M)$ into the group $U(2)$ is given by the Cayley transform

$$\frac{i}{2}\tilde{x} \rightarrow Z = \frac{1 + \frac{i}{2}\tilde{x}}{1 - \frac{i}{2}\tilde{x}} = \underset{\sim}{z}\underset{\sim}{q} + z_4 \ \left(= \frac{1 - \frac{x^2}{4} + \underset{\sim}{x}\underset{\sim}{q}}{1 + \frac{x^2}{4} - ix^0} \right) \ \left(e^{2i\tau} = \frac{1 + \frac{x^2}{4} + ix^0}{1 + \frac{x^2}{4} - ix^0} \right). \quad (1.11)$$

It is easily verified that for $x \in \mathbb{R}^4$

$$ZZ^* = |\underset{\sim}{z}|^2 + |z_4|^2 + (\bar{z}_4 \underset{\sim}{z} - z_4 \bar{\underset{\sim}{z}} + \bar{\underset{\sim}{z}} \times \underset{\sim}{z})\underset{\sim}{q} = 1 \quad (1.12a)$$

$$\left(\text{as} \quad \left(1 - \frac{x^2}{4}\right)^2 + \underset{\sim}{x}^2 = \left(1 + \frac{x^2}{4}\right)^2 + x_0^2 \right)$$

or, since $(\bar{z}_4 \underline{z} - z_4 \underline{\bar{z}})\underline{\bar{z}} x \underline{z} = 0$,

$$|\underline{z}|^2 + |z_4|^2 = 1, \quad \bar{z}_4 \underline{z} - z_4 \underline{\bar{z}} = 0 = \underline{\bar{z}} \times \underline{z} \quad \text{so that} \quad \bar{\underline{z}} = \frac{\underline{z}}{z^2} \tag{1.12b}$$

The inverse formula $\frac{i}{2}\underline{x} = (Z-1)(Z+1)^{-1}$, only makes sense for $\det(Z+1) = 1+2z_4+z^2 \neq 0$.

Proposition 1.1 The G-action on $\bar{M}$ is given by

$$g: Z \to Z_g = (u_1 Z + v u_2)(v^* u_1 Z + u_2)^{-1} \tag{1.13}$$

where *)

$$\begin{aligned} &v v^* < 1, \quad v^* v < 1, \\ &u_1 u_1^* = (1 - v v^*)^{-1}, \quad u_2 u_2^* = (1 - v^* v)^{-1}. \end{aligned} \tag{1.14}$$

It is defined globally, since for any $Z \in U(2)$

$$\det(v^* u_1 Z + u_2) \neq 0. \tag{1.15}$$

Proof. The transformation law (1.13) is a consequence of (1.9) and (1.11) for

$$g_c \equiv \begin{pmatrix} u_1 & v u_2 \\ v^* u_1 & u_2 \end{pmatrix} = S g \, {}^t S, \quad S = \frac{1}{\sqrt{2}} \begin{pmatrix} 1 & 1 \\ -1 & 1 \end{pmatrix} = {}^t S^{-1} \tag{1.16a}$$

(${}^t S$ standing for the transposed of the matrix S), or

$$\begin{aligned} &u_1 = \tfrac{1}{2}(A+B+C+D), \qquad u_2 = \tfrac{1}{2}(A+D-B-C) \\ &v u_2 = \tfrac{1}{2}(B+D-A-C), \qquad v^* u_1 = \tfrac{1}{2}(C+D-A-B). \end{aligned} \tag{1.16b}$$

In this compact picture the $G^* x G$-invariant metric tensor is diagonal:

$$\beta_c \equiv S \beta \, {}^t S = \begin{pmatrix} -1 & 0 \\ 0 & 1 \end{pmatrix}. \tag{1.17}$$

---

*) We use tne notation B 0 if the matrix B is hermitian and all its eigenvalues are positive.

Eq.(1.14) then reflects the invariance property

$$g_c^* \beta_c g_c = \beta_c .$$

To prove (1.15) we use the identity

$$TT^+ = (\det T)\,\mathbb{1}, \quad \text{for } T^+ = \varepsilon\, {}^t T \varepsilon^{-1}, \quad \varepsilon = \begin{pmatrix} 0 & 1 \\ -1 & 0 \end{pmatrix} = -\varepsilon^{-1} \qquad (1.18)$$

valid for any 2x2 matrix T. It implies

$$\det(v^* u_1 Z + u_2) = \det v^* \det u_1 \det Z + \det u_2 + \operatorname{tr}(v^* u_1 Z u_2^+).$$

On the other hand, we find from (1.14) $|\det u_1| = |\det u_2| \geq 1 > |\det v^*|$, which yields (1.15), since $|\det Z| = 1$ for $Z \in U(2)$.

Finally, we verify the implication $Z^*Z=1 \Rightarrow Z^*_g Z_g = 1$ for Z_g given by (1.13):

$$Z_g^* Z_g = (u_2^* + Z^* u_1^* v)^{-1} (Z^* u_1 u_1^* Z + u_2^* v^* v u_2 + Z^* u_1^* v u_2 +$$
$$+ u_2^* v^* u_1 Z + u_2^* v^* v u_2)(u_2^* u_2 + u_2^* v^* u_1 Z + Z^* u_1^* v u_2 +$$
$$+ Z^* u_1^* v v^* u_1 Z)^{-1} (u_2^* + Z^* u_1^* v) = 1.$$

In deriving the last equaiton we have used (1.14) which gives

$$u_2^* v^* v u_2 = u_2^* u_2 - 1,$$
$$Z^* u_1 v v^* u_1 Z = Z^* u_1^* u_1 Z - 1 \quad (\text{for } Z^* Z = 1).$$

The following statement gives a matrix (or complex quaternion) characterization of the analyticity domain of a vector of the form $\varphi(z)|0\rangle$ where φ is a local covariant field, $|0\rangle$ is the Poincaré invariant vacuum vector in a QFT satisfying energy positivity.

Proposition 1.2. The forward tube

$$T_+ = \{\zeta = x + iy \in \mathbb{C}^4;\ y^0 > |\underset{\sim}{y}|,\ x \in \mathbb{R}^4\} =$$
$$= \left\{ i\tilde{\zeta} = i\tilde{x} - \tilde{y};\ \tilde{y} = \begin{pmatrix} y^0 - y^3 & -y^1 + iy^2 \\ -y^1 - iy^2 & y^0 + y^3 \end{pmatrix} > 0 \right\} \qquad (1.19)$$

is mapped by (the analytic continuation of) the Cayley transforms (1.11) into the "unit ball" of complex 2x2 matrices:

$$\mathcal{B}: Z^* Z = \bar{z} z + (\bar{z}_4 \underset{\sim}{z} - z_4 \bar{\underset{\sim}{z}} + \underset{\sim}{z} \times \bar{\underset{\sim}{z}}) \underset{\sim}{q} < 1 \qquad (1.20a)$$

or $z \in \mathcal{B}$ iff

$$|z|^2 + |\bar{z}_4 \underset{\sim}{z} - z_4 \bar{\underset{\sim}{z}} + \underset{\sim}{z} \times \bar{\underset{\sim}{z}}| < 1. \qquad (1.20b)$$

Sketch of the proof. One first proves that the tube domain (1.19) is an orbit of the conformal action g: $\frac{i}{2}\tilde{\zeta} \to \frac{i}{2}\zeta_g = (A\frac{i}{2}\tilde{\zeta}+B)(C\frac{i}{2}\tilde{\zeta}+D)^{-1}$ and then verifies (1.20) for $\frac{i}{2}\tilde{\zeta} = -1$ (Z=0). To verify the implication $Z^*Z = 1 \Rightarrow Z_g^*Z_g = 1$ for Z_g given by (1.13) we use the identities $u_2^*v^* vu_2 = u_2^*u_2 - 1$, $Z^*u_1^*vv^*u_1 Z = Z^*u_1^*u_1 Z - 1$ (for $Z^*Z = 1$) which follow from (1.14).

1C. A distinguished complex 0-curvature metric on $\bar{M}$.

We recall that the (local) conformal properties of Minkowski space with metric

$$dx^2 = dx^\mu \eta_{\mu\nu} dx^\nu = d\underset{\sim}{x}^2 - dx_0^2 \quad (= d\tilde{x}\, d\underset{\sim}{x}) \qquad (1.21)$$

do not differ from those of a conformally flat space with metric tensor

$$g_{\mu\nu}(x) = \Omega^2(x)\eta_{\mu\nu}, \qquad \Omega(x) \neq 0 \text{ for all } x \in \mathbb{R}^4. \qquad (1.22)$$

(Indeed, the light-cone (causal) structure in the tangent space at each point is the same in both cases.)

The general O(4) invariant conformally flat metric (1.22) that extends to $\bar{M}$ is

$$\chi\left(\frac{1+\frac{x^2}{4}+ix^0}{1+\frac{x^2}{4}-ix^0}\right)\frac{dx^2}{(1+\frac{x^2}{4}-ix^0)^2} = \chi(z^2)\,dz^2 \tag{1.23}$$

where $dz^2 = \left(\sum_\alpha\right) dz_\alpha^2 = dZdZ^+$ and χ is an arbitrary smooth (non-vanishing) function on the unit circle.

Indeed, from (1.5) and (1.11) we find

$$\frac{1}{4}dx^2 = d\frac{i}{2}\tilde{x}\, d\frac{i}{2}\underset{\sim}{x} = \frac{4dZdZ^+}{(z^2+2Z_4+1)^2} . \tag{1.24}$$

On the other hand, any scalar O(4)- invariant function of z should be a function of

$$z^2 \left(= e^{2i\tau}\right) = \frac{1+\frac{x^2}{4}+ix^0}{1+\frac{x^2}{4}-ix^0} . \tag{1.25}$$

Eq.(1.23) follows from (1.24), if we note that for Z given by (1.11)

$$\left(1+\frac{x^2}{4}-ix^0\right)\left(1+z^2+2Z_4\right) = 4. \tag{1.26}$$

There are two distinguished choices of the conformal factor $\chi(z^2)$. One (that is being commonly made - see, e.g. [P1-3, S1-3,T2]) is determined by the requirement that the metric on $\bar{M}$ is (real and) invariant under the maximal compact subgroup S(U(2)xU(2)) of G. It is $\chi(z^2) = \frac{c}{z^2}$ (with c a non-zero real). The second,

$$\chi(z^2) = 1, \tag{1.27}$$

has zero Riemann curvature tensor (it leads to a complex valued metric form (1.23) and has not been considered).

The (traceless) Weyl curvature tensor vanishes identically for a conformally flat space-time:

$$C^{\kappa}{}_{\lambda\mu\nu} = R^{\kappa}{}_{\lambda\mu\nu} + \tfrac{1}{2}\left(R_{\lambda[\mu}\delta^{\kappa}_{\nu]} + g_{\lambda[\mu}R^{\kappa}_{\nu]}\right) - \tfrac{1}{6} R\, g_{\lambda[\mu}\delta^{\kappa}_{\nu]} = 0. \tag{1.28}$$

(Here $g_{\mu\nu}$ has the form (1.22) and brackets stand for antisymmetrization:

$$g_{\lambda[\mu}\delta^{\kappa}{}_{\nu]} = g_{\lambda\mu}\delta^{\kappa}_{\nu} - g_{\lambda\nu}\delta^{\kappa}_{\mu},$$

$R_{\lambda\nu} := R^{\mu}{}_{\lambda\mu\nu}$ is the Ricci tensor, $R = R^{\lambda}{}_{\lambda}$ is the scalar curvature.) Therefore, for a conformally flat space, the vanishing of the Riemann curvature $R^{\kappa}{}_{\lambda\mu\nu}$ is equivalent to the vanishing of the Ricci tensor $R_{\mu\nu}$, which gives

$$\Omega\,\partial_\mu\partial_\nu\Omega - 2\,\partial_\mu\Omega\,\partial_\nu\Omega + \tfrac{1}{2}\eta_{\mu\nu}(\partial\Omega)^2 = 0. \tag{1.29}$$

Taking the trace of (1.29) we find that the conformal factor Ω for a Ricci flat space should satisfy the free wave equation

$$\Box\,\Omega(x) = 0. \tag{1.30}$$

It is easily verified that

$$\Omega(x) = \left(1 + \frac{x^2}{4} - i x^0\right)^{-1} \tag{1.31}$$

satisfies (1.29) (and hence, also (1.30)).

These results are summarized (and sharpened) by the following

<u>Proposition 1.3.</u> The condition (1.29) for vanishing curvature has exactly two SO(4)-invariant solutions (each determined up to a factor): $\Omega_{\pm} = \left(1+\frac{x^2}{4} \pm i x^0\right)^{-1}$ or $\mathcal{X}_+ = \frac{1}{z^4}$, $\mathcal{X}_- = 1$ in the z-picture.

<u>Proof.</u> Writing (1.29) in the compact picture for $\mathcal{X}(z^2) = \Omega^2(z^2)$, $\eta_{\alpha\beta} = \delta_{\alpha\beta}$ we have: $2\Omega'(\Omega + z^2\Omega')\delta_{\alpha\beta} + 4(\Omega\Omega'' - 2\Omega'^2) z_\alpha z_\beta = 0.$

This equation has two solutions: $\Omega = \frac{c}{z^2}$ and $\Omega = c$.

Our choice (1.27) (1.31) corresponds to the use of analytic functions of z in the compact picture QFT. The second solution (Ω_+) corresponds to the metric form $d\bar{z}^2$ and would have involved instead analytic functions of $\bar{z}$.

1D. Non-parallelizable "flat frame bundle" on $\bar{M}$.

The geometry of $\bar{M}$ equipped with the complex valued flat metric (1.23) (1.27) can be also characterized by a "canonical frame bundle" and related to the Cartan connection form on U(2).

The group manifold U(2) has a left invariant Lie algebra valued 1-form

$$\theta = Z^{-1} dZ = q_\alpha \theta^\alpha \ (\in u(2)) \tag{1.32a}$$

where for $Z = q\, z$, θ^α are given by

$$z^2\theta^1 = z_3 dz_2 - z_2 dz_3 + z_4 dz_1 - z_1 dz_4, \quad z^2\theta^2 = z_1 dz_3 - z_3 dz_1 + z_4 dz_2 - z_2 dz_4,$$

$$z^2\theta^3 = z_2 dz_1 - z_1 dz_2 + z_4 dz_3 - z_3 dz_4, \quad z^2\theta^4 = z_\alpha dz_\alpha = \tfrac{1}{2} dz^2 \tag{1.32b}$$

and satisfy (as a consequence of $d\theta = -\theta \wedge \theta$)

$$d\theta^4 = 0, \qquad d\theta^i = -\varepsilon_{ijk}\, \theta^j \wedge \theta^k. \tag{1.33}$$

The tetrad θ^α is associated with the real S(U(2)xU(2) invariant metric on U(2): $\sum_\alpha (\theta^\alpha)^2 = \frac{dz^2}{z^2}$. If we introduce the 0-curvature Cartan connection forms on U(2)

$$\omega^k{}_4 = \theta^k \ (= -\omega_4{}^k), \quad \omega^i{}_j = -\varepsilon_{ijk}\, \theta^k, \tag{1.34}$$

then Eqs.(1.33) would imply, in view of the structure equation $\mathcal{T}^\alpha = d\theta^\alpha + \omega^\alpha{}_\beta \wedge \theta^\beta$, the presence of a non-zero torsion $\mathcal{T}^\alpha = \omega^\alpha{}_4 \wedge \theta^4$. If, however, we introduce the bundle*) of fundamental 1-forms

$$e_\alpha = \sqrt{z^2}\,\theta^\alpha, \quad \text{such that} \quad e_1^2 + e_2^2 + e_3^2 + e_4^2 = dz^2, \tag{1.35}$$

then both the torsion and the curvature would vanish:

$$de_\alpha + \omega_{\alpha\beta} \wedge e_\beta = 0, \quad \text{for} \quad \omega_{k4} = \theta^k, \ \omega_{ij} = -\varepsilon_{ijk}\theta^k, \tag{1.36}$$

$$d\omega_{\alpha\beta} + \omega_{\alpha\sigma} \wedge \omega_{\sigma\beta} = 0 \tag{1.37}$$

(ω coinciding with the Cartan connection). Thus, substituting the globally defined left invariant 1-forms θ by the tetrad e_α, we have traded non-zero torsion (or, alternatively, non-zero curvature for the Riemannian connection $\Gamma^i{}_k = \varepsilon_{ijk}\theta^j$, $\Gamma^i{}_4 = 0$) to parallelization on $\bar{M}$.

2. FREE ZERO-MASS FIELDS ON $\bar{M}$

2A. Hermitian scalar field

A free real scalar massless quantum field φ_M in Minkowski space is characterized by the (analytically continued) 2-point function

$$\langle \varphi_M(\zeta_1)\, \varphi_M(\zeta_2)\rangle_0 = \frac{1}{4\pi^2(\zeta_1-\zeta_2)^2} \quad \text{for} \quad \zeta_1 \in T_-, \quad \zeta_2 \in T_+, \tag{2.1}$$

T_+ being the forward tube (1.19) while T_- is the backward tube ($-T_- = T_+$).

As noted in the proof of Proposition 1.2 the tube domain T_+ (as well as T_-) is a homogeneous space of the conformal group G under the action (1.9). Moreover, the linear span of vectors

---

*) e_α can be regarded as functions on the double covering $\bar{\bar{M}}$ of $\bar{M}$; if we stick to the manifold $\bar{M}$, we should regard them as a local section of a frame bundle with structure group U(1).

$$\Phi_M(\zeta) = \varphi_M(\zeta)|0\rangle \qquad (2.2)$$

carries a lowest weight (positive energy) unitary representation U(g)

$$(U(g)\Phi_M)(\zeta) = \{det(C\tfrac{i}{2}\tilde{\zeta}+D)\}^{-1}\Phi_M(\zeta_g),\ \tfrac{i}{2}\tilde{\zeta}_g = (A\tfrac{i}{2}\tilde{\zeta}+B)(C\tfrac{i}{2}\tilde{\zeta}+D)^{-1}, \qquad (2.3)$$

g being given by (1.7).

Remark. To see that the tranformation law (2.3) always makes sense it is sufficient to verify that the detrminant in the denominator does not vanish for special conformal transformations

$$g = \begin{pmatrix} 1 & 0 \\ \frac{i}{2}\underset{\sim}{c} & 1 \end{pmatrix},\ \frac{i}{2}\tilde{\zeta}_g = \frac{i}{2}\tilde{\zeta}\left(\frac{i}{2}\underset{\sim}{c}\frac{i}{2}\tilde{\zeta}+1\right)^{-1},\quad \zeta_g = \frac{\zeta + \frac{\zeta^2}{4}c}{1+\frac{1}{2}c\zeta+\frac{1}{16}c^2\zeta^2}.$$

Indeed, if the imaginary part of the determinant $\mathrm{Im}\det(\frac{i}{2}\underset{\sim}{c}\frac{i}{2}\tilde{\zeta}+1) = \frac{1}{2}(cy+\frac{c^2}{4}xy)$ vanishes for $y^0 > |y|$ then the vector $c+\frac{c^2}{4}x$ must be spacelike. Therefore, either $c^2 \neq 0$ and $c^2 Re\, det(\frac{i\underset{\sim}{c}}{2}\frac{i\tilde{\zeta}}{2}+1) = (c+\frac{c^2}{4}x)^2 - \frac{c^4y^2}{16} > 0$ or $c = 0$ and the determinant is 1.

The transformations (2.3) are known to leave invariant the wave equation

$$\Box\Phi_M(\zeta) \equiv \left(\Delta_{\underset{\sim}{\zeta}} - \frac{\partial^2}{\partial\zeta^{0\,2}}\right)\Phi_M(\zeta) = 0 \qquad (2.4)$$

(for d = 1 only) although the d'Alembert operator $\Box$ is not conformally invariant.

Proposition 2.1. Every solution of the wave equation (2.4) is mapped by

$$\Phi_M(\zeta) \to \Phi(z) = \frac{4\pi}{1+2z_4+z^2}\Phi_M(\zeta(z)) \qquad (2.5)$$

where z belongs to the "unit ball" $\mathcal{B}$ (1.20) and

$$\frac{i}{2}\tilde{\zeta}(z) = \frac{Z-1}{Z+1} = \frac{z^2-1+2\underset{\sim}{q}\underset{\sim}{z}}{1+2z_4+z^2} \qquad (2.6)$$

into a solution of the 4-dimensional (complex) Laplace equation

$$\Delta_4 \Phi(z) = \left(\Delta_{\underset{\sim}{z}} + \frac{\partial^2}{\partial z_4^2}\right)\Phi(z) = 0 \quad (z \in \mathcal{B}). \tag{2.7}$$

The map (2.5) intertwines between the representation U_M (2.3) and

$$[U(g)\Phi](z) = \{\det(v^* u_1 Z + u_2)\}^{-1} \Phi(z_g) \tag{2.8}$$

where the transformation law $Z \to Z_g$ is defined by (1.13) ($Z=zq$) and g is expressed in terms of the 2x2 matrices u_1, u_2, and v (satisfying (1.14)) by (1.16).

Remark. The numerical coefficient in (2.5) is chosen in such a way that the 2-point function (2.1) is transformed into

$$\langle \Phi(z_1) | \Phi(z_2) \rangle = \frac{1}{(z_1 - z_2)^2} \qquad \left(\text{for } \frac{z_1}{z_1^2} \in \mathcal{B} \text{ and } z_2 \in \mathcal{B}\right). \tag{2.9}$$

Proof. Setting $\zeta_4 = -i\zeta^0$ we define the translation generators T_α (α = 1,2,3,4) in the z-picture by

$$-2\pi \frac{\partial}{\partial \zeta_\alpha} \Phi_M(\zeta) = \frac{2}{1 + \frac{1}{4}\zeta^2 + \zeta_4} \left(T_\alpha \Phi(z)\right)_{z = z(\zeta)} . \tag{2.10}$$

This gives

$$\underset{\sim}{T} = \frac{1}{2}\underset{\sim}{z}\left(z\frac{\partial}{\partial z} + 1 + \frac{\partial}{\partial z_4}\right) - \frac{1}{4}(1 + 2z_4 + z^2)\frac{\partial}{\partial \underset{\sim}{z}} , \tag{2.11a}$$

$$T_4 = -\frac{1}{4}(1 + 2z_4 + z^2)\frac{\partial}{\partial z_4} + \frac{1 + z_4}{2}\left(z\frac{\partial}{\partial z} + 1 + \frac{\partial}{\partial z_4}\right) \tag{2.11b}$$

$$= \frac{1}{4}(1 - z^2)\frac{\partial}{\partial z_4} + \frac{1 + z_4}{2}\left(z\frac{\partial}{\partial z} + 1\right); \tag{2.11c}$$

hence,

$$\underset{\sim}{T}^2 = \frac{1}{4}\underset{\sim}{z}^2\left(z\frac{\partial}{\partial z} + 2 + \frac{\partial}{\partial z_4}\right)\left(z\frac{\partial}{\partial z} + 1 + \frac{\partial}{\partial z_4}\right) + \frac{1}{16}\left(1+2z_4+z^2\right)^2 \Delta_{\underset{\sim}{z}} -$$
$$-\frac{1}{4}\left(1+2z_4+z^2\right)\underset{\sim}{z}\frac{\partial}{\partial \underset{\sim}{z}}\left(1+z\frac{\partial}{\partial z} + \frac{\partial}{\partial z_4}\right) - \frac{3}{8}\left(1+2z_4+z^2\right)\left(z\frac{\partial}{\partial z}+1+\frac{\partial}{\partial z_4}\right),$$

$$T_4^{\,2} = \frac{1}{2}(1+z_4)^2\left\{1+\frac{1}{2}\left(z\frac{\partial}{\partial z}+\frac{\partial}{\partial z_4}+3\right)\left(z\frac{\partial}{\partial z}+\frac{\partial}{\partial z_4}\right)\right\} +$$
$$+\frac{1}{4}\left(1+2z_4+z^2\right)\left\{\frac{1}{4}\left(1+2z_4+z^2\right)\frac{\partial^2}{\partial z_4^2} - \left(z\frac{\partial}{\partial z}+1+\frac{\partial}{\partial z_4}\right)\left(\frac{1}{2}+z_4\partial_4+\partial_4\right)\right\}$$

so that

$$T^2 \equiv \underset{\sim}{T}^2 + T_4^{\,2} = \frac{1}{16}\left(1+2z_4+z^2\right)^2\Delta_4 \qquad \left(\Delta_4 = \frac{\partial^2}{\partial z^2}\right). \tag{2.12}$$

The intertwining property of the map (2.5) is also straightforward We shall only note that the z-picture counterpart of Segal's conformal Hamiltonian

$$H_\zeta = \frac{\partial}{\partial \zeta_4} - \frac{1}{2}\zeta_4(1+\zeta\partial) + \frac{1}{4}\zeta^2\frac{\partial}{\partial \zeta_4} =$$
$$= i\left\{\frac{\partial}{\partial \zeta^0} + \frac{1}{2}\zeta^0(1+\zeta\partial) + \frac{1}{4}\zeta^2\frac{\partial}{\partial \zeta^0}\right\} \tag{2.13}$$

(the rotation generator in the 6-4 plane) is

$$H_z = -1 - z\partial\ . \tag{2.14}$$

In deriving (2.14) we have used that the dilation generator $-1-\zeta\partial$ corresponds to $\zeta T - 1 = 2T_4 + z\frac{\partial}{\partial z} + \frac{\partial}{\partial z_4} + 1$ in the z-picture and

$$\left(1+\frac{1}{4}\zeta^2+\zeta_4\right)\frac{1+z_4}{2} - \frac{1}{2}\zeta_4 = 1.$$

<u>Corollary.</u> The compact picture free massless scalar field

$$\varphi(z) = \frac{4\pi}{1+2z_4+z^2}\,\varphi_M(x(z)) \qquad (zq \equiv Z \in U(2)) \tag{2.15}$$

satisfies the Laplace equation

$$\Delta_4\,\varphi(z) = 0. \tag{2.16}$$

Proposition 2.2. The free field $\varphi(z)$ with 2-point function (2.9) can be expanded in the form

$$\varphi(z) = \sum_{n=1}^{\infty} \left\{ \frac{1}{z^2} a_n\left(\frac{z}{z^2}\right) + a_n^*(z) \right\} \tag{2.17a}$$

where $a_n^{(*)}$ are homogeneous harmonic polynomilas of degree n-1:

$$a_n^{(*)}(z) = a_n^{(*)\alpha_1 \dots \alpha_{n-1}} z_{\alpha_1} \dots z_{\alpha_{n-1}} ; \tag{2.17b}$$

here $a_n^{(*)\alpha_1 \dots \alpha_{n-1}}$ are symmetric traceless tensors. The a_n satisfy the following discrete basis commutation relations for creation and annihilation operators:

$$[a_n(z_1), a_m^*(z_2)] = \delta_{mn} H_{n-1}(z_1, z_2); \tag{2.18}$$

here H_k is the unique O(4)-invariant harmonic polynomial of degree k in both z_1 and z_2, normalized by the condition

$$H_{n-1}(z_1, z_2) = n \quad \text{for} \quad z_1 z_2 = 1 = z_1^2 z_2^2, \quad n=1,2,\dots . \tag{2.19}$$

We have

$$H_k(z_1, z_2) = (z_1^2 z_2^2)^{\frac{k}{2}} C_k^1(\hat{z}_1, \hat{z}_2) \qquad \left(\hat{z} = \frac{z}{z^2}\right). \tag{2.20}$$

where C_k^1 are the (hyperspherical) Gegenbauer polynomilas with generating function

$$(1 - 2t\xi + t^2)^{-1} = \sum_{k=0}^{\infty} t^k C_k^1(\xi). \tag{2.21}$$

Energy positivity implies

$$a_n(z)|0\rangle = 0 = \langle 0| a_m^*(z), \tag{2.22}$$

$|0\rangle$ being the unique conformally invariant vacuum.

Sketch of the proof. All properties of a free field are read from its 2-point function. On the other hand, the 2-point function (2.9) is derived from (2.17-22):

$$\langle \varphi(z_1)\varphi(z_2)\rangle_0 = \frac{1}{z_1^2}\sum_{n=1}^{\infty}\left[a_n\left(\frac{z_1}{z_1^2}\right), a_n^*(z_2)\right] = \frac{1}{z_1^2}\sum_{k=0}^{\infty}\left(\frac{z_2^2}{z_1^2}\right)^{\frac{1}{2}k} C_k^1(\hat{z}_1\hat{z}_2) =$$

$$= \frac{1}{z_1^2}\left(1 - 2\frac{z_1 z_2}{z_1^2} + \frac{z_2^2}{z_1^2}\right)^{-1} = \frac{1}{(z_1-z_2)^2} \quad \text{for } \left|\frac{z_2^2}{z_1^2}\right| < 1.$$

We note that in the analyticity domain (1.20) of the vector

$$\Phi(z) = \varphi(z)|0\rangle = \sum_{n=1}^{\infty} a_n^*(z)|0\rangle \tag{2.23}$$

we have $|z^2| < 1$; similarly, the vector function $\langle 0|\varphi(z)$ is analytic in the conformal reflection of $\mathcal{B}$, where $|z^2| > 1$. Therefore, we should assume $|z_2^2| < 1 < |z_1^2|$, thus automatically falling in the convergence domain of the expansion of the 2-point function.

Positivity. The 1-particle subspace of the Fock space for the field $\varphi(z)$ is the Hilbert-space closure $\mathcal{H}_1$ of the direct sum $\bigoplus_n \mathcal{H}_1^{(n)}$, where $\mathcal{H}_1^{(1)} = \mathbb{C}\, a_1^*|0\rangle$, $\mathcal{H}_1^{(n+1)} = \{|F\rangle = f_{\alpha_1\ldots\alpha_n} a_{n+1}^{*\,\alpha_1\ldots\alpha_n}|0\rangle\}$ of K-finite vectors - K being the maximal compact subgroup S(U(2)xU(2)) of G. Here $f_{\alpha_1\ldots\alpha_n}$ runs over all (rank n) symmetric traceless tensors. $\mathcal{H}_1^{(n)}$ and $\mathcal{H}_1^{(m)}$ being orthogonal for $n \neq m$, to verify the positivity of the inner product in $\mathcal{H}_1$ it suffices to study its properties in each of the finite dimensional subspaces $\mathcal{H}_1^{(n)}$. On the other hand, the positivity of $\langle F|F\rangle$ in $\mathcal{H}_1^{(n+1)}$ is made obvious by the explicit formula $\langle F|F\rangle = 2^n \sum_{\alpha_1\ldots\alpha_n} \bar{f}_{\alpha_1\ldots\alpha_n} f_{\alpha_1\ldots\alpha_n}$ which follows from (2.18) and (2.22). Indeed, if $f_{\alpha_1\ldots\alpha_n}$ is a symmetric traceless tensor, then

$$\frac{1}{n!}\bar{f}_{\alpha_1\ldots\alpha_n}\frac{\partial}{\partial \bar{z}_{\alpha_1}}\cdots\frac{\partial}{\partial \bar{z}_{\alpha_n}} H_n(\bar{z}, z) = 2^n \bar{f}_{\alpha_1\ldots\alpha_n} z_{\alpha_1}\ldots z_{\alpha_n}. \tag{2.24}$$

(Proof: the general form of the right hand side of (2.24) is deduced from O(4)-covariance, the tracelessness of $\bar{f}_{\alpha_1\ldots\alpha_n}$, and the fact that $H_n(\bar{z},z)$ - and hence its n-th derivative(2.24) - is harmonic in z. The coefficient 2^n can be evaluated setting $\bar{f}_{\alpha_1 \dots \alpha_n} = \bar{\zeta}_{\alpha_1} \dots \bar{\zeta}_{\alpha_n}, \bar{\zeta}^2 = 0$ and applying (2.20) (2.21) for $z_1 = \bar{\zeta}$.)

Eq. (2.24) allows to present the canonical commutation relations in a purely algebraic form:

$$[a_1, a_1^*] = 1, \qquad [a_{2\alpha}, a_2^{*\beta}] = 2\,\delta_\alpha^{\ \beta},$$

$$[a_{3\alpha_1\alpha_2}, a_3^{*\beta_1\beta_2}] = \frac{2^2}{2!}\left(\delta_{\alpha_1}^{\beta_1}\delta_{\alpha_2}^{\beta_2} + \delta_{\alpha_1}^{\beta_2}\delta_{\alpha_2}^{\beta_1} - \frac{1}{2}\delta_{\alpha_1\alpha_2}\delta^{\beta_1\beta_2}\right) \text{ etc.} \tag{2.25}$$

Conversely, the positivity of the inner product in $\mathcal{H}_1^{(n)}$ is made obvious by (2.25).

Eqs.(2.18-20) or (2.25) should be regarded as a canonical form for the Heisenberg algebra of a free massless field in 4 dimensions.

Remark. It follows from the conformal invariance of the Laplace equationn that if $a_n(z)$ is harmonic, so is its "Kelvin transform" $\frac{1}{z^2}\, a_n\!\left(\frac{z}{z^2}\right)$. Moreover, the homogeneous harmonic polynomials that appear in the expansion (2.17) form a complete set of harmonic functions on $\bar{M} = U(2)$. The space of homogeneous harmonic polynomials of degree n-1 (isomorphic to $\mathcal{H}_1^{(n)}$) has dimension n^2 - equal to the number of independent components of a symmetric traceless tensor of rank n-1 in four dimensions, $n^2 = \binom{n+2}{3} - \binom{n}{3}$. We also note that setting $a_n^* = a_{-n}$ we can write (2.17) in a form, reminiscent to the 2-dimensional models:

$$\varphi(z) = \sum_{n=\pm1,\pm2,\dots} a_n(\hat{z})\,(z^2)^{-\frac{1}{2}(n+1)} \quad \text{where } [a_n(\hat{z}), a_m(\hat{z})] = n\,\delta_{n+m}. \tag{2.17c}$$

2B. A Weyl spinor field

A free 2-component (Weyl) spinor field $\psi_M(x)$ is characterized by its (analytic) 2-point function

$$\langle \Psi_M(\zeta_1) | \Psi_M(\zeta_2)\rangle = \langle \psi_M(\zeta_1)\psi_M(\zeta_2)\rangle_0 = -i\,\underset{\sim}{\partial}\,\frac{1}{4\pi^2\zeta_{12}^2} =$$

$$= \frac{i\,\underset{\sim}{\zeta}_{12}}{4\pi^2\zeta_{12}^4} = \frac{-q\,\zeta_{12}}{2\pi^2\zeta_{12}^4} \quad \text{for } \zeta_{12} = \zeta_1 - \zeta_2 \in T_-,\ (\zeta_{12})_4 = -i\zeta_{12}^0. \tag{2.26}$$

It is, in particular, a solution of the Weyl equation

$$\tilde{\partial}\psi_M(x) = 0 \quad \text{for} \quad \tilde{\partial} = \tilde{\sigma}^\mu \partial_\mu = -\partial_0 - \underline{\sigma}\,\underline{\partial}\,, \tag{2.27a}$$

$$\partial_\mu \psi_M^*(x)\tilde{\sigma}^\mu \quad (\equiv \psi_M^* \overleftarrow{\tilde{\partial}}) = 0 = \Psi_M(\zeta)\frac{\overleftarrow{\partial}}{\partial\zeta} q^+ . \tag{2.27b}$$

A counterpart of Proposition 2.1 can be formulated as follows.

Proposition 2.3. The vector function

$$\Psi(z) = \frac{\pi}{2}\left(\frac{4}{1+2z_4+z^2}\right)^2 \Psi_M(\zeta(z))\,(1+zq^+) \tag{2.28}$$

admits an analytic continuation in the domain $\mathcal{B}$ (1.20) and satisfies as a consequence of (2.7) the "quaternionic Weyl equation"

$$\Psi(z)\overleftarrow{\partial} q \equiv \partial_\alpha \Psi(z) q_\alpha = 0. \tag{2.29}$$

The 2-point function (2.26) is transformed in the "compact picture" (2.28) into

$$\langle \Psi(z_1) | \Psi(z_2) \rangle = \langle \psi(z_1)\,\psi^*(z_2)\rangle_0 =$$

$$= 2\,\frac{z_{12}\,q^+}{z_{12}^4} = -q^+\partial\,\frac{1}{z_{12}^2}\,, \quad z_{12} = z_1 - z_2 . \tag{2.30}$$

The map $\Psi_M(\zeta) \to \Psi(z)$ intertwines between two realizations of the positive-energy elementary representation $[\frac{3}{2}; \frac{1}{2}, 0]$ of G.

Proof. In the case of spinors Eq.(2.10) is replaced by

$$-2\pi \frac{\partial}{\partial\zeta^\alpha}\Psi_M(\zeta) = \Psi(z)\overleftarrow{T}_\alpha^{(\frac{1}{2})}\,\frac{1+zq}{4}(1+2z_4+z^2)\Big|_{z=z(\zeta)} \tag{2.31a}$$

or

$$-\pi \frac{\partial}{\partial\zeta^\alpha}\Psi_M(\zeta) = \Psi(z)\overleftarrow{T}_\alpha^{(\frac{1}{2})}\Big|_{z=z(\zeta)}\;\frac{1+\frac{1}{2}\zeta q^+}{1+\zeta_4+\frac{1}{4}\zeta^2}\,, \tag{2.31b}$$

where

$$\overleftarrow{T}_\alpha^{(\frac{1}{2})} = -\overleftarrow{\partial}_\alpha \tfrac{1}{4}(1+2z_4+z^2) + (\overleftarrow{\partial} z + \overleftarrow{\partial}_4 + 1)\tfrac{1}{2}(z_\alpha + \delta_{\alpha 4}) + $$
$$+ \tfrac{1}{4}(1+zq)q_\alpha^+ , \qquad (2.32)$$

so that

$$\overleftarrow{T}_\alpha^{(\frac{1}{2})} \frac{1+zq}{2} q_\alpha^+ = \overleftarrow{\partial} q \tfrac{1}{8}(1+2z_4+z^2)(1+zq^+). \qquad (2.33)$$

The proposition also follows from the relation between the 2-point functions (2.26) and (2.30) which is derived from (1.26) and

$$\tfrac{1}{4}\left[\zeta_1(z_1) - \zeta_2(z_2)\right]^2 = 4\left(\frac{\underset{\sim}{z}_1}{1+2z_{14}+z_1^2} - \frac{\underset{\sim}{z}_2}{1+2z_{24}+z_2^2}\right)^2 + $$
$$+\left(\frac{z_1^2-1}{1+2z_{14}+z_1^2} - \frac{z_2^2-1}{1+2z_{24}+z_2^2}\right)^2 =$$
$$= \sum_{a=1}^{2} \frac{1-2z_{a4}+z_a^2}{1+2z_{a4}+z_a^2} - \frac{(z_1^2-1)(z_2^2-1) + 4\underset{\sim}{z}_1\underset{\sim}{z}_2}{(1+2z_{14}+z_1^2)(1+2z_{24}+z_2^2)} =$$
$$= \frac{2(z_1-z_2)^2}{(1+2z_{14}+z_1^2)(1+2z_{24}+z_2^2)},$$

$$(1+z_1q^+)\tfrac{i}{2}\left(\underset{\sim}{\zeta}_1(z_1) - \underset{\sim}{\zeta}_2(z_2)\right)(1+z_2q^+) = (z_1q^+-1)(z_2q^++1) - (z_1q^++1)(z_2q^+-1) =$$
$$= 2(z_1-z_2)q^+.$$

The intertwining property of the map $\Psi_M \to \Psi$ is exhibited, in particular, by the two forms of the conformal Hamiltonian (cf.(2.13) (2.14))

$$H_\zeta^{(\frac{1}{2})} = (1+\tfrac{1}{4}\zeta^2)\frac{\partial}{\partial\zeta_4} - \tfrac{1}{2}\zeta_4\left(\tfrac{3}{2} + \zeta\frac{\partial}{\partial\zeta}\right) + \tfrac{1}{4}\underset{\sim}{\zeta q}, \qquad (2.34a)$$

$$H_z^{(\frac{1}{2})} = -\frac{3}{2} - z\frac{\partial}{\partial z} . \tag{2.34b}$$

The compact picture 2-component spinor field

$$\psi(z) = \frac{\pi}{2}\left(\frac{4}{1+2z_4+z^2}\right)^2 (1+zq^+)\,\psi_M(x(z)) \tag{2.35}$$

satisfies the Weyl equation

$$q\partial\,\psi(z) = 0 \left(= \Delta_4 \psi(z)\right) \tag{2.36}$$

and has the following expansion in terms of creation and annihilation operators:

$$\psi(z) = \sum_{n=1}^{\infty} \left\{ \frac{q^+z}{z^4}\, b_{n+\frac{1}{2}}\left(\frac{z}{z^2}\right) + c^*_{n+\frac{1}{2}}(z) \right\} \tag{2.37}$$

where $b_{n+\frac{1}{2}}$ and $c^*_{n+\frac{1}{2}}$ are 2-component homogeneous harmonic polynomials of degree n-1, such that $(H_z + n+\frac{1}{2})\, b_{n+\frac{1}{2}}(z) = 0$ etc.,

$$q^+\frac{\partial}{\partial z} b_{n+\frac{1}{2}}(z) = 0 = q\frac{\partial}{\partial z} c^*_{n+\frac{1}{2}}(z),$$

$$b^*_{n+\frac{1}{2}}(z)\overleftarrow{\frac{\partial}{\partial z}} q = 0 = c_{n+\frac{1}{2}}(z)\overleftarrow{\frac{\partial}{\partial z}} q^+ . \tag{2.38}$$

The only non-trivial canonical anticommutation relations are

$$\left[b_{n+\frac{1}{2}}(z_1),\, b^*_{n+\frac{1}{2}}(z_2)\right]_+ = \frac{qz_1}{z_1^2}\, q^+\frac{\partial}{\partial z_2} H_n(z_1,z_2) = \left[c_{n+\frac{1}{2}}(z_1),\, c^*_{n+\frac{1}{2}}(z_2)\right]_+ \tag{2.39}$$

where H_k is given by (2.20).

<u>Proposition 2.4</u>. The expansion (2.37) together with the canonical anticommutation relations (2.39) and the vacuum property

$$b_{n+\frac{1}{2}}(z)|0\rangle = 0 = c_{n+\frac{1}{2}}(z)|0\rangle \tag{2.40}$$

(implied by the energy positivity) allows to reconstruct the 2-point function (2.30).

Proof. Indeed, using again (2.21) we find

$$\langle \psi(z_1)\, \psi^*(z_2) \rangle = q^+ \partial_2 \sum_{n=1}^{\infty} H_{n-1}(z_1, z_2)\, (z_1^2)^{-n} =$$

$$= q^+ \partial_2 \frac{1}{(z_1 - z_2)^2} = 2\, \frac{z_{12}\, q^+}{z_{12}^4}$$

in accord with (2.30).

Hermitian conjugation and positivity. The positivity of the inner product, implicitly defined by (2.39) and (2.40), can again be verified - as in the scalar field case - in a purely algebraic way. Indeed, Eq. (2.38) just says that the vector $b^*_{n+\frac{1}{2}}(z)|0\rangle$ transforms under the n(n+1) dimensional representation $\left(\frac{n-1}{2}, \frac{n}{2}\right)$ of SU(2)xSU(2), its algebraic expression being

$$\left(b^*_{n+\frac{3}{2}}\right)^{\bar{A}}_{\alpha_1 \ldots \alpha_n} \left(q_{\alpha_n}\right)_{\bar{A} B} = 0 \tag{2.41}$$

(α_n is summed from 1 to 4, the spinor index $\bar{A}$, from 1 to 2). If $S_{\alpha_1 \ldots \alpha_n \bar{A}}$ is a (complex valued) spin-tensor that is symmetric and traceless in $\alpha_1 \ldots \alpha_n$ and satisfies $q^{+A\bar{A}}_{\alpha_1} S_{\alpha_1 \ldots \alpha_n \bar{A}} = 0$ (i.e. also transforms under the irreducible representation $\left(\frac{n-1}{2}, \frac{n}{2}\right)$ of SU(2)xSU(2)) then

$$\left\| S_{\alpha_1 \ldots \alpha_n \bar{A}}\, b^{*\bar{A}}_{n+\frac{3}{2}\, \alpha_1 \ldots \alpha_n} |0\rangle \right\|^2 =$$

$$= 2^{n+1} (n+1)\, \bar{S}^{\bar{A}}_{\alpha_1 \ldots \alpha_n} S_{\alpha_1 \ldots \alpha_n \bar{A}} \quad (>0). \tag{2.42}$$

The multiindex form of the anticommutation relations (2.39) (i.e., the spin ½ counterpart of (2.25)) is

$$[b_{\frac{3}{2}\bar{B}}, b^{*\bar{A}}_{\frac{3}{2}}]_+ = 2\delta^{\bar{A}}_{\bar{B}}, \quad [b_{\frac{5}{2}\beta\bar{B}}, b^{*\bar{A}}_{\frac{5}{2}\alpha}]_+ = 8\delta_{\alpha\beta}\delta^{\bar{A}}_{\bar{B}} - 2q_{\beta\bar{B}C}q^{+C\bar{A}}_{\alpha}, \text{ etc.}$$

$$[c_{\frac{5}{2}\alpha A}, c^{*B}_{\frac{5}{2}\beta}]_+ = 8\delta_{\alpha\beta}\delta_A^{B} - 2q^{+B\bar{C}}_{\beta}q_{\alpha\bar{C}A} \text{ etc.} \tag{2.43}$$

The field (2.37) and its conjugate

$$\psi^*(z) = \sum_{n=1}^{\infty}\left(b^*_{n+\frac{1}{2}}(z) + c_{n+\frac{1}{2}}\left(\frac{z}{z^2}\right)\frac{q^+z}{z^4}\right) \tag{2.44}$$

have analytic K-finite matrix elements for $z^2 \neq 0$. The analytic field operators are related as follows under hermitian conjugation:

$$(\psi(z))^* = \psi^*\left(\frac{\bar{z}}{\bar{z}^2}\right)\frac{q\bar{z}}{\bar{z}^4} \tag{2.45a}$$

$$(\psi^*(z))^* = \frac{q\bar{z}}{\bar{z}^4}\,\psi\left(\frac{\bar{z}}{\bar{z}^2}\right). \tag{2.45b}$$

We leave it to the reader to verify that the star operation so defined is involutive and is equivalent to the conditions

$$(b_\nu(z))^* = b^*_\nu(\bar{z}), \quad (c_\nu(z))^* = c^*_\nu(\bar{z}) \tag{2.45c}$$

(for $\nu = n+\frac{1}{2}$, n=1,2,...).

3. COMPOSITE CONFORMAL FIELDS AND LIGHT-CONE OPE

3A. U(1)-current algebra

The compact picture electromagnetic current $J_\alpha(z)$ is a conserved vector field on $\bar{M}$ of dimension 3. It has a generalized Fourier-Laurent expansion of the form

$$J_\alpha(z) = \sum_{n=-\infty}^{\infty} Q_{n\alpha}(z) \quad \text{where} \quad \partial_\alpha Q_{n\alpha}(z) = 0 \tag{3.1}$$

and $Q_{n\alpha}$ are series of polynomials in z_β and z^{-2} homogeneous in z. of degree -n-3; in particular,

$$Q_{0\alpha}(z) = 2\Big\{ Q\frac{z_\alpha}{z^4} + {}^*\Sigma_{\alpha\beta}\frac{z_\beta}{z^4} +$$

$$+ \sum_{\nu=1}^{\infty}(z^2)^{-\nu-2}\big(z_\alpha\, S^{(\nu)}_{\alpha_1\ldots\alpha_{2\nu}}\, z_{\alpha_1}\ldots z_{\alpha_{2\nu}} + f_{\alpha\alpha_1\ldots\alpha_{2\nu+1}}\, z_{\alpha_1}\ldots z_{\alpha_{2\nu+1}}\big) \qquad (3.2)$$

Here Q is the charge operator satisfying for an electron field ψ

$$[\psi(z), Q] = \psi(z), \qquad [Q, \psi^*(z)] = \psi^*(z); \qquad (3.3)$$

for a current composed out of a free Weyl spinor field,

$$J_\alpha(z) = :\psi^*(z)\, q_\alpha\, \psi(z):$$

we have

$$Q = \tfrac{1}{2}(b^*b - c^*c) + \tfrac{1}{8}(b^*_\alpha b_\alpha - c^*_\alpha c_\alpha) + \ldots \; ; \qquad (3.4)$$

$\Sigma_{\alpha\beta}$ and ${}^*\Sigma_{\alpha\beta}$ are the dual Lorentz generators

$$\Sigma_{\alpha\beta} = \tfrac{1}{2}(b^* q_{\alpha\beta} b + c\,\tilde{q}_{\alpha\beta} c^*) + \cdots \qquad (3.5a)$$

$$(q_\alpha q^+_\beta = \delta_{\alpha\beta} + q_{\beta\alpha} = \delta_{\alpha\beta} - q_{\alpha\beta}\,, \quad q^+_\beta q_\alpha = \delta_{\alpha\beta} + \tilde{q}_{\alpha\beta})$$

$${}^*\Sigma_{\alpha\beta} = \tfrac{1}{2}\varepsilon_{\alpha\beta\gamma\delta}\Sigma_{\gamma\delta} = \tfrac{1}{2}(c\,\tilde{q}_{\alpha\beta} c^* - b^* q_{\alpha\beta} b) + \ldots ; \qquad (3.5b)$$

the defining property for $\Sigma_{\alpha\beta}$ in this case is the commutation rule

$$\tfrac{1}{2}[\psi(z), \Sigma_{\alpha\beta}] = (z_\beta\partial_\alpha - z_\alpha\partial_\beta + \tfrac{1}{2}\tilde{q}_{\alpha\beta})\,\psi(z). \qquad (3.6)$$

$S^{(\nu)}$ is a rank 2ν symmetric traceless tensor (that transforms under the $(2\nu+1)^2$-dimensional representation (ν,ν) of SU(2)xSU(2)); $f^{(\nu)}$ transforms under the representation $(\nu+1,\nu) + (\nu,\nu+1)$ of dimension, $2(2\nu+3)(2\nu+1)$.

Similar expansions can be written for all $Q_{n\alpha}$; $Q_{-3\alpha}$ is the first among the negative index terms that involves a part regular for z=0, that is the constant 4-vector "creation operator" $b^* q_\alpha c^*$.

The operator valued coefficients of the series $Q_{n\alpha}$ generate an infinite graded Lie algebra $\mathfrak{A}$.

There are (at least) two major complications as compared to the 2-dimensional current algebra (i.e. the direct sum to two Heisenberg algebras in the case of a U(1)xU(1) chiral symmetry). First of all, each level of the 4-dimensional current algebra is infinite dimensional. In particular, the 0-th level gives rise to an infinite dimensional subalgebra of $\mathfrak{A}$, which can be approximated by the series of imbedded compact Lie algebras $u(\frac{(N+1)(N+2)(N+3)}{3}) \oplus u(\frac{(N+1)(N+2)(N+3)}{3})$ (N=0,1,2,..) in the case of the Weyl field current (3.4). Secondly, it can be shown that the current $J_\alpha(z)$ and the unit operator do not span $\mathfrak{A}$; rather, an infinite ladder of conserved conformal tensors is needed in order to expand the current commutator.

3B. A light cone current-field OPE.

The 1-dimensional OPE algebra, described in Chapter II, appears in an appropriate limit of the 4-dimensional current-field OPE.

Consider, for the sake of simplicity, a charged scalar field $\varphi(z)$ of (possibly anomalous) dimensions d. Its conformal-invariant 3-point function with a current $J_\alpha(z)$, consistent with a standard Ward identity,

$$\langle J_\alpha(z_1)\varphi(z_2)\varphi^*(z_3)\rangle = \\ = ie\left(\frac{1}{z_{13}^2}\overleftrightarrow{\partial}_{1\alpha}\frac{1}{z_{12}^2}\right) z_{23}^2 \langle\varphi(z_2)\varphi^*(z_3)\rangle, \tag{3.6}$$

suggests that the light-cone bilocal operator

$$B_d(z,\varepsilon;\kappa) = \lim_{\substack{z_1\to z+\varepsilon\kappa \\ z_2=z,\ \kappa^2=0}} z_{12}^2\, z_{12}^\alpha\, J_\alpha(z_1)\varphi(z_2) \tag{3.7}$$

exists; moreover,

$$\langle B_d(z,\varepsilon;\kappa)\varphi^*(0)\rangle = \\ = \frac{-ie\, z^2}{z^2+2\varepsilon\kappa z}\left(1+\frac{z^2}{z^2+2\varepsilon\kappa z}\right)\langle\varphi(z)\varphi^*(0)\rangle. \tag{3.8}$$

This indicates that $B_d(z, \varepsilon ; \kappa)$ has an OPE of the type (II.4.6) starting with a term proportional to φ :

$$B_d(z,\varepsilon;\kappa) = -ie\,\Gamma(d)\int_0^1 du \left\{ \frac{u(1-u)^{d-3}}{\Gamma(d-2)} + \frac{(1-u)^{d-2}}{\Gamma(d-1)} \right\} \varphi(z+\varepsilon u\kappa) + \cdots \tag{3.9}$$

(The weight in (3.9) is obtained from (3.8) using the integral representation (II.3.11).) If we are allowed to assume that

$$\left(e^{2\pi i H} - e^{2\pi i d}\right) B_d\,(z,\varepsilon;\kappa)|0\rangle = 0 \tag{3.10}$$

(where $H = J_{60}$ is the "second quantized conformal Hamiltonian" - cf. (2.14) and (2.34)) then the OPE (3.9) would only involve fields of dimension d+n (n=0,1,2,...).

We can define a primary field, as a field φ whose commutator with the conformal (traceless) stress energy tensor $T_{\alpha\beta}$ is homogeneous in φ, in any number of space-time dimension. We, however, do not know whether there are non-trivial (i.e. non-free) conformal QFT models in 4 dimensions with infinite conformal families (in particular, with an infinite ladder of conserved tensor currents). More generally, we do not know the analogue of the "fusion rules" (and hence of the "minimal theories") of ref. [B1] for higher than two dimensions.

Acknowledgments

A short version of these lectures was first given while I was enjoying the hospitality of the Arnold-Sommerfeld-Institut für Mathematische Physik in Clausthal. The main part of these notes was written while I was visiting the International School for Advanced Studies in Trieste, Collège de France in Paris, Faculté des Sciences Mirand in Université de Dijon and Institut des Hautes Etudes Scientifiques in Bures-sur-Yvette (France) during the winter of 1986. It is a pleasure to thank my hosts and colleagues in all these institutions for their hospitality. I would also like to thank V. Kac, G. Mack, G. Sotkov and A. Zamolodchikov for acquainting me with their unpublished work and useful discussions at various stages of this work.

References

A1 M. Ademollo, L. Brink, A. d'Adda, R. D'Auria, E. Napolitano, S. Sciuto, E. Del Giudice, P. di Vecchia, S. Ferrara, F. Gliozzi, R. Musto, R. Pettorino, Supersymmetric string and colour confinement, Phys. Lett. 62B (1976) 105-110; -, ..., - , J.H. Schwarz, Dual string with U(1) colour symmetry, Nucl. Phys. B111 (1976) 77-110.

A2 D. Altschüler, H.P. Nilles, String models with lower critical dimensions, compactification and nonabelian symmetries, Phys. Lett. 154B (1985) 135-140.

A3 H. Aratyn, A.H. Zimerman, On covariant formulation of free Neveu-Schwarz and Ramond string models, Phys. Letters 166B (1986) 130-134

A4 H. Aratyn, H.H. Zimerman, Gauge invariance of the bosonic free field string theory, Phys. Lett. 168B (1986) 75-77

B1 A.A. Belavin, A.M. Polyakov, A.B. Zamolodchikov, Infinite conformal symmetry in two dimensional quantum field theory, Nucl. Phys. B241 (1984) 333-380.

B2 M.A. Bershadsky, V.G. Knizhnik, M.G. Teitelman, Superconformal symmetry in two dimensions, Phys. Lett. 151B (1985) 31-36.

B3 N.N. Bogolubov, A.A. Logunov, I.T. Todorov, "Introduction to Axiomatic Quantum Field Theory," Benjamin, Reading, Mass. 1975.

C1 C. Callan, D. Friedan, E. Martinec, M. Perry, Strings in background fields, Nucl. Phys. B262 (1985) 593-609.

C2 A. Casher, F. Englert, H. Nicolai, A. Taormina, Consistent superstrings as solutions of the D=26 bosonic string theory, Phys. Lett. 162B (1985) 121-126.

C3 S. Coleman, D. Gross, R. Jackiw, Fermion avatars of the Sugawara model, Phys. Rev. 180 (1969) 1359-1366.

C4 N.S. Craigie, V.K. Dobrev, I.T. Todorov, Conformally covariant composite operators in quantum chromodynamics, Ann. Phys. (N.Y) 159 (1985) 411-444.

D1 G.F. Dell'Antonio, Y. Frishman, D. Zwanziger, Thirring model in terms of currents: solution and light-cone expansion, Phys. Rev. D6 (1972) 988-1007.

D2 P. Di Vecchia, V.G. Knizhnik, J.L. Petersen, P. Rossi, A supersymmetric Wess-Zumino Lagrangian in two dimensions, Nucl. Phys. B253 (1985) 701-726;

P. Di Vecchia, J.L. Petersen, H.B. Zheng, N=2 extended superconformal theories in two dimensions, Phys. Lett. 162B (1985) 327 332.

D3 V.K. Dobrev, V.B. Petkova, S.G. Petrova, I.T. Todorov, Dynamical derivation of vacuum operator product expansion in Euclidean conformal quantum field theory, Phys. Rev. D13 (1976) 886-912.

D4 Vl. D. Dotsenko, Critical behaviour and associated conformal algebra of the Z_3 Potts model, Nucl. Phys. B235 (1984) 54-74.

D5 Vl. S. Dotsenko, V.A. Fateev, Conformal algebra and multipoint correlation functions in 2D statistical models, Nucl. Phys. B240 (1984) 312-348; Four-point correlation functions and the operator algebra in 2D conformal invariant theories with central charge $c \leq 1$, Nucl. Phys. B251 (1985) 691-734; Operator algebra of two-dimensional conformal theories with central charge $c \leq 1$, Phys. Lett. 154B (1985) 291-295.

E1 H. Eichenherr, Minimal operator algebras in superconformal field theory, Phys. Lett. 151B (1985) 26-30.

F1 B.L. Feigin, D.B. Fuchs, Representations of the Virasoro algebra, Moscow preprint 1984.

F2 S. Ferrara, R. Gatto, A. Grillo, Conformal algebra in two spacetime dimensions and the Thirring model, Nuovo Cim. 12A (1972) 959-968.

F3 I.B. Frenkel, Two constructions of affine Lie algebra representations and Boson-fermion correspondence in quantum field theory, J. Funct. Anal. 44 (1981) 259-327.

F4 I.B. Frenkel, V.G. Kac, Basic representations of affine Lie algebras and dual resonance models, Inventiones Math. 62 (1980) 23-66.

F5 D. Friedan, Introduction to Polyakov's string theory, in: 1982 Les Houches Summer School, J.B. Zuber, R. Stora eds. Les Houches, Session 39, Recent Advances in Field Theory and Statistical Mechanics (North Holland 1984) pp. 839-867; On two-dimensional conformal invariance and the field theory of strings, Phys. Lett. 162B (1985) 102-108.

F6 D. Friedan, E. Martinec, S.H. Shenker, Covariant quantization of superstrings, Phys. Lett. 160B (1985) 55-61; Conformal invariance, supersymmetry and string theory, Princeton Univ. preprint (November 1985).

F7 D. Friedan, Z. Qui, S. Shenker, Conformal invariance, unitarity and critical exponents in two dimensions, Phys. Rev. Lett. 52 (1984) 1575-1578, and contribution in [VOMP] .

F8 D. Friedan, Z. Qui, S. Shenker, Superconformal invariance in two-dimensions and the tricritical Ising model, Phys. Lett. 151B (1985) 37-43.

F9 S. Fubini, A. Hanson, R. Jackiw, New approach to field theory, Phys. Rev. D7 (1973) 1732-1760.

G1 I.M. Gel'fand, D.B. Fuchs, Cohomology of the algebra of vector fields on a circle, Funk. Anal. i Prilozh. 2 (1968) 92-93; The cohomology of the Lie algebra of tangent vector fields on a smooth manifold, I and II, ibid.3 n. 3 (1969) 32-52 and 4 N. 2 (1970) 23-32 (English transl.: Funct.Anal. Appl. 3 (1969) 194-224 and 4(1970) 110-119).

G2 J.L. Gervais, A. Neveu, Non-standard critical statistical models from Liouville theory, Nucl. Phys. B257 [FS 14] (1985) 59-76.

G3 F. Ghiozzi, Ward like identities and twisting operators in dual resonance models, Nuovo Cim. Lett. 2(1969) 846-850.

G4 P. Goddard, Kac-Moody and Virasoro algebras: representations and applications, Cambridge Univ. preprint DAMTP 85-21.

G5 P. Goddard, A. Kent, D. Olive, Virasoro algebra and coset space models, Phys. Lett. 152B (1985) 88-93; Unitary representations of the Virasoro and super-Virasoro algebras, Commun. Math. Phys. 103 (1986) 105-119.

G6 P. Goddard, W. Nahm, D. Olive, Symmetric spaces, Sugawara's energy momentum tensor in two dimensions and free fermions, Phys. Lett. 160B (1985) 111-116.

G7 P. Goddard, D. Olive, Kac-Moody algebras, conformal symmetry and critical exponents, Nucl. Phys. B257 (1985) 226-240; Algebras lattices and strings, in [VOMP] pp. 51-96.

G8 P.Goddard, D. Olive, A. Schwimmer, The heterotic string and a fermionic construction of the E_8-Kac-Moody algebra, Phys. Lett.157B (1985) 393-399.

G9 R. Goodman, N.R. Wallach, Projective unitary positive energy representations of Diff(S^1), J. Funct. Anal. 63 (1985) 299-321.

I1 C. Itzykson, J.B. Zuber, Two-dimensional conformal invariant theories on a torus, Saclay preprint PhT 85-019 (January 1986). submitted to Nucl. Phys. B [FS]

K1 V.G. Kac, Contravariant form for infinite dimensional Lie algebras and superalgebras, Lecture Notes in Physics 94 (Springer, Berlin 1979) pp. 441-445.

K2 V.G. Kac, "Infinite Dimensional Lie Algebras: An Introduction" (Birkhauser, Boston 1983)

K3 V.G. Kac, I.T. Todorov, Superconfomal current algebras and their unitary representations, Commun. Math. Phys. 102 (1985) 337-347.

K4 V.G. Kac, M. Wakimoto, Unitarizable highest weight representations of the Virasoro, Neveu-Schwarz and Ramond algebras (to be published in Lecture Notes in Physics).

K5 A. Kent, Conformal invariance and current algebra, Chicago preprint EFI 85-62 (October 1985).

K6 V.G. Knizhnik, Covariant fermionic vertex in superstrings, Phys. Lett. 160B (1985) 403-407.

K7 V.G. Knizhnik, A.B. Zamolodchikov, Current algebra and Wess-Zumino model in two dimensions, Nucl. Phys. B247 (1984) 83-103.

M1 S. Mandelstam, Dual resonance models, Phys. Rep. 13 (1974) 259-353.

N1 D. Nemeschanski, S. Yankielowicz, Critical diemsnions of string theories in curved space, Phys. Rev. Lett. 54 (1985) 620-623.

N2 Yu. A. Neretin, Unitary representations with highest weight of the group of diffeomorphisms of a circle, Funk. Anal. i Prilozh. 17(3) (1983) 85-96 (Transl.: Funct. Anal. Appl. 17 (1983) 235 236).

N3 A. Neveu, J.H. Schwarz, Factorizable dual model of pions, Nucl. Phys. B31 (1971) 86-112.

N4 A. Neveu, P.C. West, Gauge symmetries of the free supersymmetric string theories, Phys. Lett. 165B (1985)63-70.

N5 A. Neveu, H. Nicolai, P. West, New symmetries and ghost structure of covariant string theories, Phys. Lett. 167B (1986) 307-314.

O1 W. Ogura, A. Hosoya, Kac-Moody algebra and nonlinear sigma model, Phys. Lett. 164B (1985) 329-332.

O2 D. Olive, Kac-Moody algebras: an introduction for physicists, Imperial College preprint Imperial /TP/84-85/14; Kac-Moody and Virasoro algebras in local quantum phsyics, Imperial /TP/84-85/33, London; Kac-Moody algebras in relation to quantum physics, ICTP Spring School on Supergravity and Supersymmetry, SMR 170-5 Trieste (April 1986).

P1 S.M. Paneitz, I.E. Segal, Analysis of space-time bundles I and II, J. Funct. Anal. 47 (1982) 78-142 and 49 (1982) 335-414.

P2 S.M. Paneitz, Analysis of space-time bundles III, J. Funct. Anal. 54 (1983) 18-112.

P3 A.Z. Petrov, New Methods in the General Theory of Relativity, (Nauka, Moscow 1966) (English transl.: Einstein Spaces (Pergamon Press, Oxford, 1969)).

R1 P. Ramond, Dual theory for free fermions, Phys. Rev. D3 (1971) 2415-2418.

R2 W. Rühl, B.C. Yunn, The transformation behaviour of fields in conformally covariant quantum field theory, Fortschr. d. Physik 25 (1977) 83-99.

S1 I. Segal, Mathematical Cosmology and Extragalactic Astronomy (Academic Press, N.Y. 1976).

S2 I.E. Segal, H.P. Jakobsen, B.Ørsted, S.M. Paneitz, N. Speh, Covariant chronogeometry and extreme distances: Elementary particles Proc. Nat. Acad. Sci. USA 78 (1981) 5261-5265.

S3 I.E. Segal, Covariant chronogeometry and extreme distances, III Macro-micro relations, Int. J. Theor. Phys. 21 (1982) 852-869.

S4 G.M. Sotkov, I.T. Todorov, V. Yu. Trifonov, Quasiprimary composite fields and operator product expansions in 2-dimensional conformal models, ISAS preprint 9/86/EP.

T1 Ch.B. Thorn, Computing the Kac determinant using dual model techniques and more about the no-ghost theorem, Nucl. Phys. B248 (1984) 551-569.

T2 I.T. Todorov, Conformal description of spinning particles, ISAS Preprint 1/81/EP, Trieste (to be published as ISAS Lecture Notes, Springer).

T3 I.T. Todorov, Local field representations of the conformal group and their applications, Lectures presented at ZiF, November - December 1983, in Mathematics + Physics, Lecture Notes on Recent Results, Vol. 1, Ed. L. Streit (World Scientific, Singapore, Philadelphia 1985) pp 195-338.

T4 I.T. Todorov, Current algebra approach to conformal invariant two-dimensional models, Phys. Lett. 153B (1985) 77-81; Algebraic approach to conformal invariant 2-dimensional models, Bulg. J. Phys. 12 (1985) 3-19.

T5 I.T. Todorov, Infinite Lie algebras in 2-dimensional conformal field theory, ISAS-ICTP Lecture Notes, ISAS Preprint 2/85/EP. Trieste (World Scientific, to be published).

T6 I.T. Todorov, M.C. Mintchev, V.B. Petkova, Conformal Invariance in Quantum Field Theory (Scuola Normale Superiore, Pisa 1978).

U1 A. Uhlmann, The closure of Minkowski space, Acta Phys. Polon. 24 (1963) 295-296; Some properties of the future tube, preprint KMU-HEP 7209 Leipzig,(1972).

VOMP Vertex Operators in Mathematics and Physics, Ed. by J. Lepowski, S. Mandelstam, I.M. Singer (Springer N.Y. 1985).

V1 M.A. Virasoro, Subsidiary conditions and ghosts in dual resonance models, Phys. Rev. D1 (1970) 2933-2966.

Z1 A.B. Zamolodchikov, V.A. Fateev, Nonlocal (parafermi) currents in 2-dimensional conformal QFT and self-dual critical points in Z_N-symmetric statistical systems, Zh. Eksp. Teo. Fiz. 89 (1985) 380-399. [Transl.: Sov. Phys. JETP 62 (1985) 215-225]

Z2 A.B. Zamolodchikov, V.A. Fateev, Representations of the algebra of "parafermi currents" of spin 4/3 in 2-dimensional conformal field theory. Minimal models and the 3-critical Z_3-Potts model, Landau Institute of Theoretical Physics, Moscow 1985 (to be published)

Lecture Notes in Physics

Vol. 237: Nearby Molecular Clouds. Proceedings, 1984. Edited by G. Serra. IX, 242 pages. 1985.

Vol. 238: The Free-Lagrange Method. Proceedings, 1985. Edited by M.J. Fritts, W.P. Crowley and H. Trease. IX, 313 pages. 1985.

Vol. 239: Geometrics Aspects of the Einstein Equations and Integrable Systems. Proceedings, 1984. Edited by R. Martini. V, 344 pages. 1985.

Vol. 240: Monte-Carlo Methods and Applications in Neutronics, Photonics and Statistical Physics. Proceedings, 1985. Edited by R. Alcouffe, R. Dautray, A. Forster, G. Ledanois and B. Mercier. VIII, 483 pages. 1985.

Vol. 241: Numerical Simulation of Combustion Phenomena. Proceedings, 1985. Edited by R. Glowinski, B. Larrouturou and R. Temam. IX, 404 pages. 1985.

Vol. 242: Exactly Solvable Problems in Condensed Matter and Relativistic Field Theory. Proceedings, 1985. Edited by B.S. Shastry, S.S. Jha and V. Singh. V, 318 pages. 1985.

Vol. 243: Medium Energy Nucleon and Antinucleon Scattering. Proceedings, 1985. Edited by H.V. von Geramb. IX, 576 pages. 1985.

Vol. 244: W. Dittrich, M. Reuter, Selected Topics in Gauge Theories. V, 315 pages. 1986.

Vol. 245: R.Kh. Zeytounian, Les Modèles Asymptotiques de la Mécanique des Fluides I. IX, 260 pages. 1986.

Vol. 246: Field Theory, Quantum Gravity and Strings. Proceedings, 1984/85. Edited by H.J. de Vega and N. Sánchez. VI, 381 pages. 1986.

Vol. 247: Nonlinear Dynamics Aspects of Particle Accelerators. Proceedings, 1985. Edited by J.M. Jowett, M. Month and S. Turner. VIII, 583 pages. 1986.

Vol. 248: Quarks and Leptons. Proceedings, 1985. Edited by C.A. Engelbrecht. X, 417 pages. 1986.

Vol. 249: Trends in Applications of Pure Mathematics to Mechanics. Proceedings, 1985. Edited by E. Kröner and K. Kirchgässner. VIII, 523 pages. 1986.

Vol. 250: Lie Methods in Optics. Proceedings 1985. Edited by J. Sánchez Mondragón and K.B. Wolf. XIV, 249 pages. 1986.

Vol. 251: R. Liebmann, Statistical Mechanics of Periodic Frustrated Ising Systems. VII, 142 pages. 1986.

Vol. 252: Local and Global Methods of Nonlinear Dynamics. Proceedings, 1984. Edited by A.W. Sáenz, W.W. Zachary and R. Cawley. VII, 263 pages. 1986.

Vol. 253: Recent Developments in Nonequilibrium Thermodynamics Fluids and Related Topics. Proceedings, 1985. Edited by J. Casas-Vázquez, D. Jou and J.M. Rubí. X, 392 pages. 1986.

Vol. 254: Cool Stars, Stellar Systems, and the Sun. Proceedings, 1985. Edited by M. Zeilik and D.M. Gibson. XI, 501 pages. 1986.

Vol. 255: Radiation Hydrodynamics in Stars and Compact Objects. Proceedings, 1985. Edited by D. Mihalas and K.-H. A. Winkler. VI, 454 pages. 1986.

Vol. 256: Dynamics of Wave Packets in Molecular and Nuclear Physics. Proceedings, 1985. Edited by J. Broeckhove, L. Lathouwers and P. van Leuven. VIII, 187 pages. 1986.

Vol. 257: Statistical Mechanics and Field Theory: Mathematical Aspects. Proceedings, 1985. Edited by T.C. Dorlas, N.M. Hugenholtz and M. Winnink. VII, 328 pages. 1986.

Vol. 258: Wm. G. Hoover, Molecular Dynamics. VI, 138 pages. 1986.

Vol. 259: R.F. Alvarez-Estrada, F. Fernández, J.L. Sánchez-Gómez, V. Vento, Models of Hadron Structure Based on Quantum Chromodynamics. VI, 294 pages. 1986.

Vol. 260: The Three-Body Force in the Three-Nucleon System. Proceedings, 1986. Edited by B.L. Berman and B.F. Gibson. XI, 530 pages. 1986.

Vol. 261: Conformal Groups and Related Symmetries – Physical Results and Mathematical Background. Proceedings, 1985. Edited by A.O. Barut and H.-D. Doebner. VI, 443 pages. 1986.

I. T. Todorov

Conformal Description of Spinning Particles

1986. VIII, 74 pages. (Trieste Notes in Physics).
Soft cover DM 35,-.
ISBN 3-540-16890-7

Contents: Introduction. - A Guide to the List of References. - The Conformal Group of a Conformally Flat Space Time and Its Twistor Representations. - Twistor Flag Manifolds and SU (2,2) Orbits. - Classical Phase Space of Conformal Spinning Particles. - Twistor Description of Classical Zero Mass Fields. - Appendix: A. Clifford Algebra Approach to Twistors. Relation to Dirac Spinors. - References. - Subject Index.

The concept of twistors introduced into mathematical physics by R. Penrose has by now reached its maturity and has become a valuable tool, in particular in the context of conformal structures. These lectures introduce the student to the concept of the conformal group in conformally flat space time and its twistor representation, and present some twistor flag manifolds in detail. This results in a systematic study of positive energy conformal orbits, interpreted as classical phase spaces of "conformal particles", which have dilation invariant mass spectrum. The twistor description of the phase space of conformal spinning particles and of classical zero mass fields is presented in some detail and should be of interest not only to graduate students but also to researchers in mathematics and mathematical physics.

Springer-Verlag
Berlin Heidelberg New York
London Paris Tokyo

Springer

MIX
Papier aus verantwortungsvollen Quellen
Paper from responsible sources
FSC® C105338

If you have any concerns about our products,
you can contact us on
ProductSafety@springernature.com

In case Publisher is established outside the EU,
the EU authorized representative is:
Springer Nature Customer Service Center GmbH
Europaplatz 3, 69115 Heidelberg, Germany

Printed by Libri Plureos GmbH
in Hamburg, Germany